GEOLOGICAL SOCIETY SPECIAL PUBLICATION NO. 176

# Deformation of Glacial Materials

EDITED BY

## ALEX J. MALTMAN, BRYN HUBBARD
## & MICHAEL J. HAMBREY

(University of Wales, Aberystwyth)

2000

Published by

The Geological Society

London

# THE GEOLOGICAL SOCIETY

The Geological Society of London was founded in 1807 and is the oldest geological society in the world. It received its Royal Charter in 1825 for the purpose of 'investigating the mineral structure of the Earth' and is now Britain's national society for geology.

Both a learned society and a professional body, the Geological Society is recognized by the Department of Trade and Industry (DTI) as the chartering authority for geoscience, able to award Chartered Geologist status upon appropriately qualified Fellows. The Society has a membership of 9099, of whom about 1500 live outside the UK.

Fellowship of the Society is open to those holding a recognized honours degree in geology or cognate subject and who have at least two years' relevant postgraduate experience, or who have not less than six years' relevant experience in geology or a cognate subject. A Fellow with a minimum of five years' relevant postgraduate experience in the practice of geology may apply for chartered status. Successful applicants are entitled to use the designatory postnominal CGeol (Chartered Geologist). Fellows of the Society may use the letters FGS. Other grades of membership are available to members not yet qualifying for Fellowship.

The Society has its own publishing house based in Bath, UK. It produces the Society's international journals, books and maps, and is the European distributor for publications of the American Association of Petroleum Geologists, (AAPG), the Society for Sedimentary Geology (SEPM) and the Geological Society of America (GSA). Members of the Society can buy books at considerable discounts. The publishing House has an online bookshop (http://bookshop.geolsoc.org.uk).

Further information on Society membership may be obtained from the Membership Services Manager, The Geological Society, Burlington House, Piccadilly, London W1V 0JU, UK. (Email: enquiries@geolsoc.org.uk: tel: +44 (0)207 434 9944).

The Society's Web Site can be found at http://www.geolsoc.org.uk/. The Society is a Registered Charity, number 210161.

Published by The Geological Society from:
The Geological Society Publishing House
Unit 7, Brassmill Enterprise Centre
Brassmill Lane
Bath BA1 3JN, UK

(*Orders*: Tel. +44 (0)1225 445046
Fax +44 (0)1225 442836)
Online bookshop: http://bookshop.geolsoc.org.uk

The publishers make no representation, express or implied, with regard to the accuracy of the information contained in this book and cannot accept any legal responsibility for any errors or omissions that may be made.

**British Library Cataloguing in Publication Data**
A catalogue record for this book is available from the British Library.

ISBN 1-86239-072-X

Typeset by Aarontype Ltd, Bristol, UK

Printed by Whitstable Printers Ltd, UK

**Distributors**

*USA*
  AAPG Bookstore
  PO Box 979
  Tulsa
  OK 74101-0979
  USA
*Orders*: Tel. +1 918 584-2555
         Fax +1 918 560-2652
         E-mail: *bookstore@aapg.org*

*Australia*
  Australian Mineral Foundation Bookshop
  63 Conyngham Street
  Glenside
  South Australia 5065
  Australia
*Orders*: Tel. +61 88 379-0444
         Fax +61 88 379-4634
         E-mail: *bookshop@amf.com.au*

*India*
  Affiliated East-West Press PVT Ltd
  G-1/16 Ansari Road, Daryaganj,
  New Delhi 110 002
  India
*Orders*: Tel. +91 11 327-9113
         Fax +91 11 326-0538
         E-mail: *affiliat@nda.vsnl.net.in*

*Japan*
  Kanda Book Trading Co.
  Cityhouse Tama 204
  Tsurumaki 1-3-10
  Tama-shi
  Tokyo 206-0034
  Japan
*Orders*: Tel. +81 (0)423 57-7650
         Fax +81 (0)423 57-7651

Deformation of Glacial Materials

Glacier de Saleina, Valais, Switzerland – a small, fast-flowing and heavily-crevassed valley glacier. (Photo: M. J. Hambrey).

# Contents

It is recommended that reference to all or part of this book should be made in one of the following
ways:

MALTMAN, A. J., HUBBARD, B. & HAMBREY, M. J. (eds) 2000. *Deformation of Glacial Material.*
Geological Society, London, Special Publications, **176**, 000–000.

LAWSON, W. J., SHARP, M. J. & HAMBREY, M. J. 2000. Deformation histories and structural
assemblages of glacier ice in a non-steady flow regime. *In*: MALTMAN, A. J., HAMBREY, M. J. &
HUBBARD, B. (eds) *Deformation of Glacial Material.* Geological Society, London, Special
Publications, **176**, 000–000.

# Referees

The following are acknowledged as referees of papers in this volume:

R. Alley
I. Baker
D. Blankenship
M. Bennett
S. Carr
D. Cowan
D. Evans
S. Fitzsimons
N. Glasser
J. Glen
H. Gudmundsson
M. Hambrey
S. Hanmer
P. Harrison
J. Hart
P. Herbst

B. Hubbard
P. Hudleston
J. Knight
P. Knight
D. Kohlsdedt
D. Lawson
W. Lawson
B. Marmo
F. Ng
T. Payne
S. Pedersen
E. Pettit
E. Phillips
P. Porter
B. Rea
J. Rose

A. Russell
E. Rutter
M. Sharp
M. Siergert
A. Smith
R. Souchez
D. Sugden
T. Thorsteinsson
J.-L. Tison
J. van der Meer
J. Walder
W. Warren
G. Williams
C. Wilson

# Deformation of glacial materials: introduction and overview

ALEX J. MALTMAN, BRYN HUBBARD & MICHAEL J. HAMBREY

*Institute of Geography and Earth Sciences, University of Wales,*
*Aberystwyth, Ceredigion SY23 3DB, UK*
*(e-mail: ajm@aber.ac.uk; byh@aber.ac.uk; mjh@aber.ac.uk)*

The flow of glacier ice can produce structures that are striking and beautiful. Associated sediments, too, can develop spectacular deformation structures and examples are remarkably well preserved in Quaternary deposits. Although such features have long been recognized, they are now the subject of new attention from glaciologists and glacial geologists. However, these workers are not always fully aware of the methods for unravelling deformation structures evolved in recent years by structural geologists, who themselves may not be fully aware of the opportunities offered by glacial materials. This book, and the conference from which it stemmed, were conceived of as a step towards bridging this apparent gap between groups of workers with potentially overlapping interests.

Glaciologists have long been aware of the remarkable structures developed in flowing ice. Nineteenth century Alpine mountaineers and natural scientists, such as the Swiss naturalist, Agassiz ('The Father of Ice Ages'), and Forbes and Tyndall from Britain, described a range of structures in glaciers, and were clearly impressed by the similarity with deformation structures in rocks. The first half of the twentieth century saw few advances in glaciological thinking, but renewed interest in structures followed the formulation of a flow law for ice (e.g. Nye 1953; Glen 1955) and its application to glaciers (e.g. Nye 1957). Since then, as outlined by Hambrey & Lawson (this volume), there has been numerous studies of deformation in glaciers. Many of these studies link glacier structures to measured deformation rates, notably in the classic case studies of Allen *et al.* (1960) and Meier (1960) in North America. However, few glaciologists have applied the structural geological concepts of progressive and cumulative deformation to glaciers. Where such an approach has been adopted (primarily within valley glaciers and ice caps), new insights have emerged concerning ice deformation in relation to the development of foliation, folds, and crevasses and other faults (e.g. Hudleston 1976; Hambrey 1977; Hambrey &

Milnes 1977; Hooke & Hudleston 1978; Lawson *et al.* 1994). However, these concepts remain to be applied to deformation at the scale of ice sheets, where analogous structures are at least an order of magnitude larger. Significantly, these structures may well contain information with the potential for assessing the long-term dynamic behaviour and stability of their host ice masses.

Striking deformation structures are also produced at a wide variety of scales in the sediments associated with ice. Fine examples appear, for example, in the works of Brodzikowski (e.g. *in* Jones & Preston 1989) and in the volumes by Ehlers *et al.* (1995*a, b*). In fact, of all the various kinds of geological deformation structures, among the very first to be described were features ascribed to the action of ice. Two great geological pioneers were involved: both Sir Charles Lyell (1840) and the visionary Henry Sorby (1859) interpreted the disturbance of deposits on the coast of eastern England as being due to the movement of icebergs. Lyell (1840) then expanded his ideas on the deformation of the glacial deposits of Norfolk, and by the end of the century a variety of structures in Europe and North America had been interpreted as the result of glacial processes. A history of these early studies of deformation of sediments by glaciers is given in Aber *et al.* (1989).

Despite this long pedigree of research, there is commonly disagreement on the actual mechanisms involved in generating glacigenic structures. The dominant concept this century, until a decade or so ago, involved the notion of ice bulldozing into sediments and generating various 'ice-thrust' structures as a result. Thus, most deformation of glacial sediment was envisaged as being 'made at or near glacial termini' (Flint 1971, p. 121). It was the discovery that some glaciers and parts of major ice-sheets rest not on bedrock, but on a layer of sediment (e.g. Engelhardt *et al.* 1978; Boulton 1979) that launched the 'deformable bed' hypothesis and a new significance to deformation structures in glacial sediments.

*From*: MALTMAN, A. J., HUBBARD, B. & HAMBREY, M. J. (eds) *Deformation of Glacial Materials*. Geological Society, London, Special Publications, **176**, 1–9. 0305-8719/00/$15.00 © The Geological Society of London 2000.

Even so, a common approach today among Quaternary specialists is to see the structures as an aid to deducing glacial environments and directions of ice movement (e.g. Evans *et al.* 1999), risky though this is without a knowledge of the geometries, kinematics, and physical conditions of the deformation. On the other hand, understanding such aspects of deformation has been a growing theme of structural geology over the last 50 years or so, as the subject has become less descriptive (e.g. Twiss & Moores 1992). These efforts, however, have almost exclusively been restricted to rocks. As noted above, few structural geologists have taken an interest in structures in glaciers. The same is largely true for glacial sediments, and concerted efforts at com-bining ideas developed for rocks with glacial sediments, such as those of Banham (1975), Aber *et al.* (1989), Warren & Croot (1994) and Harris *et al.* (1997), are sparse. Yet analysing structures in Quaternary sediments can have an important practical advantage over studying those in lithified rock: the sediment can readily be scraped away to reveal the full three-dimensional arrangement of the structures. Indeed, some of the most spectacular structures appear in working sand and gravel pits, where large-scale sections are constantly changing.

Because of the importance of sub-glacial deforming sediments to the motion of many ice masses, it is appropriate to examine the deformation of ice and glacial sediments together. In some cases, sediments shearing beneath an ice sheet are best regarded as being in continuum with overlying debris-rich ice (e.g. Hart 1998). Thus the division of the papers included in this volume into separate sections on ice and sediments is largely for convenience. We have attempted to head each section with a paper that provides some insight and overview. Some papers deal with the linkage between ice and sediments, and there are other papers that could have been placed in more than one group. While some of the papers are explicitly contemporary reviews, others combine review with new findings or point strongly to future work.

In many ways, the papers included here confirm the continuing existence of a gap between different groups of workers – especially in terms of approaches, methods and terminology. Some papers that were refereed by, say, a Quaternary specialist and a structural geologist received conflicting reviews, with contrasting opinions on the practices used and the clarity of the terms. Hence, a number of articles herein represent a working compromise. We have attempted to produce a balanced volume, but are keenly aware that something of a schism continues. Our aim is that by collecting together papers on similar subjects from workers with a range of backgrounds and approaches, new possibilities and collaborations may open. We hope the mixture presented here will provide a basis for more integrated approaches in the future.

## Ice deformation

The volume opens with four papers concerned with processes of ice deformation (relationships between ice deformation and structural development being considered in the second section). Despite the small number of papers in this section, a broad range of approaches to investigating ice deformation is presented. These include results from large- and small-scale ice coring projects, laboratory analysis of ice character (including isotopic, gas and chemical composition and ice crystallography), and laboratory analysis of ice rheology, using both triaxial deformation apparatus and a novel **trifuge apparatus.**

The first paper in the section, by **Souchez et al.**, addresses a complex picture of ice formation by freeze-on and deformation near the base of the Greenland Ice Sheet as revealed in a number of basal-ice core sections. Souchez and his team at Brussels are ideally placed to write such a review (based on both published and new information), since they have been working for some years on the physical properties of the basal sections of many of the world's most important ice cores. The data summarised in this contribution relate mainly to the basal sections of 3 cores located in central Greenland (Dye 3 GRIP and GISP2). Interpretation of the gas, stable isotope and chemical composition of these core sections indicates that their silt-laden basal ice layers are largely composed of ice that was formed prior to ice-sheet advance. Souchez *et al.* argue that such silt-rich ice is incorporated without a phase change into the advancing ice sheet, and that it is subsequently distributed and mixed tectonically with clean (firn-derived) glacier ice over some metres to tens of metres at the base of the ice sheet.

The second paper in the section, by **Tison & Hubbard**, presents new information based on a series of eight short ice cores recovered from along a flow-line at Glacier de Tsanfleuron, Switzerland. This paper supplements an earlier classification of the ice facies present in these cores (Hubbard *et al.* 2000) with a wealth of ice crystallographic data. Significantly, the locations of the ice cores allow a glacier-wide flow-line model of crystallographic evolution to be reconstructed. This progression involves a sequence of four crystallographic units, from initial ice development within 20 m of the surface in the accumulation area of the glacier, to strongly metamorphosed ice located within c. 10 m of the glacier bed. The crystallography of this basal ice reflects a steady-state balance between processes of grain growth and strain-related processes of grain-size reduction.

**Baker** *et al.* report a set of constant strain-rate compression tests on individual ice crystals at a variety of orientations, comparing the strain response of undoped ice with that of ice doped with $H_2SO_4$ at 6.8 ppm. While the test corroborates earlier results indicating that the presence of very low concentrations of $H_2SO_4$ reduces both peak stress and subsequent flow stress (Trickett *et al.* 2000), the present contribution goes on to demonstrate that doping does not significantly affect the stress exponent in the flow law for that ice.

Finally, **Irving** *et al.* report a series of ice deformation experiments conducted using the Cardiff Geotechnical Centrifuge in which ice blocks are rotated at a centripetal acceleration of 80 g in a beam centrifuge to recreate the true self-weight stress field that drives real ice masses. Even though the laboratory apparatus is still being perfected (particularly in terms of maintaining a constant, controlled temperature through the sample), the preliminary results presented in this paper are encouraging. These indicate that the technique is viable and that it may have significant potential in terms of testing viscosity differences between different ice types. In particular, ice containing 10% by volume of sand strained an order of magnitude slower ($10^{-7}$ s$^{-1}$) than did clean ice ($10^{-6}$ s$^{-1}$) in these experiments.

## Glacier flow and structures

Glacier flow is manifested in a variety of structures, commonly reflecting deformation on decadal time-scales in temperate valley glaciers to millenial time-scales in polar ice sheets (e.g. Hambrey 1994, chapter 2; Paterson 1994, chapter 9). The 'primary structure', sedimentary stratification (derived from the accumulation of snow and ice) is subject to considerable modification or even obliteration during glacier flow. It is commonly overprinted by the 'secondary' or deformational structure called foliation, either through folding and transposition of the initial layering, or as a completely new structure resulting from shearing. Extensional flow results in the development of crevasses or tensional

veins, ultimately giving rise to crevasse traces or even a new foliation if these layers are subject to severe cumulative deformation, as below an icefall. In compressive flow regimes, thrusts may develop, especially in polythermal glaciers. All these structures are analogous to those in deformed rocks and can be used to generate models of deformation in fold-and-thrust belts. Like rock structures, those in ice reflect long-term deformation (or cumulative strain). Recognizing the significance of structures in glaciers has implications for understanding not only ice dynamics on all scales, but also how debris is incorporated and ultimately deposited by the ice.

The papers in this section focus primarily on the structures observed englacially and at the surface of glaciers, and how they relate to the flow of ice. We start with a review of structural styles and deformation fields in glaciers by **Hambrey & Lawson**. This paper outlines the historical development of this topic, beginning with the nineteenth century pioneers but concentrating on developments since the 1950s, when the physics of glacier flow was elucidated. Deformation rates and histories of various glaciers are described, together with measurements indicating measurements which indicate that ice experiences a polyphase deformation history. The importance of cumulative strain in discussing structural development is highlighted, and comparisons are made between ice structures and structures in deformed rocks. The significance of structural glaciology to the way in which debris is incorporated, transported and finally deposited by glaciers is also evaluated. Pointers are offered to future work on linking ice structures to deformation, especially through modelling approaches.

In a paper on deformation histories and structural assemblages of glacier ice, **Lawson** *et al.* evaluate a range of velocity and strain-rate data from the world's most intensively studied surge-type glacier, the Variegated Glacier in Alaska. The strain histories of ice at different positions on this non-steady-state glacier, following the well-documented 1982–83 surge, are compared. The histories of accumulation of cumulative strain are complex, and can mask the effects of large, transient strain events. Substantial cumulative strain can be 'undone', and the cumulative strain signal may be unrepresentative of earlier strain-rates. Structural relationships do not always reflect the complexity of deformation histories, and it is clear that brittle structures in particular can be reactivated several times as they pass through the glacier.

From the large scale, we turn to a paper by **Wilson** on how deformation is localized in anisotropic ice; this aspect is studied by means of an experimental study involving a series of creep tests. The effects of initial *c*-axis preferred orientation and the inclination of primary layering, during both plane strain compression and combined simple shear-compression are evaluated. The author found that significant variations in both the strain rate and in the development of microstructures occurred during the plane-strain compression experiments. In the combined compression and shear experiments, the minimum shear strain rates vary according to anisotropy in the ice, and deviate from the normal power flow law for isotropic ice.

**Marmo & Wilson** provide an analysis of the stress distribution and deformation history associated with one type of structure, a set of boudins, formed in an outlet glacier of the East Antarctic ice sheet. These boudins arise out of stretching of frozen water-filled crevasses. By documenting the geometrical evolution of the boudins and comparing this with surface strain-rates, it can be determined how these structures form in relation to the stress distribution, as derived from two-dimensional finite-difference modelling. The demonstration that stress is re-fracted across ice-rheological boundaries has implications for analysing planar and linear fabric in rocks.

Next, a paper by **Hubbard & Hubbard** offers a new approach to the understanding of glacier structures using a high-resolution, three-dimensional finite difference model. The model is used to reconstruct the velocity field of Haut Glacier d'Arolla in the Swiss Alps. From this field, it is possible to predict the generation, passage and surface expression of a variety of structures. By means of flow vectors and strain ellipses, the evolution of crevasses, crevasse traces and stratification (mapped from aerial photographs) can be examined. The authors also use the model to track ice that is formed in the accumulation area and then subsequently modified by burial and englacial transport, before being re-exposed in the ablation area. Although this application of the modelling technique is still at an early stage of development, it offers considerable scope to understand better the evolution of a wide variety of structures, including foliation, folds, boudins and a variety of fractures, as highlighted in the first paper in this section.

**Ximenis *et al*.** then report an excellent field study of the kind of glacier-wide structural development that fully complements Hubbards' modelling approach. Ximenis *et al*. deal with folding in Johnsons Glacier, a small cirque-shaped tidewater glacier in the South Shetland Islands of Antarctica. This glacier is unique in having excellent marker horizons in the form of tephra layers, and a cliff section allowing the three-dimensional structure to be observed. The glacier is characterized by converging flow, and several large folds develop axes sub-parallel to the flow-lines. These folds thus result from transverse shortening in response to reducing channel cross-sectional width, and from an increase in differential flow-rates between the centre of the glacier and the margins. Similar structures have been reported from valley glaciers in Svalbard (Norwegian High-Arctic) by Hambrey *et al*. (1999).

The final paper in this section is by **Herbst & Neubauer**, who report on research at the well-known Pasterze Glacier (Pasterzenkees) in the Austrian Alps. This paper explains how a wide range of structures in the glacier, including foliation, shear zones, folds, thrusts and other faults develop, and how they compare with similar structures in rocks. These structures, combined with an understanding of the kinematics of the glacier, are used to develop the analogy with a model of an extensional allochthon, formed on top of an orogenic wedge. The East Carpathian orogen is used as the example.

## Subglacial deformation

Subglacial sediment deformation probably represents the single most actively studied glaciological process over the past 15 years or so. However, many aspects of the actual mechanisms that contribute to this sediment deformation are still poorly understood – principally because of the extreme difficulties involved in both observing the process first-hand and recreating the boundary conditions of those processes in the laboratory. Partly as a result of these difficulties, the very nature of the deformation itself remains unclear. The debate is principally argued out in terms of whether the deformation can be regarded as essentially viscous (where the rate of sediment deformation varies with applied stress) or essentially plastic (where the sediment is unable to sustain any stress above that at which the material fails). Recent research by workers such as Iverson (e.g. Iverson *et al*. 1997, 1998, 1999; Iverson 1999) and Tulaczyk (Tulaczyk *et al*. 1998, 2000*a*, *b*; Tulaczyk 1999) on the details of sediment deformation, and Fischer *et al*. (1998) and Alley (e.g. 1996) on sediment hydrology reflect a shift in the balance of opinion away from a simple (commonly viscous-based) approach towards a more complex (and probably realistic) analysis involving multiple plastic failure, non-steady pore-water pressures, and the role of ploughing of large clasts protruding from the base of the overriding ice (or the roughness of the ice itself). The inclusion of these (perhaps non-steady) processes into dynamic models of ice-mass motion provides a major stimulus to current research on subglacial deformation.

The papers presented in this section fully reflect these contemporary issues: a review of the role of continuity in the presence and character of a deformable subglacial sediment layer; an analysis of the basal sediment-basal ice continuum from an outlet glacier of the East Antarctic ice sheet; a theoretical treatment (based on a viscous deformation model) of the relationships between sediment loading and flow-induced landforms; a presentation of field evidence from Iceland indicating that basal motion during a surge is concentrated close to the ice-sediment interface; radar evidence from the Antarctic interior that indicates the presence of extensive areas of water-saturated subglacial sediments, and a report of a suite of laboratory tests investigating the relationships between pore-water pressure, permeability and strength.

The first of these articles is by **Alley**, who presents a broad review of our current understanding of the physical character of subglacial sediments. Alley concludes that many of the glaciologically important aspects of deforming subglacial sediments are ultimately controlled by till production, continuity and, to some extent, pre-existing surface geology, rather than the finer details of their rheological character.

**Fitzsimons et al.** then present investigations of ice deformation undertaken beneath Suess Glacier, Antarctica, where a 25 m long subglacial tunnel has been painstakingly constructed by chainsaw to gain direct access to the ice-bed interface. This remarkable facility has allowed basal ice velocity to be measured over an extended period. Results indicate that fine-grained amber ice containing relatively high solute concentrations deforms more readily than adjacent clean ice. Further, sediment-laden ice is characterized by more brittle deformation features than cleaner ice, suggesting different modes of failure in this $-17°C$ ice.

Next, **Hindmarsh & Rijsdijk** evaluate whether a viscous model of sediment deformation is consistent with field observations of the nature and scale of loading instabilities within glacigenic sediments. The application of Rayleigh–Taylor theory to layered sediments that are assumed to be characterized by a pre-existing density inversion is found by the authors to result in features that are consistent with those observed in the field. Further, Hindmarsh & Rijsdijk's model indicates that variations in

layer thickness are likely to exert a stronger control over the scale of the resulting instability form than are variations in sediment viscosity.

**Fuller & Murray** report painstaking field evidence from the forefield of recently surged Hagafellsjökull Vestari, Iceland. Detailed analysis of macro-scale and micro-scale structures in recently exposed flutes and drumlins suggests that surge deformation is confined to the upper 16 cm of the glacier's subglacial till layer. Detailed field-based evidence also indicates that the (surge phase) coupling between the glacier base and the subglacial sediment layer was strong and that ploughing by clasts entrained within, but protruding from, the ice was widespread.

In the penultimate paper in this section, **Siegert** reports evidence from the remarkable SPRI-TUD airborne radar database of the East Antarctic ice sheet relating to the nature of the ice–bed interface. Although the presence and character of now well-documented sub-ice lakes have been inferred from flat basal radar returns in earlier papers (e.g. Siegert *et al.* 1996; Siegert & Ridley 1998), this contact is here supplemented by a frozen ice–bed interface (weak and scattered radar return) and an ice-saturated subglacial sediment interface. The latter radar return is similar to, but less flat than, that produced at an ice–lake interface. This significant development points to the potential for ice radar not only to map areas of presumably deformable (and deforming) subglacial sediments, but also to identify large-scale structures within those sediments.

Finally, **Hubbard & Maltman** report on a suite of laboratory investigations of the dynamic permeability of subglacial sediments recovered from Haut Glacier d'Arolla, Switzerland, and a (late Pleistocene) glacierized beach cliff section in South Wales. Results from these experiments indicate that, although large variations in permeability did not occur during deformation (to total strains of <20%), pore-water pressure did exert a major influence over permeability, whether dynamic or static. The nature of this relationship between permeability and effective pressure is best described by an inverse power relationship above a base permeability such that permeability increased dramatically (several orders of magnitude) at effective pressures of less than *c.* 150 kPa.

## Glaciotectonic structures

Before considering the individual papers in this section we contrast terminologies and approaches used. Since the mention of 'glacial tectonics' by Slater (1926), the term has evolved into 'glaciotectonics' and become well established. It can, however, cause confusion in interdisciplinary work because of the way that structural geologists use 'tectonics'. In structural geology the term has a dual meaning. Originally it referred to the 'architecture' (its meaning translated from the original Greek) or configuration of rock masses, as in papers describing the 'tectonics' or 'tectonic style' of a particular region. This would seem to be the usage perpetuated in 'glaciotectonics', but modern usage in structural geology tends to use 'tectonic' to refer to the origin of the forces that caused the deformation. Tectonic forces originate from physico-chemical changes within the Earth, contrasting strongly with gravity-based forces that drive glacial motion (Maltman 1994). Hence the recent habit of some workers to drop the prefix glacio- is dangerous if structural geologists and glaciologists are to communicate clearly. 'Tectonic detachment', for example, has been reported by glaciologists as a process occurring within ice, but the term might puzzle structural geologists since tectonic forces are unlikely to operate within glaciers. Owen (1989) described some structures in sediments in the Himalayas that are partly the product of tectonic forces associated with crustal uplift, and others that are due to ice movement, yet both are referred to as tectonic. To avoid this confusion, all structures directly associated with glacial processes should be referred to as glaciotectonic, a practice we have followed in this volume.

'Glaciotectonic' refers to the direct deformation and structures resulting from the movement or loading of glacier-ice, outside the ice itself. The definition and limitations of the term have been discussed by Aber *et al.* (1989). In line with their suggestions, the term as employed here does not involve structures within glacier ice, and structures resulting from primary depositional processes (such as till fabric) are excluded, as are the effects of freeze/thaw and iceberg grounding. Lithospheric adjustments to changing ice loads can induce deformation, but these effects are also excluded, as the role of ice here is indirect. Glaciotectonic structures include a wide range of features conventionally studied by structural geologists and the larger scale effects can grade into major landforms (e.g. Aber *et al.* 1989; Warren & Croot 1994; Van der Wateren 1995). The structures themselves include a wide variety of folds and faults, on a range of scales, together with a host of related features.

Almost all glaciotectonic structures, unlike those normally encompassed by structural geology, involve only one deformation mechanism:

frictional grain-boundary sliding, sometimes called independent particulate flow (e.g. Maltman 1994). In this, the mineral particles themselves undergo negligible deformation but simply slide past each other; in most glacial deposits the mineral grains are stronger than any bonding between them. Aggregates of clays may be deformed, and, particularly in frozen sediments, any brittle lithic clasts or bedrock that are involved may undergo cataclasis (grain breakage). However, the deformation mechanisms central to most crustal structures, that is the various modes of crystal plasticity and diffusion mass transfer, are unlikely to be significant in the relatively cold and rapid conditions of glaciotectonic deformation. One significant aspect of this difference between glacial sediments and deforming rocks is that terms for the resulting structures have to be used carefully if confusion is to be avoided. Mylonite, for example, is a structural geological term that by definition applies to materials that deformed dominantly by plastic, intracrystalline, processes such as dislocation creep and dynamic recrystallization (e.g. Hippert & Hongn 1998). Yet the term 'mylonitic' has been applied to glacial deposits that almost certainly deformed chiefly by grain-boundary sliding rather than any form of crystal plasticity.

Differences in approach between structural geology and glacial geology are particularly marked in the microscopic study of deformation features, and these differences are reflected in some of the papers in this section. Even the subject is named differently: 'micromorphology', a term used in this volume, is foreign to structural geologists, who tend to talk about 'microstructure'. Basic working terms such as fabric, texture, matrix and structure are minefields of different meanings.

These differences in the structural and Quaternary approaches seem fundamentally to result from their differing heritages. Since Sorby's pioneering work in the middle of the nineteenth century (Sorby 1859), structural geology has drawn largely on materials science and metallurgy (e.g. Kameyama *et al.* 1999), and there has been strong emphasis on the motion geometry of rock particles in response to deforming stresses, in particular the notion of employing 'shear sense indicators'. In contrast, the field concerned with micromorphology of glacial sediments has existed for little more than about twenty years, and is still in the process of cataloguing the features and evolving the most useful terminology. So far, the subject has drawn more on the concepts and terminology of soil science (e.g. Fitzpatrick 1984) and less on the experience of deformed rocks, or for that matter the approaches of soil (geotechnical) engineering. This has the advantage of enabling biological and other high-level constituents that are not normally present in lithified rocks to be dealt with. It also provides a detailed terminology, based on the term 'plasma' with various prefixes, for the fine-grained, typically clayey material that structural geology tends to have neglected. However, it does reduce the emphasis on deformation, not normally an important aspect of agricultural soils. The result is that, at the present time, microstructural studies are rather impenetrable to Quaternary geologists, some of whom use a terminology quite alien to structural geologists. Therefore, there seems a particularly pressing need in the field of microscopic studies of glaciotectonism for greater commonality of approach and terminology. However, it seems premature to attempt a unison here; we have not attempted to lay down yet another set of working definitions and impose editorial prejudices. While any particularly arcane terminology has hopefully been eliminated, authors' preferences have been honoured. Consequently, there is some duality of usage in the following papers and the exact meaning has to be deduced from the context.

The section on glaciotectonics begins with two reviews that illustrate the differences in approach to microscopic scale analyses and that may help coordinate them in future. Each uses a simple terminology and approach that should make the reviews fully accessible. **Menzies** reviews the state-of-the-art in Quaternary studies that draw from the soil science heritage, providing a qualitative classification of deformation structures in glacial sediments and then attempting to link this taxonomy to the different glacial environments. The review minimizes specialist terminology and introduces some terms common to structural geology. Menzies makes it clear that his work is still provisional; deformed glacial sediments have not been surveyed comprehensively at the microscopic scale and features may remain unrecognized. The ideal goal of this kind of work is to seek criteria at the microscopic scale for recognizing particular lithofacies. However, Menzies emphasises that the complexities of nature are likely to preclude this in practice. Rather, microstructural criteria will have to be used together with other features to diagnose particular glacial settings.

**Van der Wateren et al.** review aspects of a structural geological topic that has advanced greatly in recent years: that of diagnosing the kinematics of the deformation. They explain how various microstructural features can lead to a better understanding of deformation move-

ments and magnitudes, but here also, the analyses are best used in conjunction with other kinds of observations. Structural geology has by now evolved methods for dealing with shear sense in flowing materials to a considerable degree of sophistication (e.g. Passchier 1998), with details of the relative roles of different strain symmetries being hotly debated (e.g. Lin et al. 1999). No doubt, particularly complex symmetries exist in deforming glacial materials, through such things as non-planar boundaries and accompanying heterogeneous volumetric strains, but at this stage of the subject's development in glacial geology, Van der Wateren et al. felt it more instructive to restrict their review largely to simple shear.

A structural geological topic that has by now matured is that of recognising and deciphering multiple or polyphase deformation, common, for example, in the rocks that form the mountain belts of the world. The contribution by **Phillips & Auton** illustrates how such concepts can be used to unravel sequences of events in glacial deposits. The paper also shows how a more coordinated terminological approach can be achieved. It deals explicitly with 'micromorphology' yet employs conventional structural geology terminology of S1, S2, for successive generations of fabrics. Structural geological terms such as Riedel shears, boudinage, and pressure shadows are used as appropriate alongside terms derived from Van der Meer (1993) such as lattisepic and unistrial plasmic fabrics, quite foreign to structural geology. Such capitalizing on the strengths of the different approaches in order to employ the most effective terms and concepts must be the way forward if we are to gain before long a more unified approach to microscopic studies of deformed glacial sediments.

Structural geological advances in recent decades on fold-and-thrust belts have already been employed in glacial geology (e.g. Croot 1987; Hambrey et al. 1997) and glaciology (e.g. Hambrey et al. 1999), but the article by **Huuse & Lykke-Andersen** extends such adaptations further. All the glaciotectonic structures they report are submarine and are interpreted from high-resolution multi-channel seismic data, giving unprecedented quality of resolution for such structures. Moreover they are documented on a scale of kilometres that is normally simply too large to observe on land. The profiles they derive compare closely with on-land fold-and-thrust belts (e.g. Macedo & Marshak 1999), and the rooting of the thrusts into an underlying detachment zone is comparable to the architecture of accretionary prisms forming at convergent plate margins (e.g. Morgan & Karig 1995).

The two final papers in this section illustrate the interface between glaciotectonic structures and landforms. **Fowler** provides a dynamic-numerical analysis of the flow processes that may give rise to that still controversial feature – drumlins. **Graham & Midgley** outline a case study of moraines in North Wales, in which the interplay between deformation processes and structures, and particularly the role of thrusting, is explored. Their paper also nicely reflects the pedigree of the studies reported in this volume. It was in the same locality in North Wales that Charles Darwin in 1842 showed that, not only must ice have once extended over southern Britain, but also that the associated processes had much influence on today's landscape.

This volume represents the product of an interdisciplinary conference, entitled the 'Deformation of Glacial Materials', held at the apartments of the Geological Society of London in Burlington House in September 1999. The editors would first and foremost like to thank the conference participants, who contributed so fully to the success of both the conference and this volume. We would also like to thank the staff of Burlington House and the staff of the Geological Society Publishing House respectively for their assistance in organizing the meeting and in producing this volume. In addition, we acknowledge the efforts of the many referees who gave up their time to read and comment on the manuscripts contained in this volume. Finally, we gratefully acknowledge the support of the sponsors of the conference – the Tectonics Studies Group of the Geological Society, the International Glaciological Society, the Quaternary Research Association, and our commercial sponsors, Lasmo plc and Badley Ashton & Associates Ltd.

# References

ABER, J. S., CROOT, D. G. & FENTON, M. M. 1989. *Glaciotectonic landforms and structures.* Kluwer, Dordrecht.

ALLEN, C. R., KAMB, W. B., MEIER, M. F. & SHARP, R. P. 1960. Structure of the lower Blue Glacier, Washington. *Journal of Geology*, **68**, 601–625.

ALLEY, R. B. 1996. Towards a hydrological model for computerized ice-sheet simulations. *Hydrological Processes*, **10**, 649–660.

BANHAM, P. 1975. Glacitectonic structures: a general discussion with particular reference to the contorted drift of Norfolk. *In*: WRIGHT, A. E. & MOSELEY, F. (eds) *Ice Ages, Ancient and Modern.* Geological Journal Special Issue. Seel House Press, Liverpool, 69–94.

BOULTON, G. S. 1979. Processes of glacier erosion on different substrates. *Journal of Glaciology*, **23**, 15–38.

CROOT, D. G. 1987. Glacio-tectonic structures – a mesoscale model of thin-skinned thrust sheets. *Journal of Structural Geology*, **9**, 797–808.

EHLERS, J., GIBBARD, P. L. & ROSE, J. 1995a. *Glacial deposits of Great Britain and Ireland.* A. A. Balkema, Rotterdam.

——, KOZARSKI, S. & GIBBARD, P. 1995b. *Glacial deposits in north-east Europe.* A. A. Balkema, Rotterdam.

ENGELHARDT, H. F., HARRISON, W. D. & KAMB, B. 1978. Basal sliding and conditions at the glacier bed as revealed by bore-hole photography. *Journal of Glaciology*, **20**, 469–508.

EVANS, D. J. A., SALT, K. E. & ALLEN, C. S. 1999. Glacitectonized sediments, Barrier Lake, *Kanaskis* Country, Canadian Rocky Mountains. *Canadian Journal of Earth Sciences*, **36**, 395–407.

FISCHER, U. H., IVERSON, N. R., HANSON, B., HOOKE, R. L. & JANSSON, P. 1998. Estimation of hydraulic properties of subglacial till from ploughmeter measurements. *Journal of Glaciology*, **44**, 517–522.

FITZPATRICK, E. A. 1984. *Micromorphology of soils.* Chapman & Hall, London.

FLINT, R. F. 1971. *Glacial and Quaternary Geology.* John Wiley, New York.

GLEN, J. W. 1955. The creep of polycrystalline ice. *Proceedings of The Royal Society of London, Ser. A*, **228**.

HAMBREY, M. J. 1977. Foliation, minor folds and strain in glacier ice. *Tectonophysics*, **39**, 397–416.

——1994. *Glacial Environments.* UCL Press, London.

—— & MILNES, A. G. 1977. Structural geology of an Alpine glacier (Griesgletscher, Valais, Switzerland). *Eclogae Geologicae Helvetiae*, **70**, 667–684.

——, BENNETT, M. R., DOWDESWELL, J. A., GLASSER, N. F. & HUDDART, D. 1999. Debris entrainment and transfer in polythermal valley glaciers. *Journal of Glaciology*, **45**.

——, HUDDART, D., BENNETT, M. R. & GLASSER, N. F. 1997. Genesis of 'hummocky moraines' by thrusting in glacier ice: Evidence from Svalbard and Britain. *Journal of the Geological Society, London*, **154**, 623–632.

HARRIS, C., WILLIAMS, G., BRABHAM, P., EATON, G. & MCCARROLL, D. 1997. Glaciotectonized Quaternary sediments at Dinas Dinlle, Gwynedd, North Wales, and their bearing on the style of deglaciation in the eastern Irish Sea. *Quaternary Science Reviews*, **16**, 109–127.

HART, J. K. 1998. The deforming bed debris-rich basal ice continuum and its implications for the formation of glacial landforms (flutes) and sediments (melt-out till). *Quaternary Science Reviews*, **17**, 737–754.

HIPPERT, J. F. & HONGN, F. D. 1998. Deformation mechanisms in the mylonite/ultramylonite transistion. *Journal of Structural Geology*, **20**, 1435–1448.

HOOKE, R. I. & HUDLESTON, P. 1978. Origin of foliation in glaciers. *Journal of Glaciology*, **20**, 285–299.

HUBBARD, B., TISON, J.-L. & JANSSENS, L. 2000. Ice-core evidence for the thickness and character of clear facies basal ice: Glacier de Tsanfleuron, Switzerland. *Journal of Glaciology*, **46**, 140–150.

HUDLESTON, P. J. 1976. Recumbent folding in the base of the Barnes Ice Cap, Baffin Island, Northwest

Territories, Canada. *Geological Society of America Bulletin*, **87**, 1684–1692.

IVERSON, N. R. 1999. Coupling between a glacier and a soft bed: II. Model results. *Journal of Glaciology*, **45**, 41–53.

——, BAKER, R. W., HOOKE, R. L., HANSON, B. & JANSSON, P. 1999. Coupling between a glacier and a soft bed: I. A relation between effective pressure and local sheer stress determined from till elasticity. *Journal of Glaciology*, **45**, 31–40.

——, —— & HOOYER, T. S. 1997. A ring-shear device for the study of till deformation: Tests on tills with contrasting clay contents. *Quaternary Science Reviews*, **16**, 1057–1066.

——, HOOYER, T. S. & BAKER, R. W. 1998. Ring-shear studies of till deformation: Coulomb-plastic behavior and distributed strain in glacier beds. *Journal of Glaciology*, **44**, 634–642.

JONES, M. E. & PRESTON, R. M. F. (eds) 1989. *Deformation of Sediments and Sedimentary Rocks*. Geological Society, London, Special Publications, 29.

KAMEYAMA, M., YUEN, D. A. & KARATO, S. I. 1999. Thermal-mechanical effects of low-temperature plasticity (the Peierls mechanism) on the deformation of a viscoelastic shear zone. *Earth and Planetary Science Letters*, **168**, 159–172.

LAWSON, W., SHARP, M. & HAMBREY, M. J. 1994. The structural geology of a surge-type glacier. *Journal of Structural Geology*, **16**, 1447–1462.

LIN, S., JIANG, D., WILLIAMS, P. F., DEWEY, J. F., HOLDSWORTH, R. E. & STRACHAN, R. A. 1999. Discussion on transpression and transtension zones. *Journal of the Geological Society, London*, 1045–1050.

LYELL, C. 1840. The boulder formation or drift and associated freshwater deposits composing the mud cliffs of east Norfolk. *London and Edinburgh Philosophical Magazine and Journal of science, Series 3*, **16**, 345.

MACEDO, J. & MARSHAK, S. 1999. Controls on the geometry of fold-thrust belt salients. *Geological Society of America Bulletin*, **111**, 1808–1822.

MALTMAN, A. J. 1994. Introduction and overview. *In*: MALTMAN, A. (ed.) *The Geological Deformation of Sediments*. Chapman and Hall, London, 1–35.

MEIER, M. F. 1960. *Mode of flow of Saskatchewan Glacier, Alberta, Canada*. U.S. Geological Survey Professional Paper, Report No. **351**.

MORGAN, J. K. & KARIG, D. E. 1995. Kinematics and a balanced and restored cross-section across the toe of the eastern Nankai accretionary prism. *Journal of Structural Geology*, **17**, 31–45.

NYE, J. F. 1953. The flow law of ice from measurements in glacier tunnels, laboratory experiments, and the Jungfraufirn borehole experiment. *Proceedings of the Royal Society of London*, **A219**, 477–489.

——1957. The distribution of stress and velocity in glaciers and ice sheets. *Proceedings of the Royal Society of London*, **A239**, 113–133.

OWEN, L. A. 1989. Neotectonics and glacial deformation in the Karakoram Mountains and nanga Parbat Himalaya. *Tectonophysics*, **163**, 227–265.

PASSCHIER, C. W. 1998. Monoclinic model shear zones. *Journal of Structural Geology*, **20**, 1121–1137.

PATERSON, W. S. B. 1994. *The Physics of Glaciers*. Pergamon.

SIEGERT, M. J. & RIDLEY, J. K. 1998. An analysis of the ice-sheet surface and subsurface topography above the Vostok Station subglacial lake, central East Antarctica. *Journal of Geophysical Research–Solid Earth*, **103**, 10 195–10 207.

——, DOWDESWELL, J. A., GORMAN, M. R. & MCINTYRE, N. F. 1996. An inventory of Antarctic sub-glacial lakes. *Antarctic Science*, **8**, 281–286.

SLATER, G. 1926. Glacial tectonics as reflected in disturbed drift deposits. *Proceedings of the Geologists' Association*, **37**, 392–400.

SORBY, H. C. 1859. On the contorted stratification of the drift of the coast of Yorkshire. *Proceedings of the Geological and Polytechnic Society, West Riding, Yorkshire, 1849–1859*, 220–224.

TRICKETT, Y. L., BAKER, I. & PRADHAN, P. M. S. 2000. The orientation dependance of the strength of single ice crystals. *Journal of Glaciology*, **46**, 41–44.

TULACZYK, S. 1999. Ice sliding over weak, fine-grained tills: dependence of ice-till interactions on till granulometry. *In*: MICKELSON, D. M. & ATTIG, J. W. (eds) *Glacial processes past and present*. Geological Society of America Special Papers, **337**, 159–177.

——, KAMB, B. & ENGELHARDT, H. 2000*a*. Basal mechanics of Ice Stream B. II. Plastic undrained-bed model. *Journal of Geophysical Research*, **105**, 483–494.

——, —— & ——2000*b*. Basal mechanics of Ice Stream B: I. Till mechanics. *Journal of Geophysical Research*, **105**, 463–481.

——, ——, SCHERER, R. & ENGELHARDT, H. F. 1998. Sedimentary processes at the base of a West Antarctic ice stream: Constraints from textural and compositional properties of subglacial debris. *Journal of Sedimentary Research*, **68**, 487–496.

TWISS, R. J. & MOORES, E. M. 1992. *Structural Geology*. W. H. Freeman, New York.

VAN DER MEER, J. J. F. 1993. Microscopic evidence of subglacial deformation. *Quaternary Science Reviews*, **12**, 553–587.

VAN DER WATEREN, F. M. 1995. *Structural geology and sedimentology of push moraines – processes of soft sediment deformation in a glacial environment and the distribution of glaciotectonic styles*. Mededelingen Rijks Geologische Dienst, **54**.

WARREN, W. P. & CROOT, D. G. (eds) 1994. *Formation and deformation of glacial deposits*, A. A. Balkema, Rotterdam.

# Ice Deformation

(Overleaf) Basal ice in Taylor Glacier, Dry Valleys, Antarctica. Sediment has frozen to the base of the glacier by regelation processes, giving rise to layers containing varying concentrations of debris. At the lower left of the photograph, debris is being released to form basal till; clean glacier ice is visible at the top right. (Photo: M. J. Hambrey).

# Basal ice formation and deformation in central Greenland: a review of existing and new ice core data

R. SOUCHEZ[1], G. VANDENSCHRICK[2], R. LORRAIN[1] & J.-L. TISON[1]

[1] *Université Libre de Bruxelles, Département des Sciences de la Terre et de l'Environnement,*
*CP 160/03, B-1050 Brussels, Belgium (e-mail: glaciol@ulb.ac.be)*
[2] *Université Catholique de Louvain, Département GEO, Bâtiment Mercator,*
*B-1348 Louvain-La-Neuve, Belgium*

**Abstract:** In this paper we review, and supplement, existing data to investigate the character, origin and deformation of the basal silty ice of the centre of the Greenland ice sheet as revealed in the Dye3, GRIP and GISP2 cores.

A major process of basal silty ice formation in the central part of Greenland is incorporation of relict non-glacial ice at the base of the ice sheet during its development. The evidence for this can be found in a stable isotope composition study, both in $\delta D$ and $\delta^{18}O$, in a total gas content and gas composition study, in a comparison of the dielectric conductivity profile and chemical profiles. Ice crystallographic investigations and the study of the isotopic compositions in Nd, Sr and Pb of the mineral particles embedded in the silty ice help to clarify the situation.

The processes proposed to explain the stacked sequence developed in the basal silty ice of these central areas are folding resulting in the interbedding of silty ice and glacial ice, and flow-induced mixing related to circular motion of ice in bedrock depressions with flow separation accompanied by some entrainment of the underlying ice.

Within the last ten years, several review papers on basal ice have been published in the glaciological literature (Hubbard & Sharp 1989; Knight 1997; Alley *et al.* 1997). They indicate the various possible processes of formation and deformation of this ice. Within this context, a multi-parametric study can give important information. Such a multi-parametric approach implies the acquisition of stable isotope profiles, both in $\delta D$ and $\delta^{18}O$, total gas content and gas composition profiles and/or dielectric conductivity profile and chemical profiles. The investigation of the mineralogy and the isotopic composition of the particles embedded in the ice can also help to decipher the record stored in the basal part of ice sheets and glaciers.

There is a major difference between situations encountered through deep drillings in the central part of the Greenland Ice Sheet and situations studied along its margins. This paper is largely a review of the results obtained by our group in the study of basal ice retrieved by three deep drillings performed along the main ice divide of the Greenland Ice Sheet, more specifically by two of them for which stable isotope and gas composition profiles are available.

## Basal ice from Central Greenland

Three deep ice cores have penetrated basal silty ice along the main ice divide: the Dye 3 core (65°19'N, 43°82'W), the GRIP core (72°34'N, 34°34'W) and the GISP2 core (72°58'N, 38°48'W) about 28 km downglacier from GRIP (Fig. 1). At Dye 3 more than 25 m of silty ice was recovered at the bottom of the drill hole, which reached a depth of 2037.8 m. The temperature at the bottom was about −12°C. At GRIP, more than 6 m of basal ice was recovered and the temperature at the bottom was recorded at −9°C. The deep drilling was stopped at 3028.7 m depth. At GISP2, 13.1 m of silty ice was penetrated before reaching the ice–bedrock interface. The bed was reached at a drilled depth of 3053.4 m and 1.55 m of bedrock material was obtained before drilling was terminated (Gow & Meese 1996). The bottom temperature is −9°C, like at GRIP. Thus, the basal temperature at these three sites is significantly (at least 6°C) below the pressure melting point.

Debris particles embedded in the basal ice at the three sites are mainly silt- and fine sand-sized with an occasional lithic particle up to 2 cm in

*From*: MALTMAN, A. J., HUBBARD, B. & HAMBREY, M. J. (eds) *Deformation of Glacial Materials*. Geological Society, London, Special Publications, **176**, 13–22. 0305-8719/00/$15.00 © The Geological Society of London 2000.

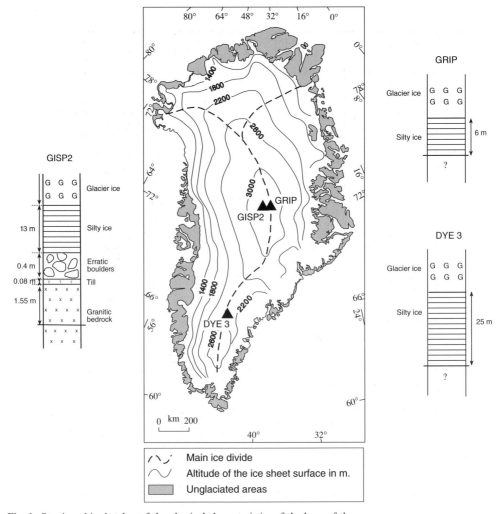

**Fig. 1.** Stratigraphic sketches of the physical characteristics of the base of the cores.

diameter. The dispersed silty and sandy particles are evenly distributed in the silty ice and diffuse banding is frequent. At GISP2, the top 3 m or so of layers of silty ice are intermixed with clear ice layers ranging in thickness from a few millimeters to more than 10 cm (Gow *et al.* 1997, plate 1). At GRIP, the upper 1.2 m of the silty ice contains interbedded silty and clear ice layers (unit 2 in Tison *et al.* 1998, fig. 3). The lower 5 m exhibit only diffuse banding like the lowest 10 m at GISP2. At Dye 3 there is no interbedding with clear ice layers in the upper part of the sequence. The sediment content is much higher in the GISP2 silty ice than in the GRIP silty ice: sediment loads in the ice range from 0.3% to 0.65% by weight at GISP2 (Gow & Meese 1996; Gow *et al.* 1997) and 0.01% to 0.2% at GRIP.

The dirt content has a mean value of 0.03% by weight at Dye 3 (Souchez *et al.* 1998), but is more variable (0.02–1.2% by weight).

Basal ice from the ice divide is fairly different from that exposed along the margins of the Greenland Ice Sheet. In the marginal zone, basal ice typically contains higher concentrations of coarser debris particles, and exhibits strong evidence for melting and refreezing processes (Sugden *et al.* 1987; Clausen & Stauffer 1988; Souchez *et al.* 1993).

## Overriding of pre-existing non-glacial ice

The stable isotope composition of the silty ice at GRIP, obtained by measuring both the $\delta D$ and

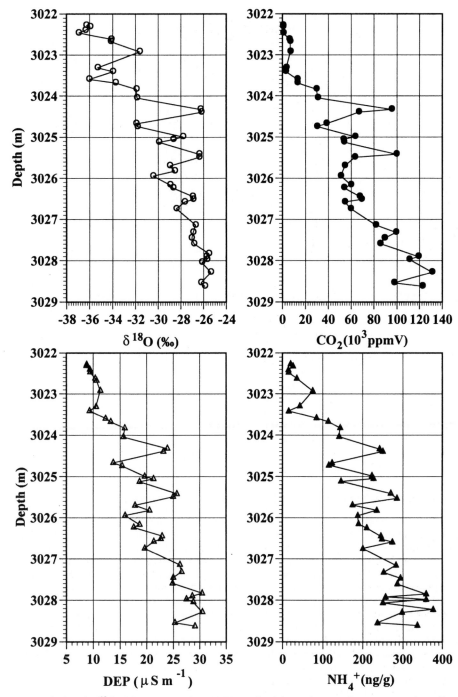

**Fig. 2.** Variations in $\delta^{18}O$, $CO_2$ concentrations, DEP conductivity and ammonium concentrations along the silty ice profile of the GRIP core.

the $\delta^{18}O$ values continuously along the ice core on samples 3.4 cm long, shows striking features (Souchez 1997). $\delta^{18}O$ or $\delta D$ values reach levels never observed anywhere else in the whole core. For example, $\delta^{18}O$ values as high as $-25‰$ are recorded at the base (Fig. 2), while the least negative values in the whole core above, those from the supposed Eemian ice, are around $-32‰$. The local mean annual surface temperature deduced from such a high $\delta^{18}O$ value using the relationship developed for Central Greenland by Johnsen *et al.* (1992) is $-16.9°C$, a value much too high for an extensive ice sheet to exist. Moreover, if the basal ice originated at or near the ground surface in the absence of the ice sheet, a $\delta^{18}O$ value of $-25‰$ would be obtained for climatic conditions similar to those prevailing during the formation of the glacier ice just above the silty ice, taking into account the current $\delta^{18}O$ change with elevation and the isostatic rebound (Souchez *et al.* 1994). This rationale is based on the use of the present-day isotope–temperature spatial relationship. If the isotope–temperature temporal relationship obtained by Jouzel *et al.* (1997) and by Boyle (1997) is taken into account, another estimate is obtained. The internal temperature of the Central Greenland Ice Sheet determined from borehole thermometry (Cuffey *et al.* 1995) implies that the Last Glacial Maximum was 15–20°C colder than the Holocene while the isotopic approach based on the present-day $\delta^{18}O$–temperature relationship from Johnsen *et al.* (1992) would give about 12°C. The shifted $\delta^{18}O$–temperature relationship is a consequence of cooler tropical temperatures. Using the temporal relationship (0.37‰ per °C), the new deduced surface temperature for a $\delta^{18}O$ value of $-25‰$ at the base of the silty ice would be $-4.8°C$, a value, that certainly does not allow the existence of an ice sheet. This interpretation is based on the assumption that the initial $\delta^{18}O$ composition of the ice has not subsequently been modified at the ice–bedrock interface by processes such as melting and refreezing. This assumption is supported by ice properties presented below.

Gas composition analyses of the silty ice give further arguments for its origin at or near the ground surface in the absence of the ice sheet. Extremely high levels (Fig. 2) of carbon dioxide (as high as 130 000 ppmv) and methane (as high as 6000 ppmv) and extremely low levels of oxygen (as low as 3% in volume) have been detected. Diffusion from the ground into the ice and within the ice cannot explain the situation: for the time scale involved, the diffusion length is much too short if one takes the most commonly accepted values for diffusion coefficients of

gases in the ice (Souchez *et al.* 1995a). A striking similarity exists between the gas profiles and the $\delta^{18}O$ profile of the ice itself (Fig. 2). Higher $\delta$ values are related to higher $CO_2$ or $CH_4$ concentrations and lower total gas contents and oxygen concentrations. The similarity between gas and ice parameters is a strong argument in favor of an explanation based on a mixing process (Souchez *et al.* 1995a, b). A mixing model was devised and the characteristics of the local ice component involved in the mixing process defined. These characteristics fit well with ice formed in a marshy environment, possibly within a peat deposit in a permafrost environment. A present-day analogue can be found in the gas distribution in ground ice studied in Alaska by Kvenvolden & Lorenson (1993) and Rasmussen *et al.* (1993).

The dielectric profile of the silty ice from the GRIP core shows a particular pattern (Fig. 2). Di-electric profile conductivities (DEP) reach values in the bottom part of the silty ice higher than the highest measured in the whole core that are observed during the isotopically warm periods ($15–25\,\mu S\,m^{-1}$), with a maximum of $33\,\mu S\,m^{-1}$. Tison *et al.* (1998) have shown that, by contrast with the situation developed in the whole core, the main control on DEP conductivities in the silty ice must fulfil two conditions. First, it must be able to introduce defects in the ice lattice that enhance the DEP signal and, second, it must be associated with the gaseous phase in the ice because of strong correlation between the DEP profile and gas properties. Previous work on the GRIP core (Moore *et al.* 1994; Wolff *et al.* 1995) has indicated the critical role of ammonium in driving the DEP conductivity term linked with intracrystalline defects. Ammonium could have initially existed in the $NH_3$ gaseous form and has thus been regarded as a plausible candidate for explaining the DEP conductivity profile in the silty ice. Tison *et al.* (1998) have shown not only that ammonium is the main control on the dielectric properties in the silty ice, but also that ammonium is strongly linked to oxalate, less to formate and not to acetate. The close association of oxalate and ammonium has been recently explained (Legrand *et al.* 1998) as being one of the fingerprints of ornithogenic soils. This suggests the existence of considerably deglaciated areas, with significant plant cover and efficient biological activity in close vicinity. It therefore strongly corroborates the findings deduced from the stable isotope and gas compositions that part of the ice in the basal sequence originated in a periglacial environment before the settling of the present-day ice sheet.

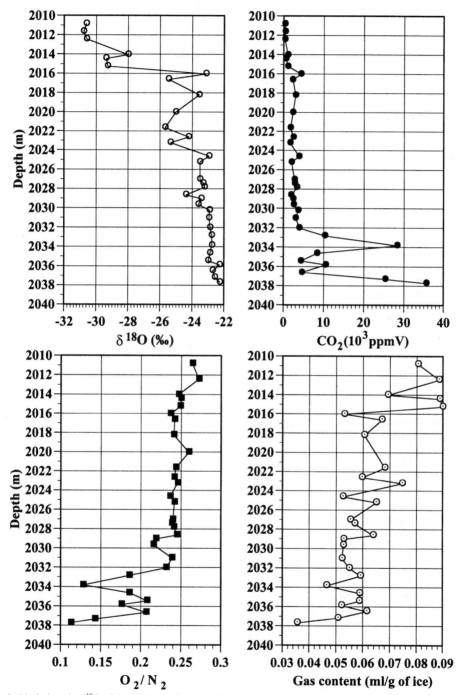

**Fig. 3.** Variations in $\delta^{18}O$, $CO_2$ concentrations, $O_2/N_2$ ratio and total gas content in the basal ice from Dye 3. Unit 2 (as defined in Souchez *et al.* 1988) extends between 2012.50 and 2030 m depth.

At Dye 3 the basal part of the silty ice also shows much higher $\delta^{18}O$ values than anywhere else in the core (Fig. 3). A difference of 5.5‰ is present between the warmest ice in the core (−28‰) and the basal part of the silty ice (−22.5‰). Since the site is also located near the main ice divide of the ice sheet and currently exhibits a mean temperature lower than −14°C, the temporal $\delta^{18}O$–temperature relationship from Boyle (1997) mentioned above can also be used. A local mean annual surface temperature of −6.4°C is computed from the isotopic data, a value much too high for an extensive ice sheet to exist. One can thus consider that, like in GRIP, the basal part of the silty ice at Dye 3 is made of ice formed at or near the ground surface and represents a local member. This was not considered in a previously published paper devoted to the Dye 3 silty ice (Souchez *et al.* 1998). Indeed the δ-shift cannot be the result of ice accretion as developed in Souchez *et al.* (1998).

Gas composition studies are not yet as detailed in the Dye 3 silty ice as in GRIP; the $CH_4$ profile for example has still to be established. $CO_2$ concentrations reach values higher than 40 000 ppmv and the $O_2/N_2$ ratio can be as low as 0.02 (Fig. 3). There is a close relationship between δ values and gas parameters in the higher part of the silty ice, between 2012.5 m and 2030 m: higher δ values are correlated with higher $CO_2$ concentrations, lower $O_2/N_2$ ratios and lower total gas concentrations (unit 2 in Souchez *et al.* 1998). This striking similarity with the GRIP core provides a strong argument in favour of a mixing process between glacial ice (unit 1 in Souchez *et al.* 1998) and the top of the local member. The seven bottom metres of silty ice at Dye 3 represent this local member with quite constant higher δ values and highly variable $CO_2$ concentrations (unit 3 in Souchez *et al.* 1998).

Concerning the GISP2 core, the δ profiles and the gas composition profiles of the silty ice are not yet available. It is therefore premature to tell that the silty ice at GISP2 does not result from ice accretion and/or ice diagenesis at the base of the ice sheet. However, some information can be derived from a comparative isotopic study of the mineral particles present in the silty ice. The particles embedded in the silty ice of the GRIP core have been studied for their Nd, Sr and Pb isotopic compositions. Similar analyses were performed for comparison on the particles embedded in the silty ice of the GISP2 core and on the subglacial till and granitic bedrock from the GISP2 rock core (Weis *et al.* 1997). The till unit (Fig. 1) present in the GISP2 rock core just above the granitic bedrock, and the till

particles embedded in the GRIP basal ice, show similar Sr and Nd isotopic compositions, different from the local granitic bedrock. Conversely, the silt particles embedded in the GISP2 basal ice have Sr and Nd isotopic compositions similar to those of the local granitic bedrock. This suggests that subglacial bedrock erosion occurred locally, perhaps on topographically high bedrock knolls, where the till sheet was not present. The presence of the till unit near GRIP, 28 km upglacier, implies that the till must have a significant areal extent in Central Greenland. We think that the till layer overlying bedrock at GISP2 and the mixing of glacier ice with local ice formed in the absence of the Greenland Ice Sheet at GRIP date back from the build up of the present ice sheet. This supports the hypothesis that the Greenland Ice Sheet in the Summit area did not result from in situ growth from local snow banks. The probable origin of a dolerite erratic boulder found in the subglacial till at GISP2 from East Greenland indicates progression of the ice sheet from the East. The information retrieved from the comparative study of the isotopic compositions of the particles indicates a different situation at GISP2 from that at GRIP. The results from stable isotope and gas composition studies of the silty ice would be particularly helpful within this context.

## Deformation by folding and flow-induced mixing

There are significant differences in crystallographic properties between the silty and the clear ice layers within the upper part of the basal debris zone at Summit. The coarse-grained texture of the clear, debris-free ice contrasts strikingly with the fine-grained nature of the silt-bearing ice. Additionally, the coarse-grained clear-ice exhibits a multi-maximum orientation of its *c*-axes as opposed to the vertical clustering of *c*-axes exhibited by the silty ice (Tison *et al.* 1994; Gow & Meese 1996). Since the glacial ice textures and fabrics bear the imprint of annealing recrystallization at relatively high temperatures, it seems that the same process has also extended into the clear ice. The single maximum fabric of the silt-bearing ice indicates the existence of significant shear deformation. Present-day stress conditions at Summit are unlikely to favour active simple shear in the basal ice layers. At the ice divide, horizontal shear stress should be considerably reduced in favour of vertically compressive stress. The possible effect of sloping bedrock implying some simple shear flow at a large scale is precluded from radio-echo-soundings in the area (Hempel & Thyssen 1992). The observed fabrics in the silty ice must therefore partly result from previous stress conditions. At GRIP, the location of the single maximum fabric within a vertical girdle can be understood by overriding of the site by an ice sheet in progression, as a result of fluctuating simple shear stress on the changing bedrock topography (Tison *et al.* 1994).

A study of the composition of the clear ice layers present in the upper part of the silty ice sequence at GRIP gives further information on their origin (Fig. 4). While the chlorine and nitrate concentrations in the silty ice are in general at the level of $40\,ng\,g^{-1}$ and $25\,ng\,g^{-1}$ respectively, a major peak is present in the record: it is developed in the thickest clear ice layer interbedded with the silty ice in the upper part of the sequence. In this clear ice layer, the concentration in chlorine reaches $399\,ng\,g^{-1}$, while the concentration in nitrate reaches $96\,ng\,g^{-1}$. This is to be compared with Cl concentrations of $218\,ng\,g^{-1}$ and $NO_3$ concentrations of $102\,ng\,g^{-1}$ present in the glacier ice just above the silty ice. Such high concentrations are accompanied by (1) a higher total gas content, (2) quasi atmospheric $CO_2$, $CH_4$ and $O_2$ concentrations, and (3) low mean $\delta^{18}O$ values (averaged over the size of the chemical samples), although there is no strict, fine-scale correspondence. The interbedded clear ice layer has slightly higher gas concentrations than glacier ice above the silty ice but one has to take into account post-interbedding diffusion of gases from neighbouring silty ice layers. Such a picture leaves little doubt that clear ice layers interbedded in the basal sequence are glacial ice layers. The interbedding is presumably the result of folding.

Thickening of the basal layer could occur by flow divergence and concentration around subglacial obstructions in zones of longitudinal compression. Deformation is more pronounced close to the bed where simple shear is important

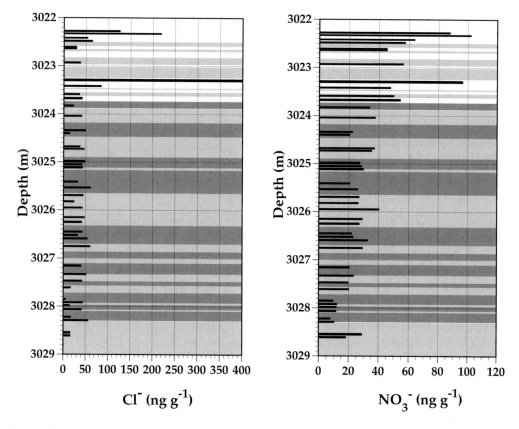

**Fig. 4.** Chlorine and nitrate concentration profiles in the GRIP silty ice (analyses done at LGGE, Grenoble, France). Bar width shows individual sample thickness. The gray scale in the background is a qualitative estimate of the debris load, through visual inspection of the cores in transmitted light.

and the viscosity of the ice reduced by higher temperatures (Hubbard & Sharp 1989). Folds can also be generated under simple shear where it acts on an initially shallow undulatory layering (Hudleston 1977). Subsequent shear strain will tend to tighten these fold structures until they become isoclinal and banding again develops parallel to the bed. If folding can explain the interbedding of clear and silty ice layers in the upper part of the basal ice at GRIP and probably at GISP2, no visible evidence that this process occurs exists either in the lower, main part of the silty ice at Summit, or over the complete sequence of basal ice at Dye 3.

A flow-induced mixing process was considered to explain the situation at GRIP (Souchez *et al.* 1995*a, b*) and at Dye 3 (Souchez *et al.* 1998). Figure 2 shows the variations with depth of several parameters in the GRIP silty ice: $\delta^{18}O$, which is a property of the ice, $CO_2$, which is a gas parameter, and ammonium, which is a compound present in the ice matrix. The co-variations of such different parameters imply an explanation based on a flow-induced mixing process. If, at Dye 3 the top part of the silty ice is considered (Fig. 3), there is also a very good correlation along a mixing curve between $\delta^{18}O$ and $CO_2$ (unit 2 in Souchez *et al.* 1998). Very large $\delta$ fluctuations are present in GRIP and in Dye 3 before tapering out in the base of the sequence. This can be considered as a signature of the proposed deformation process.

It is possible to understand these properties using the results obtained by Gudmundsson (1997). Ice within a trough rotates slowly with the upper part of this circular motion in the general flow direction of the ice. For a critical roughness, calculated by Gudmundsson for different boundary conditions including no-slip, there is flow separation accompanied by some entrainment of the underlying ice. The resolutions of the radio-echo-soundings at GRIP or at Dye 3 are, however, not clearly sufficient to compare a known roughness to that required to cause flow separation at a metric scale. The presence of relict ice from a stage when the ice sheet was not present at the site calls nevertheless for some preservation mechanism. Such preservation is better explained by flow-induced mixing through circular motion than by folding.

### Physical conditions at the bed

The present-day thermal situation at the three sites is a temperature well below the pressure melting point at the base. The question therefore arises as to whether such basal conditions were continuously present in the past or whether phase changes were a likely phenomenon at some time so that their possible effects must be considered?

Melting and refreezing have occurred in the Holocene, and are presently occurring in the marginal zone of the Greenland Ice Sheet. Basal ice from the marginal zone has a co-isotopic composition, both in $\delta D$ and $\delta^{18}O$, indicative of freezing of a limited water reservoir (Jouzel & Souchez 1982; Souchez & Jouzel 1984; Sugden *et al.* 1987; Souchez *et al.* 1988; Knight *et al.* 1994). Gas composition analysis shows either a depletion or an enrichment in the most soluble gases, $CO_2$ and $O_2$, in accordance with selective melting or with partial freezing (Souchez *et al.* 1993). Such physical conditions were presumably also frequent in the marginal zone in the past. We thus believe that if entrainment of relict ice took place in the marginal zone of a progressing ice sheet, then ice composition would show the influence of phase changes. Neither at GRIP, nor at Dye 3 is there evidence for this.

The deuterium excess at GRIP and Dye 3 is quite constant in the silty ice profile and similar to the values obtained in the glacier ice above the basal sequence. No trend in deuterium excess was observed, precluding any significant ice–water phase change.

Glacial ice above the basal sequence at GRIP and at Dye3 has $\delta$ values implying temperatures intermediate between that of the Last Glacial Maximum and that of the Holocene for a well-developed ice sheet. The gas content of about 0.1 and $0.09 \, cm^3 \, g^{-1}$ respectively is a further argument for a substantial ice body, since total gas content is a good indicator of the elevation of the site of formation of the glacial ice considered (Raynaud & Lebel 1979).

The physical conditions at the bed therefore imply entrainment by a cold-based growing ice sheet. Entrainment at cold glacier beds is now considered in the recent literature. The existence of thin water films at the interfaces between ice and mineral particles at temperatures well below the pressure melting point (Dash *et al.* 1995) is presumably the driving mechanism.

### Conclusion

The silty ice present at the base of the central part of the Greenland Ice Sheet cannot have been formed in a situation where ice domes are developed as they are today. A previous configuration of the ice sheet is required in order to produce the necessary shear for entrainment and deformation. This does not mean that the silty ice was formed under marginal conditions since the basal ice along the margins of the ice sheet has markedly different characteristics.

Incorporation into the basal part of the ice sheet of relict non glacial ice by overriding during ice sheet development appears to be an important process of basal ice formation in the central part of the Greenland Ice Sheet. This can be demonstrated by stable isotope composition analyses and by total gas content and gas composition analyses. For the GRIP core, DEP conductivity and chemical profiles provide additional supporting arguments. Crystallographic investigations and the study of the isotopic composition of particles embedded in the ice help to decipher the complexity of the situation.

Two main processes of basal ice deformation occur: folding, which clearly results in interbedding of silty ice and glacial ice in the upper part of the sequence, and flow-induced mixing related to circular motion within troughs, depending on a critical bedrock roughness, at the metric scale. Such processes are likely to be active well below the pressure melting point since there is no evidence that phase changes occurred there in the past at such a scale.

This work is a contribution to the Greenland Ice Core Project (GRIP) organized by the European Science Foundation. We thank the GRIP participants and supporters for their co-operative effort. We also thank the national science foundations in Belgium, Denmark, France, Germany, Iceland, Italy, Switzerland and the United Kingdom, as well as the XII Directorate of the E.C. Thanks are also due to M. de Angelis and M. Legrand from LGGE, Grenoble for the chlorine and nitrate analyses. J.-L. Tison is Research Associate at the National Foundation for Scientific Research (Belgium).

# References

ALLEY, R. B., CUFFEY, K. M., EVENSON, E. B., STRASSER, J. C., LAWSON, D. E. & LARSON, G. J. 1997. How glaciers entrain and transport sediment at their beds: physical constraints. *Quaternary Science Reviews*, **16**, 1017–1038.

BOYLE, E. A. 1997. Cool tropical temperatures shift the global $\delta^{18}$O-T relationship: An explanation for the ice core $\delta^{18}$O-borehole thermometry conflict? *Geophysical Research Letters*, **24**, 273–276.

CLAUSEN, H. B. & STAUFFER, B. 1988. Analyses of two ice cores drilled at the ice-sheet margin in West Greenland. *Annals of Glaciology*, **10**, 23–27.

CUFFEY, K. M., CLOW, G. D., ALLEY, R. B., STUIVER, M., WADDINGTON, E. D. & SALTUS, R. W. 1995. Large Arctic temperature change at the Wisconsin-Holocene glacial transition. *Science*, **270**, 455–458.

DASH, J. G., FU, H. & WETTLAUFER, J. 1995. The premelting of ice and its environmental consequences. *Report on Progress in Physics*, **58**, 115–167.

GOW, A. J. & MEESE, D. A. 1996. Nature of basal debris in the GISP2 and Byrd ice cores and its relevance to bed processes. *Annals of Glaciology*, **22**, 134–140.

——, ——, ALLEY, R. B., FITZPATRICK, J. J., ANANDAKRISHNAN, S. G. A. & ELDER, B. C. 1997. Physical and structural properties of the Greenland Ice Sheet Project 2 ice core – A review. *Journal of Geophysical Research*, **102**(C12), 26 559–26 575.

GUDMUNDSSON, G. H. 1997. Basal-flow characteristics of a non-linear flow sliding frictionless over strongly undulating bedrock. *Journal of Glaciology*, **43**, 80–89.

HEMPEL, L. & THYSSEN, F. 1992. Deep radio echo soundings in the vicinity of GRIP and GISP2 drill sites, Greenland. *Polarforschung*, **62**, 11–16.

HUBBARD, B. & SHARP, M. 1989. Basal ice formation and deformation: a review. *Progress in Physical Geography*, **13**, 529–558.

HUDLESTON, P. J. 1977. Progressive deformation and development of fabric across zones of shear in glacial ice. *In*: SAXENA, S. K. & BHATTACHARIJS (eds) *Energetics of Geological Processes*. Springer-Verlag, New-York, 121–150.

JOHNSEN, S. J., CLAUSEN, H., DANSGAARD, W., FUHRER, K., GUNDESTRUP, N., HAMMER, C., IVERSEN, P., JOUZEL, J., STAUFFER, B. & STEFFENSEN, J. 1992. Irregular glacial interstadials recorded in a new Greenland ice core. *Nature*, **359**, 311–313.

JOUZEL, J. & SOUCHEZ, R. 1982. Melting-refreezing at the glacier sole and the isotopic composition of the ice. *Journal of Glaciology*, **28**, 35–42.

——, ALLEY, R. B. CUFFEY, K. M., DANSGAARD, W., GROOTES, P., HOFFMANN, G., JOHNSEN, S. J., KOSTER, R. D., PEEL, D., SHUMAN, C. A., STIEVENARD, M., STUIVER M. & WHITE, J. 1997. Validity of the temperature reconstruction from water isotopes in ice cores. *Journal of Geophysical Research*, **102** (C12), 23 471–26 487.

KNIGHT, P. G. 1997. The basal ice layer of glaciers and ice sheets. *Quaternary Science Reviews*, **16**, 975–993.

——, SUGDEN, D. E. & MINTY, C. 1994. Ice flow around large obstacles as indicated by basal ice exposed at margin of the Greenland ice sheet. *Journal of Glaciology*, **40**, 359–367.

KVENVOLDEN, K. A. & LORENSON, T. D. 1993. Methane in permafrost – preliminary results from coring at Fairbanks, Alaska. *Chemosphere*, **26**, 609–616.

LEGRAND, M., DUCROZ, F., WAGENBACH, D., MULVANEY, R. & HALL, J. 1998. Ammonium in coastal antarctic aerosol and snow: Role of polar ocean and penguin emissions. *Journal of Geophysical Research*, **103**, 11 043–11 056.

MOORE, J. C., WOLFF, E. W., CLAUSEN, H. B., HAMMER, C. U., LEGRAND, M. R. & FUHRER, K. 1994. Electrical response of the Summit-Greenland ice core to ammonium, sulphuric acid and hydrochloric acid. *Geophysical Research Letters*, **21**, 565–568.

RASMUSSEN, R. A., KHALIL, M. A. K. & MORAES, F. 1993. Permafrost methane content: 1. Experimental data from sites in Northern Alaska. *Chemosphere*, **26**, 591–594.

RAYNAUD, D. & LEBEL, B. 1979. Total gas content and surface elevation of polar ice sheets. *Nature*, **281**, 289–291.

SOUCHEZ, R. 1997. The buildup of the ice sheet in central Greenland. *Journal of Geophysical Research*, **102** (C12), 26 317–26 323.

—— & JOUZEL, J. 1984. On the isotopic composition in dD and $d^{18}O$ of water and ice during freezing. *Journal of Glaciology*, **30**, 369–372.

——, BOUZETTE, A., CLAUSEN, H., JOHNSEN, S. J. & JOUZEL, J. 1998. A stacked mixing sequence at the base of the Dye 3 core. *Geophysical Research Letters*, **25**, 1943–1946.

——, JANSSENS, L., LEMMENS, M. & STAUFFER, B. 1995b. Very low oxygen concentration in basal ice from Summit, Central Greenland. *Geophysical Research Letters*, **22**, 2001–2004.

——, —— & CHAPPELLAZ, J. 1995a. Flow-induced mixing in the GRIP basal ice deduced from the $CO_2$ and $CH_4$ records. *Geophysical Research Letters*, **22**, 41–44.

——, ——, TISON, J.-L., LORRAIN, R. & JANSSENS, L. 1993. Reconstruction of basal boundary conditions at the Greenland Ice Sheet margin from gas composition in the ice. *Earth and Planetary Science Letters*, **118**, 327–333.

——, LORRAIN, R., TISON, J.-L. & JOUZEL, J. 1988. Co-isotopic signature of two mechanisms of basal-ice formation in arctic outlet glaciers. *Annals of Glaciology*, **10**, 163–166.

——, TISON, J. L., LORRAIN, R., LEMMENS, M., JANSSENS, L., STIÉVENARD, M., JOUZEL, J., SVEINBJÖRNSDOTTIR, A. & JOHNSEN, S. J. 1994. Stable isotopes in the basal silty ice preserved in the Greenland Ice Sheet at Summit; environmental implications. *Geophysical Research Letters*, **21**, 693–696.

SUGDEN, D. E., KNIGHT, P. G., LIVESEY, N., LORRAIN, R. D., SOUCHEZ, R. A., TISON, J.-L. & JOUZEL, J. 1987. Evidence of two zones of debris entrainment beneath the Greenland Ice Sheet. *Nature*, **328**, 238–241.

TISON, J.-L., SOUCHEZ, R., WOLFF, E., MOORE, J. C., LEGRAND, M. M. & DE ANGELIS, M. 1998. Is a periglacial biota responsible for enhanced dielectric response in basal ice from Summit-Greenland ice core? *Journal of Geophysical Research*, **103**, (D15), 18885–18894.

——, THORSTEINSSON, T., LORRAIN, R. D. & KIPFSTHUL, S. 1994. Origin and development of textures and fabrics in basal ice at Summit, Central Greenland. *Earth and Planetary Science Letters*, **125**, 421–437.

WEIS, D., DEMAIFFE, D., SOUCHEZ, R., GOW, A. J. & MEESE, D. A. 1997. Ice sheet development in Central Greenland: implications from the Nd, Sr and Pb isotopic compositions of basal material. *Earth and Planetary Science Letters*, **150**, 161–169.

WOLFF, E. W., MOORE, J. C., CLAUSEN, H. B., HAMMER, C. U., KIPFSTUHL, J. & FUHRER, K. 1995. Long-term changes in the acid and salt concentrations of the Greenland Ice Core Project ice core from electrical stratigraphy. *Journal of Geophysical Research*, **100** (D8), 16 249–16 263.

# Ice crystallographic evolution at a temperate glacier: Glacier de Tsanfleuron, Switzerland

JEAN-LOUIS TISON[1] & BRYN HUBBARD[2]

[1] *Département des Sciences de la Terre et de l'Environnement, Faculté des Sciences, Université Libre de Bruxelles, CP 160/03 50 Avenue F.D. Roosevelt. 1050 Bruxelles, Belgium (e-mail: jtison@ulb.ac.be)*
[2] *Centre for Glaciology, Institute of Geography and Earth Sciences, University of Wales, Aberystwyth, Ceredigion SY23 3DB, Wales, UK (e-mail: byh@aber.ac.uk)*

**Abstract:** Ice crystallographic measurements have been made on eight cores retrieved from temperate Glacier de Tsanfleuron, Switzerland. Cores are aligned approximately along a flow-parallel transect, allowing a stratigraphic model of crystallographic evolution at the glacier to be constructed.

Results indicate the presence of four crystallographic units at the glacier. Unit 1 composed of homogeneous, fine-grained ice with a uniform fabric, is located within *c*. 20 m of the ice surface in the accumulation area of the glacier. Crystal growth within this unit occurs in the absence of significant stresses, and its rate is closely described by an Arrhenius-type relationship. Unit 2 ice, characterized by the local development of coarser crystals, forms after some decades of Arrhenius growth, marking the initial influence of processes of dynamic recrystallisation. Unit 3 ice, characterized by an abrupt increase in minimum crystal size, occurs at a depth of *c*. 33 m throughout the glacier. In the accumulation area, this increase coincides with the first evidence of systematic fabric enhancement, interpreted in terms of dynamic recrystallisation. Unit 4 ice, characterized by large, interlocking grains with a multi-modal girdle fabric, develops within *c*. 10 m of the glacier bed. Here, the measured minimum crystal size is consistent with a steady-state balance between Arrhenius processes of grain growth and strain-related processes of grain-size reduction. These changes are interpreted in terms of the effects of intense, continuous deformation in this basal zone.

Ice crystal size, shape and orientation exert a strong control over bulk ice rheology. Significantly, investigations of ice cores recovered from polythermal ice masses (showing a basal layer of finite thickness at the pressure melting point) indicate that some of these crystallographic properties may vary systematically with depth. In particular, progressive crystal alignment at high total shear strains may result in 'strain softening' and a corresponding increase in strain rate response to imposed stress. This effect has been reported for the Camp Century ice core, Greenland (Shoji & Langway 1984), the Barnes Ice Cap core, Canada (Hooke *et al.* 1988), the Law Dome ice core, Antarctica (Russell-Head & Budd 1979; Thwaites *et al.* 1984). Strain rate also responds to variations in ice crystal size. In general, research has indicated that fine-grained ice is stronger than coarse-grained ice (Hambrey & Müller 1978), although isolating these influences from other inter-dependent factors, such as crystal shape and included debris concentration, is difficult (e.g. Fisher & Koerner 1986; Hooke *et al.* 1988). Re-interpreting borehole inclination measurements at Dye 3 in terms of strain-rate enhancement (Dahl Jensen & Gundestrup 1987), Thorsteinsson *et al.* (1999) recently demonstrated that, although the major features of the deformation at Dye 3 are explained by anisotropy and temperature, the residual enhancement correlates well with dust and soluble-ion concentration divided by crystal size. These authors suggest that most of the 'excess deformation' may be due to impurities or crystal size.

Crystallographic variability therefore plays an important role in controlling rates and patterns of ice mass deformation. However, although crystallographic investigations were first initiated at temperate glaciers (e.g. Perutz & Seligman 1939), and extensive crystallographic studies of temperate glaciers took place in the 1950s to the 1980s (reviewed by Budd 1972, and extended by e.g. Meier *et al.* 1974; Hambrey & Müller 1978; Hambrey *et al.* 1980), relatively little is known

*From*: MALTMAN, A. J., HUBBARD, B. & HAMBREY, M. J. (eds) *Deformation of Glacial Materials*. Geological Society, London, Special Publications, **176**, 23–38. 0305-8719/00/$15.00 © The Geological Society of London 2000.

about the internal crystallographic structure of
temperate valley glaciers. Budd & Jacka (1989)
pointed out that most data have been concentrated on near-surface observations, stating that:
'... so far, a comprehensive flow-line study of
deep ice properties for a temperate glacier is not
available'. To our knowledge this gap has not
been addressed yet, probably partly because
of the increased recent interest in the rheological properties of deep polar ice cores. In this
paper we report on the internal crystallographic
evolution of a temperate alpine glacier (Glacier
de Tsanfleuron, Switzerland) as revealed in a
series of eight, medium-length ice cores aligned
approximately along a flow-line.

## Field site, methods and ice type classification

Fieldwork was conducted at Glacier de Tsanfleuron, Switzerland, a $c.\,4\,km^2$ temperate glacier
located between $c.\,2420\,m$ a.s.l. and $c.\,2900\,m$
a.s.l., that flows over Cretaceous and Tertiary
limestones (Fig. 1). It can be considered as a
plateau glacier with a small discrete ice tongue
on its eastern side, which has experienced continuous retreat since the Little Ice Age (see
1855–1860 moraine ridge in Fig. 1). The glacier
has been extensively studied, particularly in
terms of ice–bedrock interface processes including the character and formation of its basal ice
layers (Tison & Lorrain 1987; Sharp *et al.* 1989,
1990; Hubbard & Sharp 1995; Hubbard &
Hubbard 1998).

Eight vertical ice cores were recovered using
an adapted RAND corer powered by a portable generator. Core locations were aligned
approximately along a flow line, with five cores
recovered from the ablation area of the glacier in
1996, and the remaining three cores from further
up-glacier in 1997 (Fig. 1). In total, 170 m of ice
core was recovered (Table 1). Since the total
length of each core was limited to approximately
45 m only the 1996 cores, located above relatively
thin marginal ice, reach the glacier bed. The 1997
cores, located some kilometres up-glacier, end
englacially. Individual core segments, which were
typically 75 mm in diameter and up to 0.6 m
long, were logged, bagged and stored on site
in a freezer maintained at $c.\,-30°C$. The core
remained in these freezers during transport from
the Alps and during storage until the ice was
analysed in a cold laboratory at $c.\,-18°C$.

Ice crystallographic measurements were performed on selected core sections, chosen on the
basis of visual inspection of the whole cores
in transmitted light, as described below. Thin-
sections were cut in the vertical plane, using the
standard microtoming procedure (Langway
1958) or a diamond-wire sawing technique, for
debris-laden or brittle ice (Tison 1994). Thin-
sections were typically $700\,\mu m$ thick. Crystal
sizes were measured at 20 mm depth intervals
throughout the sections, using the linear intercept method (Pickering 1976). However, since
these were measured on vertical thin-sections at
the given resolution, only single linear traverse

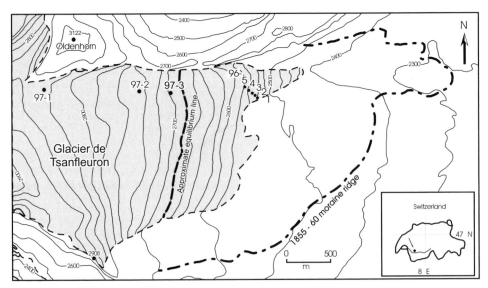

**Fig. 1.** Glacier de Tsanfleuron, Switzerland, and the locations of the ice cores drilled in 1996 (96-1 to 96-5) and
1997 (97-1 to 97-3). The stippled line delineates the present-day ice margin. Contour interval is 20 m.

**Table 1.** *Ice core lengths*

| Core no. | Core length (m) |
|----------|-----------------|
| 96-1 | 3.47 |
| 96-2 | 9.70 |
| 96-3 | 13.00 |
| 96-4 | 19.40 |
| 96-5 | 44.84 |
| 97-1 | 39.86 |
| 97-2 | 19.99 |
| 97-3 | 19.28 |

Cores prefixed by 96 were drilled near the frontal margin of the glacier in 1996, extending from the glacier surface to the bed. Cores prefixed by 97 were drilled in the accumulation area of the glacier in 1997, terminating englacially. Core locations are given in Fig. 1.

**Table 2.** *Ice type classification*

| Ice type | Description |
|----------|-------------|
| 0 | Incomplete closure of inter-grain spaces (snow and firn) |
| 1 | Medium or high density of small bubbles, or the intercalation of such ice with thin ($\leq 6$ cm) layers of any other ice type |
| 2 | Medium or high density of medium bubbles |
| 3 | Low density of medium bubbles |
| 4 | Medium/high density of large bubbles |
| 5 | Low density of large bubbles |
| 6 | Bubble-free and debris-poor |
| 7 | Bubble-free and debris-rich |

could be performed at each level. Following Pickering (1976) and Jacka (1984), we adopt a multiplication factor of 1.75 in order to standardize our crystal size measurements with those of the alternative techniques of count per unit area and maximum core intercept.

From the two longest cores (one in the accumulation area and the other in the ablation area), thin sections were selected for c-axes measurements on a Universal stage using the procedure outlined by Langway (1958). The resulting data were plotted on the lower hemisphere of a Schmidt net using the refraction corrections supplied by Kamb (1962). All measurements were performed on vertical thinsections and readings were rotated in the horizontal plane. For logistic reasons, no azimuthal orientation consistency exists between individual fabric diagrams. The maximum error in c-axis determination is 5° (Langway 1958).

Ice cores were visually logged in transmitted light and classified at a length resolution of 10 mm for bubble density and size. Core classification, reported in Hubbard et al. (2000), adopted eight material categories that were considered to reflect a progressive evolution from snow through to metamorphosed clear facies ice and debris-rich basal ice (Table 2). Categories were labelled Types 0 to 7: Type 0 material is snow and firn, characterized by the incomplete closure of incorporated gas bubbles; Type 1 ice is characterized by a medium (opaque, light grey) or high (opaque, white) density of small ($<1$ mm in diameter) bubbles or the fine-scale ($\leq 6$ cm) intercalation of such ice with any other ice type; Type 2 ice is characterized by a medium or high density of medium-sized (1 to 5 mm in

diameter) bubbles; Type 3 ice is characterized by a low density (transparent, grey) of medium-sized bubbles; Type 4 ice is characterized by a medium or high density of large bubbles ($>5$ mm in diameter); Type 5 ice is characterized by a low density (scattered) of large bubbles, and Type 6 ice is characterized by no included bubbles. Type 7 ice is bubble-free and debris-rich. All of the core material fell into these eight ices types. Classification of the ice-marginal cores (Hubbard et al. 2000) (Fig. 2a) indicates a general down-core progression from ice containing a high concentration of fine bubbles (Types 1 and 2) at the top of each core, through ice containing a progressively lower concentration of larger bubbles (Types 3 to 5), to a c. 10 m thick layer of bubble-free, clear facies ice (Type 6) towards the base of each of the cores. This layer is underlain only by a c. $10^{-1}$ m thick layer of debris-rich, Type 7 basal ice, that contacts the substrate at the base of each core (Fig. 2a). Hubbard et al. (2000) interpret this pattern of down-core gas expulsion in terms of strain-induced metamorphism involving pervasive grain-boundary melting and refreezing. Such a process may have significant implications for the crystallographic character of the ice involved. Below, we report on this crystallography, as well as the stratigraphic classification of the 1997 cores.

## Results

### Ice type classification: 1997 cores

In contrast to the 1996 cores, the 1997 cores (Fig. 2b) include a surface layer of Type 0 snow and firn. The thickness of this layer increases upglacier, from c. 1 m in core 97-3 to almost 8 m in 97-1. Below this surface layer, the cores are

**(a)**

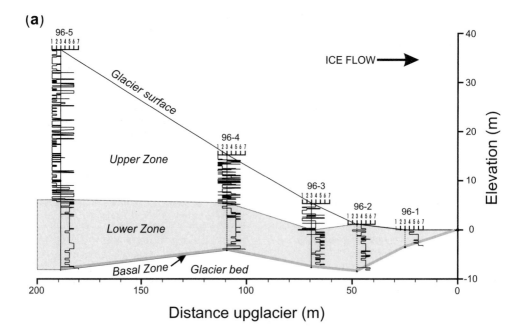

**(b)**

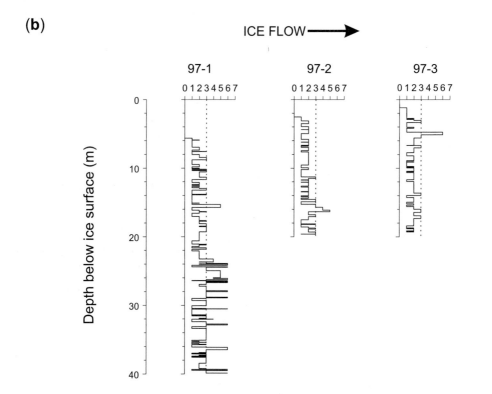

**Fig. 2.** Core stratigraphy: (**a**) 1996 cores (after Hubbard *et al.* 2000) and (**b**) 1997 cores. Ice types are plotted on a semi-quantitative scale from 0 to 7 that reflects the progressive evolution from snow-firn (Type 0) to metamorphosed clear facies (Type 6) and debris-rich basal ice (Type 7).

dominated by ice of Types 1 to 3 with very little Type 4 to 6 ice present until a depth of c. 25 m (only reached in core 97-1). From here to the base of core 97-1 (at c. 45 m), ice is predominantly of Types 1 to 3 although numerous thin layers of clear, Type 6 ice are present. This pattern is similar to that observed in the uppermost sections of the 1996 (ablation area) cores (Fig. 2a), suggesting a possible stratigraphic overlap between the basal 15 m of 97-1 and the uppermost 35 m of 96-5.

## Ice crystal size and texture

Ice crystal size has been measured on 326 thin sections from all eight cores (Figs 3 to 6). Mean grain diameters range from 2 to 69 mm, with crystal size generally increasing down-core, and, correspondingly, from Type 1 to Type 6 ice. The uppermost sections of the accumulation area cores are composed of ice that is homogeneous, with a polygonal texture and crystal size varying between 1.8 and 2.6 mm. The ice contains no visible traces of ice lenses or coarse crystal layers, although these were observed in surface pits further upstream in the central part of the accumulation area. The thickness of this layer decreases downglacier, from c. 20 m in 97-1, to c. 16 m in 97-2, to c. 2 m in 97-3. There is no evidence for the layer in the ice-marginal, 1996 cores.

Crystal size dispersion increases steadily between 20 and 35 m in core 97-1. Although the minimum crystal size remains similar to that recorded in the uppermost 20 m of the core, it shows a slight increasing trend with depth from the surface down to 34 m which is also observed in cores 97-2 (down to 16 m), 97-3 (down to 19 m with an interruption in the 10–15 m depth range) and 96-5 (down to 29 m). This is particularly clear on the thin-section photographs sequences A 28–A 59–A 71b (Fig. 3) and B 8–B 22–B 41 (Fig. 4). The baseline of minimum crystal size sharply shifts towards larger values (10–20 mm) within a few m depth at c. 36 m and c. 33 m in cores 97-1 and 96-5 respectively. The maximum crystal size increases simultaneously at these locations. A similar behaviour characterises core 97-3 between the depths of 10 and 15 m. Close inspection of the thin sections involved reveals that the increase in the smaller mean crystal size values actually reflects two crystal arrangements: a real increase of the minimum crystal size (as in thin section photograph A 87b in Fig. 3) and a minor population of smaller crystals distributed in pockets and narrow bands at the boundary of larger crystals, often showing undulose extinctions (Fig. 7, A 95 and C 39).

While the unit described above extends to the base of core 97-1 (c. 40 m deep), a further change is apparent from a depth of c. 35 to c. 38 m in core 96-5 (Fig. 5). Here, the minimum crystal size steadily decreases at the same time as the crystal size range narrows down to 10–30 mm. Ice texture also changes in this basal section, from primarily smoothed rounded contours (Fig. 5, lower half of V 72) to jagged interlocked crystal boundaries (Fig. 5, V 134 and V 205). It should be noted, in this regard, that the minimum crystal size value of 10 mm cited above could have been partially biased by including several occurrences of the same grain along a given intercept line. A similar unit is also present in cores 96-1 and 96-3, located closer to the glacier margin (Figs 1 and 6).

Many of the cores are characterized by a variety of ice structures, in particular the ice-marginal, 1996 cores. Core 96-5, for example, is characterized by several good examples of crystal discontinuities (e.g. at c. 3 m; Fig. 5, thin-section V 9), sometimes underlined by a single thin layer of microcrystals (e.g. c. 31 m; Fig. 5, thin-section V 102), that are possible evidence for thrusting in the ice mass. Crystallographic evidence for folding also appears to be present in core 96-5 at 21 m depth, consistent with observations of recumbent folds on the flanks of a crevasse located c. 10 m from the location of core 96-5.

## Ice crystal fabric

Over 3000 crystal orientations have been measured to produce fabrics from 48 thin sections, mainly from the two longest end-of-transect cores, 96-5 and 97-1. Fabrics in the uppermost 25 m of core 97-1 are generally uniform, with the exception of one observation of a single maximum about the vertical (Fig. 3, A 42b). This lack of general alignment is present not only in the fine-grained ice, but also in the coarser ice located between 22 and 25 m where crystal-size dispersion initiates. A slight departure from the uniform state in the fine-grained ice occurs close to 30 m as c-axes tend to concentrate in a wide girdle (Fig. 3, A 71 fine). Although the smaller number of axes (a function of the increased crystal size) reduces its significance, a similar pattern might arise in the coarse-grained ice fraction at the same level in the core (Fig. 3, A 71 coarse). Finally, the lowermost section of core 97-1 shows a moderate enhancement of the patterns described above, the coarse grained ice now showing c-axes organized in a multiple maximum fabric located in a girdle centred on the vertical (Fig. 3, A 87 coarse).

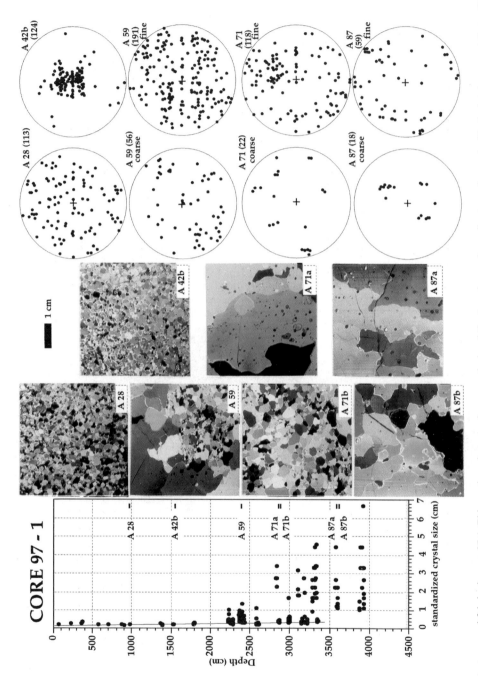

**Fig. 3.** Ice textures and fabrics in core 97-1. Thin section photographs are vertical and the 1-cm bar scale applies to each sample. Vertical striae are artefacts of the thin-sectioning procedure (in most cases due to irregular penetration of the diamond wire saw). Fabric diagrams are plotted in the horizontal plane. Depth, identification and height of each vertical section (black bar) are indicated to the right of the crystal-size profile. The thin solid line visualizes the trend in minimum crystal size.

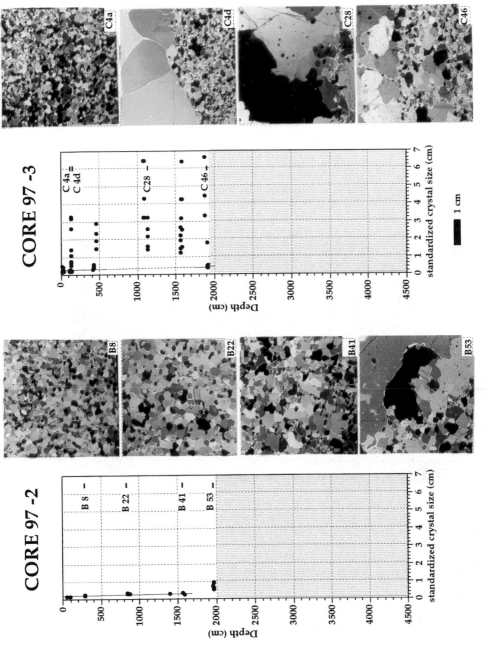

Fig. 4. Ice textures in cores 97-2 and 97-3. Conventions are as in Fig. 3.

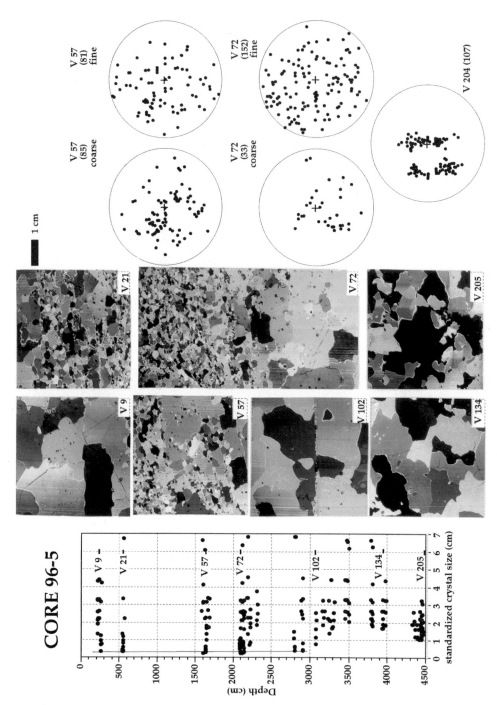

**Fig. 5.** Ice textures and fabrics in core 96-5. Conventions are as in Fig. 3.

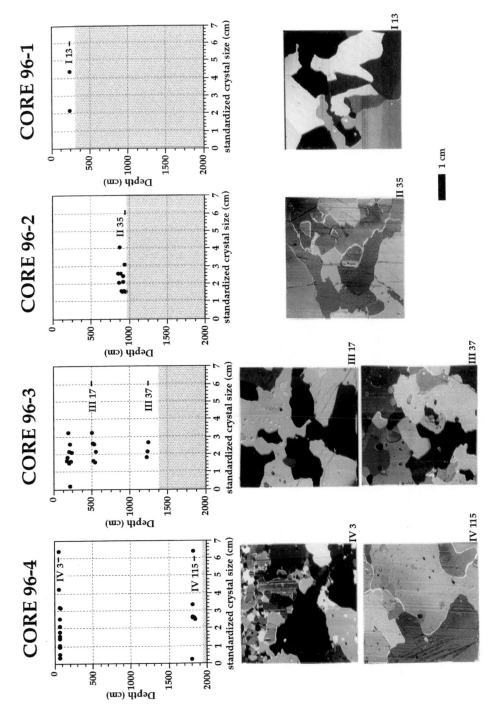

**Fig. 6.** Ice textures in cores 96-4, 96-3, 96-2 and 96-1. Conventions are as in Fig. 3.

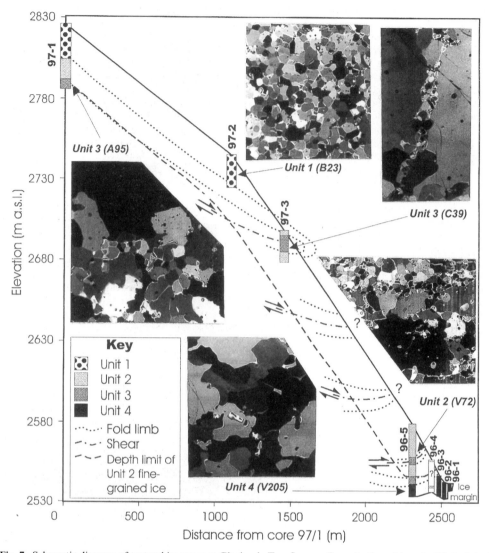

**Fig. 7.** Schematic diagram of textural ice types at Glacier de Tsanfleuron. Core stratigraphic units (Units 1 to 4) are described in the text. Photomicrographs characterize typical features of the various units. The equilibrium line altitude, estimated on the basis of late-summer snow-line position, is *c.* 2680 m a.s.l. (see Fig. 1).

The uppermost layers of the ice-marginal core 96-5 are characterized by a contrast between the fine-grained bubbly ice, which has a uniform fabric (Fig. 5, V 57 fine and V 72 fine), and the coarse-grained layers (Fig. 5, V 57 coarse and V 72 coarse), which have a tendency to concentrate in a wide maximum, possibly a girdle enclosing the vertical. Such a girdle fabric, however, has developed strongly towards the base of core 96-5 (Fig. 5, V 204), characterized by distinct multiple maxima on each thin-section analysed. Although occasional measurements

might reflect separated sections through the same grain in a given thin-section, the effect has been kept to a minimum level by combining (in sample V 204, for example) several thin-sections from the same core (thus with no azimuthal offset), and distant from each other by at least three times the maximum grain size.

## Discussion

The results presented above indicate the existence of four basic textural units at Glacier de

Tsanfleuron. The stratigraphy of these units is presented in Fig. 7.

*Unit 1* (stippled core symbols) is composed of uniformly fine-grained ice, with polygonal crystals. It generally shows a uniform fabric, but may also develop a single maximum orthogonal to the glacier surface. This unit crop out at the surface of the accumulation area cores, and thins down-glacier, forming the uppermost 22 m, 16 m and 2 m of cores 97-1, 97-2 and 97-3 respectively.

*Unit 2* (grey core shading) is characterized by an increased range of crystal sizes and a slight regular increase in the minimum crystal size. The first half of this unit shows a similar fabric to that of the directly overlying Unit 1 ice in the 1997 accumulation area cores, at least in the fine grained fraction (Fig. 3, A 59 fine). The coarser fraction (Fig. 3, A 59 coarse) may slightly diverge from uniformity. This trend towards a weak fabric organization seems to occur in the lower half of the unit, both for the fine and the coarse fractions (Fig. 3, A 71). Crystallographic foliation further develops in Unit 2 ice by the time of its arrival in the ice-marginal, 1996 cores. Here, coarse-grained layers, within which a girdle fabric has developed (Fig. 5, V 57 and V 72), frequently disrupt layers of bubbly, fine-grained ice with a predominantly uniform fabric (similar to Unit 1 ice).

*Unit 3* (black core shading) is characterized by a distinct increase in the size of its smallest constituent ice crystals (from *c.* 2.5 to *c.* 10 mm in core 97-1), and the first signs of significant fabric development. A clear transition towards this unit appears at a similar depth of just over 30 m in the two longest cores, despite their diverse locations (core 96-5 at the ice margin, and core 97-1 in the accumulation area). However, it also occurs in alternation with Unit 2 both higher up in core 96-5 (22–23 m deep) and the middle section of core 97-3 (10–15 m deep). While Unit 3 ice extends to the base of core 97-1, it is overprinted by Unit 4 ice at the ice-margin.

*Unit 4* (gravel-texture core symbols) occurs below a depth of *c.* 38 m in core 96-5, and is characterized by a marked increase in minimum grain size and a marked decrease in size variability relative to the overlying, Unit 3 ice. Unit 4 ice crystals are jagged and interlocked and characterized by a strong multi-modal fabric.

The stratigraphy (Fig. 7), the relatively uniform, small grain-size and the polygonal crystal shapes strongly suggest that Unit 1 is composed of firn and recently formed ice. This interpretation is consistent with the unit's high density of small bubbles (Fig. 2b) and the observed ice fabric (A 28 and A 42b in core 97-1; Fig. 3). Both uniform and single maximum fabrics perpendi-

cular to the glacier surface have indeed been described by several authors for firn in temperate glaciers (Perutz & Seligman 1939; Schytt 1958; Kizaki 1962; Fabre *et al.* 1972), although the origin of the single maximum fabric is still debated in terms of a depositional or recrystallized origin.

Unit 2 ice, located between 22 and 34 m from the glacier surface, is characterized by a considerable dispersion of crystal sizes superimposed on the minimum grain-size baseline described above. This dispersion could result from percolation and refreezing processes in the firn or from stress induced recrystallization. In the former case, well-defined layers of coarse crystals would be expected, and in the latter case, some fabric enhancement would be expected. As described above, however, no coarse grain melt layers are visible in the first 22 m of core 97-1 (Unit 1). This absence could be related to local conditions, since melt layers were observed, further upstream, closer to the central flow line. These should then be visible at greater depths in core 97-1. However although occasional coarse-grained layers with straight boundaries were observed at 26 m and 30 m in this core, pockets of coarser crystals are generally the rule (Fig. 8). Furthermore, detailed examination of textural relationships (Fig. 8) shows fine crystals included in larger ones, and some elongation of the crystals in the finer matrix, both suggesting a certain amount

Scale: ▬ = 1 cm

**Fig. 8.** Ice texture of sample A 60 (depth range: 24.09 to 24.18 m in core 97-1) illustrating grain to grain textural relationships in Unit 2.

of recrystallisation and deformation. This is also consistent with the weak girdle fabrics of samples A 59 coarse and A 71 in Fig. 3.

While the increasing crystal size in Unit 2 may reflect the early stages of dynamic recrystallisation at low stresses, the trend in the baseline of minimum mean grain size for polygonal crystals of Units 1 and 2 (first 33 m of the accumulation area) probably results from crystal growth at negligible deviatoric stress. The normal grain growth law may describe such a process, where crystal size is considered as a function of time and temperature (Paterson 1994, p. 23):

$$D^2 = D_0^2 + kt \qquad (1)$$

where $D^2$ = mean cross-sectional area (mm²) at time $t$, and $D_0^2$ = initial cross section area (mm²). $k$, the temperature dependence of the rate of crystal growth (mm² a⁻¹) is given by the Arrhenius relation:

$$k = k_0 \exp(-Q/RT) \qquad (2)$$

where $Q$ is the activation energy, $R$ the gas constant, $T$ the absolute temperature and $k_0$ a constant. Paterson (1994, p. 25 and Fig. 2.5) shows that for polar firn, in the $-13°C$ to $-53°C$ range, $\ln k$ plots as a linear function of $T^{-1}$, giving a constant value for the activation energy of 42.4 kJ mol⁻¹. Li (1994) and Jacka & Li (1994) carried out laboratory tests to examine ice crystal growth (under no stress) over the temperature range $-0.1°C$ to $-50°C$, clearly demonstrating that $Q$ is not independent of temperature above $-13°C$ (Jacka & Li 1994, table 2 and fig. 2). We can, however, combine these results with precipitation data from the area (Bezinge & Bonvin 1973) to check the consistency of the observed crystal sizes with Arrhenius-type growth. Bezinge & Bonvin (1973) give a yearly mean precipitation value ($A_{mean}$) of 1510 mm water equivalent (w.e.) at a meteorological station located at 2870 m a.s.l. on the flanks of the Oldenhorn (Fig. 1), for the 1960–1970 decade. This fits well with the range of extreme values (1020 to 2050 mm w.e.) recorded since the station was set up (Bezinge & Bonvin 1973). Since:

$$A_{mean} = \frac{z}{t} \qquad (3)$$

and

$$t = \frac{D^2 - D_0^2}{k} \qquad (4)$$

then

$$k = \frac{A_{mean}(D^2 - D_0^2)}{z} \qquad (5)$$

where $z$ is the ice depth in m water equivalent. Mean minimum crystal size increases from

1.8 mm to 3.3 mm between the surface and 33.5 m depth in core 97-1 (Fig. 3). This corresponds to an increase in surface area from 2.54 mm² to 8.55 mm², i.e. 6 mm² ($D^2 - D_0^2$). Using a mean firn density of 615 kg m⁻³ ($z = 20.6$ m water equivalent for an observed thickness of 33.5 m), gives $k = 4.4 \times 10^{-1}$ mm² a⁻¹. Referring to the calculated values in Jacka & Li (1994, table 2) gives a temperature range of $-1°C$ to $-3°C$. Although we do not have a record of ice temperatures with depth in the area, these slightly negative values are compatible with the accumulation area of an Alpine glacier at 3000 m a.s.l. It should be noted that the strong effect of temperature on the activation energy for crystal growth (and therefore on the crystal growth rate) in the near-zero temperatures could easily lead to growth rate of 31.5 mm² a⁻¹ (at $-0.1°C$) (Jacka & Li 1994), i.e. to crystals a few centimetres in size in less than ten years. This could partially explain the development, during warmer years, of the coarse-grained ice population that lacks any significant fabric in Unit 2 of the accumulation area. Finally, another approach to test the coherence of our crystal texture data with a normal grain growth process, is to measure the number of faces per grain. As an example, results from the measurements on 64 crystals in thin-sections A 71b (core 97-1) and B 41 (core 97-2) give 5.14 ($\sigma = 0.83$) and 4.95 ($\sigma = 0.95$) faces respectively. Both estimates are close to the expected value of 5.12 for normal growth or foam texture (Mullins 1956).

Numerous local crystallographic contrasts have developed in Unit 2 ice by the time it appears in the 1996 cores at the frontal margin of the glacier. These fabrics may be broadly classified into those in fine-grained ice, that generally show a uniform distribution (e.g. Fig. 5, V 57 and V 72 fine), and those in coarse-grained ice, that are concentrated in a girdle (e.g. Fig. 5, V 57 coarse and V 72 coarse). The latter category is also characterized by evidence of offset crystal boundaries (e.g. Fig. 5, V 102) and local folding, strongly indicating deformation-related development during the passage from the accumulation area. In contrast, the fine grained ice could be formed by a number of processes, including surviving as a remnant of the original (Unit 1) texture, forming as crevasses fills, or acquiring its characteristics by strain-related polygonization and recrystallization. Formation as crevasse fills may be discounted given the high frequency and sub-horizontal layering of this fine-grained ice. Several arguments also mitigate against formation as strain related textures.

(a) Polygonization is more likely to be a dominant process in cold ice. Deformation in temperate ice mostly occurs accompanied by grain boundary migration, intra-crystalline slip and new grain nucleation at the boundaries of coarser, highly-strained grains (Wilson & Zhang 1994). This results in dispersed small pockets of fine-grained ice, rather than in continuous thicker layers, as were visible in Unit 2 (e.g. Fig. 5, top of V 72 and V 57).

(b) The fine-grained ice is predominantly of Type 1 (Table 2, Fig. 2), i.e. showing a medium to high density of small bubbles. It is difficult to conceive that such an ice type would result from polygonization or recrystallization of (large-grained) bubble-poor ice Types 4 to 6.

(c) If the crystals in the fine-grained ice were to result from dynamic recrystallization, they should be characterized by $c$-axis organization at $c.\,45°$ to the axis of compression ('easy glide' basal plane parallel to shear directions) (Kamb 1972; Alley 1992), at least in the early stages of deformation. Alternatively if these crystals were to result from polygonization, they should show occurrence of more nearest-neighbour grains with small misorientation than would occur uniformly in a sample (Alley 1992). The observed uniform fabric (Fig. 5, V 57, V 72 fine) is not in accordance with these predictions.

Remnants of the original (Unit 1) texture is thus the most likely candidate for the fine-grained ice in Unit 2 although a certain amount of strain accumulation through grain boundary sliding should not be discarded, especially in the ablation area. This process would indeed produce extremely slow crystallographic fabric development and no shape fabric. It is worthwhile noting that this textural sub-unit only survives down to about 30–35 m both in cores 97-1 and 96-5. It therefore appears that the occurrence of this 'primary' texture is limited to the upper zone of the glacier with no or negligible deviatoric stresses (dashed line in Fig. 7).

Unit 3 ice occurs at about 33 m depth in cores 97-1 and 96-5, and is characterized by a definite fabric enhancement (Fig. 3, A 87) and an increase in mean minimum crystal size. Although this mean minimum crystal size sometimes reflects a homogeneous population dominated by crystals between 10 and 20 mm in size (Fig. 3, A 87b), more often it is the expression of a minority of smaller crystals occurring as aggregates (Fig. 7, A 95) or fine layers (Fig. 7, C 39). However, crystal size within the unit is highly variable at the scale of decimetres, suggesting that the strain field is correspondingly heterogeneous. Thus, zones of uniform, large crystals (up to 70 mm) are considered to reflect low strain rates, while zones of smaller crystals are interpreted in terms of active deformation.

This pattern is coherent with the ice type alternations presented in Fig. 2.

Unit 4 is characterized by a multi-modal girdle fabric in ice that is uniformly coarse grained with jagged, interlocked crystal contours. Jacka & Li (1994) demonstrated that, during tertiary creep, ice crystals evolve to a constant size that is determined by a balance between crystal growth (as a function of temperature) with time, and crystal change as a result of temperature-dependent deformation (ice creep). These authors showed that the temperature effect in these two processes are similar, and consideration of the activation energies for the two processes indicates that it may be appropriate to cancel them, yielding a dependence of equilibrium crystal size on stress alone, of the type:

$$D_e^2 \propto \frac{1}{\tau^3} \qquad (6)$$

Figure 9 (after Jacka & Li 1994) plots the square of the equilibrium steady-state crystal diameter, as a function of the octahedral shear stress, for a range of ice creep tests at several different temperatures (black dots). A straight line has been drawn with a slope of $-3$ to visualize the

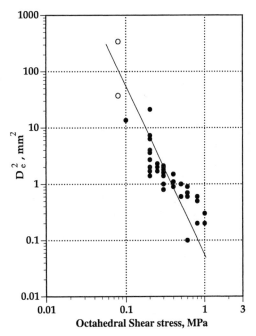

**Fig. 9.** Plot of the square of the equilibrium steady-state crystal diameter as a function of the octahedral shear stress for a range of ice-creep tests at several different temperatures (black symbols) from Jacka & Li (1994). Open circles show observed minimum and maximum $D_e^2$ at a depth of 44 m in core 96-5.

trend of Equation 6. Unit 4 crystal size at a depth of 44 m in core 96-5 falls in the range 6.1 to 18.5 mm (using direct measurements from the linear intercept method, without the 1.75 correction, in order to be consistent with Jacka & Li's (1994) values). This provides us with a range of 37.21–342.25 mm$^2$. Using the measured local slope at the site of core 96-5 ($\alpha = 11°55'$) one can calculate a rough estimate of the basal shear stress at the bottom of the core (which reached the ice–bedrock interface at a depth of 44.84 m) to be $\tau_{xy} = 0.08$ MPa. Although the fabric patterns from core 96-5 suggest that some 3D vertical compression should be considered in the calculation of the octahedral shear stress in the ablation area, plotting the $D_e^2$ and $\tau_{xy}$ values measured/calculated above as open circles in Fig. 9 agrees well with the experimental data from Jacka & Li (1994). These field data therefore provide important support for Jacka & Li's (1994) model, particularly since the conditions under which Unit 4 ice forms at Glacier de Tsanfleuron (low stresses and long time periods) are difficult to recreate and assess in the laboratory.

In Units 3 and 4 where structured fabrics develop, the pattern is generally one of a small girdle in which the c-axes are organised in multiple maxima. However, the orientation of the girdle differs between the accumulation and the ablation area. In the first case (e.g. Fig. 3, A 87 coarse) the girdle is centred on the vertical, indicating that the controlling stress field is close to vertical compression. In the ablation area, the symmetry axis of the girdle is rotated towards the horizontal, such that the girdle now *includes* the vertical direction (e.g. Fig. 5, V 57 coarse, V 72 coarse and V 204). This reflects the increasing contribution of simple shear superimposed on a vertical compressive stress, as experimentally demonstrated by Kamb (1972, figs 16 and 18). Such a stress regime is consistent with the inclination of the local crystallographic layering and the presence of recumbent fold traces (with their symmetry planes at a low-angle to the horizontal), on the crevasses walls in the ablation area of the glacier. This is unusual in the ablation zone of a temperate glacier, where basal shear stress and longitudinal compression normally combine to form conspicuous oblique layering. In contrast, the situation at Tsanfleuron Glacier may reflect its relatively thin, plateau-like geometry.

## Conclusions

The spatial distribution and stratigraphy of four units identified in eight cores at Glacier de Tsanfleuron provide evidence of a consistent pattern of ice crystallographic evolution through the glacier (Fig. 2).

Uniformly fine-grained ice with a uniform fabric (Unit 1) forms in the accumulation area of the glacier and survives within *c.* 20 m of the ice surface. Through time (and therefore depth), ice crystal size within this unit increases in the absence of significant stresses under an Arrhenius growth process, within a temperature range of −1°C to −3°C.

After downglacier movement, perhaps over a few decades, coarser ice crystals begin to develop, increasing the grain-size range of the ice (Unit 2). This increase is interpreted in terms of the initiation of dynamic recrystallisation, although local events of Arrhenius growth at temperatures closer to the pressure melting point (*c.* −0.1°C) or inheritance of melt layers originating upstream, are not precluded. By the time this ice has passed into the ablation area of the glacier, the coarser crystals have developed a multiple maximum fabric, typical of high cumulative strain, but the smaller crystals still display a uniform fabric. The properties of the latter (continuous layers of bubble-rich type 1 ice with polygonal crystals characterized by uniform fabrics) suggest that they are remnants of the original primary layering (Unit 1).

Minimum crystal size increases abruptly at a depth of *c.* 33 m (Unit 3) in both the accumulation area and the ablation area of the glacier. In the accumulation area, this increase coincides with the evidence of systematic fabric enhancement, perhaps marking the local dominance of processes of dynamic recrystallization. Later, as the proportion of larger, recrystallized grains increases, the process of nucleation of new grains at the margins of highly strained coarser crystals dominates. In the ablation area, Unit 3 ice is characterized by the progressive expulsion of high densities of small bubbles from the ice. However, the persistence of crystal size foliation in Unit 3 ice suggests that deformation in this intermediate zone of the glacier occur along discrete layers, a few decimetres apart.

Because the glacier has been in general recession since 1855–1860, the ice at the margin is stagnant and favours overriding by the ice flowing from upstream. This occurs along discrete shear zones and results in recumbent folding and the insertion of Unit 3 ice into the shallower Unit 2 in the ablation area (Fig. 7, core 97-3 between 10 and 15 m; Fig. 4; Fig. 7, core 96-5 between 22 and 23 m; Fig. 5). Figure 10 (V 73, located 10 cm below V 72 in Fig. 5) illustrates the microstructures associated with this process: dotted lines marked *s* show a repeated, consistent structural

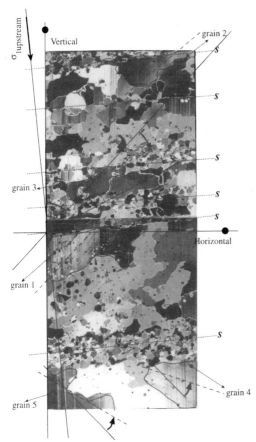

**Fig. 10.** Vertical thin-section photograph of V 73 (core 96-5 in the ablation area) showing microstructural relationships between layering, and crystals' internal deformation patterns.

Grains 3, 4 and 5 are in a later stage of the process, and have already been partially rotated under cumulative strain. Note that grain 3 is surrounded by pockets of fine-grained crystals and shows internal crystal growth, both suggesting locally active recrystallization.

Within 10 m of the glacier base, mean grain size increases, grain-size dispersion decreases, and crystals develop a jagged, interlocked structure and a strong multi-modal, girdle fabric (Unit 4). These changes are interpreted in terms of the effects of intense, continuous deformation near the glacier bed. The measured crystal sizes are consistent with a steady state balance between Arrhenius processes of grain growth and a strain-related reduction in grain size (Jacka & Li 1994).

We thank L. Janssens, R. Lorrain, A. Khazendar, R. Prytherch, L. Plewes and R. Cloke for fieldwork assistance, and C. J. L. Wilson and E. Rutter for helpful comments on the manuscript. This work was funded by NERC Grant GR9/2026. J.-L. Tison is Research Associate at the Fonds National de la Recherche Scientifique (F.N.R.S.).

direction that commands both the alternation of the fine-grained/coarse-grained layers and the suturing of previous recrystallization events in Unit 2. This layering can be considered as perpendicular to the principal compression axis of the stress field developed upstream (Fig. 10, $\sigma_{1\text{upstream}}$). A few grains (labelled 1 to 5) show clear traces of internal slip boundaries, similar to those observed by Wilson & Zhang (1994) in experimental runs. Grains 1 to 3 dip upglacier, and grains 4 and 5 dip downglacier. Grains 1 and 2 show the highest angle to the vertical and are located at 45 to the $\sigma_{1\text{upstream}}$ direction. This suggests that these 2 grains have their 'easy glide' basal plane broadly in equilibrium with the upstream stress field, and are just in the process of responding, by internal slip, to the new stress configuration developing in the ablation area.

## References

ALLEY, R. B. 1992. Flow-law hypotheses for ice-sheet modeling. *Journal of Glaciology*, **38**, 245–256.

BEZINGE, A. & BONVIN, G. 1973. Images du Climat sur les Alpes Pennines. *In: Réunion de la Section Glaciologie de la Société Hydrotechnique de France*, Société Hydrotechnique de France, Grenoble, 26.

BUDD, W. F. 1972. The development of crystal orientation fabrics in moving ice. *Zeitschrift für Gletscherkunde und Glazialgeologie*, **8**, 65–105.

—— & JACKA, T. H. 1989. A review of ice rheology for ice sheet modelling. *Cold Regions Science and Technology*, **16**, 107–144.

DAHL JENSEN, D. & GUNDESTRUP, N. 1987. Constitutive properties of ice at Dye 3 Greenland. *In: The Physical Basis of Ice Sheet Modelling–Vancouver, Symposium*. International Association of Hydrological Sciences Publication, **170**, 31–43.

FABRE, J.-P., PETIT, J.-R. & VALLON, M. 1972. *Carottage continu de 187 m sur le glacier de la Vallée Blanche*. Résultats Préliminaires, Laboratoire de Glaciologie, Grenoble.

FISHER, D. A. & KOERNER, R. M. 1986. On the special rheological properties of ancient microparticle-laden Northern Hemisphere ice as derived from bore-hole and core measurements. *Journal of Glaciology*, **32**, 501–510.

HAMBREY, M. J. & MÜLLER, F. 1978. Structures and ice deformation in the White Glacier, Axel Heiberg Island, Northwest Territories, Canada. *Journal of Glaciology*, **20**, 41–67.

——, MILNES, A. G. & SIEGENTHALER, H. 1980. Dynamics and structure of Griesgletscher, Switzerland. *Journal of Glaciology*, **25**, 215–228.

HOOKE, R.LEB., GAO, X. Q., JACKA, T. H. & SOUCHEZ, R. A. 1988. Rheological contrast between Pleistocene and Holocene ice in Barnes Ice Cap, Baffin ISLAND, N. W. T., Canada: a new interpretation. *Journal of Glaciology*, **34**, 364–365.

HUBBARD, B. & HUBBARD, A. 1998. Bedrock surface roughness and the distribution of subglacially precipitated carbonate deposits: implications for formation at Glacier de Tsanfleuron, Switzerland. *Earth Surface Processes and Landforms*, **23**, 261–270.

—— & SHARP, M. 1995. Basal ice facies and their formation in the Western Alps. *Arctic and Alpine Research*, **27**, 301–310.

——, JANSSENS, L., TISON, J.-L. & SPIRO, B. Ice core evidence for the thickness and character of Clear facies basal ice at Glacier de Tsanfleuron, Switzerland. *Journal of Glaciology*, **46**, 140–150.

JACKA, T. H. 1984. Laboratory studies on relationships between ice crystal size and flow rate. *Cold Regions Science and Technology*, **10**, 31–42.

—— & LI JUN 1994. The steady state crystal size of deforming ice. *Annals of Glaciology*, **20**, 13–18.

KAMB, B. 1962. Refraction corrections for universal stage measurements. I. Uniaxial crystals. *The American Mineralogist*, **47**, 227–245.

——1972. Experimental recrystallization of ice under stress. *In*: edited by HEARD, H. C., BORG, T. Y., CARTER, N. L. & RALEIGH, C. B. (eds) *Flow and fracture of rocks*. AGU Geophysical Monographs, **16**, 211–241.

KIZAKI, K. 1962. Ice fabric studies on Hamna Ice Fall and Honhörbrygga Glacier, Antarctica. *The Antarctica Record*, **16**, 54–74.

LANGWAY, C. C. JR. 1958. *Ice fabrics and the Universal stage*. Cold Regions Research and Engineering Laboratory Technical Report, **62**.

LI JUN. 1994. *Interrelation between flow properties and crystal structure of snow and ice*. PhD Thesis, University of Melbourne.

MEIER, M. F., KAMB, B., ALLEN, C. R. & SHARP, R. P. 1974. Flow of Blue Glacier, Olympic Mountains, Washington, USA. *Journal of Glaciology*, **13**, 187–212.

MULLINS, W. W. 1956. Two-dimensional motion of idealized grain boundaries. *Journal of Applied Physics*, **27**, 900–904.

PATERSON, W. S. B. 1994. *The Physics of Glaciers*. Elsevier Science Ltd, Oxford.

PERUTZ, M. F. & SELIGMAN, G. A. 1939. Crystallographic investigation of glacier structure and the mechanism of glacier flow. *Proceedings of the Royal Society of London*, **A950**, 335–360.

PICKERING, F. B. 1976. *The Basis of Quantitative Metallography*. Metals and Metallurgy Trust for the Institute of Metallurgical Technicians, Whetsone.

RUSSELL-HEAD, D. S. & BUDD, W. F. 1979. Ice-sheet flow properties derived from bore-hole shear measurements combined with ice core studies. *Journal of Glaciology*, **24**, 117–130.

SCHYTT, V. 1958. Snow studies at Maudheim – Snow studies inland – The inner structure of the ice-shelf at Maudheim as shown by core drilling. *Norwegian–British–Swedish Antarctic Expedition 1949–1952*, 151.

SHARP, M., GEMMEL, J. & TISON, J.-L. 1989. Structure and stability of the former subglacial drainage system of the Glacier de Tsanfleuron, Switzerland. *Earth Surface Processes and Landforms*, **14**, 119–134.

——, TISON, J.-L. & FIERENS, G. 1990. Geochemistry of subglacial calcites: implications for the hydrology of the basal water film. *Arctic and Alpine Research*, **22**, 141–152.

TANAKA, H. 1972. On preferred orientation of glacier and experimentally deformed ice. *Journal of the Geological Society of Japan*, **78**, 659–675.

THORSTEINSSON, T., WADDINGTON, E. D. TAYLOR, K. C., ALLEY, R. B. & BLANKENSHIP, D. D. 1999. Strain-rate enhancement at Dye 3 Greenland. *Journal of Glaciology*, **45**, 338–345.

THWAITES, R. J., WILSON, C. J. L. & MCCRAY, A. P. 1984. Relationship between bore-hole closure and crystal fabrics in Antarctic ice core from Cape Folger. *Journal of Glaciology*, **30**, 171–179.

TISON, J.-L. 1994. Diamond-wire saw cutting technique for investigating textures and fabrics of debris-laden ice and brittle ice. *Journal of Glaciology*, **40**, 410–414.

—— & LORRAIN, R. 1987. A mechanism of basal ice layer formation involving major ice-fabrics changes. *Journal of Glaciology*, **33**, 47–50.

WILSON, C. J. L. & ZHANG, Y. 1994. Comparison between experiment and computer modelling of plane strain simple shear ice deformation. *Journal of Glaciology*, **40**, 46–55.

# The effect of H$_2$SO$_4$ on the stress exponent in ice single crystals

I. BAKER, Y. L. TRICKETT & P. M. S. PRADHAN

*Thayer School of Engineering, Dartmouth College, Hanover, NH 03755, USA*
*(e-mail: Ian.Baker@Dartmouth.edu)*

**Abstract:** Sulphuric acid is a naturally occurring contaminant in ice. Recently, it was demonstrated that, during compression tests at $-20°C$ and a strain rate of $1 \times 10^{-5}\,s^{-1}$, as little as 0.1 ppm H$_2$SO$_4$ reduced both the peak strength and the subsequent flow stress of ice single crystals that deform primarily by basal slip. In the present work, compression tests were performed at $-20°C$ at a variety of strain rates on both undoped ice single crystals and ice single-crystals containing 6.8 ppm sulphuric acid of various orientations again deforming primarily by basal slip. The results show that sulphuric acid dramatically decreases both the peak stress and the subsequent flow stress of ice single crystals at all strain rates. In contrast, the stress exponent was determined to be 1.89–1.97 and was unaffected by the dopant. This value is similar to values found by previous workers for undoped ice crystals.

Here we consider how the presence of sulphuric acid (H$_2$SO$_4$), a natural contaminant in ice, affects the stress exponent in ice single crystals. It is well established that hydrofluoric acid and hydrochloric acid reduce both the peak strength and the subsequent flow stress of ice crystals (Jones 1967; Jones & Glen 1968, 1969; Naka-mura & Jones 1970, 1973), and recently it was shown that at $-20°C$ and an axial strain rate of $1 \times 10^{-5}\,s^{-1}$, H$_2$SO$_4$ has a similar effect (Trickett *et al.* 2000*a*). The softening effect on the peak stress was shown to be related to the square root of the concentration of sulphuric acid, up to 11.5 ppm. This observation is important because H$_2$SO$_4$ occurs naturally in ice, the result of aerosols produced both by volcanic activity and by marine biological emissions. For example, concentrations of sulphate ions up to 300 ppb have been recorded (Lorius *et al.* 1969). Mulva-ney & Wolff (1994) have compiled data for the major impurites in the Antarctic ice sheet and, although there are some uncertainties in the calculation, they noted that the sulphate ion concentration appears to decline slowly with distance from the coast. These impurities, if located in the water veins in the ice, may explain the high d.c. electrical conductivity of Antarctic ice compared to laboratory grown ice (Paren & Walker 1971). The present study concerns the effect of H$_2$SO$_4$ on the stress exponent.

In their investigation of the creep of ice single crystals, Higashi *et al.* (1965) concluded that the plastic behaviour of ice resembles that of metals, and the following equation was used to describe the steady-state creep rate:

$$\dot{\varepsilon} = A\sigma^n \exp\left(\frac{-Q}{RT}\right) \qquad (1)$$

where $\sigma$ is the applied stress, $T$ is the absolute temperature, $n$ is the stress exponent, $A$ is a constant and $Q$ is the activation energy for creep. Values of $n$ and $Q$ that have been obtained for ice single crystals are summarized in Table 1. The value of $n$ relates the sensitivity of the velocity of mobile dislocations in ice crystals to change in stress. The smaller is the value of $n$ the less sensitive the mobility of dislocations is to a change in stress. The activation energy, $Q$ is controlled by point defects such as positive ions and $L$ defects, which affect the motion of dislocations.

In this paper we present the results of a study in which ice single crystals of different orientations containing 6.8 ppm H$_2$SO$_4$, deforming by basal slip, were strained at a variety of rates at $-20°C$. Similar experiments were performed on undoped ice single crystals. It is demonstrated that even though H$_2$SO$_4$ softens ice it does not change the stress exponent associated with the peak stress.

## Experimental details

Both the undoped and doped crystals were made from distilled and deionized water with a ph $\approx 7$ and a resitivity $>18\,M\Omega$. Details of the preparation of ice

*From:* MALTMAN, A. J., HUBBARD, B. & HAMBREY, M. J. (eds) *Deformation of Glacial Materials.* Geological Society, London, Special Publications, **176**, 39–45. 0305-8719/00/$15.00 © The Geological Society of London 2000.

**Table 1.** *Values of stress exponent, n, and activation energy for creep, Q, for single crystal ice under a variety of different testing conditions obtained by different researchers (in chronological order)*

| Researchers | $n$ | $Q$ (kJ/mole) | Testing conditions |
|---|---|---|---|
| Kamb (1961) | 2.5 | | Torsion ($-2.9°C$) Artificial crystals |
| Readey & Kingery (1964) | 2.0 | $60 \pm 6.3$ | Tension (0 to $-42°C$) Artificial crystals |
| Higashi *et al.* (1964) | 1.53 | 66.8 | Tension ($-15$ to $-40°C$) Natural crystals |
| Higashi *et al.* (1965) | 1.58 | 66.4 | Bending ($-4.8$ to $-40°C$) Natural crystals |
| Jones & Glen (1968) | 2.4 <br> 2.4 | 75.6 <br> 43.3 | Compression ($-10$ to $-50°C$) Compression ($-55$ to $-80°C$) Artificial crystals |
| Muguruma (1969) | 1.3 <br> 1.7 | 46.2 <br> 63 | Chemically polished Mechanically polished Compression ($-10$ to $-30°C$) Artificial crystals |
| Nakamura & Jones (1973) | 1.8 | $75.6 \pm 8.4$ | Tension ($-5$ to $-46°C$) Artificial crystals |
| Jones & Brunet (1978) | $1.95 \pm 0.04$ <br> $2.07 \pm 0.08$ | $70 \pm 2$ <br> $70 \pm 2$ | Compression ($-0.2°C$) Compression ($-20°C$) Artificial crystals |

single crystals with and without sulphuric acid have been given elsewhere (Trickett *et al.* 2000*b*).

In order to determine the stress exponent, constant strain-rate compression tests were performed at $-20°C \pm 0.2°C$ at a variety of axial strain rates from $1 \times 10^{-6} \text{s}^{-1}$ to $1 \times 10^{-3} \text{s}^{-1}$. The strain rate could be set to $\pm 2\%$ on the system used for testing. For the crystal orientations used here, the peak stress is reproducible to $\pm 5\%$ (Trickett *et al.* 2000*b*).

The three orientations of crystals for the undoped ice used are schematically shown in Fig. 1. In Fig. 1a, the basal plane was tilted 33° from the top surface and

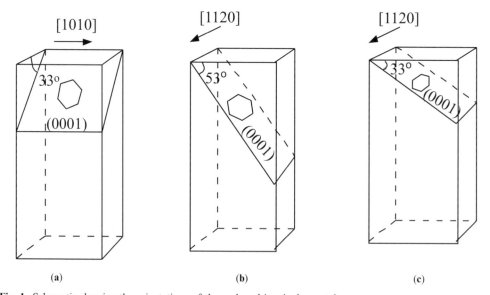

**Fig. 1.** Schematic showing the orientations of the undoped ice single crystals.

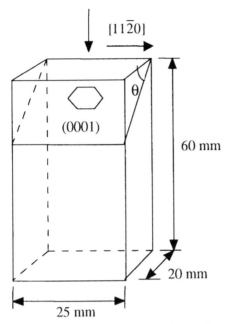

**Fig. 2.** Schematic showing the orientations of the H₂SO₄-doped ice single crystals.

the (10$\bar{1}$0) plane was on the side. For the crystal shown in Fig. 1b, the basal plane was tilted 53° from the top surface, with the (11$\bar{2}$0) plane on the front, while the crystal shown in Fig. 1c had the basal plane tilted 33° from the top surface, with the (11$\bar{2}$0) plane on the front. Therefore the values of $\theta$, the angle between the c-axis and the loading direction, are 33°, 53°, and 33° respectively.

The orientations used for the sulphuric acid-doped ice are shown in Fig. 2. Theta values of 3° and 25° were

used. The crystals were doped with 6.8 ppm H₂SO₄. For both undoped and doped ice crystals, basal slip was the primary deformation mode.

## Results and discussion

The resolved shear stress-strain curves for the undoped ice single crystals obtained at various strain rates are illustrated in Figs 3–5. They all show behaviour typical for ice deforming by basal slip, that is, with increasing strain the stress increases linearly to a peak value before falling, initially rapidly, and then more slowly toward a plateau stress. The pronounced drop in flow stress after the peak is presumably due to dislocation multiplication.

The relationship between the (log) critical resolved shear stress (CRSS), (for the peak stress) and the (log) shear strain rate is shown in Fig. 6. The shear strain rate $\dot{\gamma}$ was calculated by dividing the shear strain on the basal slip plane, $\gamma$, by the time, where $\gamma$ for compression is given by (Schmid & Boas 1950)

$$\gamma = \{-\cos\theta_0 + [(l_0/L)^2 - \sin^2\theta_0]^{1/2}\}/\cos\lambda_0$$

where $l_0$ is the original length of the specimen, $L$ is the deformed length of the specimen, $\lambda_0$ is the angle between the loading direction and the slip direction at the start of the test, $\theta_0$ is the angle between the loading direction and the normal to the slip plane at the start of the test).

The slope of the fitted lines, which is the stress exponent, $n$ is $1.89 \pm 0.17$ for orientation (a), $1.92 \pm 0.11$ for orientation (b), and $1.97 \pm 0.08$ for orientation (c) (where $\pm$ indicates one standard deviation). Thus, within the measurable

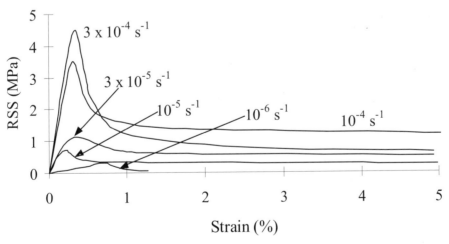

**Fig. 3.** Typical resolved shear stress-strain curves at different axial strain rates (indicated) of undoped ice single crystals with orientation (a) at −20°C.

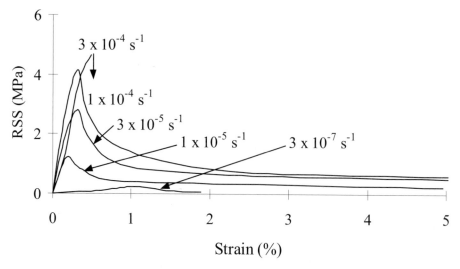

**Fig. 4.** Typical resolved shear stress-strain curves at different axial strain rates (indicated) of undoped ice single crystals with orientation (b) at −20°C. The downward arrow on the test performed at $3 \times 10^{-4}\,\text{s}^{-1}$ indicates fracture.

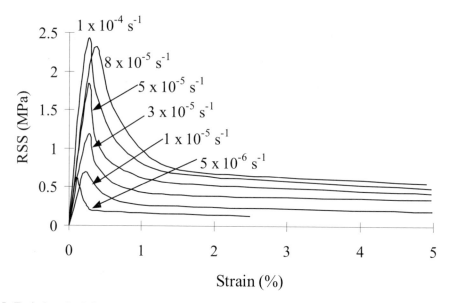

**Fig. 5.** Typical resolved shear stress-strain curves at different axial strain rates (indicated) of undoped ice single crystals with orientation (c) at −20°C.

error, there is no orientation effect on $n$. These values of $n$ agree reasonably well with the value of $n = 2.07 \pm 0.08$ at −20°C obtained by Jones & Brunet (1978). They carried out constant strain-rate compression tests on single crystals with similar orientations, that is, the angle $\theta$ between the $c$-axis and the compressive axis ranged from 35° to 55°.

The value of $n$ obtained by other workers was typically somewhat different; see Table 1. For example, Jones & Brunet (1978) measured $n$ at various temperatures and found that it varied slightly from 1.95 at −0.2°C to 2.07 at −20°C in constant strain-rate compression tests. In contrast, Nakamura & Jones (1973) carried out constant strain-rate tension tests on ice single

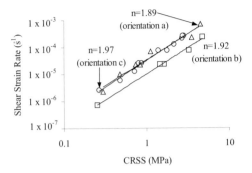

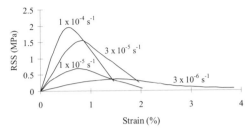

**Fig. 6.** The dependence of the shear strain rate on the CRSS at $-20°C$ for undoped ice single crystals with the three orientations **(a)** (triangles), **(b)** (squares) and **(c)** (circles) shown in Fig. 1. The error bars, which are estimated to be $\pm5\%$ of the CRSS and $\pm2\%$ of the shear strain rate, are not shown since they are generally smaller than the symbols used to depict the data. The slope of each line, which is the stress exponent, $n$, is indicated in each case.

crystals. Their $n$ value varied only slightly from 1.4 at $-26°C$ to 1.5 at $-11°C$. Their specimens had the $c$-axis tilted around $45°$ from the tensile axis. Similarly, Higashi *et al.* (1964) obtained $n = 1.53$ at $-15°C$ using the same deformation mode and orientation as Nakamura & Jones.

Figure 7 shows resolved shear stress–strain curves at a variety of strain-rates for ice single crystals doped with 6.8 ppm $H_2SO_4$ with $\theta = 25°$ at $-20°C$. Interestingly, even at the highest axial strain-rate used of $1 \times 10^{-3}\,s^{-1}$ the ice crystals were still ductile. In contrast, pure ice specimens with a similar orientation fail by brittle fracture at the peak stress at the slower axial strain rate of $3 \times 10^{-4}\,s^{-1}$, as shown in Fig. 4. This observation suggests that sulphuric acid increases the ductility of ice single crystals.

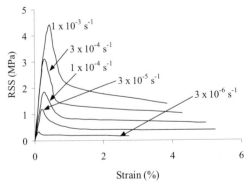

**Fig. 7.** Typical resolved shear stress-strain curves at various axial strain rates (indicated) for $H_2SO_4$-doped ice single crystals with the $c$-axis at $25°$ to the loading direction and at $-20°C$. The doping level is 6.8 ppm.

**Fig. 8.** Typical resolved shear stress-strain curves at various axial strain rates (indicated) for $H_2SO_4$-doped ice single crystals with the $c$-axis at $3°$ to the loading direction and at $-20°C$. The doping level is 6.8 ppm.

Another set of resolved shear stress-strain curves for $H_2SO_4$-doped ice single crystals but with the $c$-axis tilted $3°$ from the loading direction are shown in Fig. 8. The concentration of $H_2SO_4$ in these crystals is also 6.8 ppm Although cracking was observed in all these specimens, none of them failed by brittle fracture. The unusual shape of curves is due to the orientation (Trickett *et al.* 2000a, b).

In Fig. 9 the $n$ values for ice crystals doped with 6.8 ppm are shown to be $1.89 \pm 0.21$ and $1.94 \pm 0.26$, for orientations of $25°$ and $3°$ respectively; that is, essentially the same. These values fall within the range of values for the undoped ice ($n = 1.89-1.97$), which are shown on the same figure and are within the standard deviations of the slopes. Therefore, the value of $n$ for

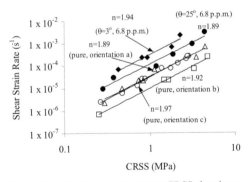

**Fig. 9.** Graph of strain rate versus CRSS showing a comparison between the $n$ values (slope of each line, which is the stress exponent) of undoped ice and $H_2SO_4$-doped ice single crystals. The open symbols are for the three undoped ice crystals orientations: **(a)** (triangles), **(b)** (squares) and **(c)** (circles), as shown in Fig. 1. The filled symbols are for $H_2SO_4$-doped crystals: $\theta = 3°$ (diamonds) and $\theta = 25°$ (circles). The error bars, which are estimated to be $\pm5\%$ of the CRSS and $\pm2\%$ of the shear strain rate, are not shown since they are generally smaller than the symbols used to depict the data.

$H_2SO_4$-doped ice single crystals is not significantly different from that for pure ice. Although the CRSS appears to depend on the crystallographic orientation for $H_2SO_4$-doped ice (Trickett *et al.* 2000*a*), no obvious orientation dependence of *n* was observed in either the undoped ice or the $H_2SO_4$-doped ice single crystals.

Although sulphuric acid does not affect the stress exponent of ice single crystals, as noted previously (Trickett *et al.* 2000*a*), it does soften ice. This is evident here by the larger shear strain rate for a given CRSS (Fig. 9), a behaviour consistent with Glen's model (1968). According to Glen, proton disorder presents the major obstacle to dislocation glide in ice $I_h$ crystals. Since adjacent (0001) glide planes are connected through randomly oriented hydrogen bonds, in order for a dislocation to move the hydrogen bonds have to be appropriately reoriented. This bond reorientation occurs by the movement of ions ($H_3O^+$ and $OH^-$) and Bjerrum defects (D type and L type). Thus, Glen's model predicts that impurities that introduce additional point defects into ice should increase the mobility of dislocations. For example, an HF molecule introduces one L-defect and the $H_3O^+$ concentration also increases (Granicher 1963). Thus, Jones (1967) and Jones & Glen (1969) demonstrated that very small amounts of dissolved HF could significantly reduce the strength of ice single crystals. Since $SO_4^{2-}$ occurs as an interstitial ion in the lattice but $F^-$ is present as a substitutional one, the softening mechanism of $H_2SO_4$ may be a little different from HF. However, the $H_2SO_4$ presumably also increases the dislocation mobility due to the addition of two $H^+$ ions from each $H_2SO_4$ molecule, thus decreasing the strength. The $H_2SO_4$ may also increase the grown-in dislocation density (Trickett *et al.* 2000*a*), as has been shown for Hf and HCl doping of ice crystals (Jones & Gilra 1972*a, b*; Oguro 1988).

Since according to Glen's model (1968), $H_2SO_4$ doping produces easier dislocation motion this would imply improved ductility by allowing plastic flow to dissipate local stress concentrations that would otherwise lead to cracking. The increase in ductility of ice crystals due to $H_2SO_4$ may also be simply related to the reduction in the strength – an inverse relationship between strength and ductility is well documented for metallic alloys.

The implications of the softening effect of sulphuric acid on ice single crystals for the flow of natural polycrystalline ice are not clear. It has been shown that sulphuric acid is present in natural ice at the triple junctions where three grains meet (Mulvaney *et al.* 1988; Wolff *et al.* 1991; Fukazawa *et al.* 1998). A substantial concentration of sulphuric acid (greater than 1 M in areas of $1\,\mu m^2$) has been found, using energy dispersive spectroscopy (EDS) in a scanning electron microscope (SEM), as a solid phase at the triple junctions of ice from Dolleman Island held at $-160°C$ (Mulvaney *et al.* 1988). Whether a significant amount of $H_2SO_4$ remains in the lattice in polycrystalline ice and how $H_2SO_4$ affects the deformation of polycrystalline ice are both unknown and the subjects of current studies.

## Conclusions

The value of the stress exponent, *n*, at 20°C for ice single crystals doped with 6.8 ppm $H_2SO_4$ was found to be $1.89 \pm 0.21$ at $\theta = 25°$ and $1.94 \pm 0.26$ at $\theta = 3°$ where $\theta$ is the orientation between the basal plane and the compression axis. For undoped single crystal ice of three different orientations, *n* values ranged from 1.89 to 1.97. The insignificant variance of *n* for different orientations and between pure ice and doped ice indicates that *n* does not depend on the crystallographic orientation or $H_2SO_4$ doping.

Grant OPP-9526454 from the National Science Foundation and Grant DAA-H04-96-1-0041 from the Army Research Office supported this research. H. J. Frost is acknowledged for his useful comments.

## References

FUKAZAWA, H., SUGIYAMA, K. & MAE, S. 1998. Acid ions at triple junction of Antarctic ice observed by Raman scattering. *Geophysical Research Letters*, **25**, 2845–2848.

GRANICHER, H. 1963 Properties and lattice imperfections of ice crystals and the behaviour of $H_2O$-HF solid solutions. *Physics Kondensierten Materie*, **1**, 1–12.

GLEN, J. W. 1968. The effect of hydrogen disorder on dislocation movement and plastic deformation of ice. *Physics Kondensierten Materie*, **7**, 43–51.

HIGASHI, A., KOINUMA, S. & MAE, S. 1964. Plastic yielding in ice single crystals. *Japan Journal of Applied Physics*, **3**, 610–616.

——, —— & ——1965. Bending creep of ice single crystals. *Japan Journal of Applied Physics*, **4**, 575–582.

JONES, S. J. 1967. Softening of ice crystals by dissolved fluoride ions. *Physics Letters*, **25A**, 366–377.

—— & BRUNET, J. G. 1978. Deformation of ice single crystals close to the melting point. *Journal of Glaciology*, **21**, 445–455.

—— & GILRA, N. K. 1972*a*. Increase of dislocation density in ice by dissolved hydrogen fluoride. *Applied Physics Letters*, **20**, 319–320.

—— & ——1972b. X-Ray Topographic Study of Dislocations in Pure and HF-Doped Ice. *Philosophical Magazine*, **27**, 457–472.

—— & GLEN, J. W. 1968. The mechanical properties of single crystals of ice at low temperatures. *International Association of Scientific Hydrology Publications*, **79**, 326–340.

—— & —— 1969. The effect of dissolved impurities on the mechanical properties of ice crystals. *Philosophical Magazine*, **19**, 13–24.

KAMB, W. B. 1961. The glide direction in ice. *Journal of Glaciology*, **3**, 1097–1106.

LORIUS, C., BAUDIN, G., CITTANOVA, J. & PLATZER, R. 1969 Impuretes solubles contenues dans la glace de l'Antarctique. *Tellus*, **21**, 136–148.

MUGURUMA, J. 1969. Effects of surface condition on the mechanical properties of ice crystals. *Journal Applied Physics (J. Phys. D)*, Ser. 2, **2**, 1517–1525.

MULVANEY, R. & WOLFF, E. W. 1994. Spatial variability of the major chemistry of the Antarctic ice sheet. *Annals of Glaciology*, **20**, 440–446.

——, —— & OATES, K. 1988. Sulphuric acid at grain boundaries in Antarctic ice. *Nature*, **331**, 247–249.

NAKAMURA, T. & JONES, S. J. 1970. Softening effect of dissolved hydrogen chloride in ice crystals. *Scripta Metallurgica*, **4**, 123–126.

—— & ——1973. Mechanical properties of impure ice crystals. *In*: WHALLEY, E., JONES, S. J. &

GOLD, L. W. (eds) *Physics and Chemistry of Ice*. Papers presented at the Symposium on the Physics and Chemistry of Ice, Ottawa, Canada, August 14–18 1972. Ottawa, Royal Society of Canada, 365–369.

OGURO, M. 1988. Dislocations in artificially grown single crystals of ice. *In*: HIGASHI, A. (ed.) *Lattice Defects in Ice Crystals*. Hokkaido University Press, Sapporo, Japan, 27–47.

PAREN, J. G. & WALKER, J. C. F. 1971. Influence of Limited Solubility on the Electrical and Mechanical Properties of Ice. *Nature*, **230**, 77–79.

READEY, D. W. & KINGERY, W. D. 1964. Plastic deformation of single crystal ice. *Acta Metallurgica*, **12**, 171–178.

SCHMID, E. & BOAS, W. 1950. *Plasticity of Crystals*. F. A. Hughes and Co. Ltd, London.

TRICKETT, Y. L., BAKER, I. & PRADHAN, P. M. S. 2000a. The effects of sulphuric acid on the mechanical properties of ice single crystals. *Journal of Glaciology*, in press.

——, —— & ——2000b. The orientation dependence of the strength of ice crystals. *Journal of Glaciology*, **46**, 41–44.

WOLFF, E. W., MULVANEY, R. & OATES, K. 1988. The location of impurities in Antarctic ice. *Annals of Glaciology*, **11**, 194–197.

# Physical modelling of the rheology of clean and sediment-rich polycrystalline ice using a geotechnical centrifuge: potential applications

DUNCAN H. B. IRVING, BRICE R. REA & CHARLES HARRIS

*Department of Earth Sciences, Cardiff University, Cardiff, CF10 3YE, UK*
*(e-mail irvingd@cf.ac.uk)*

**Abstract:** A self-weight stress gradient is developed through the body of a glacier such that strain rates are highest in the lowest few metres as described by Glen's flow law and other stress- and temperature-dependent relationships. Conventional laboratory technology limits the size and complexity of physical models of glacier ice, particularly in the complicated basal ice layers. A geotechnical centrifuge can be used to replicate such stress regimes in a controlled environment using a scaled model of the field 'prototype' that is subjected to an accelerational field that is a factor $N$ greater than that of the Earth, $g$. The development of a technique employing a geotechnical centrifuge as a testbed for such physical models is described. Strain rates of $10^{-6}$–$10^{-7}$ s$^{-1}$ are calculated for models of low and moderate stress, high temperature ice. Relationships between the physical models and glacial systems suggest a scaling of the effects of transient creep by $1:N$, diffusion creep by $1:N^2$–$1:N^3$ and power law creep by $1:1$. Preliminary results demonstrate the potential applications of the technique in the fields of glaciology and glacial geomorphology, in particular where low stresses and high temperatures are key characteristics of a glacial system and in systems containing several stratigraphic units.

The studied mechanical behaviour of ice has generally been conducted using triaxial and uniaxial rigs in which stresses are applied to samples along one or more axes. In the case that the confining stress is greater than the applied stress, the deformation of the sample is generally expressed as a conical pair of shear zones as shown schematically in Fig. 1a. In this configuration, the departures from the true stress field in a glacier can be summarized as follows.

- The sample size limits the grain size which can be used in the sample (Baker 1981) before grain size becomes volumetrically significant. Stress is localized in the larger grains and the resultant shear strain is modified by this effect.
- The size of the shear zone is governed by the end platens and the sample length.
- The shear zones are conical and hence have a point discontinuity.

Appropriate laboratory and mathematical methods have minimized the effects of the last two factors to allow testing at small grain sizes to be conducted. However, grain sizes in excess of 10 mm are common in ice bodies at high homologous temperatures as a result of dynamic recrystallisation (Paterson 1994).

A more effective laboratory technique is now introduced that allows large models of appropriate geometry to be tested under the correct stress regime. This method employs a geotechnical beam centrifuge to apply an acceleration so that self weight stress increases linearly with depth through a large (maximum $0.5 \times 0.7 \times 0.2$ m) model in a controlled environment. Such a model size allows spatially varying composition and structure as observed in the field (known as the *prototype*) to be replicated in the laboratory as shown in Fig. 1b.

This paper presents a methodology developed for the rheological testing of ice on a geotechnical centrifuge and the microphysical considerations which must be applied in relating a scaled physical model to field measurements. Centrifuge methods are appropriate to the refinement of current rheological models of ice. Sample size and fundamental theoretical considerations have dictated the nature of laboratory studies of ice rheology to date. This study has identified the need for an extension of the current knowledge to incorporate the following factors.

- Low shear stress and temperatures above $-10°C$. This promotes deformation by diffusion creep in addition to power-law creep.

*From:* MALTMAN, A. J., HUBBARD, B. & HAMBREY, M. J. (eds) *Deformation of Glacial Materials.* Geological Society, London, Special Publications, **176**, 47–55. 0305-8719/00/$15.00 © The Geological Society of London 2000.

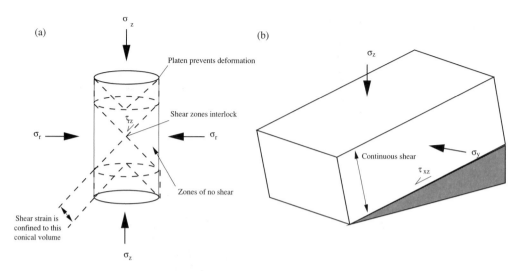

**Fig. 1.** Schematic comparison of (**a**) biaxial and (**b**) ideal stress configurations in mechanical testing. Triaxial testing develops shear, $\tau_{rz}$ in response to ($\sigma_z$) ancl radial ($\sigma_r$) stresses in a discrete conical zone governed by the platen width and the sample length (a). The ideal configuration applies a shear stress, $\tau_{xz}$, in response to an applied load, $\sigma_z$, and a confining stress, $\sigma_y$, which increases linearly with depth to give a planar shear zone through the sample (b).

However, the diffusional component is rarely considered in numerical models of ice subjected to low shear stresses.

- The effect of grain size, which is important in the transient creep component and the diffusion creep component. Models subjected to low stresses will display a rheology which is strongly controlled by the size of grains used in the model construction. Models displaying power-law creep are not affected by the grain size used within the ranges observed in glacial systems (1–100 mm).
- Low stresses, which are dissipated by diffusion creep more readily than by power-law creep. Large grain sizes and high temperatures exacerbate this effect.

## Centrifuge technology

The geotechnical beam centrifuge is a unique physical modelling tool which has been used by geotechnical engineers for some years, but is only now gaining a profile in Earth Sciences (Rea *et al.* in press; Harris *et al.* in press). In many Earth surface phenomena the role of self-weight body forces is crucial in explaining observed behaviour; i.e. where *in situ* stresses change with depth and the response of the material may alter in response to these stress differences.

A model fixed to the end of a centrifuge arm that is then rotated ('flown') rapidly, experiences

an inertial radial acceleration field, which, as far as the model is concerned, is equivalent to a gravitational acceleration field but many times stronger than Earth's gravity field (Taylor 1995). The model experiences zero stress at the surface but this increases with depth, producing a stress gradient that is directly comparable to field prototypes. The ratio of the stress gradient in the model to that in the prototype is equal to the $g$ level at which the model is being flown. This is conventionally expressed as the dimensionless parameter $N$. This effect is shown in Fig. 2. The angular velocity of the centrifuge is such that the angular acceleration gives a centrifugal force that replicates the required stress at the base of the model.

## Scaling

In a scaled model (Fig. 2) subjected to an inertial acceleration field which is $N$ times Earth's gravity, the vertical stress in the model, $\sigma_{zm}$ at model depth $h_m$, is:

$$\sigma_{zm} = \rho N g h_m \qquad (1)$$

where $\rho$ is the density and $g$ is the acceleration due to gravity. If the stress at the base of the model is the same as that at the base of the prototype then the thickness of the prototype is $Nh_m$. For example, a model with a thickness of 0.1 m subjected to a centripetal acceleration of

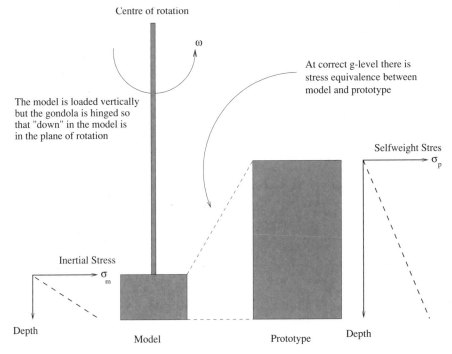

**Fig. 2.** The scaling of a physical model on a rotating centrifuge with the field prototype. At a certain angular frequency, $\omega$, the centrifugal force through the model will give the same surficial ($\sigma = 0$) and basal ($\sigma_{zm} = \sigma_{zp}$) stresses. $N$ is the factor required to scale the stress gradient in the model to that in the prototype.

$100g$ will have a stress gradient 100 times that in a prototype of equal density whose thickness is 10 m. Hence the macroscopic linear dimension scales as $1 : N$.

In common with other polycrystalline materials, ice can deform by a number of mechanisms; including power-law creep, diffusion creep and plastic flow. The degree to which each operates is governed by the grain size, stress and temperature. The manner in which each mode of deformation can be scaled between the model and the prototype is of key importance to accurate physical modelling. Plastic flow is not considered here but as this is a laboratory methodology with models being tested from an initially artificial fabric, the effects of transient creep are considered. Each of these mechanisms is controlled by very different parameters summarized below.

*Power-law creep*

Power-law creep rate is a function of dislocation distribution within a polycrystalline medium. In an isotropic medium the shape of grains and the distribution of the size and orientation

of dislocations is independent of the grain size of the medium (Mellor & Testa 1969; Baker 1981). Dislocation geometry is a function of applied stress, $\sigma$, only and is described by the relationship

$$\dot{\varepsilon} \approx \frac{D_{sd}\mu b}{kT}\left(\frac{\sigma}{\mu}\right)^3 \qquad (2)$$

where $D_{sd}$ is the coefficient of self-diffusion, $\mu$ is the shear strength, $b$ is the Burgers vector, $k$ is the Boltzmann constant and $T$ is the temperature (Ranalli 1987).

Over the grain size range of 0.1 mm to 0.1 m, there is no reported variation in the parameters included in Equation 2 (Duval & LeGac 1980; Jacka 1984) so under equal applied shear stresses, a model constructed of ice with a grain size of 1 mm will exhibit power-law creep at the same rate as one with a grain size of 100 mm. This scale-invariance implies that the scaling factor for the power-law component of strain is $N = 1$ and if power-law creep is the exclusive mechanism in the deformation of a model then scaling of grain size is unnecessary as hypothesized by Smith (1995). The effect of dynamic recrystallization upon the grain size will give rise to a change in the microphysical description of the dislocation configuration but this is not considered in the macroscopic domain.

## Diffusion creep

Diffusion creep in ice is described by the Nabarro–Herring model (Ranalli 1987; Hooke 1998). Nabarro–Herring creep occurs by the motion of atoms through the intracrystal lattice. Diffusivity is strongly dependent on grain size as the ratio of the surface area of a crystal to its volume governs the path which dislocations take through a grain. Grain boundary diffusion is described by the Coble model and becomes the dominant mode at temperatures below $-50°C$ (Hooke 1998). The Nabarro–Herring relationship with grain size and applied stress has the form

$$\dot{\varepsilon} \approx \frac{D_{sd}\mu\Omega}{kTd^2}\left(\frac{\sigma}{\mu}\right) \qquad (3)$$

where $\Omega$ is the atomic volume (Ranalli 1987). This has been expressed using an Arrhenius temperature relationship by Duval *et al.* (1983) as

$$\dot{\varepsilon} \approx \frac{B}{d^l}\sigma^n \qquad (4)$$

where $B$ is a temperature dependent strength parameter and $d$ is the $2 \leq l \leq 3$ is a fabric related power where $l \to 2$ for isotropic ice (Baker 1981). In both cases creep has at least a $d^{-2}$ dependence and given that $d$ is a macroscopic parameter, the effect of scaling the grain size as $Nd$ will give rise to a scaling factor of $N^{-2}$–$N^{-3}$.

## Transient creep

At temperatures above $-10°C$ the activation energy for creep increases from 65 to 140 kJ mol$^{-1}$, which is over twice the self-diffusion activation energy for oxygen or hydrogen (Barnes *et al.* 1971; Weertman 1973). This excludes crystal climb as the rate limiting factor (Duval *et al.* 1983). Crystal drag is the most likely rate-limiting factor at shear stresses below 0.1 MPa (Hooke 1998, chapter 4) and this mechanism is controlled by the available surface area of ice grains. Grain size is a macroscopic parameter so the transient creep component has a scaling factor of $N$.

## Scaling effects

To achieve the accurate replication of a prototype ice body in the laboratory, consideration must be given to the stress, temperature and grain size regimes, which can be summarized as follow:

- At low shear stresses (below 0.1 MPa), temperatures close to the triple point and grain sizes greater than 10 mm the effects of diffusion creep are several orders of magnitude greater than the other creep components (Baker 1981; Irving 1999).
- At higher stresses, power-law creep becomes the dominant form of creep and the effects of grain size scaling become negligible (Mellor & Testa 1969).

It is often the case in geotechnical problems that the scaling is such that there is time compression, which is obviously advantageous if a costly and lengthy series of field tests may be avoided. In the case of non-accelerating polycrystalline deformation the above scaling relations have no time-dependent components other than the rate of decay of the transient component. The advantage of using a beam centrifuge over more conventional methods are the size and complexity of the model which may be tested rather than reduction in time necessary to measure deformations.

## Centrifuge Method

The centrifuge used in this study was the Cardiff Geotechnical Centrifuge, a 2.3 m beam centrifuge capable of rotating packages of $1 \times 1 \times 1$ m at accelerations of $100g$. The gondola that carries the package is pivoted on the end of the beam so it can swing outwards and the model becomes oriented such that the vertical the direction in the model is along a radius of the centrifuge locus. The difference in $g$ level between the top and bottom of a 200 mm thick model is less that 1% (Irving 1999) so the accelerational field can be considered constant through the model. Figure 3 shows the layout of the box, model, cooling systems and instrumentation described below.

### Model construction

Models were constructed following the method of Jellinek & Brill (1956), whereby ice grains were sieved through a 10 mm sieve into a mould of 150 mm depth, 760 mm length and either 250 or 500 mm in breadth. Water, chilled to 0°C was added to a the mould to bond the grains and the models were chilled slowly over a period of several days to $-7°C$. Coarse sand was mixed with the ice grains before the chilled water was added to give a sediment–rich (10%) ice model

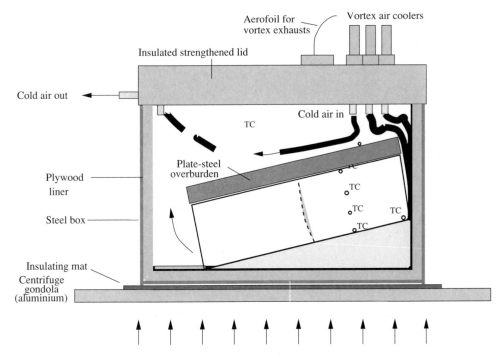

**Fig. 3.** Side view of centrifuge strongbox and model. The centrifuge box is constructed of 20 mm steel plate with a 20 mm wooden liner. The sample length is 760 mm in the downslope direction, up to 200 mm in thickness and 1 × 500 mm or 2 × 250 mm in breadth. The coolers circulate cold, dry air around the model.

(all percentages refer to volumetric content). It was observed that the sediment was fully dispersed within the model with localized grain-to-grain contacts at triple junctions.

## Environmental controls

Temperature and stress were the two parameters that could be controlled in the centrifuge tests. The model was placed in a strongbox on a slope of 15° which maximized the basal shear stress possible in the confines of the box. The angular velocity of the centrifuge governs the stress through the model but after initial tests a method was developed which allowed the attachment of a 60 mm steel overburden to

the surface of the model which increased the basal shear stress to 106 kPa (Rea *et al.* in press). Temperature control was achieved using vortex coolers which delivered a 250 l min$^{-1}$ flow of air at −20°C. Testing was conducted over a period of several days with intermittent rest periods to allow the centrifuge to cool.

## Instrumentation

A range of instrument was deployed with varying success in such a harsh environment. temperature was measured continuously using K-type thermocouples in and around the model. Deformation was initially measured at the surface of the model using linear voltage displacement transformers, and within the model using ductile solder thread markers (Irving 1999). Neither of these produced satisfactory data sets as a result of sample preparation and high g. They were superseded with a grid of 1 mm diameter, 20 mm long hollow plastic markers frozen into the side of the model (Irving 1999) and columns of 5 mm diameter, 20 mm high plastic markers drilled into the model extending from the surface (Rea *et al.* in press). Image analysis allowed the time development of the marker grids and columns to be calculated.

## Behaviour of Models

Several tests were conducted which were used to refine the methodology (Irving 1999). A second series of tests was begun with an overburden of steel plate to increase the basal shear stress (Rea *et al.* in press). The recorded strain rates are presented in Table 1. The tests are summarised as follows:

- Two models of isotropic polycrystalline ice, one pure and the other containing a 10% coarse sand fraction were tested (Irving 1999). The model dimensions were 760 mm in the downslope direction, 500 mm wide and 170 mm in thickness. The 15° slope and 80g accelerational field induced basal shear stresses of 25 kPa in the clean ice model and 27 kPa in the sediment-rich model. They were flown for a total of 48 and 70 ks respectively and average strain rates of $10^{-6}\,s^{-1}$ were calculated. The temperatures were kept within a range of $[-6, -1°C]$ with a mean value of $-2.5°C$. For the purposes of interpretation it was assumed that this value could be used as a constant. The time development was recorded and the strain rates decreased at a decelerating rate although a minimum value was not attained. In the sediment-rich model the creep was slower and there appeared to

**Table 1.** *Calculated strain rates from preliminary tests*

| Model description | Basal shear stress (kPa) | Duration (ks) | Strain rate ($s^{-1}$) |
|---|---|---|---|
| Bubble-rich | 106 | 52.2 | $10^{-6}$ |
| Massive | 106 | 52.2 | $10^{-6}$ |
| Massive | 25 | 48 | $10^{-7}$ |
| Massive, 10% sand | 27 | 72 | $10^{-7}$ |

The strain rates in the low stress tests were the minimum achieved during testing. Those in the 106 kPa tests were the mean values over the duration of the test and do not provide information about the time development of the strain rate.

be 5% compaction during the first 30–40 ks of the test.

- The effects of increasing the basal shear stress with the use of steel overburden were investigated using two models of isotropic ice, one massive and the other with a high bubble content. Rea *et al.* (in press) describes the test procedure, the key differences from the first tests series being that the models were 250 mm in breadth and 130 mm in thickness with an additional 60 mm of steel attached to the surface by ridges moulded onto the underside of the steel. The total deformation was measured using marker columns drilled into the ice. Strain rates of $10^{-7}$–$10^{-6}\,s^{-1}$ were calculated. Time development data were not obtained due to instrumentation failure so it is not possible to speculate on the degree to which the transient deformation had decayed or whether a minimum creep rate was reached.

The downslope deformation recorded after 14.5 hours in each of the tests is shown in Fig. 4. Compaction (slope-perpendicular deformation) was measured, but was negligible in the clean ice models and transient in the sediment-rich model. The uppermost markers are missing from the marker positions in Fig. 4a as the majority of them had melted out by this stage in the testing. The extent of the deformation is greater in the samples that were subjected to the overburden (Fig. 4b) but not to the extent predicted by a power-law rheology as there would be a transient component superposed on this mode. Of the unburdened samples, the deformation in the sediment-rich ice is mechanically impeded by the presence of the sand grains although the retardation by a factor of 80% is greater than that reported by Hooke *et al.* (1972).

The time development of the deformation in the unburdened samples reported by Irving (1999) suggests that a steady-state deformation was not achieved and hence the deformation was by a superposition of transient creep and diffusional creep due to the low stress. The higher stresses in the overburden samples require a power-law deformation and the duration of the test suggests that a steady-state of deformation was reached.

Sources of inaccuracy were identified as inadequate refrigeration techniques. The refrigeration system was not capable of dissipating heat generated by the rotation of the centrifuge through an enclosed air-mass so the temperature fluctuated throughout the tests. Problems were also encountered due to the formation of ice over the surfaces of some displacement markers

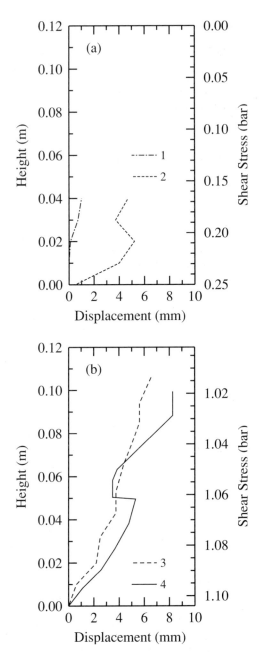

**Fig. 4.** Low stress (**a**) and high stress (overburden applied) (**b**) deformation curves after 14.5 hours at 80g. Curves 1 and 2 are measured from side-mounted deformation markers. Curve 1 is from a sample of ice containing 10% sediment by volume and curve 2 is from a sample of clean ice. Both samples were bubble poor. Curves 3 and 4 are measured from column markers in the centre lines of the models. Curve 3 is from a clean, bubble-poor sample and curve 4 is from a clean sample with bubble-poor ice from its base to 0.05 m and bubble-rich above.

making their exact locations difficult to detect. Hence these results are preliminary but demonstrate the potential of the centrifuge method.

## Discussion

It was anticipated that the conventional geotechnical centrifuge methodology physically constraining a field prototype would, with appropriate scaling considerations lead to a model which would deform in a predictable manner. However, the complexity of the scaling issues in terms of both composition and superposed modes of deformation became apparent after the initial tests (Irving 1999). It became more appropriate to construct a well-constrained model and calculate the physical properties of a prototype from the behaviour of the model.

The simpler case is the two models in which the basal shear stress was increased with overburden. As argued above, it could be assumed that all deformation was in the form of power-law creep with an unquantifiable decaying transient component. In the absence of time-development data, it is not possible to extract the two components from the total deformation but from the scaling relationships described above, the transient component of the deformation would scale as $1:N$ and the power-law component as $1:1$. The sensitivity of the transient effect to grain size allows this effect to be amplified using the centrifuge and has a potential application in modelling the effects of transient creep in ice bodies.

The temperatures and geometries were kept as constant as possible over the series of tests. Hence, the similar extent of deformation in the three clean samples suggests that the high temperature at which the tests were carried out has a greater positive effect on diffusional creep (Fig. 4a) than on a dislocational creep (Fig. 4b) even though the latter was achieved at higher shear stresses. This difference is amplified by the scaling effects described above whereby a power-law rheology is unaffected by the grain size of the ice whereas diffusional creep displays a sensitivity between the second and third power to grain size at high temperature

This has strong implications for further modelling, both numerical and physical, of ice bodies which are subjected to low stresses and high temperatures. Examples of such environments where this low-stress modelling method is applicable are:

● the firnification zone in glaciers where deformation has yet to reach a steady-state (Paterson 1994);

- cirque glaciers in climatically marginal zones, the mass-balance fluctuations of which preclude a steady-state rheology (Haeberli & Hoelzle 1995);
- proto-rock glaciers evolving from talus (Azizi & Whalley 1995).

The models which were subjected to lower stresses were considered to be displaying a combination of diffusion creep and power-law creep in addition to the initial transient creep component. The scaling relationships described for each of these components are poorly constrained in the absence of exclusively steady-state creep so the relative magnitude of each component cannot be extracted. The grain size in the model exerts the greatest control on its rheology and the poor constraint on the scaling factor ($N^2$–$N^3$) prevent rigorous interpretation of the rheology in the context of a field prototype. However, in a model where grain size is not scaled with the model thickness, the amplification of the diffusion component relative to the power-law component permits diffusion creep to be investigated once a steady-state is achieved.

## Conclusions

The tests described demonstrate the rheological effects which can be investigated using centrifuge modelling. A large sample size and a linear stress field allow the behaviour of ice grains of realistic sizes to be modelled and a continuous stress regime to be applied through a model. The addition of overburden permits a wide range of ice thicknesses to be investigated. The tests presented above demonstrate that, close to the melting point and at moderate stresses (106 kPa), polycrystalline ice with a large grain size deforms by power-law creep with little, or no diffusional component. Models tested at lower stresses disiplayed a combination of the two deformation modes (Irving 1999). The extent of the deformation at both stress levels was comparable and this is attributed to a scaling effect which amplifies low-stress deformation at large grain size and high temperature. This effect warrants further consideration in low-stress, high-temperature ice bodies. Constructing and deforming such a model is only made possible using a large testbed and linear stress regime and these are the key features of the method.

Thus, this method has significant advantages over triaxial rig tests in both practical and numerical sophistication. This experimental technique is in the developmental stage and improvements in data acquisition and temperature control will allow tests of increased sophistication and duration with more thermal control to be conducted.

The potential applications for such a method beyond quantifying the rheology of models at various grain sizes, sediment fractions, shear stresses and temperatures include the following.

- Models containing compositional heterogeneities such as varying debris or water content as found in basal layers of glaciers. This could be a extended in a simple manner to structurally sophisticated models including two-layer systems such as rock glaciers and multi-layer systems such as tectonized basal ice.
- Theoretical models of subglacial quarrying (Iverson 1991) have, to date, not been calibrated through any physical data. Field constraints are likely to ensure that obtaining such data will remain impractical making scaled physical modelling the best alternative.
- Subglacial hydrology may also be modelled using this technique. For example, development of the variable pressure axis (VPA) described by Hubbard et al. (1955) may be replicated with an ice overburden atop a sediment bed with a controlled flow of water, at varying pressures, through the bed.
- Meltwater migration and fractionation processes (Souchez & Lorrain 1991), which are stress gradient and temperature dependent, are readily modelled. Indeed, much work has already been completed on the migration of water through granular materials (Depountis et al. 1999) in centrifuge models.

## References

AZIZI, F. & WHALLEY, W. B. 1995. Finite element analysis of the creep containing thin ice bodies. *In*: *Proceedings of the Vth International Offshore and Polar Engineering Conference*. ISOPE.

BAKER, R. W. 1981. Textural and crystal-fabric anisotropies and the flow of ice masses, *Science*, **211**, 1043–1044.

BARNES, P., TABOR, D. & WALKER, J. C. F. 1971. The friction and creep of polycrylstalline ice'. *Proceedings of the Royal Society London*, **324**, 127–155.

DEPOUNTIS, N., DAVIES, M. C. R., BURKART, S., HARRIS, C., THOREL, L., REZZOUG, A., KONI, D., MERRIFIELD, C. & CRAIG, W. H. 1999. Scaled centrifuge modelling of capillary rise. *In*: YOUNG, R. N. & THOMAS, H. R. (eds) *2nd BGS International Geoenvironmental Engineering Conference*, Thomas Telford, London, 264–271.

DUVAL, P. & LEGAC, H. 1980, Does the permanent creep-rate of polycrystalline ice increase with crystal size? *Journal of Glaciology*, **25**, 151–157.

——, ASHBY, M. & ANDERMAN, I. 1983. Rate-controlled processes in the creep of polycrystalline ice. *Journal of Physical Chemistry*, **87**, 4066–4074.

HACBERLI, W. & HOELZLE, M. 1995. Application of inventory data for characteristics of and regional climate-change effects on mountain glaciers: a pilot study with the European Alps. *Annals of Glaciology*, **21**, 206–212.

HARRIS, C., REA, B. R. & DAVIES, M. C. R. Geotechnical centrifuge modelling of gelifluction processes: Validation of a new approach to periglacial slope studies. *Annals of Glaciology*, **31**, in press.

HOOKE, R. L. 1998. *Principles of Glacier Mechanics.* Prentice Hall.

——, DAHLIN, B. B. & KAUPER, M. T. 1972. Creep of ice containing dispersed fine sand. *Journal of Glaciology*, **11**, 327–336.

HUBBARD, B., SHARPE, M. J., WILLIS, I. C., NIELSEN, M. K. & SMART, C C. 1995. Borehole water-level variations and the structure of the subglacial hydrological system of Haut Glacier d'Arolla, Valais, Switzerland. *Journal of Glaciology*, **31**, 572–583.

IRVING, D. H. B. 1999. *The Application of Geotechnical Centrifuge Modelling to the Deformation of Permafrost.* PhD thesis, Cardiff University.

IVERSON, N. R. 1991. Potential effects of subglacial water-pressure fluctuations on quarrying. *Journal of Glaciology*, **37**, 27–36.

JACKA, T. H. 1984. Laboratory studies on relationships between ice crystal size and flow rate. *Cold Regions Science and Technology*, **10**, 31–42.

JELLINEK, H. H. G. & BRILL, R. 1956. Viscoelastic properties of ice. *Journal of Applied Physics*, **27**, 1198–1209.

MELLOR, M. & TESTA, R. 1969. Creep of ice under low stress. *Journal of Glaciology*, **8**, 147–152.

PATERSON, W. S. 1994. *The Physics of Glaciers,* third edn. Pergamon Press, Oxford.

RANALLI, G. 1987. *Rheology of the Earth.* Allen and Unwin.

REA, B., IRVING, D., HUBBARD, B. & MCKINLEY, J., Preliminary investigations of centrifuge modelling of polycrystalline ice deformation. *Annals of Glaciology*, in press.

SMITH, C. C. 1995. Cold regions engineering. *In:* TAYLOR, R. N. (ed.) *Geotechnical Centrifuge Technology.* Blackie, London, 264–292.

SOUCHEZ, R. A. & LORRAIN, R. D. 1991. *Ice Composition and Glacier Dynamics.* Springer-Verlag, New York.

TAYLOR, R. N. 1995. Centrifuges in modelling: Principles and scale effects. *In:* TAYLOR, R. N. (ed.) *Geotechnical Centrifuge Technology.* Blackie, London, 19–33.

WEERTMAN, J. 1973. Creep of ice. *In:* WHALLEY, E. S., JONES, S. Y. & GOLD, L. W. (eds) *Physics and Chemistry of Ice.* Royal Society of Canada, 320–337.

# Glacier Flow and Structures

(Overleaf) View down on the icefall of Glacier de Saleina, Valais, Switzerland. As the ice approaches a rock step, it accelerates and clean-cut transverse crevasses develop normal to the maximum extending stress. As the icefall steepens, linking crevasses develop, fracturing the ice more intensely, and producing leaning towers of ice (*séracs*) which collapse to form ice-rubble. (Photo: M. J. Hambrey).

# Structural styles and deformation fields in glaciers: a review

MICHAEL J. HAMBREY[1] & WENDY LAWSON[2]

[1] Centre for Glaciology, Institute of Geography and Earth Sciences, University of Wales, Aberystwyth SY23 3DB, UK (e-mail: mjh@aber.ac.uk)

[2] Department of Geography, University of Canterbury, Christchurch, New Zealand (e-mail: w.lawson@geog.canterbury.ac.nz)

**Abstract:** Early structural glaciological research focused on analysis of particular structures or on mapping of structural features at particular glaciers. More recently, glacier structures have been interpreted in the context of deformation rates and histories measured or estimated using a range of techniques. These measurements indicate that glacier ice experiences complex, polyphase deformation histories that can include a wide range of types, rates and orientations of strain. Deformation styles in glacier ice resemble those in rocks, but occur at a much faster rate, allowing direct measurements to be undertaken, and providing potentially useful models of rock deformation. Structural analysis in the context of measured deformation shows that a wide range of structures (e.g. folds, foliations, boudins, shear zones, crevasses and faults) develop in response to complex strain environments, but strain does not necessarily result in the generation of structures. In the future, three-dimensional numerical modelling may be able to interpret and predict deformation histories and structural development.

The aim of this paper is to provide a review of the current state of knowledge in structural glaciology for the non-specialist, drawing on concepts developed for structural geology, particularly strain analysis in deformed rocks (e.g. Ramsay 1967; Hobbs et al. 1976; Ramsay & Huber 1983, 1987). In particular, we explore ways to improve our understanding of debris transfer, and deposition by, glaciers. In addition, the way in which numerical modelling may elucidate deformation histories in ice is highlighted.

The manner in which ice deforms in glaciers, and produces a variety of ductile and brittle structures, resembles that in rocks. Despite this, relatively few studies use glaciers as analogues of rock deformation, as structural glaciology has lagged behind developments in structural geology. We therefore hope that future application of structural geological principles will aid understanding of the dynamics of ice masses. This paper focuses on englacially formed structures exposed at the surface of valley glaciers, rather than the basal ice layer, the latter having been the subject of several recent investigations and reviews (e.g. Hubbard & Sharp 1989; Alley et al. 1998; Knight 1998; Lawson et al. 1998).

## Historical background

Interest in glacier structures dates back to the early days of glaciology. Early workers, such as Tyndall (1859) and Forbes (1900) compared foliation, the ubiquitous layering in glaciers, with geological structures. The nineteenth century was a time when the fundamental physical principles of glacier flow were being determined, and both those authors used ice structures to support their ideas about deformation in glaciers.

The first half of the twentieth century saw few significant advances in our understanding of glacier flow and ice deformation. However, in the 1950s a combination of borehole deformation measurements in glaciers, the formulation of a flow law for ice, and the establishment of a sound, physically-based theory for glacier flow (e.g. Glen 1955; Nye 1952, 1953, 1957), stimulated interest in glaciers, and led to the first detailed systematic studies of glacier structures (Table 1). Early contributions included those of Allen et al. (1960) on Blue Glacier, Washington and Meier (1960) on Saskatchewan Glacier, Alberta, in which the three-dimensional orientation of all the major structures were

From: MALTMAN, A. J., HUBBARD, B. & HAMBREY, M. J. (eds) Deformation of Glacial Materials. Geological Society, London, Special Publications, **176**, 59–83. 0305-8719/00/$15.00 © The Geological Society of London 2000.

**Table 1.** *Representative investigations of structures in glaciers since the 1950s*

| Glacier | Region | Author(s) | Date | Topic |
|---------|--------|-----------|------|-------|
| Twin Glacier | Alaska | Leighton | 1951 | Ogives |
| Pasterzenkees | Austria | Schwarzacher & Untersteiner | 1953 | Foliation |
| Outlet glaciers, SW Vatnajökull | Iceland | King & Ives | 1954 | Ogives |
| Pasterzenkees | Austria | Untersteiner | 1955 | Foliation |
| Greenland Ice Sheet | Thule, NW Greenland | Bishop | 1957 | Shear moraines |
| Malaspina Glacier | Alaska, USA | Sharp | 1958 | Folding of moraines |
| Glaciar Universidad | Central Chile | Lliboutry | 1958 | Ogives |
| Blue Glacier | Washington, USA | Kamb | 1959 | Ice petrofabrics in relation to structure |
| Blue Glacier | Washington, USA | Allen *et al.* | 1960 | Overall structure |
| Blue Glacier | Washington, USA | Sharp | 1960 | Overall structure |
| Saskatchewan Glacier | Alberta, Canada | Meier | 1960 | Overall structure; strain-rates |
| Vesl-Skautbreen | Norway | Grove | 1960 | Overall structure |
| Austerdalsbreen | Norway | King & Lewis | 1961 | Ogives |
| Various | Switzerland & France | Fisher | 1962 | Ogives |
| Greenland Ice Sheet | Thule, NW Greenland | Swinzow | 1962 | Shear zones |
| Burroughs Glacier | Alaska, USA | Taylor | 1962 | Overall structure |
| Various | Greenland, Iceland, Alaska | Atherton | 1963 | Ogives |
| Fox & Franz Josef Glaciers | New Zealand | Gunn | 1964 | Overall structure |
| Gulkana Glacier | Alaska, USA | Rutter | 1965 | Foliation |
| Mer de Glace | France | Vallon | 1967 | Overall structure |
| Ferrar Glacier | Antarctica | Souchez | 1967 | Shear moraines |
| Hintereisferner & Langtaufererjochferner | Austria | Ambach | 1968 | Crevasse formation |
| Gulkana Glacier | Alaska, USA | Ragan | 1969 | Structures below icefall |
| Kaskawulsh Glacier & Meserve Glacier | Alaska, USA Antarctica | Holdsworth | 1969 | Crevasse formation |
| Various | Svalbard | Boulton | 1970 | Debris/structure relationships |
| Kaskawulsh Glacier | Yukon, Canada | Anderton | 1970 | Structures at confluence; strain-rates |
| Isfallsglaciären | Sweden | Jonsson | 1970 | Overall structure & fabric |
| Bjørnbo Gletscher | East Greenland | Rutishauser | 1971 | Folding of surge-type moraines |
| Ice Cap | Ellesmere I., Canada | Souchez | 1971 | Structure of ice-cored moraines |
| Bering Glacier | Alaska, USA | Post | 1972 | Folding of moraines in piedmont glacier |
| Barnes Ice Cap | Baffin I., Canada | Hooke | 1973 | Foliation in relation to flow |
| Various | Switzerland | Hambrey & Milnes | 1975 | Boudinage |
| Various, Okstindan | Norway | Hambrey | 1975 | Origin of foliation |
| Charles Rabots Bre | Norway | Hambrey | 1976 | Overall structure |
| Barnes Ice Cap | Baffin I., Canada | Hudleston | 1976 | Folding |
| Griesgletscher | Switzerland | Hambrey & Milnes | 1977 | Structural evolution |
| Various | Switzerland; Axel Heiberg I., Canada | Hambrey | 1977 | Foliation, minor folds & strain |
| Barnes Ice Cap (num. model) | Baffin I., Canada | Hooke & Hudleston | 1978 | Foliation; relation to strain |
| White Glacier | Axel Heiberg I., Canada | Hambrey & Müller | 1978 | Overall structure; strain-rates |
| [Theoretical] | | Posamentier | 1978 | Ogives |
| Austerdalsbreen & Berendon Glacier | Norway Br. Columbia, Canada | Eyles & Rogersen | 1978*a, b* | Medial moraine formation |
| Griesgletscher | Switzerland | Hambrey *et al.* | 1980 | Structure and strain fields |

| Glacier | Region | Author(s) | Date | Topic |
|---|---|---|---|---|
| Mer de Glace | France | Lliboutry & Reynaud | 1981 | Ogives |
| Barnes Ice Cap | Baffin I., Canada | Hudleston | 1983 | Shear zones and strain patterns |
| Ventisquero Soler | N. Patagonian Icefield, Chile | Aniya & Naruse | 1987 | Thrusts and debris entrainment |
| Variegated Glacier | Alaska, USA | Sharp et al. | 1988 | Tectonic processes during surge |
| George VI Ice Shelf | Antarctica | Reynolds & Hambrey | 1988 | Overall structure from satellite imagery & photos |
| Wordie Ice Shelf | Antarctica | Reynolds | 1988 | Overall structure from satellite imagery |
| Glaciers at Phillips Inlet | Ellesmere I., Canada | Evans | 1989 | Debris-structure relationships |
| Several glaciers | N. Sweden | Hudleston | 1989 | Folds & veins in ice & rocks |
| Ice Stream B | West Antarctica | Vornberger & Whillans | 1990 | Crevasse development |
| Lambert Glacier – Amery Ice Shelf System | East Antarctica | Hambrey | 1991 | Satellite interpretation of ice structures |
| Variegated Glacier | Alaska, USA | Pfeffer | 1992 | Foliation formation during surge |
| Lambert Glacier – Amery Ice Shelf System | East Antarctica | Hambrey & Dowdeswell | 1994 | Satellite interpretation of ice structures & dynamics |
| Variegated Glacier | Alaska, USA | Lawson et al. | 1994 | Structural geology of surge-type glacier |
| Variegated Glacier | Alaska, USA | Lawson | 1996 | Structural evolution through several surge cycles |
| Bakaninbreen | Svalbard | Hambrey et al. | 1996 | Thrusting in surge-type glacier |
| Hessbreen | Svalbard | Hambrey & Dowdeswell | 1996 | Overall structure of surge-type glacier |
| Various | Russian High-Arctic | Dowdeswell & Williams | 1997 | Satellite interpretation of surge-type moraines |
| Bakaninbreen | Svalbard | Murray et al. | 1997 | Structural interpretation using GPR |
| Worthington Glacier | Alaska, USA | Harper et al. | 1998 | Crevasse formation in relation to strain fields |
| Kongsvegen | Svalbard | Glasser et al. | 1998 | Overall structure & debris relationships |
| East Antarctic ice sheet | Framnes Mts., East Antarctica | Marmo & Wilson | 1998 | Deformation in ice masses |
| Various | Svalbard | Hambrey et al. | 1999 | Debris entrainment in relation to structure |
| Holmströmbreen | Svalbard | Boulton et al. | 1999 | Moraine formation |

documented. Several of these early studies attempted to relate structures to measured strain rates, but with limited success. It was later recognized that certain structures, notably folds and foliation, were more likely to be related to cumulative strain (reflecting long-term strain history), calculated either from measured velocity distributions (Hambrey & Milnes 1977) or from numerical modelling (Hudleston & Hooke 1980; Hudleston 1983). Recent research has focused on how structures develop in surge-type glaciers (e.g. Sharp et al. 1988; Lawson et al. 1994; Murray et al. 1997; Hambrey & Dowdeswell 1996), and on applying concepts developed for valley glaciers to large ice streams, ice shelves and ice caps (e.g. Reynolds 1988; Hambrey &

Dowdeswell 1994; Dowdeswell & Williams 1998) using satellite imagery.

During the past three decades it has been established that deformation processes in glaciers can explain how debris is transported and deposited. Early workers examined the relationship between particular structures and debris features (e.g. Goldthwait 1951; Souchez 1967, 1971). However, few have attempted to explore these links in detail. Pioneering work by Boulton (1967, 1970, 1978) on debris-entrainment and depositional processes has provided a framework for identifying such processes in the geological record. However, the connection between debris transport and structures (such as foliation, folding and thrusting) on a glacier-wide scale is only now receiving detailed attention (Hambrey *et al.* 1999).

## Glaciers as analogues of rock deformation

### Comparison between glacier ice and deformed rocks

Glacier ice can be regarded as a metamorphic rock which is deforming under temperatures very close to, or below the pressure melting point (in temperate and cold glaciers respectively) (Hambrey & Milnes 1977). The residence time for ice in valley glaciers varies from a few decades at the centreline of a temperate tidewater glacier to a few centuries at the margins of a cold or polythermal glacier. During this time, a mass of snow in the accumulation area undergoes burial, lithification and diagenesis into glacier ice. During subsequent flow the ice undergoes the equivalent of regional high-grade metamorphism, partial melting and recrystallisation, fracture and flow. Deeper ice is subjected to ductile deformation, producing such structures as folds and foliation, while near-surface ice is mainly affected by fracturing (manifested by crevassing). The deeper ice is exposed progressively by melting in the ablation area as the terminus is approached. Completion of this deformation cycle in a typical alpine valley glacier takes place at a rate six orders of magnitude faster than in compressive mountain belts such as the Alps (Hambrey & Milnes 1977). Hence, glaciers provide opportunities for measuring deformation on the human time-scale, potentially allowing inferences to be made concerning deformation in orogenic belts.

In large cold ice masses, such as the ice sheets of Greenland and Antarctica, the rates of deformation are generally two orders of magnitude less than in temperate valley glaciers (except in ice streams), and the residence time of ice is measured in hundreds of thousands of years. However, similar structures appear in these ice masses as in temperate glaciers, indicating that, although deformation is slower, the end products are the same.

### Polyphase deformation

As in rocks, ice is subject to 'polyphase' deformation, that is structural development may result from several phases of deformation. Polyphase deformation implies events separated in time. However, 'early' structures form high up the glacier whilst 'late' structures are forming below. Thus, within the entire deforming mass of ice, temporal separation of deformation phases is not feasible. For example in Griesgletscher, Switzerland, the earliest structure, stratification ($S_0$), is folded to produce an axial-planar foliation ($S_1$), that is subsequently cross-cut by fractures (crevasses) and tensional veins (crevasse traces) ($S_2$), followed by a later phase of folding and foliation development ($S_3$). As $S_0$ forms in the accumulation area, $S_3$ is forming downstream. Thus one 'phase' runs continuously into another in some areas, and is clearly separable in others. Furthermore, periods of homogeneous deformation may occur between two 'phases', leaving no structural imprint, whilst the timing of the transition from one 'phase' to another can vary from place to place (Hambrey & Milnes 1977). There is now limited recognition that this concept is applicable to rocks. For example, it has been shown in the Appalachians that separate episodes of deformation, inferred from analysis of structures at one location, occurred at the same time in different locations as deformation swept through the rocks (Gray & Mitra 1993).

Recognition of 'phases' of deformation in glaciers is readily achieved by detailed structural mapping. Commonly three or four phases are recognizable but, beyond that, the relatively coarse crystal structure and recrystallization tend to hide other 'phases'. In contrast, many more phases of deformation may be deciphered in metamorphic rocks, as early structures are preserved at the microscopic scale and be visible in thin-sections.

### Cumulative strains in glaciers

The concept of cumulative (or finite) strains in rocks underpins the discipline of structural geology (e.g. Ramsay 1967; Hobbs *et al.* 1976: Ramsay & Huber 1983, 1987), yet has only been applied to glaciers in a few cases (e.g. Hambrey

& Milnes 1977; Hooke & Hudleston 1978; Lawson *et al.* this volume). In glaciers, strain-rates may be used to determine stresses (*via* Glen's flow law for ice), and attempts have been made to link strain-rate tensors to the development of foliation and crevasses.

The reasons why it is difficult to relate structural development to strain-rate are that: (i) structures may develop at depth, and continue to evolve with ice-flow; (ii) the orientation of structures is a reflection of the strain history they have experienced (i.e. how increments of strain add up); (iii) a structure may be reactivated at strain-rates lower than, or at a different orientation to, those required for their formation (Lawson *et al.* this volume). In other words, it is important to consider the sum total of successive increments of strain, since measurement taken at any one time or in any one place may be fortuitous with regard to structural development (Hambrey *et al.* 1980).

Glaciers typically have zones which are subject to (i) simple shearing, as at the ice margins, where the strains are non-coaxial; and (ii) pure shearing, as in zones of longitudinal compression in mid-glacier, where the strains are coaxial. Hooke & Hudleston (1978) have illustrated the different ways in which pre-existing structural inhomogeneities respond to these different types of strain path (Fig. 1).

Cumulative strain in rocks is generally depicted graphically by means of the strain ellipse or ellipsoid. Milnes & Hambrey (1976) described a method of determining strain ellipses for glaciers using ice velocity and flow-line data, and applied it to Griesgletscher, a valley glacier in Switzerland, to interpret a wide range of structures at the ice surface (Hambrey & Milnes 1977; Hambrey *et al.* 1980) (Fig. 2). Hooke & Hudleston (1978) and Hudleston (1983) examined numerically the change in shape and

orientation of the cumulative strain ellipse on flow-lines through a longitudinal section of the Barnes Ice Cap (Fig. 3). Lawson (1990) determined surface-parallel cumulative strain ellipses from the change in shape of moraine features on successive aerial photographs.

A comprehensive structural analysis of a glacier requires determination of the cumulative strain history, but this requires derivation of the velocity field of the entire glacier. Furthermore, the analysis has to rely on the assumption of steady-state flow during the residence time of ice in the glacier, as no velocity measurements over sufficiently long periods have been undertaken. Alternatively, robust modelling approaches need to be developed.

## Deformation in glaciers

Deformation occurs as a result of the gravity-driven flow of ice by creep, and by the super-imposed effects of basal motion in the form of sliding over a deformable bed. The deformation of ice resulting from creep occurs at a rate proportional to the third power of the shear stress, and is given by Glen's Flow Law (Glen 1955). The flow law has been generalized for glaciers by Nye (1957):

$$\dot{\varepsilon}_{ij} = A\tau_{ij}^n$$

where $\dot{\varepsilon}_{ij}$ represents the shear-strain rate, $A$ and $n$ are constants, and $\tau_{ij}$ is the shear stress. The constant $A$ is temperature-dependent, its value falling as the ice becomes colder, indicating that cold ice deforms less readily than warm ice. The exponent $n$ also varies, but is generally regarded as being close to 3. More complex forms of the flow law have been proposed, but the original form of the equation gives a satisfactory approximation in most cases.

Superimposed on the relatively steady-state deformation fields, fixed in space by the three-dimensional geometry of the glacier, are the dynamic deformation effects of strain caused by variations in basal sliding rates at various timescales. In general, regional steady-state glacier deformation patterns consist broadly of longitudinally extensional strain in the accumulation area, and longitudinally compressive strain in the ablation area (e.g. Hambrey *et al.* 1980). Local deformation effects, resulting from changes in bed-gradient (e.g. Harper *et al.* 1998) (Fig. 3), valley bending and glacier confluence (e.g. Gudmundsson *et al.* 1997) are superimposed on this pattern. Shear strain is concentrated near shear boundaries at the glacier margin and bed. Shear-strain rate is approximately coaxial

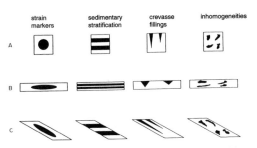

**Fig. 1.** Cumulative strain in glaciers showing the effect of pure shearing and simple shearing on the shape of structural inhomogeneities in glacier ice (from Hooke & Hudleston 1978, reproduced with permission of the International Glaciological Society).

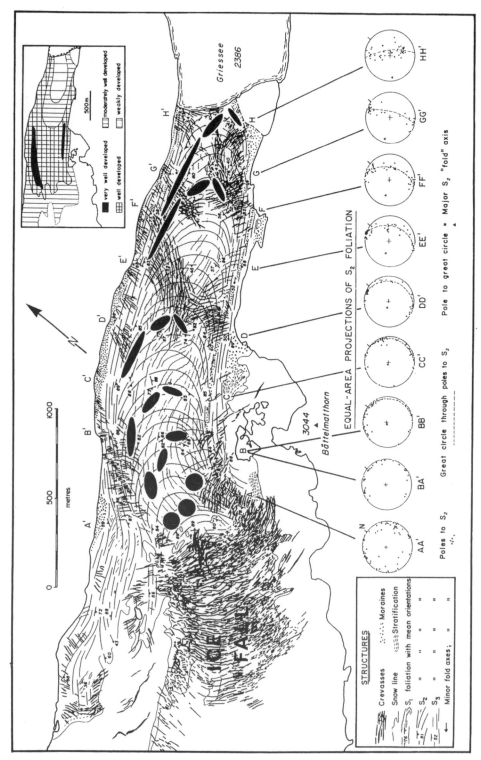

**Fig. 2.** Structural map of Griesgletscher, Switzerland, illustrating the relationship between foliation and the cumulative strain ellipse, starting with arbitrary circles below the icefall. Schmidt lower-hemisphere equal-area projections are of S₂ foliation. Inset shows qualitative assessment of 'strength' of S₂ foliation. (From Hambrey *et al.* 1980, reproduced with permission of the International Glaciological Society).

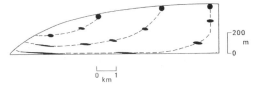

**Fig. 3.** Change in the strain ellipse on different flow paths through a longitudinal profile through the Barnes Ice Cap, based on finite-element modelling (from Hooke & Hudleston 1978, reproduced with permission of the International Glaciological Society).

along the glacier centre-line near the surface, and increasingly non-coaxial towards the margins and bed.

Various techniques have been used to measure rates of deformation in glaciers. Nye (1959) calculated principal surface strain rates from the relative motion of arrays of five stakes forming a 'strain element' on the glacier surface. This technique (with variants) is the standard method for measuring surface strain rates (e.g. Holdsworth 1969; Meier et al. 1974; Hambrey & Müller 1978; Sharp et al. 1988). Principal surface strain rates have also been estimated from velocity field gradients derived from the motion of velocity markers (e.g. Harper et al. 1998; Holdsworth 1969; Lawson et al. this volume), and more recently using remote sensing techniques (e.g. Gudmundsson et al. 1997). However, the velocity-gradient method involves inherently greater sampling errors (Vaughan 1993), although these can be mitigated by filtering (e.g. Gudmundsson et al. 1997). Deformation has also been measured directly using wire

strain meters (e.g. Evans et al. 1978; Raymond & Malone 1986).

Vertical shear-strain rates can be calculated from relative vertical displacements of markers in a borehole (Gudmundsson et al. 1997). In many cases, vertical strain rate is assumed to be linear or constant with depth (e.g. Raymond et al. 1987). Increasingly, numerical models of glacier flow are able to predict distributions of principal stresses in three dimensions (e.g. Hubbard et al. 1998). These models have considerable potential for elucidating strain patterns.

Surface strain rates have been determined for numerous glaciers at a range of scales (examples in Table 2). In general, the calculated strain rates reflect deformation at a scale that may include discontinuous deformation, for example across a crevasse. Maximum values of both compressive and extending strain rates are typically in the range $0.1–0.2\,a^{-1}$, but during surges values may reach two orders of magnitude greater than in steady-state glaciers.

Many authors have indicated that deformation in glaciers is spatially inhomogeneous at various scales. Hudleston (1977) demonstrated that discrete zones a few centimetres thick of enhanced vertical shearing occur in the Barnes Ice Cap in association with a distinctive crystal fabric. Marmo & Wilson (1998) found transverse shear strain concentrated in zones up to 3 km wide in the outlet glaciers from the Framnes Mountains of East Antarctica. Hulbe & Whillans (1997) identified zones of enhanced shearing within Ice Stream B in West Antarctica which they attributed to the variations of ice crystal fabric.

**Table 2.** *Examples of maximum strain-rates measured in valley glaciers*

| Glacier | Strain-rate $(a^{-1})$ | Reference | Comments |
|---|---|---|---|
| Saskatchewan Glacier, Alberta, Canada | 0.15 | Meier (1960) | Temperate |
| White Glacier, Axel Heiberg Island, Canada | 0.16 | Hambrey & Müller (1978) | Polythermal |
| Griesgletscher, Switzerland | 0.18 | Hambrey et al. (1980) | Temperate; base of ice-fall |
| Variegated Glacier, Alaska, USA | >36.5 | Raymond et al. (1987) | Temperate; during 1982–83 surge; regional strain from velocity distribution |
| Variegated Glacier, Alaska, USA | >100 | Sharp et al. (1988) | Direct measurements between stakes |
| Unteraargletscher, Switzerland | c. 0.2 | Gudmundsson et al. (1997) | Temperate; confluence of two tributaries |
| Unteraargletscher, Switzerland | 0.05* | Gudmundsson et al. (1997) | Vertical strain-rate |
| Worthington Glacier, Alaska, USA | 0.07 | Harper et al. (1998) | Temperate |

These are horizontal or surface-parallel values, except *.

## Structures in glaciers and their relationship to strain

Structures in glacier ice may be classified as 'primary' and 'secondary' (Hambrey 1994) as in rocks (e.g. Hills 1963; Hobbs *et al.* 1976). The most important primary structure is stratification which represents the original annual layering in the snow pack before firnification and transformation to glacier ice. Other primary structures include ice lenses derived from frozen water bodies and layers of superimposed ice. Secondary structures are of two types, brittle and ductile. Brittle structures are represented by crevasses (open fractures), and closed fractures and faults. Ductile structures result from the creep of ice, which can affect primary structures to varying degrees. Folding, foliation and crevasse traces are typical products of ductile deformation, whilst other minor structures such as boudinage and mylonite zones may also occur. The relationship between secondary structures and strain is considered below.

### Crevasses and other fractures

The relationship between crevasses and flow in valley glaciers is well-known from the theoretical work of Nye (1952). Crevasses generally form perpendicular to the maximum extending strain-rate, although the critical value for fracture is highly variable. Fracture orientation is largely influenced by morphology of the glacier-filled channel. Thus, crevasses develop especially in zones of extending flow, for example in an ice-fall, on the outside of a bend, when flowing over an uneven bed, or when the ice accelerates as it enters the sea or a lake (Frontispiece).

In surge-type glaciers, deformation associated with plug-type flow during the surge phase leads to cross-cutting longitudinal and transverse crevasses. These in turn are superimposed on the variably-orientated crevasses associated with quiescent-phase deformation (Lawson 1996).

When crevasses close, such as when they pass into a compressive flow regime, healed zones comprising coarse clear ice (contrasting with the adjacent coarse bubbly ice) remain; these represent a type of 'crevasse trace'. As ice continues to deform these traces may become arcuate, reflecting differential flow between the middle of the glacier and the margins. In so doing, the crevasse traces become more attenuated. Closed crevasses commonly display strike-slip displacements of a few centimetres to several metres (Fig. 4a), indicating passage into a different strain regime.

**Fig. 4.** Crevasse traces. (**a**) Healed water-filled crevasses showing strike-slip displacements, Hessbreen, a surge-type glacier in Svalbard. (**b**) Transverse crevasse traces, mainly of the extensional vein type, in the lower part of the icefall on Griesgletscher, Switzerland.

Another type of 'crevasse trace' is the extensional vein which forms as the lateral and vertical continuation of open crevasses, or as features in their own right (Hambrey 1976; Hambrey & Müller 1978) (Fig. 4b). Crystals are elongate and aligned perpendicular to the crack. Generally crevasse traces form normal to the maximum extending strain rate by nucleation of vein crystals along a fracture. Occasionally crevasse traces show oblique crystal growth with respect to the fracture, suggesting that vein growth has taken place in a rotational strain regime. These structures are equivalent to the crack-seal mechanism recognised in deformed rocks (e.g. Ramsay & Huber 1983).

Brittle failure in glacier ice is also represented by the thrust fault (Fig. 5). In zones of longitudinal compression, especially near the snout of a glacier, recumbent folds in basal ice may be initiated by protuberances in the bed (Hudleston 1976). The lower limb of the fold may become

**Fig. 5.** Thrusts in Variegated Glacier immediately following 1982–83 surge, showing small-scale displacements and folding of stratification.

attenuated and is replaced by a narrow zone of shearing or a discrete fracture or thrust. Such fractures may extend to the glacier surface, but many die out within the ice body (i.e. they are 'blind thrusts'). Thrusts are commonly well-developed towards the snouts of land-based cold and polythermal glaciers. Typical polythermal glaciers have terminal and marginal zones frozen to the bed, and a temperate, sliding zone in the interior. The frozen mass serves as an obstacle to flow, so the sliding ice overrides it, creating a thrust-fault (Clarke & Blake 1991; Hambrey et al. 1997). Thrusting is facilitated by the presence of structural weaknesses, such as favourably-orientated crevasse traces (Hambrey & Müller 1978) or previous generations of thrusts

(Lawson *et al.* 1994). Thrusting is also a feature of surge-type glaciers. The passage of the surge front through a glacier is accompanied by active thrusting, as in temperate Variegated Glacier, Alaska during its several month-long surge in 1982–83 (Sharp *et al.* 1988), and in polythermal Bakaninbreen, Svalbard, during its extended surge phase from about 1985–1995 (Hambrey *et al.* 1996; Murray *et al.* 1997).

## Folding

Folding of all amplitudes takes place in glaciers in response to changes in channel geometry and differential flow. Folds are commonly linked to, and have an axial planar relationship with, foliation. In many glaciers, stratification becomes folded as ice passes from a broad accumulation basin into a narrow tongue. Folds include the following styles: 'similar' (with attenuated limbs and thickened hinges; Fig. 6a), 'isoclinal' (parallel, attenuated limbs; Fig. 6b), 'parallel' (limbs and hinges of uniform thickness), chevron (zig-zag shape) and 'intrafolial' (hinges severed from limbs) (see Ramsay 1967 for descriptions). Folds are best observed at the glacier surface at the centimetre- to metre-scale. Fold axes plunge at angles that range from steep to shallow and, with their axial planes, are commonly parallel to ice flow. Flow-parallel minor folds, are commonly 'parasitic' on larger fold struc-

tures. Larger-scale folds are commonly difficult to define on the ground in the absence of distinctive marker beds, but may be evident from their vergence (defined by their asymmetry) or in aerial photographs.

Folds with axial planes transverse to flow form in glaciers that have zones of strong longitudinal compression. Minor folds are common below icefalls, particularly if there is a pre-existing longitudinal foliation. Their axes may also be parallel to near-vertical arcuate foliation, which also forms in the ice-fall. Fold styles are similar to those associated with longitudinal foliation.

Piedmont glaciers, in which longitudinal compression is particularly pronounced, have folds with amplitudes of several kilometres, comprising both longitudinal foliation and moraines, e.g. the Bering and Malaspina glaciers in Alaska (see cover photograph). However, in such glaciers the pattern of folding is complicated by surges (Sharp 1958; Post 1972). Ramberg (1964) successfully modelled non-surge type folding of moraines in piedmont glaciers using a centrifuge.

Surge-type behaviour gives rise to a distinctive style of fold structure, defined by moraines (e.g. Rutishauser 1971; Post 1972; Driscoll 1980; Lawson 1996). These are in the form of teardrop-shaped or bulbous structures with their fold axes parallel to the valley sides. They are the result of short-lived periods of enhanced discharge of ice from tributaries into the main

**(a)**

**(b)**

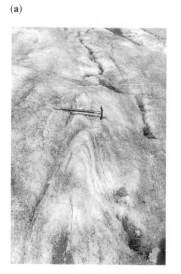

**Fig. 6.** Metre-scale folding in glacier ice. (**a**) 'Similar' fold in fine-grained and coarse bubbly ice foliation with vertical axial plane, Vadret del Forno, Switzerland. (**b**) Isoclinal fold with gently upglacier-dipping axial plane, involving layers of debris derived from a basal position, Thompson Glacier, Axel Heiberg Island, Canada.

glacier, or the steady discharge of ice from non-surge-type tributaries into a quiescent-phase surge-type glacier. Such structures are useful in identifying surge-type glaciers, even when there have been no documented surges (Fig. 7).

## Foliation

Foliation is the pervasive planar structure found in almost all glaciers, comprising discontinuous layers of coarse bubbly, coarse clear and fine

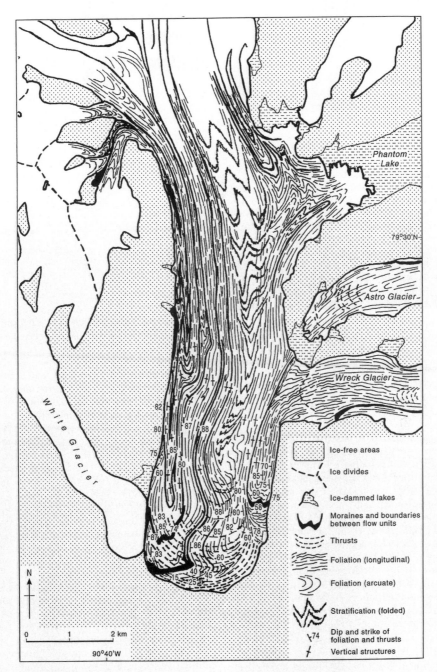

**Fig. 7.** Generalized structural map of lower Thompson Glacier, Axel Heiberg Island, showing possible surge-related folding of medial moraines, longitudinal foliation, thrusts and stratification.

**(a)**

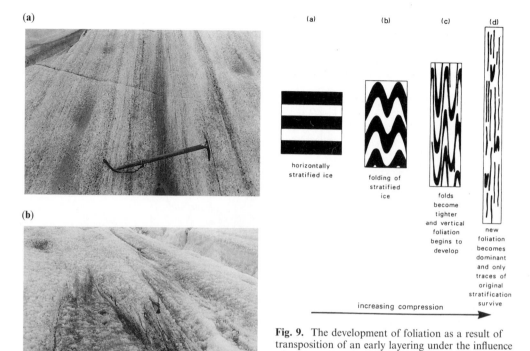

**(b)**

**Fig. 8.** Types of foliation. **(a)** Typical foliation with intercalated layers of coarse bubbly, coarse clear and fine grained ice, Austre Brøggerbreen, Svalbard. **(b)** Open folding of stratification with axial plane foliation, Austre Lovénbreen, Svalbard. The ice axe is inserted along the fold axis, which is parallel to flow; view upglacier.

**Fig. 9.** The development of foliation as a result of transposition of an early layering under the influence of prolonged compression (from Hambrey 1994, fig. 2.22).

ice (Allen *et al.* 1960) (Fig. 8a). It is commonly mistaken for stratification, but even cursory observation will demonstrate that it is the product of deformation. One view is that foliation is inherited from stratification or other structural inhomogeneities by 'transposition', both in simple and pure shear regimes (Fig. 9). This process, well-known from metamorphic rocks (Hobbs *et al.* 1976), is also common in glaciers (Hambrey 1977; Hooke & Hudleston 1978; Lawson *et al.* 1994). However, not all foliation is derived from pre-existing structures. Several instances of foliation being a completely new structure have been reported (e.g. Lawson 1990; Pfeffer 1992; Hambrey *et al.* 1999); indeed, in polythermal glaciers it forms a clear axial planar relationship with stratification (Fig. 8b), superficially resembling the folding/slaty cleavage relationship observable in low-grade metamorphic rocks, but being defined by variations in ice

type. The exact mechanism for producing the planar elements in such axial-plane foliations is unknown.

As noted above, two main types of foliation develop in glaciers: longitudinal and transverse (or arcuate). Longitudinal foliation develops as ice converges from a wide accumulation, or series of basins, into a narrow tongue, in the same simple-shear regime as folding with flow-parallel axes. Foliation is most pronounced where folding is tightest, such as at the glacier margin or at the confluence zone of two ice-flow units. The relationship between foliation and strain rates is often unclear at the place of measurement. It has been demonstrated that longitudinal foliation develops approximately parallel to the maximum shear strain-rate tensor, such as at the glacier margins or at the confluence of two flow units, but not if it has been subsequently transported passively into a new stress regime (e.g. Meier 1960; Anderton 1970; Hambrey & Müller 1978). The cumulative strain ellipse, orientated originally at 45° to the foliation, rotates gradually towards parallelism with the structure, as long as the foliation remains in this type of simple shearing regime. Thus in marginal areas, where cumulative strains may be up to several hundred percent, the long axis of the cumulative strain ellipse may essentially be parallel to the foliation

(Hambrey & Milnes 1977; Hambrey *et al.* 1980) (Fig. 2). In three-dimensions, one would expect the longitudinal foliation to attain parallelism with the *xz* plane of the strain ellipsoid.

In contrast, pervasive transverse foliation is developed from crevasse traces rather than stratification, although it can also develop from local transposition associated with folding around transverse thrust faults. Typically, as these extensional structures pass into a zone of longitudinal compression below an icefall, the layering is intensified. The transverse foliation thus develops normal to the maximum compressive strain rate. This represents a pure shear regime, and the structure develops and remains perpendicular to the short axis of the strain ellipse during its development. In this case, the magnitude of cumulative strain and intensity of transverse foliation is much less than for longitudinal foliation (Hambrey & Milnes 1977; Hambrey *et al.* 1980) (Fig. 2).

In surge-type glaciers, the formation of pervasive longitudinal foliation is associated with deformation during quiescence, although local transverse foliation may develop around thrust faults associated with intense surge-phase shortening (Lawson *et al.* 1994). Pfeffer (1992) argued that the particular stress conditions associated with surging could produce bubble-migration foliation.

## Boudins

Two types of boudinage structure have been identified in temperate valley glaciers: competence contrast and foliation boudinage, occurring on scales of a few centimetres to several metres (Hambrey & Milnes 1975). However, the ductility contrasts between different ice types vary. In a number of Swiss glaciers, competence contrast boudinage occurs where pods of relatively debris-free ice become isolated in debris-laden basal ice, or where fine-grained ice layers become attenuated within foliated coarse bubbly ice (Fig. 10a). These relationships would suggest that the clean ice and fine-grained ice are the more competent (less ductile) components respectively. Competence contrast boudinage has also been observed in the East Antarctic ice sheet, where frozen water-filled crevasses have been extended, and the water-ice has behaved as the more competent material (Marmo & Wilson this volume).

Foliation boudinage is commonly found in association with strong longitudinal foliation, such as at the glacier margins or where two flow units combine (Hambrey & Milnes 1975) (Fig. 10b). Both symmetric and asymmetric

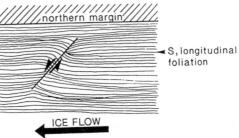

**Fig. 10.** Boudinage structures. (**a**) Competence-contrast boudinage; fine-grained ice boudins, surrounded by coarse bubbly ice, Vadrec da l'Albigna, Switzerland. (**b**) Asymmetric foliation boudinage in longitudinal foliation, Griesgletcher, Switzerland. (**c**) Schematic diagram of asymmetric foliation boudinage at the nothern margin of Variegated Glacier, Alaska; sense of displacement along crevasse trace and necking of longitudinal foliation are shown.

varieties have been identified. The latter are the most common, and are associated with fracture or shear zones which displace the ice sideways at the boudin neck and form an acute angle with the margin downstream (Fig. 10c). Two

possible explanations have been proposed for
these structures.

(1) The orientation of these boudins suggests that they are a response to longitudinal
extension and shearing (such as on the outside
of a bend). The fractures may be cracks that
formed upglacier and offered that most convenient points for failure during stretching
(Lawson *et al.* 1994).

(2) Boudin-like features may be initiated as
extensional fractures, followed by rotation and
closure in a non-coaxial strain regime, leading to
bending of foliation to form hook folds adjacent
to the boudin neck or vein (Hudleston 1989).

Boudins of these types are typically a few
metres wide and long (Hambrey & Milnes 1975;
Lawson *et al.* 1994), but 'mega-boudins', two-
orders of magnitude larger have been documented from satellite images of the Lambert Glacier,
one of Antarctica's largest glacier systems
(Hambrey & Dowdeswell 1994).

### Shear zones

A variety of shear zones, related to thrusting
and foliation, may be found in some glaciers.
Shearing has the effect of either transforming
large ice crystals, through a sub-grain stage, to
fine-grained ice, or generating larger crystals.
This depends on the balance between destructive
and constructive processes operating during the
deformation process (Hudleston pers. comm.
1999). Similar processes are represented by
mylonite zones in rocks. Zones of sheared ice
are commonly parallel to thrusts, and are best
seen when the thrusting process has been relatively recent. Particularly well-developed shear-
zones have been described from the margin of the
Greenland ice sheet, near Thule, where they are
responsible for a particular type of moraine-
formation (Swinzow 1962; Souchez 1967). Examples of recently active shear zones were observed
soon after the termination of the surge of
Bakaninbreen, Svalbard (Hambrey *et al.* 1996).
These shear zones extended across the glacier for
several tens of metres and were up to half a metre
wide. Intrafolial folds and lenses of coarse clear
ice and coarse bubbly ice were present in the
shear zone.

Shearing is evident in many cases where strike-
slip faulting and crevasse formation occur.
Shearing is indicated by the bending of layers,
particularly longitudinal foliation, towards the
fracture (Fig. 11). In such cases, ductile shearing
proceeds until the brittle strength of the ice is
exceeded, and the ice fractures.

Surge-type Variegated Glacier illustrated
spaced ductile shear zones in association with

**Fig. 11.** Shear zone associated with crevasse trace and
displacing longitudinal foliation, Glacier d'Otemma,
Switzerland.

foliation boudinage following the 1982–83 surge
(Lawson 1990). These shear zones are discontinuous and anastomosing, and affected about 10%
of the ice at the northern margin. Like the foliation boudinage, they develop to accommodate
rotational shearing and longitudinal extension.
As a result of rotation during shearing, the
foliation attains a new orientation in affected ice.
The shear zones strike parallel to the fractures and necks of foliation boudinage. Similar
zones in anisotropic rocks have been termed
'cleavage zones' (Platt & Vissers 1980).

### Ogives

Ogives are arcuate structures below ice-falls and
are only associated with valley glaciers. Two
main types of ogive have been distinguished
(Post & LaChapelle 1971): wave ogives and
Forbes bands. The former represent the passage
of thicker ice through an ice-fall in winter than
in summer, giving rise to a wave and a trough
that represents one year's movement through the
ice-fall. Forbes bands (named after the nineteenth century glaciologist) are defined by pairs
of dirty and clean ice, each also representing a
year's flow through the ice-fall. Several hypotheses have been suggested, but that of
Nye (1958) is most widely accepted. According
to him, the dirty ice is the product of debris
entrainment and concentration during the summer passage of ice through the ice-fall, the clean
ice representing passage in winter. Waddington
(1986) examined the formation of wave ogives
mathematically, and showed that although any
section of a glacier with a velocity gradient can
produce waves, they are formed in different
places and interfere with one another. Only
where the velocity gradient is large and localised, as in an icefall, do observable waves form.

**Fig. 12.** Ogives (Forbes bands) in the tongue of Bas Glacier d'Arolla. The foot of the icefall is at the upper left of the photograph.

Furthermore, the ice must pass through the critical zone in no more than six months, and no waves form if the ice takes a whole year or more to move through the icefall.

Commonly, wave ogives pass into Forbes bands downglacier (Fig. 12). The precise structural nature of ogives has not been fully defined. However, from preliminary observations on Bas Glacier d'Arolla, Switzerland, they appear to comprise transverse foliation, and it is possible that they also could originate from crevasse-traces under a longitudinal extension regime high in the ice-fall, followed by longitudinal compression and transposition of earlier longitudinal layering.

## Characteristic structural assemblages

Different parts of a glacier develop their own distinctive suites of structures, depending on whether the ice is undergoing longitudinal extension or compression, or whether it is subject to shearing, as at the margins or where two flow units combine. Because ice passes through several such deformation regimes, and that these also change through time as the dynamic state of the glacier changes, a glacier acquires considerable structural complexity. Below, we describe some examples of structural assemblages for different deformation regimes in valley glaciers.

### Extensional flow regimes

Extending flow occurs above and near the equilibrium line, or where valley morphology promotes ice-acceleration. Structures resulting from extension include crevasses and crevasse traces. Crevasses typically have a transverse orienta-

tion, but linking longitudinal crevasses create *sérac* fields. Mapped examples of crevasse fields include Blue Glacier, Washington State (Allen *et al.* 1960); Griesgletcher, Switzerland (Hambrey & Milnes 1977); Variegated Glacier, Alaska during a succession of different surges (Lawson 1996); and Kronebreen, a tidewater glacier in Svalbard (Glasser *et al.* 1998). Many glaciers in Svalbard, currently undergoing strong recession, show an abundance of crevasse traces, but no crevasses, reflecting how they have become progressively less dynamic in the last 100 years.

### Compressive flow regimes associated with the base of an ice-fall

The detailed structure beneath an ice-fall has been documented by several authors. Allen *et al.* (1960) described the transverse foliation below the ice-fall on Blue Glacier Washington, which begins as a vertical structure, but is deformed into a progressively less steep, upglacier-dipping structures that they described as 'nested spoons'. In a study of Gulkana Glacier, Alaska, Ragan (1969) noted that in the western part of the ice-fall stratification is destroyed, and that foliation begins to form well up the ice-fall. In contrast, in the eastern half, an irregular line of cone-shaped ice lobes, defined by arcuate foliation, develop. These 'cones' merge into larger and fewer arcs downglacier, until only a single arc system remains.

In a glacier-wide analysis of Griesgletcher, Hambrey & Milnes (1977) demonstrated how open crevasses and tensional veins in the upper part of an ice-fall of a wide variety of orientations become compressed in a pure shear regime at the base, and are transformed to transverse, then arcuate foliation. These tensional structures are superimposed on an earlier compositional layering, a longitudinal foliation (itself probably originating from stratification inherited from the accumulation area above the ice-fall). In the compressive zone, this early layering is folded, with axial planes developing parallel to the transverse foliation. Complete transposition is achieved in places, leaving only isolated fold hinges. The resulting transverse structure is parallel to the long axis of the strain ellipse (Fig. 2).

### Compressive flow regimes associated with terminal lobes

Piedmont glacier lobes occur where ice spreads out in unconstrained fashion after leaving the confines of a glacial trough. Fold structures, with their axial planes parallel to the ice margin,

(a)

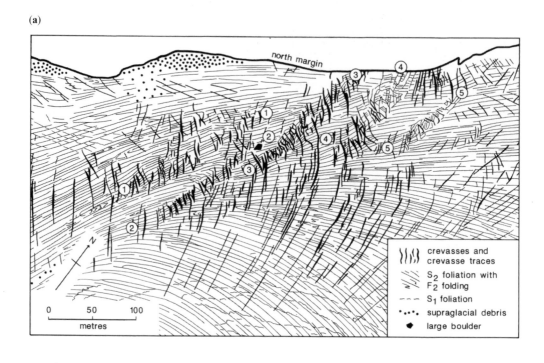

(b)                                              (c)

**Fig. 13.** Marginal shear regime at N margin of Griesgletscher, Switzerland. (**a**) Structural map drawn from aerial photograph, illustrating two foliations, crevasses and crevasse traces. Five sets of *en échelon* crevasses and crevasse traces (numbered 1–5) are present, each representing a shear zone in the ice. Note sigmoidal nature of some crevasses and rotation of foliation in the intercrevasse blocks. (**b**) Oblique view of shear zone from northern flank above glacier. (**c**) Close-up view of crevasses in shear zone and rotated longitudinal foliation.

develop on a scale of hundreds of metres to kilometres. The structures that have been folded are characteristically longitudinal foliation and medial moraines. Few such structural complexes have been mapped, but the spectacular folding in the piedmont lobes of the Malaspina and Bering glaciers in Alaska is well known, although these structures are complicated by folding of looped moraines induced by former surges (see cover photograph). Although cumulative deformation has not been modelled in these cases, it is likely that the folds have an axial planar relationship with foliation, and that the axial plane is parallel to the cumulative strain ellipsoid. Lawson *et al.* (1994, fig. 5) described a more complex variant of this type of structural assemblage from the terminal lobe of Variegated Glacier, where longitudinal compression is induced not only by emergence beyond the confines of a narrow valley, but also by abutment against the much larger Hubbard Glacier. Here, distinctive lithologies in the medial moraines were mapped and shown to define third-generation folds resulting from the passive folding of surge-generated bulb-like loops in the moraines.

## Shear regime in a non-surge type temperate glacier

Zones of simple shearing in the marginal areas and between ice-flow units of valley glaciers are commonly defined by *en échelon* crevasses, which are analogous to tension gashes in deformed rocks. In Griesgletscher, an aerial photograph illustrates some of the characteristic features of shear zones at its northern margin (derived map in Fig. 13a; general view Fig. 13b; close-up of shear zone with rotated foliation in Fig. 13c). The shear zones result from faster moving ice from above the icefall converging with, and pinching out, slow-moving ice at the margin. The structure in this part of the glacier is complex, comprising $S_1$, a longitudinal foliation folded in axial planar relationship with $S_2$, the dominant set of arcuate foliation formed from crevasse traces in the ice-fall (as referred to above). At this northern margin, $S_2$ has a trend nearly parallel to the margin. Cutting across this foliation at an acute angle are three prominent (1, 2, 3 in Fig. 13a), and other less well-defined sets of *en échelon* crevasses (in addition to other crevasses of various orientations). Some show sigmoidal characteristics indicative of subsequent rotation in a dextral shearing regime. Foliation in the inter-crevasse blocks is rotated and truncated, again indicating the same sense of shearing in most cases. These relationships are preserved

even after the *en échelon* crevasses close up, or pass into sets of *en échelon* crevasse traces (sets 4, 5; Fig. 13a).

## Shear regime in a surge-type temperate glacier

A variety of characteristic structural assemblages developed during the 1982–83 surge of Variegated Glacier (Sharp *et al.* 1988; Lawson *et al.* 1994) (Fig. 14). Marginal shear zones up to 250 m wide developed on each side of the glacier as the downglacier-propagating surge front approached. Within these shear zones, a series of margin-parallel bands of crevasses, with distinctive shear-related orientations developed. Closest to the margin, tension cracks formed. In a 50 m wide band 100–150 m from the margin, synthetic and antithetic Riedel shear fractures developed. Further away from the margin, a band of 'P' shear fractures formed in the centre of the glacier. As the surge front passed, a 40 m wide band of brecciated ice developed about 50 m from the margin, where a velocity discontinuity developed. This pattern of structural development is closely analogous to that found in wrench-fault zones (Tchalenko 1970; Wilcox *et al.* 1973).

At the end of the 1982–83 surge, there was an overall longitudinal structural zonation of Variegated Glacier, comprising three distinct tectonic zones each associated with a characteristic surge-phase deformation history (Fig. 15). In the upper part of the glacier, upstream of the location of initiation of the surge (the surge nucleus), transverse fractures were the dominant structure. Between the location of the surge nucleus, and the final position of the upper edge of surge front, both transverse and longitudinal fractures had developed (Fig. 15a). Further downglacier, longitudinal fractures, surface folds and thrust faults occurred together (Fig. 15b). The most distal tectonic zone contained an assemblage of structures characteristics of the large longitudinally compressive

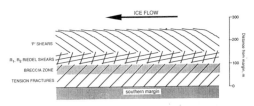

**Fig. 14.** Schematic diagram of the marginal shear zone which developed during the 1982–83 surge of Variegated Glacier, Alaska. The breccia zone was last to form. (From Lawson 1990.)

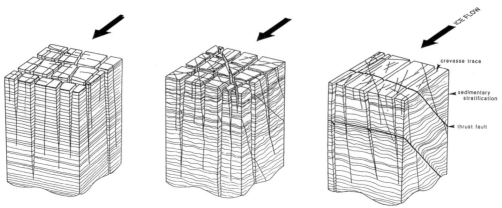

**Fig. 15.** Post surge structural assemblages at three points on the centreline of Variegated Glacier, four years after the 1982–83 surge; (**a**) at about 11 km from the head of the glacier; (**b**) at about 14 km from the head of the glacier; (**c**) in the terminal lobe. Foliations are omitted for clarity. (From Lawson 1990.)

strain-rates in the terminal lobe. These structures included overturned folds verging down-glacier with axial planar thrust faults (Fig. 15c), and a transposed foliation developed locally around the folds (Pfeffer 1992).

## Characteristic structural and deformation histories

Steady-state deformation fields and superimposed dynamic strain events during the down-glacier motion of ice result in a complex, polyphase deformation history (Table 3). Deformation histories of Variegated Glacier occurred in one to six phases of rapid strain rate during surging, depending on where the ice originated (Lawson *et al.* this volume). Structural assemblages at various points on the glacier (Fig. 15) may be simpler than that suggested by the deformation histories. Some structures are repeatedly reactivated during their passage downglacier, while others may be rotated and by a second phase of deformation. For example, transverse crevasses formed by flow-parallel extension high in the accumulation area, may be progressively rotated by vertical shearing as they move downglacier. In the lower part of the glacier they may be reactivated as thrusts, accompanied by overturned folding in flow-parallel compression at the terminus.

**Table 3.** *Examples of planar structures associated with successive deformation phases in valley glaciers*

| Glacier | $S_0$ | $S_1$ | $S_2$ | $S_3$ | $S_4$ | Source |
|---|---|---|---|---|---|---|
| *Non-surge-type glaciers* | | | | | | |
| Griesgletscher, Switzerland | Stratification | Longitudinal foliation | Crevasse traces → transverse foliation | Weak longitudinal foliation | – | Hambrey & Milnes (1977) |
| White Glacier, Canada | Stratification | Longitudinal foliation | Crevasse traces | Thrusts | – | Hambrey & Müller (1978) |
| *Surge-type glaciers* | | | | | | |
| Hessbreen, Svalbard | Stratification | Early crevasse traces | Longitudinal foliation | Thrusts | Late crevasse traces | Hambrey & Dowdeswell (1996) |
| Kongsvegen, Svalbard | Stratification | Longitudinal foliation | Thrusts | Crevasse traces | Basal ice fractures | Glasser *et al.* (1998) |
| Variegated Glacier, Alaska | Stratification | Longitudinal foliation | Transverse foliation; thrusts | – | – | Lawson *et al.* (1994) |

## Relationships between ice structure, debris transport and deposition

Although the effect of subglacial and proglacial deformation has been widely documented (e.g. Hart & Boulton 1991; Boulton 1996), the role played by ice structures in determining land-form morphology and the distribution of glacigenic sedimentary facies is little known. The importance of both high- and low-level debris-entrainment processes in explaining the character of glacial sediments has been recog-nised for three decades (e.g. Boulton 1967, 1970, 1978; Eyles & Rogerson 1978a, b; Small et al. 1979; Sollid et al. 1994), and reviewed recently by Benn & Evans (1998), but few

studies have been made concerning the rela-tionship between debris distribution and ice structure. In a series of papers on structures in polythermal glaciers and landform genesis in Svalbard (e.g. Bennett et al. 1996a, b, 1999; Boulton et al. 1999; Etzellmüller et al. 1996; Glasser et al. 1998, 1999; Hambrey & Huddart 1995; Hambrey et al. 1996, 1997, 1999; Huddart & Hambrey 1996; Sollid et al. 1994), both surge-type and non-surge-type, ending both on land and in the sea, have been examined. Apart from incorporation of debris into basal ice (see Hubbard & Sharp 1989; Knight 1998; Lawson et al. 1998), three main modes of debris entrainment have been distinguished (Hambrey et al. 1999).

**Fig. 16.** Relationships between debris and structure in valley glaciers. (**a**) Hinge zones of folded stratification, here defined by layers of angular debris, originating as rockfall material, Austre Lovénbreen, Svalbard. Nore how these features develop into a medial moraine and survive as debris trains on the proglacial area. (**b**) Subglacially derived cobbles and boulders (predominantly subangular and subrounded), thrust to surface near the snout of White Glacier, resulting from reactivation of rotated transverse crevasse trace. (**c**) Recumbent fold involving the raising of basal debris to high-level position, Thompson Glacier, Axel Heiberg Island; cliff face is about 20 m high; flow is from left to right.

(1) *Incorporation of angular rockfall material within the stratified sequence of snow and firn.* This debris takes an englacial transport path, becoming folded as the ice flows from a wide accumulation basin into a narrow channel. Folding is most intense near the glacier margins, or at flow-unit boundaries, and is associated with an axial planar foliation. The debris re-emerges near the snout as a suite of medial moraines, representing tight to open fold structures with axes parallel to flow. The resulting lines of debris are transmitted to the proglacial area in the form of regular trains of angular material, where they have sometimes been mistaken as flutes (Fig. 16a).

(2) *Incorporation of debris of both supraglacial and basal affinity within longitudinal foliation.* This is evident especially at the glacier margins and at flow-unit boundaries near the snout where deeper ice has become exposed by ablation. This foliation is the product of strong folding that also involves deformable bed material, and represents more intense deformation than (1) above. The sedimentary products of this process are 'foliation-parallel ridges' (Glasser *et al.* 1998), also sometimes mistaken for flutes.

(3) *Incorporation by thrusting.* Here, debris-rich basal ice and even rafts of subglacial sediment are uplifted into an englacial position, sometimes emerging at the ice surface, especially near the snout (Fig. 16b). This material is variable in character, typically comprising diamicton (reworked from basal till); sand and gravel (reworked from glaciofluvial sediment); or laminated muds and sands (of glaciolacustrine or glaciomarine origin). This variety is indicative of the different facies over which the ice flowed. Thrusting is facilitated in polythermal glaciers by the transition from ice that is sliding, to ice that is frozen to the bed. For these thermal and dynamic reasons, polythermal glaciers are more prone than truly cold or temperate glaciers to produce a distinctive suite of moraine-mound complexes (often described in past literature, and assigned to ice-stagnation processes, as 'hummocky moraines') (Hambrey *et al.* 1997).

A fourth mechanism of debris entrainment, not observed in Svalbard glaciers, but common in the Canadian Arctic, is by recumbent folding involving basal ice. Although not specifically linked to debris entrainment, Hudleston (1976) explained this style of folding as being initiated when ice departs from steady-state flow as, for example, when the underlying bed is uneven (Fig. 16c).

At least some of these processes are now recognised in British glacial landform assemblages dating from the Younger Dryas climatic stage (Hambrey *et al.* 1997; Bennett *et al.* 1998; Graham & Midgley this volume), and it is anticipated that they will be identified in other areas where polythermal glaciers existed.

## Implications and future work

Structural analyses of ice masses can be performed at a wide variety of scales, and all can be used to make inferences, not only of present, but also past dynamic glacier regimes. As glaciers respond to climatic amelioration, it is important to determine how glaciers have altered their behaviour on the decadal to centennial time-scale. As an example, many glaciers in Svalbard are undergoing rapid recession. They have become much less dynamic, yet ice structures record a time when the glaciers were thermally different and were subject to intense crevassing. Interpretation of these changes from the sedimentary and landform record is one of the challenges for the future.

Although detailed surface mapping yields a good understanding of the dynamic history of a glacier, it is often difficult to define large-scale structures as individual layers are hard to trace for long distances at the glacier surface. New technology, such as ground-penetrating radar (GPR) offers the prospect of defining large-scale structures by sub-surface mapping of distinct marker beds, especially debris layers. GPR, in combination with surface mapping of layers with distinctive lithologies, can be used to test hypotheses concerning large-scale folding and thrusting in valley glaciers. Such studies are vital for understanding the entrainment of sediment, its subsequent release, and the genesis of resulting landforms.

On the larger scale, structures in ice sheets record a dynamic history that is as long as the residence time of ice in the system. Typically, this is tens of thousands of years. Structures may be inferred from satellite imagery, and Landsat images in the Antarctic and Russian Arctic have already provided useful information for inferring the dynamic history of ice-drainage basins. The potential for structural information to yield results that are crucial for determining dynamic history is illustrated by the inference of 'switching off' of Ice Stream C in West Antarctica from the existence of buried crevasses (Hulbe & Whillans 1997). Here, the consequences of instability of large ice-drainage basins are likely to be catastrophic. Structural glaciology offers the prospect of assessing the probability of such events.

Recent developments in three-dimensional numerical modeling present exciting possibilities for structural glaciological analysis. In particular, the development of models that can predict principal stresses in three-dimensions at spatial resolutions that are useful for structural interpretation is already yielding predictions of stress tensors that match well with observed crevasse patterns (Hubbard *et al.* 1998). The potential for these models to evaluate deformation histories along realistic three-dimensional flow-lines will also enhance structural interpretation.

The application of satellite interferometry to the evaluation of large-scale velocity fields in glaciers and ice sheets (e.g. Unwin & Wingham 1997) also provides exciting possibilities for the calculation of deformation fields in regions where field access is difficult, or where the areas are too large for undertaking effective field measurements.

## Conclusions

(1) Glaciers can be considered as models of rock deformation, although deforming at rates that can be actually measured. The snow from which they are derived undergoes diagenesis through firn to ice, and is then subjected to deformation and recrystallisation under gravity-induced stress. The long-term effects of deformation under changing stress regimes are manifested in terms of various structures and structural assemblages, including foliation, folding, crevasses, crevasse traces, shear zones, boudins and ogives; many of these structures have analogues in rocks.

(2) Glacier ice commonly displays structures that can be linked to several 'phases' of deformation. These phases relate to movement of a 'parcel' of ice through the glacier, rather than to a strictly temporal sequence. At any one time the first phase is taking place high up the glacier, when near the snout the ice may already be experiencing its third or fourth phase, i.e. structural evolution is time transgressive.

(3) Many structures, notably folds, foliation and boudinage, do not show a clear relationship with measured strain rates. Rather they are the product of cumulative strain, and tend to show a clearer, but variable relationship with the strain ellipse. For example, foliation may develop under either a pure shear or simple shear deformation regime, but end up in both cases parallel to the long axis of the cumulative strain ellipse.

(4) Ice can be subject to strain which need not lead to the development of new structures, although the geometry of existing ones will be modified. In other cases structures can be reactivated; for example thrusts may develop out of rotated crevasse traces.

(5) There is a demonstrable relationship between ice structures and debris entrainment. Supraglacially-derived debris may be tightly folded and may define large-scale structures; foliation may be the product of tight folding that involves basal material from the deformable substrate; and thrusting may incorporate large volumes of sediment from the many and varied basal sources. Recent work has demonstrated how ice structures can explain landform development. Thus landforms associated with former ice masses have the potential to assess their dynamic and thermal characteristics.

(6) Developments in modelling strain-rate fields and cumulative deformation in glaciers offer the prospect of advances in structural glaciology. New ideas concerning the relationship of debris and structure need to be applied to, and tested in, different glaciological settings. Finally, remote sensing techniques are providing new means of assessing the stability and dynamics of the large polar ice sheets, through application of structural glaciological principles.

The authors thank numerous glaciologists and structural geologists for sharing their knowledge and insight concerning ice and rock deformation processes, especially P. J. Hudleston, A. G. Milnes, M. J. Sharp and W. H. Theakstone. For suggesting valuable improvements to the manuscript we thank N. F. Glasser and referees P. J. Hudleston and D. E. Lawson. The first author acknowledges receipt of NERC Grants GR9/02185 and GST/03/2192 for glacial geological and structural glaciological work in Svalbard respectively.

## References

ALLEY, R. B., CUFFEY, K. M., EVENSON, E. B., STRASSER, J. C., LAWSON, D. E. & LARSON, G. J. 1998. How glaciers entrain and transport basal sediment: physical constraints. *Quaternary Science Reviews*, **16**, 1017–1038.

ALLEN, C. R., KAMB, W. B., MEIER, M. F. & SHARP, R. P. 1960. Structure of the lower Blue Glacier, Washington. *Journal of Geology*, **68**, 601–625.

AMBACH, W. 1968 The formation of crevasses in relation to the measured distribution of strain-rates and stresses. *Archiv für Meteorologie, Geophysik und Bioklimatologie*, **A17**, 78–87.

ANIYA, M. & NARUSE, R. 1987. Structural and morphological characteristics of Soler Glacier, Patagonia. *Bulletin of Glacier Research*, **4**, 69–77.

ANDERTON, P. W. 1970. Deformation of surface ice at a glacier confluence, Kaskawulsh Glacier. *In*: BUSHNELL, V. C. & RAGLE, R. H. (eds) *Icefield Ranges Research Project Scientific Results*. American Geographical Society, New York & Arctic Institute of North America, Montreal, **2**, 59–76.

ATHERTON, D. 1963. Comparisons of ogive systems under various regimes. *Journal of Glaciology*, **4**, 547–557.

BENN, D. I. & EVANS, D. J. A. 1998. *Glaciers and Glaciation*. Arnold, London.

BENNETT, M. R., HAMBREY, M. J., HUDDART, D. & GHIENNE, J. F. 1996a. The formation of geometrical ridge networks ('crevases-fill' ridges), Kongsvegen, Svalbard. *Journal of Quaternary Science*, **11**, 430–438.

——, ——, —— & GLASSER, N. F. 1998. Glacial thrusting and moraine-mound formation in Svalbard and Britain: the example of Coire a' Cheudchnoic (Valley of a Hundred Hills), Torridon, Scotland. *Quaternary Proceedings*, **6**, 17–34.

——, ——, ——, —— & CRAWFORD, K. 1999. The landform and sediment assemblage produced by a tidewater glacier surge in Kongsfjorden, Svalbard. *Quaternary Science Reviews*, **18**, 1213–1246.

——, HUDDART, D., HAMBREY, M. J. & GHIENNE, J. F. 1996b. Moraine development at the high-arctic valley glacier Pedersenbreen, Svalbard. *Geografiska Annaler*, **78A**, 209–222.

BISHOP, B. C. 1957. Shear moraines in the Thule area, northwest Greenland. *US Snow, Ice & Permafrost Research Establishment, Research Report* **17**.

BOULTON, G. S. 1967. The development of a complex supraglacial moraine at the margin of Sørbreen, Ny Friesland, Vestspitzbergen. *Journal of Glaciology*, **6**, 717–736.

——1970. On the origin and transport of englacial debris in Svalbard glaciers. *Journal of Glaciology*, **9**, 213–229.

——1978. Boulder shapes and grain size distributions of debris as indicators of transport paths through a glacier and till genesis. *Sedimentology*, **25**, 773–799.

——1996. Theory of glacial erosion, transport and deposition as a consequence of subglacial sediment deformation. *Journal of Glaciology*, **42**, 43–62.

——, VAN DER MEER, J. J. M., BEETS, D. J., HART, J. K. & RUEGG, G. H. J. 1999. The sedimentary and structural evolution of a recent moraine complex, Holmstrømbreen, Spitsbergen. *Quaternary Science Reviews*, **18**, 339–371.

CLARKE, G. K. C. & BLAKE, E. W. 1991. Geometric and thermal evolution of a surge-type glacier in its quiescent state, Trapridge Glacier, Yukon Territory, Canada 1969–89. *Journal of Glaciology*, **37**, 158–169.

DOWDESWELL, J. A. & WILLIAMS, M. 1997. Surge-type glaciers in the Russian High Arctic identified from digital satellite imagery. *Journal of Glaciology*, **43**, 489–494.

DRISCOLL, F. G. 1980. Formation of the Neoglacial surge moraines of the Klutlan Glacier, Yukon Territory, Canada. *Quaternary Research*, **14** 19–30.

ETZELMÜLLER, B., HAGEN, J. O., VATNE, G., ØDEGÅRD, R. S. & SOLLID, J. L. 1996. Glacial debris accumulation and sediment deformation influenced by permafrost: examples from Svalbard. *Annals of Glaciology*, **22**, 53–62.

EVANS, D. J. A. 1989. Apron entrainment at the margins of subpolar glaciers, northwest Ellesmere Island, Canadian high arctic. *Journal of Glaciology*, **35**, 317–324.

EVANS, K., GOODMAN, D. J. & HOLDSWORTH, G. 1978. Recording wire strain meters on the Barnes Ice Cap, Baffin Isaldn, Canada. *Journal of Glaciology* **20**, 409–423.

EYLES, N. & ROGERSON, R. J. 1978a. A framework for the investigation of medial moraine formation: Austerdalsbreen, Norway and Berendon Glacier, British Columbia, Canada. *Journal of Glaciology*, **20**, 99–113.

—— & ——1978b. Sedimentology of medial moraines on Berendon Glacier, British Columbia Glacier, Canada and implications for debris transport in a glacierized basin. *Geological Society of America Bulletin*, **89**, 1688–1693.

FISHER, J. E. 1962. Ogives of the Forbes type on Alpine glaciers and a study of their origins. *Journal of Glaciology*, **4**, 53–61.

FORBES, J. D. 1900. Travels through the Alps. London, Adam & Charles Black. [New edition revised and annotated by, W. A. B. COOLIDGE].

GLASSER, N. F., BENNETT, M. R. & HUDDART, D. 1999. Distribution of glaciofluvial sediment within and on the surface of a high arctic valley glacier. Marthabreen, Svalbard. *Earth Surface Processes and Landforms*, **24**, 303–318.

——, HAMBREY, M. J., CRAWFORD, K. R., BENNETT, M. R. & HUDDART, D. 1998. The structural glaciology of Kongsvegen, Svalbard, and its role in landform genesis. *Journal of Glaciology*, **44**, 136–148.

GLEN, J. W. 1955. The creep of polycrystalline ice. *Proceedings of The Royal Society of London*, Ser. A, **228**, 519–538.

GOLDTHWAIT, R. P. 1951. Development of end moraines in east-central Baffin Island. *Journal of Geology*, **59**, 567–577.

GRAHAM, D. J. & MIDGLEY, N. G. 2000. Moraine-mound formation by englacial thrusting: the Younger Dryas moraines of Cwm Idwal, North Wales. *This volume*.

GRAY, M. B. & MITRA, G. 1993. Migration of deformation fronts during progressive deformation: evidence from detailed structural studies in the Pennsylvania Anthracite region, USA *Journal of Structural Geology*, **15**, 435–449.

GROVE, J. M. 1960. The bands and layers of Vesl-Skautbreen. *In*: LEWIS, W. V. (ed.) *Norwegian Cirque Glaciers*. Royal Geographical Society Research Series, **4**, 11–23.

GUDMUNDSSON, G. H., IKEN, A. & FUNK, M. 1997. Measurements of ice deformation at the confluence area of Unteraargletscher, Bernese Alps, Switzerland. *Journal of Glaciology*, **43**, 548–556.

GUNN, B. M. 1964. Flow rates and secondary structures of Fox and Franz Josef Glaciers, New Zealand. *Journal of Glaciology*, **5**, 173–190.

HAMBREY, M. J. 1975. The origin of foliation in glaciers: evidence from some Norwegian examples. *Journal of Glaciology*, **14**, 181–185.

——1976. Structure of the glacier Charles Rabots Bre, Norway. *Geological Society of America Bulletin*, **87**, 1629–1637.

——1977. Foliation, minor folds and strain in glacier ice. *Tectonophysics*, **39**, 397–416.

——1991. Structure and dynamics of the Lambert Glacier-Amery Ice Shelf system: implications for the origin of Prydz Bay sediments. *In*: BARRON, J., LARSEN, B. & SHIPBOARD SCIENTIFIC PARTY. *Proceedings of the Ocean Drilling Program, Scientific Results*, **119**. College Station, Texas, 61–75.

——1994. *Glacial Environments*. London, UCL Press.

—— & DOWDESWELL, J. A. 1994. Flow regime of the Lambert Glacier-Amery Ice Shelf system, Antarctica: structural evidence from Landsat imagery. *Annals of Glaciology*, **20**, 401–406.

—— & —— 1996. Structural evolution of a surge-type polythermal glacier: Hessbreen, Svalbard. *Annals of Glaciology*, **24**, 375–381.

—— & HUDDART, D. 1995. Englacial and proglacial glaciotectonic processes at the snout of a thermally complex glacier in Svalbard. *Journal of Quaternary Science*, **10**, 313–326.

—— & MILNES, A. G. 1975. Boudinage in glacier ice – some examples. *Journal of Glaciology*, **14**, 383–393.

—— & ——1977. Structural geology of an Alpine glacier (Griesgletscher, Valais, Switzerland). *Eclogae Geologicae Helvetiae*, **70**, 667–684.

—— & MÜLLER, F. 1978. Structures and ice deformation in the White Glacier, Axel Heiberg Island, Northwest Territories, Canada. *Journal of Glaciology*, **20**, 41–66.

——, BENNETT, M. R., DOWDESWELL, J. A., GLASSER, N. F. & HUDDART, D. 1999. Debris transfer in polythermal glaciers. *Journal of Glaciology*, **45**, 69–86.

——, DOWDESWELL, J. A., MURRAY, T. & PORTER, P. R. 1996. Thrusting and debris-entrainment in a surging glacier, Bakaninbreen, Svalbard. *Annals of Glaciology* **22**, 241–248.

——, HUDDART, D., BENNETT, M. R. & GLASSER, N. F. 1997. Genesis of 'hummocky moraines' by thrusting in glacier ice: evidence from Svalbard and Britain. *Journal of the Geological Society, London*, **154**, 623–632.

——, MILNES, A. G. & SIEGENTHALER, H. 1980. Dynamics and structure of Griegletscher, Switzerland. *Journal of Glaciology*, **25**, 215–228.

HARPER, J. T., HUMPHREYS, N. F. & PFEFFER, W. T. 1998. Crevasse patterns and the strain-rate tensor: a high-resolution comparison. *Journal of Glaciology*, **44**, 68–76.

HART, J. K. & BOULTON, G. S. 1991. The interrelation of glaciotectonic and glaciodepositional processes within the glacial environment. *Quaternary Science Reviews*, **10**, 335–350.

HILLS, E. S. 1963. *Elements of structural geology*. Science Paperbacks & Methuen & Co. Ltd, London.

HOBBS, B. E., MEANS, W. D. & WILLIAMS, P. F. 1976. *An Outline of Structural Geology*. John Wiley & Sons, Chichester.

HOLDSWORTH, G. 1969. Primary transverse crevasses. *Journal of Glaciology*, **8**, 107–129.

HOOKE, R. LeB. 1973. Structure and flow in the margin of the Barnes Ice Cap Baffin ISLAND, N. W. T., Canada. *Journal of Glaciology*, **12**, 423–438.

—— & HUDLESTON, P. 1978. Origin of foliation in glaciers. *Journal of Glaciology*, **20**, 285–299.

HUBBARD, A., BLATTER, H., NIENOW, P., MAIR, D. & HUBBARD, B. 1998. Comparison of a three-dimensional model for glacier flow with field data from Haut Glacier d'Arolla, Switzerland. *Journal of Glaciology*, **44**, 368–378.

HUBBARD, B. & SHARP, M. J. 1989. Basal ice formation and deformation: a review. *Progress in Physical Geography*, **13**, 529–558.

HUDDART, D. & HAMBREY, M. J. 1996. Sedimentary and tectonic development of a high-arctic thrust-moraine complex: Comfortlessbreen, Svalbard. *Boreas*, **6**, 227–243.

HUDLESTON, P. J. 1976. Recumbent folding in the base of the Barnes Ice Cap, Baffin Island, Northwest Territories, Canada. *Geological Society of America Bulletin*, **87**, 1684–1692.

——1977. Progressive deformation and development of fabric across zones of shear in glacier ice. *In*: SAXENA, S. & BHATTACHARJI, S. (eds) *Energetics of Geological Processes*. Springer-Verlag, New York, 121–150.

——1983. Strain patterns in an ice cap and implications for strain variations in shear zones. *Journal of Structural Geology*, **5**, 455–463.

——1989. The association of folds and veins in shear zones. *Journal of Structural Geology*, **11**, 949–957.

—— & HOOKE, R. LeB. 1980. Cumulative deformation in the Barnes Ice Cap, and implications for the development of foliation. *Tectonophysics*, **66**, 127–146.

HULBE, C. I. & WHILLANS, I. M. 1997. Weak bands within Ice Stream B West Antarctica. *Journal of Glaciology*, **43**, 377–386.

JONSSON, S. 1970. Structural studies of subpolar glacier ice. *Geografiska Annaler*, **52A**, 129–145.

KAMB, W. B. 1959. Ice petrofabric observations from Blue Glacier in relation to theory and experiment. *Journal of Geophysical Research*, **64**, 1891–1909.

KING, C. A. M. & IVES, J. D. 1954. Glaciological observations on some of the outlet glaciers of South-West Vatnajökull, Iceland 1954, Part II: Ogives. *Journal of Glaciology*, **2**, 646–652.

—— & LEWIS, W. V. 1961. A tentative theory of ogive formation. *Journal of Glaciology* **3**, 912–939.

KNIGHT, P. G. 1998. The basal ice layer of glaciers and ice sheets. *Quaternary Science Reviews* **16**, 975–993.

LAWSON, D. E., STRASSER, J. C., EVENSEN, E. B., ALLEY, R. B., LARSON, G. J. & ARCONE, S. A. 1998. Glaciohydraulic supercooling: a freeze-on mechanism to create stratified, debris-rich basal ice. I. Field evidence. *Journal of Glaciology*, **44**, 547–562.

LAWSON, W. 1990. *The structural evolution of Variegated, Alsaka*. PhD thesis, University of Cambridge.

——1996. Structural evolution of Variegated Glacier, Alaska, USA, since 1948. *Journal of Glaciology*, **42**, 261–270.

——, SHARP, M. & HAMBREY, M. J. 1994. The structural geology of a surge-type glacier. *Journal of Structural Geology*, **16**, 1447–1462.

——, —— & ——2000. Deformation histories and structural assemblages of glacier ice in a non-steady flow regime. *This volume.*

LEIGHTON, F. B. 1951. Ogives of the East Twin Glacier, Alaska: their nature and origin. *Journal of Geology,* **59**, 578–589.

LLIBOUTRY, L. 1958. Studies of the shrinkage after a sudden advance, blue bands and wave ogives on Glaciar Universidad (Central Chilean Andes). *Journal of Glaciology,* **3**, 216–270.

—— & REYNAUD, L. 1981. "Global dynamics" of a temperate valley glacier, Mer de Glace, and past velocities deduced from Forbes' bands. *Journal of Glaciology,* **27**, 207–226.

MARMO, B. A. & WILSON, C. J. L. 1998. Strain localisation and incremental deformation within ice masses, Framnes Mountains, East Antarctica. *Journal of Structural Geology,* **20**, 149–162.

—— & ——2000. The stress distribution related to the boudinage of a visco-elastic material: examples from a polar outlet glacier. *This volume.*

MEIER, M. F. 1960. *Mode of flow of Saskatchewan Glacier, Alberta, Canada.,* US Geological Survey Professional Paper, **351**.

——, KAMB, W. B., ALLEN, C. R. & SHARP, R. P. 1974. Flow of Blue Glacier, Olympic Mountains, Washington, USA *Journal of Glaciology,* **13**, 187–212.

MURRAY, T., GOOCH, D. L. & STUART, G. W. 1997. Structures within the surge front at Bakaninbreen, Svalbard, using ground-penetrating radar. *Annals of Glaciology,* **24**, 122–129.

MILNES, A. G. & HAMBREY, M. J. 1976. A method of estimating approximate cumulative strains in glacier ice. *Tectonophysics,* **34**, T23-T27.

NYE, J. F. 1952. The mechanics of glacier flow. *Journal of Glaciology,* **2**, 82–93.

——1953. The flow law of ice from measurements in glacier tunnels, laboratory experiments, and the Jungfraufirn borehole experiment. *Proceedings of the Royal Society of London,* **219A**, 477–489.

——1957. The distribution of stress and velocity in glaciers and ice sheets. *Proceedings of the Royal Society of London,* **239A**, 113–133.

——1958. A theory of wave formation on glaciers. *International Association of Hydrological Sciences,* **47**, 139–154.

——1959. A method of determining the strain-rate tensor at the surface of a glacier. *Journal of Glaciology,* **3**, 409–419.

PATERSON, W. S. B. 1994. *The Physics of Glaciers.* 3rd Edn. Pergamon, Oxford.

PFEFFER, W. T. 1992. Stress-induced foliation in the terminus of the Variegated Glacier, USA, formed during the 1982–83 surge. *Journal of Glaciology,* **38**, 213–222.

PLATT, J. P. & VISSERS, R. L. M. 1980. Extensional structures in anisotropic rocks. *Journal of Structural Geology,* **2**, 397–410.

POSAMENTIER, H. W. 1978. Thoughts on ogive formation. *Journal of Glaciology,* **20**, 218–220.

POST, A. 1972. Periodic surge origin of folded medial moraines on Bering piedmont glacier, Alaska. *Journal of Glaciology,* **11**, 219–226.

—— & LACHAPELLE, E. R. 1971. *Glacier Ice.* Seattle, The Mountaineers & University of Washington Press, Seattle.

RAGAN, D. M. 1969. Structures at the base of an icefall. *Journal of Geology,* **77**, 647–667.

RAMBERG, H. 1964. Note on model studies of folding of moraines in piedmont glaciers. *Journal of Glaciology,* **5**, 207–218.

RAMSAY, J. G. 1967. *Folding and Fracturing of Rocks.* McGraw-Hill, New York

—— & HUBER, M. I. 1983. *The Techniques of Modern Structural Geology. Volume 1: Strain Analysis.* Academic Press, London.

—— & ——1987. *The Techniques of Modern Structural Geology. Volume 2: Folds and Fractures.* Academic Press, London.

RAYMOND, C. F. & MALONE, S. 1986 Propagating strain anomalies during mini-surges of Variegated Glacier, Alaska, USA *Journal of Glaciology,* **32**, 178–191.

——, JOHANNESSON, T., PFEFFER, T. & SHARP, M. 1987. Propagation of a glacier surge into stagnant ice. *Journal of Geophysical Research,* **92(B9)**, 9037–9049.

REYNOLDS, J. M. 1988. The structure of the Wordie Ice Shelf, Antarctic Peninsula. *British Antarctic Survey Bulletin,* **80**, 57–64.

—— & HAMBREY, M. J. 1988. The structural glaciology of George VI Ice Shelf, Antarctica. *British Antarctic Survey Bulletin,* **79**, 79–95.

RUTISHAUSER, H. 1971 Observations on a surging glacier in East Greenland. *Journal of Glaciology,* **10**, 227–236.

RUTTER, N. W. 1965. Foliation pattern of Gulkana Glacier, Alaska Range, Alaska. *Journal of Glaciology,* **5**, 711–718.

SCHWARZACHER, N. & UNTERSTEINER, N. 1953. Zum Problem der Bänderung des Gletschereises. *Sitzungsberichte der Oesterreich Akadamie der Wissenschften Mathem-Naturio.,* Klasse Abt. IIa, **162** (1–4), 111–145.

SHARP, M. J., LAWSON, W. & ANDERSON, R. S. 1988. Tectonic processes in a surge-type glacier. *Journal of Structural Geology,* **10**, 499–515.

SHARP, R. P. 1958. Malaspina Glacier, Alaska. *Geological Society of America Bulletin* **69**, 617–646.

——1960. *Glaciers.* Congden Lectures, Oregon State System for Higher Education.

SMALL, R. J., CLARKE, M. J. & CAWSE, T. J. P. 1979. The formation of medial moraines on Alpine glaciers. *Journal of Glaciology,* **22**, 43–52.

SOLLID, J. L., ETZELMÜLLER, B., VATNE, G. & ØDEÅARD, R. S. 1994. Glacial dynamics, material transfer and sedimentation of Erikbreen and Hannabreen, Liefdefjorden, northern Spitsbergen. *Zeitschrift für Geomorphologie, Supplementband,* **97**, 123–144.

SOUCHEZ, R. A. 1967. The formation of shear moraines: an example from south Victoria Land, Antarctica. *Journal of Glaciology,* **6**, 837–843.

——1971. Ice-cored moraines in south-western Ellesmere ISLAND, N.W.T., Canada. *Journal of Glaciology,* **10**, 245–254.

SWINZOW, G. K. 1962. Investigation of shear zones in the ice sheet margin, Thule area, Greenland. *Journal of Glaciology*, **4**, 215–229.

TAYLOR, L. D. 1962. *Ice structures, Burroughs Glacier, south-east Alaska*. Columbus, Institute of Polar Studies, Ohio State University, Report No. 3.

TCHALENKO, J. S. 1970 Similarities between shear zones of different magnitudes. *Geological Society of America Bulletin*, **81**, 1625–1640.

TYNDALL, J. 1859. On the veined structure of glaciers, with observations on white seams, air bubbles and dirt bands. *Philosophical Transactions of the Royal Society of London*, **149**, 279–307.

UNTERSTEINER, N. 1955. Some observations on the banding of glacier ice. *Journal of Glaciology*, **2**, 502–506.

UNWIN, B. & WINGHAM, D. 1997. Topography and dynamics of Austfonna, Nordaustlandet, Sval-bard, from SAR interferometry. *Annals of Glaciology*, **24**, 403–408.

VAUGHAN, D. G. 1993 Relating the occurrence of crevasses to surface strain rates. *Journal of Glaciology*, **39**, 255–266.

VALLON, M. 1967. *Etude de la Mer de Glace*. Rapport du Laboratoire de Géophysique et Glaciologie de Grenoble, No. 103, 2éme partie.

VORNBERGER, P. L. & WHILLANS, I. M. 1990. Crevasse development and examples from Ice Stream B Antarctica. *Journal of Glaciology*, **36**, 3–10.

WADDINGTON, E. D. 1986. Wave ogives. *Journal of Glaciology*, **32**, 325–334.

WILCOX, R. E., HARDING, T. P. & SEELY, D. R. 1973. Basin wrench tectonics. *American Association of Petroleum Geologists, Bulletin*, 57, 74–96.

# Deformation histories and structural assemblages of glacier ice in a non-steady flow regime

WENDY J. LAWSON[1], MARTIN J. SHARP[2] & MICHAEL J. HAMBREY[3]

[1] Department of Geography, University of Canterbury, Private Bag 4800, Christchurch,
New Zealand (e-mail: w.lawson@geog.canterbury.ac.nz)
[2] Department of Earth and Atmospheric Science, University of Alberta, Edmonton,
Alberta,T6G 2E3, Canada (e-mail: martin.sharp@ualberta.ca)
[3] Centre for Glaciology, Institute of Geography and Earth Sciences, University of Wales,
Aberystwyth SY23 3DB, UK (e-mail: mjh@aber.ac.uk)

**Abstract:** Deformation histories of ice exposed at various locations on the centreline in the ablation area of surge-type Variegated Glacier were estimated using a technique based on evaluation of strain-rate fields derived from velocity gradients. The analysis indicates that ice exposed in the upper part of the ablation area a few years after the 1982–83 surge had experienced a relatively simple deformation history that included one surge-related high magnitude deformation event. Ice exposed in the terminal lobe at the same time had experienced six surge phases during its approximately 100 year-long residence time. The histories of accumulation of cumulative strain are complex, and indicate that measures of cumulative strain can mask the effects of large but transient strain events. They also demonstrate that substantial cumulative strain can be 'undone' during subsequent deformation to leave a cumulative strain signal that is unrepresentative of earlier cumulative strain states.

Structural relationships in general do not reflect the complexity of the deformation histories experienced by the ice. A preliminary analysis of the relationships between the deformation histories and structural assemblages suggests, in particular, that brittle structures are reactivated several times during the history of the ice.

As in any tectonic setting, accurate structural interpretation in glaciers requires information about deformation and deformation history. In glaciers, measurement of instantaneous strain rates over short periods of time is relatively straightforward using a range of approaches (e.g. Hambrey et al. 1980; Gudmundsson et al. 1997; Raymond & Malone 1978). However, as in other settings, structural assemblages in glaciers are likely to reflect the total strain experienced by the ice as it has travelled downglacier, rather than simply recent or local strain rates (Hudleston 1983). Measures of cumulative deformation therefore provide information that is useful for structural interpretation (Hambrey et al. 1980). Evaluating cumulative strain in glaciers is challenging, however, since the kinds of strain markers often found in geological settings are absent from glaciers, although bubble shape and orientation can be used in weakly deformed ice (Hudleston 1977). In this paper, rather than focus on local or short-term strain rates, or single measures that provide a snapshot of cumulative strain, we establish complete deformation histories for material moving through a complex deformation regime, and compare them to observed structures.

The aim of this paper is to establish deformation histories for ice exposed at various places on the surface of a glacier, and to make a preliminary analysis of the linkages between those deformation histories and structural assemblages. The analysis is carried out for Variegated Glacier, Alaska, which has relatively well-known kinematics and structural assemblages (see below). The specific objectives of the paper are (a) to establish strain rate and cumulative strain histories for ice at Variegated Glacier; (b) to analyse the nature of linkages between histories of strain rate and accumulation of cumulative strain; (c) to analyse relationships between those deformation histories and observed structural assemblages.

The results of this analysis have a range of broader implications. The comparisons between strain rate and strain accumulation histories may provide information concerning the extent to which palaeostrain analyses from cumulative strain indicators, such as deformed inclusions (Lisle 1994), are representative of strain history.

*From:* MALTMAN, A. J., HUBBARD, B. & HAMBREY, M. J. (eds) *Deformation of Glacial Materials.* Geological Society, London, Special Publications, **176**, 85–96. 0305-8719/00/$15.00 © The Geological Society of London 2000.

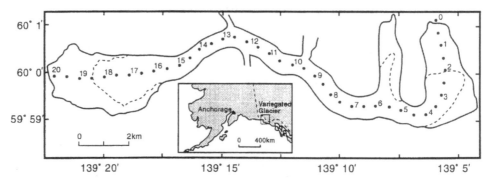

**Fig. 1.** Map of Variegated Glacier, showing location of the glacier, limits of the effects of the 1982–83 surge (dashed line), and distances along the centreline from the head of the glacier in kilometres. These distances are used as a location reference system.

In addition, the deformation histories experienced by deformed material in a non-steady state, gravity-driven surge-type glacier flow regime have direct analogies with those experienced in other tectonic settings, such as thin-skinned fold-and-thrust belts, and surge zones within those thrust belts in particular (Coward 1982).

Variegated Glacier in SE Alaska (Fig. 1) is a surge-type glacier. Glaciers of this type have a pulsating flow regime, in which long periods of quiescence alternate with regularly spaced short periods of surging, during which velocities and strain rates typically reach 10–100 times quiescent phase values (Meier & Post 1969). Surges begin at a nucleus, and the surging part of the glacier expands by the migration up- and down-glacier of the edges of the surging zone (McMeeking & Johnson 1986).

Variegated Glacier was monitored during fieldwork between 1973, the mid-point of a quiescent phase, and 1983, the end of the subsequent and penultimate surge (Kamb *et al.* 1985; Raymond & Harrison 1988). A documentary and photographic record of the glacier is also available that dates from the beginning of the twentieth century (Lawson 1997). Velocity and structural data indicate that motion during surge phases in the surging part of the glacier occurs by plug flow, involving marginal and basal detachment and a minimal amount of vertical and transverse shear (Kamb *et al.* 1985; Sharp *et al.* 1988). Motion in the quiescent part of the glacier, on the other hand, occurs by a combination of shear-dominated creep deformation and basal sliding (Bindschadler *et al.* 1977; Raymond & Harrison 1988). The long-term record indicates that Variegated Glacier surged five times in the twentieth century, at intervals ranging from 12 to 26 years, and that the behaviour of the glacier during the well-documented 1965–83 surge cycle

is broadly representative of its twentieth century kinematic behaviour (Lawson 1996, 1997).

## Method for deformation analysis

The overall methodology for the establishment of deformation histories involves the establishment of displacement paths along the centreline for a series of hypothetical ice particles, and evaluation of the deformation experienced over time by ice units between adjacent hypothetical particles.

In order to establish particle displacement paths, published velocity data (Bindschadler 1982; Kamb *et al.* 1985; Raymond *et al.* 1987; Raymond & Harrison 1988) have been assembled and interpolated in order to derive velocity fields that characterize velocity on the glacier centreline (Fig. 2). Displacement paths are calculated by establishing the downglacier movement of hypothetical ice particles through these velocity fields. Strain-rate fields (Fig. 3) are then calculated from the changing spacing of hypothetical particles, using the following algorithm:

$$\dot{e} = \frac{1}{\Delta t} \frac{l_n - l_{n-1}}{l_{n-1}} \qquad (1)$$

where $\dot{e}$ is the longitudinal flow-parallel strain rate at time $n$, $l_n$ is the spacing of two ice particles at time $n$, and $\Delta t$ is the time interval $[n - (n - 1)]$.

Long-term downglacier displacement paths are established by treating the velocity fields as generic rather than specific to the 1965–83 surge. Thus, for long-term analysis, '1965' represents year 1 of a generic 18 year-long surge cycle, rather than 1965 *per se*. When a hypothetical particle reaches year 18 in one surge cycle, it begins its movement at year 1 in the next cycle,

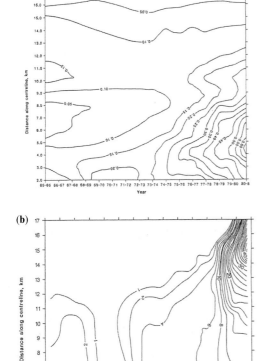

**(a)**

**(b)**

**Fig. 2.** Contoured fields of flow-parallel velocity (metres per day) at the centreline of Variegated Glacier (a) annually averaged during the 1965–81 quiescent phase; (b) monthly to bimonthly averaged during the 1982–83 surge phase. Arrows on the x-axis of (b) indicate the mid-points of periods used for averaging of velocity and strain; note that these periods vary in length. This averaging results in an apparent ending of the surge in mid-June; the surge did not actually terminate until July 5th, when velocities fell very dramatically (see Kamb et al. 1985 for details).

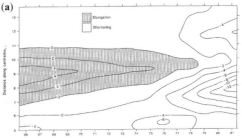

**(a)**

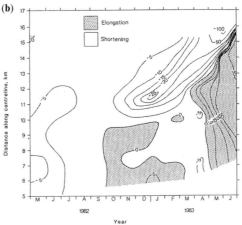

**(b)**

**Fig. 3.** Contoured fields of longitudinal flow-parallel strain rate at the centreline of Variegated Glacier (a) for the 1965–81 quiescent phase, in $10^{-5}$ per day; and (b) for the 1982–83 surge phase, in $10^{-4}$ per day.

and so on. Corresponding long-term strain rate histories (Figs 5, 6, 7) are established by treating the strain-rate fields as similarly generic rather than year-specific, and using them to determine the strain rate occurring at each point on the long-term displacement path.

Cumulative strain is calculated for inter-particle ice units with respect to some initial inter-particle length, using the following relationship:

$$e = \frac{l_n - l_0}{l_0} \qquad (2)$$

where $e$ is cumulative strain, and $l_0$ is initial length of the ice unit.

In order to calculate deformation histories for residence times of ice from burial in the accumulation area to emergence in the ablation area, two-dimensional particle paths were constructed using published mass-balance data in conjunction with long-term downglacier displacement paths derived from the surface velocity fields, as outlined above (Lawson 1990). An approximation is made that surface velocity and strain-rate fields are representative of ice motion and deformation to the maximum depth to which particles are buried within the scope of the analysis. This approximation is examined in more detail in the next section.

## Scope, scale and assumptions

Deformation histories calculated in this analysis are essentially one-dimensional, pertaining to longitudinal flow-parallel deformation on the centreline. Deformation is assumed to be co-axial. In general terms, this assumption is reasonable for deformation at a glacier centreline (e.g. Nye 1952), and velocity and deformation data indicate that it is valid at Variegated Glacier (Bindschadler *et al.* 1977; Raymond *et al.* 1987).

As indicated above, an approximation is made that surface velocity and strain-rate fields are representative of motion and deformation to the maximum depth to which particles are buried within the scope of the analysis. This maximum burial depth of ice included in this analysis is 110 m (Lawson 1990), and the maximum ice depth is over 400 m (Bindschadler *et al.* 1977). The approximation is sound during surge phases of motion, when most of the velocity is achieved by sliding (Kamb *et al.* 1985). During deformation-dominated quiescent phase flow, the differential velocity between the surface and ice at a depth of 110 m according to Nye's theory

(Paterson 1994, p. 251), is 0.01 m per day. This differential is small compared to minimum measured surface velocities in the order of 0.1 m per day (see Fig. 2). Given this small differential, the shallow burial depth compared to maximum ice depth, and the fact that 60–70% of total downglacier motion of ice through Variegated Glacier occurs during sliding-dominated surge phases of motion (Lawson 1990), the approximation of velocity and strain rate fields at depth with those of ice at the surface is considered acceptable.

The analysis incorporates that part of the glacier located between approximately 4 km and 16 km from the head of the glacier (Fig. 1). This coverage reflects that of earlier programmes of velocity measurement, and includes the majority of the section of the glacier affected by recent surges (Lawson 1997). Distances along

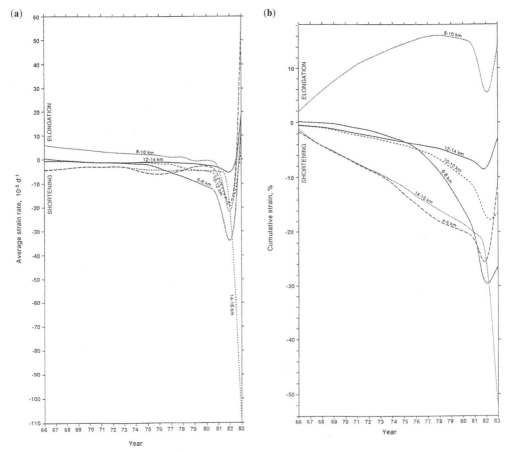

**Fig. 4.** Deformation histories experienced by six hypothetical, 2 km long ice units moving along the centreline of Variegated Glacier through the 1965–83 surge cycle. (**a**) Histories of instantaneous strain rate; (**b**) histories of accumulation of cumulative strain. The locations of the ice units at the start of the surge cycle in 1965 are indicated.

the centreline from the head of the glacier are used as a location reference system for graphical representation and discussion. The terms 'proximal', 'intermediate', and 'distal' are used relatively to indicate the longitudinal position of ice with respect to the head of the glacier.

Velocity and strain-rate fields are calculated and presented separately at different temporal averaging scales for quiescent and surge phases of motion (Figs 2, 3). For long-term analysis, the motion and deformation characteristics of the surge phase are incorporated into annual averages of velocity and strain rate. Deformation is calculated for inter-particle ice units that are initially 2 km long. Sensitivity analysis, in which deformation is calculated using different averaging time-scales and different inter-particle distances, indicates that inferences about deformation history using this type of approach are not critically dependent on the specific magnitudes of the spatial and temporal averaging scales (Lawson 1990).

Results are presented first in the form of strain rate and cumulative strain histories for six 2 km long ice units through a single surge cycle (Fig. 4). Estimates of long-term deformation histories for ice exposed at the surface at three locations in the ablation area in 1987 (the time of detailed field structural mapping; Lawson et al. 1994) are then presented, together with schematic summaries of structural assemblages observed at those locations (Figs 5–7).

## Velocity and strain-rate fields

Centreline velocity fields (Fig. 2) reflect the well-known features of the evolution and behaviour of Variegated Glacier (Kamb et al. 1985; Raymond et al. 1987; Raymond & Harrison 1988). During the quiescent phase, annually averaged velocities reached a maximum towards the end of quiescence of 0.7 m per day at a location in the upper accumulation area approximately 4–5 km from the head of the glacier (Fig. 2a). The gradual build-up to the surge during late quiescence had no discernible effect on annually averaged velocity downglacier of approximately 13 km from the head of the glacier. During the surge, monthly to bimonthly averaged velocities reached 46 m per day (Fig. 2b).

During quiescence, annually averaged strain rates reached approximately $10^{-4}$ per day (Fig. 3a). Except for an area in the central part of the glacier affected by elongation, most of the glacier experienced shortening strain throughout quiescence. This elongating area initially comprised about 20% of the total length of

the glacier, although its longitudinal extent decreased through quiescence until elongation ceased three years prior to the onset of the surge.

During the surge, monthly to bimonthly averaged strain rates reached $10^{-2}$ per day, and the strain rate field was complex, dynamic, and rapidly-evolving (Fig. 3b). During the first phase of the surge (January–June 1982), strain rates were compressive and reached magnitudes of $> -5 \times 10^{-4}$ per day (Fig. 2b). Early in the second phase (which lasted from December 1982 to July 1983), a strain-rate maximum with shortening strain rates $> 25 \times 10^{-4}$ per day developed at a location approximately 11.5 km from the head of the glacier. This compressive strain-rate peak intensified and migrated down-glacier during the second phase, resulting in shortening strain rates $> 100 \times 10^{-4}$ per day towards the end of the surge. An area of intensifying elongating strain developed upglacier of this shortening strain-rate maximum in the later months of the surge, eventually resulting in elongating strain rates in excess of $50 \times 10^{-4}$ per day in an area between 13 and 15 km on the centreline.

## Deformation histories over a surge cycle

Strain rate histories for the 1965–83 surge cycle (Fig. 4a) indicate that during the quiescent phase, five of the six ice units experienced deformation comprising relatively steady and slow shortening strain rates of up to $10 \times 10^{-5}$ per day. The sixth, the unit that began motion at 8–10 km, experienced elongation throughout quiescence. During the surge, five of the six ice units experienced a characteristic pattern of deformation comprising paired rapid shortening following by rapid elongation. The sixth, the most distal, experienced only rapid shortening during the surge, as a result of its displacement into a location down-glacier of the propagating elongating strain rate maximum. Strain rates experienced by ice units during the surge were up to two orders of magnitude larger than those occurring during the quiescent phase, and the maximum surge phase strain rate exceeded $100 \times 10^{-5}$ per day (Fig. 4a). In the context of the entire 1965–83 surge cycle, the 1982–83 surge was a high magnitude, short duration deformation event.

For five of the six ice units, histories of strain accumulation during the 1965–83 surge cycle reflect the progressive accumulation of shortening throughout quiescence (Fig. 4b). For example, by 1981, the ice unit that began the cycle located furthest upglacier, at 4–6 km from the head of the glacier, had accumulated 20%

shortening. The ice unit that started motion at 8–10 km elongated progressively through quiescence. During the surge, ice units that started motion at locations between 4 km and 14 km from the head of the glacier were first shortened then elongated. For the ice unit initially located at 8–10 km from the head of the glacier, the surge was a major but transient deformation event in terms of cumulative strain; at the end of the surge in 1983, the total cumulative elongation recorded by this ice unit was essentially the same as at the start of the surge. The ice unit that began motion in the most longitudinally-distal position, at 14–16 km from the head of the glacier, accumulated shortening throughout the entire surge cycle, until by the end of the surge

in 1983 it had accumulated a total of approximately 52% shortening during 18 years of deformation.

Overall, in the context of the 1965–83 surge cycle, the 1982–83 surge was a major deformation event in which strain rates were orders of magnitude larger than those experienced during the preceding phase of quiescence. However, in some parts of the glacier, the effect of the surge on strain accumulation was transitory.

## Deformation histories and structural assemblages in proximal ice

Ice exposed in the upper, longitudinally proximal part of the ablation area in 1987, at approximately 11 km from the head of the glacier, was initially deposited in the latter half of the preceding quiescent phase in the lower part of the accumulation area at a location approximately 8 km from the head of the glacier. It had a residence time, between burial and emergence, of approximately 10 years (Fig. 5a). During

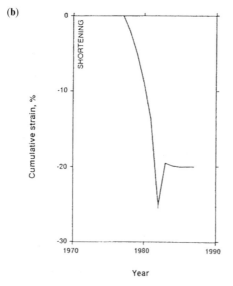

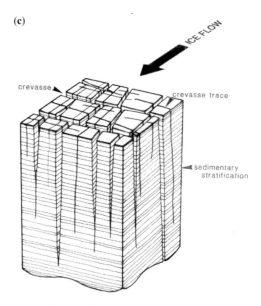

**Fig. 5.** Deformation histories and structural features of ice exposed at the surface on the centreline of Variegated Glacier at approximately 11 km from the head of the glacier. (**a**) Strain rate history of the ice, from deposition through burial to emergence; (**b**) history of accumulation of cumulative strain; (**c**) schematic block diagram summarizing structural relationships between primary sedimentary stratification, crevasses and crevasse traces (foliations omitted for clarity; see Lawson *et al.* 1994).

its residence, it first experienced four years of increasingly rapid shortening strain, followed during the surge phase by the characteristically paired intense shortening and intense elongation (Fig. 5a). In the last four years of its residence, in early quiescence, it experienced very slow shortening strain.

The history of strain accumulation for this ice involves the progressive accumulation of shortening at various rates over the 10 year residence time, except for a short period in the latter part of the surge during which the total shortening was reduced by approximately 7% from 26% to 19% (Fig. 5b). During the subsequent four years, little further strain accumulation occurred, and the total cumulative shortening of this ice when it emerged in the ablation area was 20%.

Both the strain rate and cumulative strain histories for this ice are dominated by the effects of longitudinal shortening. Cumulative shortening is recorded at all points in its history (Fig. 5b). The rapid longitudinal elongation, at rates of up to $17 \times 10^{-5}$ per day, that affected this ice for a short period during the surge phase is recorded in the cumulative strain record as a slight reduction in total shortening. The relationship between the strain rate and cumulative strain histories therefore indicates that significant strain events involving intense deformation can occur that do not substantially affect the overall pattern of progressive strain accumulation.

Ice exposed in the upper part of the ablation area in the vicinity of the centreline possesses a relatively simple set of structures (Fig. 5c; Lawson et al. 1994). The simplicity of the structural glaciology at this location is consistent with the relatively short residence time of ice exposed here. A single, well-defined set of consistently oriented transverse crevasses with relatively steep dips is superimposed on a well-defined set of longitudinally oriented crevasses. Cross-cutting relationships, in which the longitudinal crevasses predate the transverse crevasses, are simple and consistent (Fig. 5c). This structural assemblage is consistent with the development of longitudinal crevasses during the surge-phase shortening, followed by development of transverse crevasses during the immediately subsequent surge-phase elongation. The structural assemblage therefore more closely reflects the strain rate history than the cumulative strain history. In particular, it is of note that the intense but short-lived phase of elongation that has only a transitory effect on the cumulative strain history has left a clear structural imprint.

## Deformation histories and structural assemblages in intermediate ice

Ice exposed at a longitudinally intermediate location in the ablation area in 1987, at approximately 14 km from the head of the glacier, was initially deposited at approximately 5 km from the head of the glacier during the early stages of a phase of quiescence. This ice is estimated to have had a residence of 56 years between burial and emergence. During this time it experienced three surge phases.

The strain rate history of this ice began with a period of quiescent-phase shortening at rates of up to $4 \times 10^{-5}$ per day (Fig. 6a). The subsequent surges caused strain rates much higher than any that affected this ice during quiescence. All three surges produced the characteristic surge-phase strain sequence in this ice, comprising rapid shortening followed by rapid elongation. The most rapid strain affecting this ice occurred during the first period of surge-phase elongation, during which elongation rates reached $55 \times 10^{-5}$ per day. Surge-phase shortening strain rates affecting this ice reached $31 \times 10^{-5}$ per day. After the first surge, this ice experienced quiescent phase elongation for approximately 12 years at rates of up to $5 \times 10^{-5}$ per day, with the rest of the quiescent part of the deformation history involving relatively low rates of shortening.

The history of strain accumulation in this ice (Fig. 6b) is complex. Cumulative shortening was recorded throughout its history, but major fluctuations in the amount of total cumulative shortening occurred. During the first 17 years of its history, this ice accumulated 30% shortening. However, this substantial shortening was largely 'undone' during elongation over the following 14 years, such that 29 years into its history it recorded a total shortening of approximately 1%. After this initial period, the ice accumulated shortening relatively steadily over the rest of its history. This gradual shortening was punctuated by short periods of decreasing shortening that coincided with surge phase elongating strain rates. By the time this ice emerged at the surface of the glacier in the ablation area, it recorded a total shortening of 21%.

The cumulative strain rate history of this ice is significantly more complex than that of ice exposed both further upglacier and further downglacier (Fig. 7b). This complexity occurs as a result of the cumulative effects of early quiescent and surge-phase longitudinal shortening, followed by a combination of surge and quiescent phase longitudinal elongation. The relationship between the strain rate and cumulative strain histories for this ice demonstrates

**(a)**

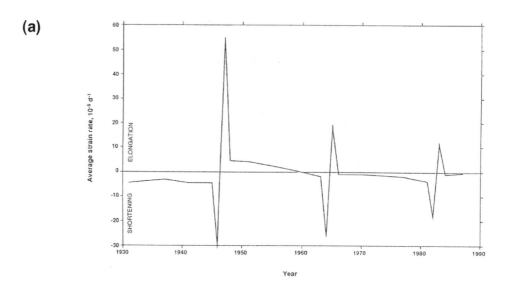

**(b)**

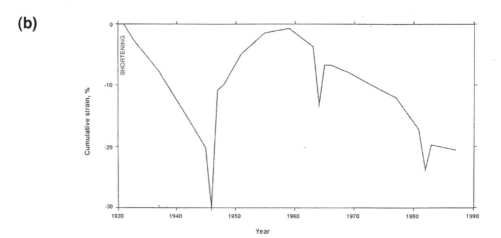

**(c)**

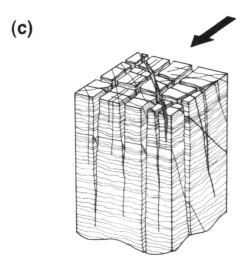

**Fig. 6.** Deformation histories and structural features of ice exposed at the surface on the centreline of Variegated Glacier at approximately 14 km from the head of the glacier. (**a**) Strain rate history of the ice, from deposition through burial to emergence; (**b**) history of accumulation of cumulative strain; (**c**) schematic block diagram summarizing structural relationships between primary sedimentary stratification, crevasses and crevasse traces (foliations omitted for clarity).

again that significant strain events may not substantially affect the overall pattern of progressive strain accumulation. The pattern of strain accumulation also indicates that large amounts of cumulative strain can be accumulated and subsequently undone, in this case such that a small amount of total strain is recorded when the ice has previously been shortened by up to nearly one third of its initial length.

Structures exposed in ice in the vicinity of the centreline at approximately 14 km from the head of the glacier are similar to those exposed further upglacier, although possessing somewhat more complex interrelationships (Fig. 6c; Lawson et al. 1994). The distinct sets of longitudinal and transverse crevasses outlined above are present. In addition, there are crevasses at other orientations with complex cross-cutting relationships. Transverse fractures have a range of dips, suggesting that some have been affected by rotational vertical simple shear. Crevasse traces are more common at this location than further upglacier. Significantly, cross-cutting relationships are relatively simple overall, and in particular they indicate that, as further upglacier, longitudinal crevasses predate transverse ones. Sedimentary stratification in this location is gently folded into overturned, downglacier-verging folds with hingelines oriented transverse to flow. The structural assemblages are consistent with the activation of both transverse and longitudinal fractures during the recent surge. The presence of crevasses traces and crevasses with oblique orientations suggests a more complex history, but the main structural imprint is of the most recent surge-induced shortening and elongation. There is no evidence for the very large accumulation and undoing of shortening that occurred early in the cumulative strain history.

## Deformation histories and structural assemblages in distal ice

Ice exposed in a longitudinally distal position in the terminal lobe in 1987, at approximately 17 km from the head of the glacier, was deposited at a location in the accumulation area upglacier of the limit of the kinematic analysis (as defined by the velocity fields shown in Fig. 2). By the time it reached 4 km from the head of the glacier, this ice was buried to a depth of 82 m (Lawson 1990). Structural analysis and limited velocity data indicate that ice upglacier of a point 4 km from the head of the glacier experiences only longitudinal elongation, and that the rate of elongation is greater during surge phases (Raymond & Harrison 1988; Lawson 1996). These data,

together with mass-balance and surge-position data, have been used to extrapolate the deformation analysis upglacier of the limit of the kinematic analysis, as shown with a dashed line in Fig. 7a (Lawson 1990). The extrapolated part of this deformation history is probably accurate in terms of sense and broad relative magnitudes of strain rate. The total residence time of ice exposed at 17 km from the head of the glacier in 1987 is estimated to be approximately 100 years.

The 100-year long deformation history of this ice includes a total of six surges, the first of which caused only elongation, and the last of which caused only shortening. Each of the intervening four surges caused the characteristic surge-phase strain sequence of rapid shortening followed by rapid elongation. For this ice, each successive surge caused annually averaged strain rates smaller than the previous surge. Strain rates during the four surges that occurred within the part of the glacier covered by the kinematic analysis reached $28 \times 10^{-5}$ per day. This ice experienced longitudinal shortening at rates of up to $8 \times 10^{-5}$ per day during three of the five phases of quiescence in its deformation history, and longitudinal elongation during the other two. The deformation history of this ice differs from the histories of ice exposed further upglacier mainly in terms of the number of surge events experienced, and of the relatively long period of longitudinal elongation in the deformation history.

The 100 year-long history of strain accumulation in this ice (Fig. 7b) contrasts with the histories of ice further upglacier, in that the former consistently records cumulative elongation. Four years after deposition, this ice was elongated by approximately 30% and 25 years after deposition it was elongated by approximately 68% (Fig. 7b). Note that this early elongation occurs in the extrapolated part of the deformation history (Fig. 7a) and that the magnitudes of total elongation therefore bear some fairly substantial uncertainties. However, as indicated above, it is likely that the broad picture is accurate. The ice remained elongated with respect to its initial state throughout its history, although the last 45 years of its residence were characterised by a gradual reduction in cumulative elongation, with minor interruptions in this overall pattern caused by surges. By the time this ice was exposed in the terminal lobe, the cumulative strain with respect to an initial state was less than 2%.

Ice exposed in the lower parts of the terminal lobe in 1987 contained a more diverse range of structures than ice further upglacier (Fig. 7c). Particularly noteworthy is the addition of thrust faults to the assemblage. These thrust faults

**(a)**

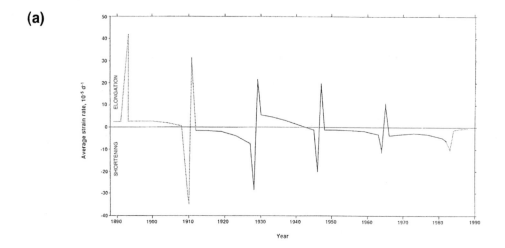

**(b)**

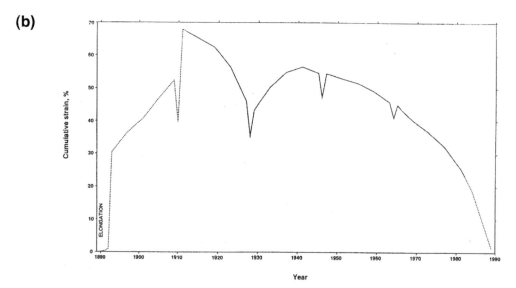

**(c)**

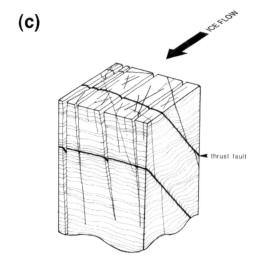

ICE FLOW

thrust fault

**Fig. 7.** Deformation histories and structural features of ice exposed at the surface on the centreline of Variegated Glacier at approximately 17 km from the head of the glacier. (**a**) Strain rate history of the ice, from deposition through burial to emergence; (**b**) history of accumulation of cumulative strain; (**c**) schematic block diagram summarizing structural relationships between primary sedimentary stratification, crevasses and crevasse traces (foliations omitted for clarity). The dashed parts of lines in (**a**) and (**b**) indicate extrapolation based on incomplete data.

are discussed in detail by Lawson *et al.* (1994). Multiple sets of closely spaced and variously oriented fractures dominate the assemblage. The overturned, downglacier-verging folds in the sedimentary stratification seen further upglacier (Fig. 6c) are better developed here, and the thrust faults are often axial planar to these folds. The thrust faults appear to be transverse fractures that have been rotated during vertical simple shear into an orientation suitable for thrusting (Lawson *et al.* 1994). The dominant set of open, recently active crevasses is longitudinal in orientation. The structural assemblage is therefore characterized by structures accommodating longitudinal shortening, including thrust faults, longitudinal crevasses, and overturned, downglacier verging folds. This assemblage is inferred to be the product of the latter part of the deformation history of the ice.

The structures and structural relationships exposed in surface ice in the lower part of the terminal lobe do not therefore reflect the complexity of the deformation history experienced by that ice. For example, cross-cutting relationships do not indicate six distinct and separate major strain events. There is no evidence for six separate phases of crevasse formation, despite the six surges that have affected this ice. The absence of evidence for the development of separate sets of crevasses in association with each surge suggests that fracture reactivation plays a significant role in strain accommodation. The reactivation of rotated transverse crevasses as thrust faults, as indicated by the structural data, suggests that structures that form initially to accommodate longitudinally elongating strain can subsequently be reactivated to accommodate longitudinally shortening strain.

## Summary and conclusions

Ice exposed at the surface in the ablation area of Variegated Glacier in 1987 had experienced a wide range of amounts of deformation. Ice located at that time in the upper part of the ablation area is estimated to have had a 10 year-long deformation history that included one surge phase strain event. In contrast, ice exposed in the lower part of the terminal lobe is estimated to have had a 100 year-long deformation history that included six surge phase strain events. Strain-rate histories for all ice included in this analysis are characterised by long periods of relatively low strain rate during quiescent phases, alternating with short periods of rapid strain rate during surge phases. Histories of strain accumulation are more complex, and indicate in particular that very large cumulative strains (up to nearly 70% in this study) can be 'undone' by subsequent deformation.

Relationships between strain rate and cumulative strain histories indicate that strain accumulation through a complex strain rate history that includes several major deformation events may be dominated by the progressive effects of background strain rates. This characteristic is illustrated in particular for ice in the terminal lobe, which had experienced a complex strain rate history including six major deformation events, but which recorded only a few per cent cumulative strain at the end of the deformation history. Overall, the nature of strain rate and cumulative strain histories, and of relationships between them, indicate that surge phase deformation events had a major but transitory effect on deformation histories in ice moving through this surge-type glacier.

The structural assemblages at all locations in the ablation area in 1987 were dominated by the effects of the most recent surge that occurred in 1982–83, even in ice that had experienced several previous surges. Where ice had experienced long deformation histories, the structural assemblages did not reflect the complexity of these deformation histories. It is inferred that reactivation of brittle structures plays a role in the maintenance of a relatively simple structural assemblage. Structures appear to be reactivated to accommodate deformation that is opposite in sense to that which formed them.

This study has indicated the potential complexity of relationships between strain rate histories and histories of the accumulation of strain, and also of the relationships between deformation histories and structural assemblages. Further research is necessary to elucidate the details of these relationships in other tectonic settings, in order to improve the accuracy of palaeostrain analysis and resulting structural interpretation. The research presented in this paper illustrates the potential of glaciers for such work.

The authors thank C. Raymond and W. Harrison for access to unpublished data. W.L. was supported during this work by a NERC Research Studentship. M.S. was supported in part by a Royal Society of London Study Award. Comments by B. Hubbard and D. Sugden helped to improve the clarity of the paper.

## References

BINDSCHADLER, R. 1982. A numerical model of temperate glacier flow applied to the quiescent phase of a surge-type glacier. *Journal of Glaciology*, **29**, 239–265.

——, HARRISON, W. D., RAYMOND, C. F. & CROS-
SON, R. 1977. Geometry and dynamics of a surge-
type glacier. *Journal of Glaciology*, **18**, 181–194.
COWARD, M. P. 1982 Surge zones in the Moine thrust
zone of NW Scotland. *Journal of Structural
Geology*, **4**, 247–256.
GUDMUNDSSON, G. H., IKEN, A. & FUNK, M. 1997.
Measurements of ice deformation at the con-
fluence area of Unteraargletscher, Bernese Alps,
Switzerland. *Journal of Glaciology*, **43**, 548–556.
HAMBREY, M. J., MILNES, A. G. & SIEGENTHALER, H.
1980. Dynamics and structure of Griesgletscher,
Switzerland. *Journal of Glaciology*, **25**, 215–228.
HUDLESTON, P. J. 1977. Progressive deformation and
development of fabric across zones of shear in
glacial ice. *In*: SAXENA, S. & BHATTACHARJI, S.
(eds) *Energetics of Geological Processes*. Springer-
Verlag, New York, 121–150.
——1983. Strain patterns in an ice cap and implications for strain variations in shear zones. *Journal
of Structural Geology*, **5**, 455–463.
KAMB, B., RAYMOND, C. F., HARRISON, W. D.,
ENGELHARDT, H. F., ECHELMEYER, K. A.,
HUMPHREY, N., BRUGMAN, M. & PFEFFER, T.
1985. Glacier surge mechanism: 1982–1983 surge
of Variegated Glacier, *Alaska Science*, **227**,
469–479.
LAWSON, W. J. 1990. *The Structural Evolution of
Variegated Glacier, Alaska*. PhD thesis, University
of Cambridge.
——1996. Structural evolution of Variegated Glacier,
Alaska, USA, since 1948. *Journal of Glaciology*,
**42**, 261–270.
——1997. Spatial, temporal and kinematic characteristics of surges of Variegated Glacier, Alaska.
*Annals of Glaciology*, **24**, 95–101.
——, SHARP, M. J. & HAMBREY, M. J. 1994. The
structural geology of a surge-type glacier. *Journal
of Structural Geology*, **16**, 1447–1462

LISLE, R. J. 1994. Palaeostrain analysis. *In*: HANCOCK,
P. L. (ed.) *Continental Deformation*. Pergamon,
Tarrytown NY, 28–42.
MCMEEKING, R. M. & JOHNSON, R. E. 1986. On the
mechanics of surging glaciers. *Journal of Glaciology*, **32**, 120–132.
MEIER, M. F. & POST, A. S. 1969. What are glacier
surges? *Canadian Journal of Earth Sciences*, **6**,
807–819.
NYE, J. F. 1952. The mechanics of glacier flow. *Journal
of Glaciology*, **2**, 82–93.
PATERSON, W. S. B. 1994. *The Physics of Glaciers*.
3rd Edition. Pergamon, Oxford.
PFEFFER, W. T. 1992. Stress-induced foliation in the
terminus of Variegated Glacier, Alaska, formed
during the 1982–1983 surge. *Journal of Glaciology*,
**38**, 213–222.
RAMSAY, J. G. & HUBER, M. I. 1983. *The Techniques
of Modern Structural Geology. Volume 1: Strain
Analysis*. Academic Press, London.
RAYMOND, C. F. 1984. Interim report on surges of
Variegated Glacier. Fourth year report on, NSF
grant number EAR-79-19530.
——1987 How do glaciers surge? A review. *Journal of
Geophysical Research*, **92**, 9121–9134.
—— & HARRISON, W. D. 1988. Evolution of
Variegated Glacier, Alaska, USA, prior to its
surge. *Journal of Glaciology*, **34**, 154–169.
—— & MALONE, S. 1986. Propagating strain anoma-
lies during mini-surges of Variegated Glacier,
USA. *Journal of Glaciology*, **32**, 178–191.
——, JOHANNESSON, T., PFEFFER, T. & SHARP, M.
1987. Propagation of a surge into stagnant ice.
*Journal of Geophysical Research*, **92**, 9037–9049.
SHARP, M. J., LAWSON, W. J. & ANDERSON, B. 1988.
Tectonic processes in a surge-type glacier – an
analogue for the emplacement of thrust sheets by
gravity tectonics. *Journal of Structural Geology*,
**10**, 499–515.

# Experimental work on the effect of pre-existing anisotropy on fabric development in glaciers

CHRISTOPHER J. L. WILSON

School of Earth Sciences, The University of Melbourne, Victoria 3010, Australia
(e-mail: c.wilson@earthsci.unimelb.edu.au)

**Abstract:** Localization of deformation in ice is known to be important at all scales in deforming glaciers. However, relatively little is known of the significance of shear localization and the influence of fabric development in anisotropic ice at microscopic scales (<mm–cm). In this experimental study, the effect of initial *c*-axis preferred orientation and the inclination of the primary layering in anisotropic ice masses, during both plane strain-compression and combined simple shear-compression, have been examined. A series of creep tests in the temperature range of −5 to −1°C over a range of shortening strains varying from 10 to 40% and compressive stresses ranging from 0 to 0.7 MPa have been undertaken. Significant variations in the strain rate and microstructural development have been observed in the plane strain-compression experiments that reflects the varying orientations of the anisotropy and its relationship to easy-glide directions in the ice mass. In the unconfined combined compression and shear experiments minimum shear stress rates vary between the variously oriented anisotropic ice masses and deviate from the normal power flow law for isotropic ice. Where annealing occurs, such ice masses preserve the pre-existing *c*-axis fabric and hence may reflect a contribution from both the recrystallized and inherited ice components.

Most workers now agree that polycrystalline ice can accommodate large amounts of intragranular plastic deformation and in so doing develops specific flow characteristics and crystallographic preferred orientations. Ice flow at the scale of ice crystals (grains), is dominated by the extreme plastic anisotropy of the ice crystal with its dominant slip system in the basal plane (Wilson & Zhang 1994). However, there is still considerable controversy as to the precise way polar ice tends to develop a preferred bulk fabric in response to its stress, strain rate, pre-existing fabric and temperature history as it ages in cold ice sheets (Hughes 1998). This means that the anisotropic behaviour during flow is also manifested at scales larger than the ice grains. In an ice sheet there is the progressive development of a fabric through an ice sheet from isotropic fabrics at the surface to a strong fabric at depth. Typical *c*-axis fabrics at depth are vertical or small-circles centred about the vertical (Gow & Williamson 1976; Thwaites *et al.* 1984; Thorsteinsson *et al.* 1997); this means that the ice becomes progressively harder to compress vertically, and easier to localize the shear horizontally.

The ice fabric, *c*-axis orientation, controls the rheological properties and therefore the deformation rate, while the temperature and strain also determine the ice fabric (Wilson 1982*a*;

Wilson & Russell-Head 1982). If the fabric distribution in an ice sheet is to be allowed to affect the flow of the ice sheet, then we also need to be able to predict the evolution of the fabric patterns. Particularly in areas where pre-existing fabric anistropies are altered by disturbances in the flowing ice mass. The deformation pattern associated with the anisotropy may also effect the pattern and characteristics of the *c*-axis fabrics we recognize in natural ice cores. It is certainly believed that it is the effect of anisotropic fabrics that contributes to the initiation of stratigraphic disturbances in the GISP2 and GRIP Greenland ice cores (Alley *et al.* 1997; Dahl-Jensen *et al.* 1997).

The reasons why *c*-axis fabrics vary in the lower levels of the GRIP and GRIP2 Greenland ice cores (Tison *et al.* 1994; Alley *et al.* 1997; Castelnau *et al.* 1998) and the observation that there is folding of the anisotropic layers begs the questions. What happens on a grain scale if a layered ice mass, composed of polycrystals or of highly anisotropic ice, is deformed in a variety of orientations, but essentially under the same bulk stress regime? What factors contribute to the variation in c-axis fabrics where polar ice is disturbed at ice divides or instabilities (e.g. Hughes 1998) and overturned folds are developed adjacent to bedrock perturbations (Fig. 1)?

*From*: MALTMAN, A. J., HUBBARD, B. & HAMBREY, M. J. (eds) *Deformation of Glacial Materials*. Geological Society, London, Special Publications, **176**, 97–113. 0305-8719/00/$15.00 © The Geological Society of London 2000.

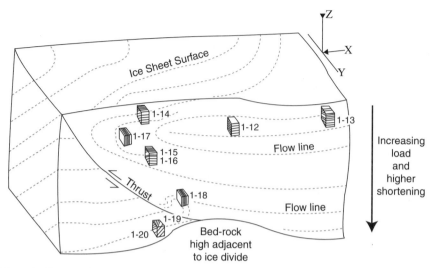

**Fig. 1.** Schematic diagram of an ice sheet containing folded flow layering adjacent to an ice divide. Flow is by pure shear with vertical compression and the orientation of the ice layering as represented by experimental Series-1 is superimposed on this model. The effect of the bed-rock high to produce a thrust is conjectural, but could be related to the processes described by Marmo & Wilson (1999).

The main problem here is that grain-scale structures are easily obliterated or reset by ongoing deformation and post-deformation grain growth through annealing processes.

In addressing these questions there are several reasons to expect shear localization to develop at a grain scale: (1) shear localization is observed at all larger scales; (2) ice is mechanically heterogeneous down to the subgrain level, due to differences in mechanical response to deformation of grains with respect to the crystal lattices (Wilson & Zhang 1994); (3) the response of most ductile materials to deformation favours the development of instabilities and hence localization of deformation (Poirier 1980; Hudleston 1983).

Although various experimental approaches have been employed to examine the problems of shearing in ice, these have focussed on polycrystalline ice aggregates (Li et al. 1996) or on highly strained aggregates (Kamb 1972; Bouchez & Duval 1982). Such experiments do not consider the role of anisotropy and its contribution to the mechanical behaviour or the final microstructure. An exception to this would be the work of Castelnau et al. (1998) who have mechanically tested and modelled strongly textured ice samples coming from a wide range of depths in the GRIP ice cores. However, a limitation of the Castelnau models is that deformation is assumed to be uniform within each grain and neighbouring grain interactions are not taken into account. Whereas, it has been pointed out by Wilson & Zhang (1994) that neighbouring grain interactions are critical in determining how an ice aggregate undergoes recrystallization.

In this study the mechanical behaviour of variously oriented samples of anisotropic ice (Fig. 2) will be compared to the behaviour of the isotropic samples in uniaxial unconfined compression tests (e.g. Li et al. 1996; Li & Jacka 1998). A flow law for ice describing such deformation has been seen as a function of the stress configuration, the total strain, an initial random crystallographic orientation and temperature and this has been established by a number of workers (Budd & Jacka 1989; Li et al. 1996). Traditionally, a flow law, Glen's Law (Glen 1955), has been applied to isotropic ice in secondary creep and a value of approximately 3 for the exponent $n$ that relates to the flow law:

$$\dot{e} = A\sigma^n$$

Where $\dot{e}$ the steady state secondary creep is related to stress ($\sigma$) and $A$ is a temperature dependent flow parameter. However, if the ice has a preferred crystallographic fabric then Glen's Law breaks down (Azuma & Goto-Azuma 1996) as the fabric produces a strong mechanical anisotropy and the strain rate is no longer independent of the magnitude of the applied stress. In such cases the flow parameter ($A$), is also based on the mean orientation of $c$-axes in the ice (Azuma & Goto-Azuma 1996). In isotropic ice experiments (e.g. Li &

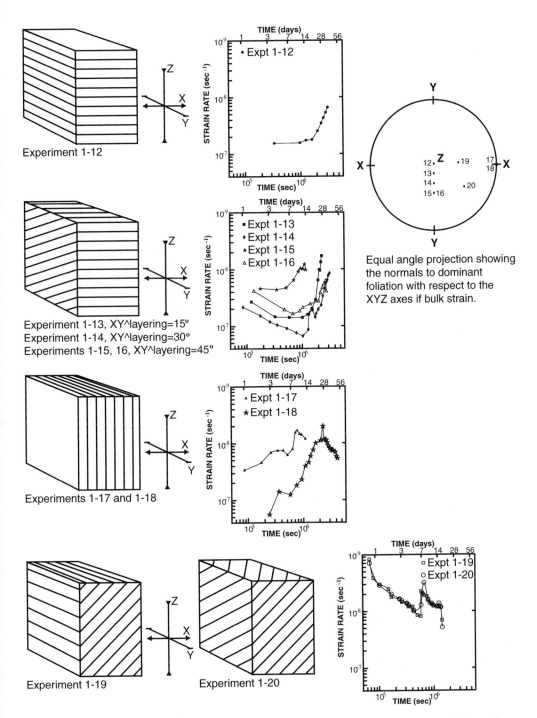

**Fig. 2.** Schematic representation of samples with respect to the bulk deformation axes *X*, *Y* and *Z*, together with strain rate versus time plots for the Series-1 experiments. The lower hemisphere of an equal area projection shows the poles to layering in experiments.

Jacka 1996) and at depth in an ice sheet, a single maximum or a small-circle girdle pattern generally develops, and this is compatible with the results of most ice flow models (Azuma & Goto-Azuma 1996). However, the contribution of inherited $c$-axis fabrics produced as a result of layer disturbances and the removal of the inherited $c$-axis fabric during deformation and subsequent grain growth is a problem that will be addressed in this paper.

## Experimental procedure

Two laboratory made ice types were used: (1) the majority of samples comprised anisotropic ice (Fig. 3) with alternating layers of elongate single crystals (area A in Fig. 3b) intergrown with fine-grained fibrous ice (area B in Fig. 3b) and (2) unlayered aggregates of polycrystalline ice (experiments 2-07, 2-09, 2-19). The platelets of anisotropic ice were prepared by alternating 5 mm layers with a film of near freezing water (0°C), and the boundary between each individual layer is defined by the presence of air bubbles (Fig. 3a). Such ice is similar to the multilayered columnar ice described by Gold (1963) and Wilson (1981). The experimental samples were cut from these slabs of layered ice, so that the layering made a precise angular relationship to the edge of the blocks (Fig. 2).

Two series of experiment were performed in order to study the distribution of deformation in the deforming anisotropic ice, and the $c$-axis fabric transitions with increasing strain in variably oriented samples. Layered samples shaped into a rectangular block measuring approximately 75 mm × 50 mm × 50 mm were used in the first series of experiments (Table 1; Fig. 2) deformed under pure-shear, plane-strain boundary conditions using the deformation apparatus described by Wilson (1982a). These are constant load (creep) experiments with initial loads of either 0.83 or 1.5 kN (Table 1). On the plane strain face ($XZ$) a set of circular strain markers were inscribed prior to deformation using the method described by (Wilson 1982a).

The second series of experiments (Table 2) used a creep deformation apparatus similar to that described by Li *et al.* (1996) that allows simultaneous application of compression and simple shear. Another nine tests were undertaken (Table 2) on rectangular oblong samples (70 mm × 15 mm × 37 mm) at 0.3 MPa to 15% strain at −5 and −2°C. Unfortunately due to a coldroom failure the final fabrics of the these samples could not be measured.

The effect of varying the inclination of this layering and hence $c$-axis distribution has a profound affect on the activation of intracrystalline slip processes and on the deformation rate. In the Series-1 experiments this was achieved by varying the initial orientation of the anisotropy, and the crystallographic preferred orientation patterns, to the bulk shortening ($Z$) and extension ($X$) axes (Fig. 2). Plane strain deformation was achieved by maintaining a constant length of the sample parallel to the ($Y$) axis (Fig. 2). These results were then compared to the Series-2 unconfined, combined compression and shear tests.

## Preferred orientation of starting material

Two distinct $c$-axis preferred orientation patterns occur in the columnar ice, the fine columnar grains have random $c$-axis orientation distribution (e.g. Fig. 4; experiments 1-14 or 1-18 and Fig. 5c, open-dots), whereas, the coarse grains intergrown with these have strong concentrations (Fig. 4; experiments 1-14 or 1-18, solid-dots). These patterns of preferred orientation are reproducible in both the $XZ$ and $YZ$ sections (Fig. 3d and e; experiments 1-13). Where strong $c$-axis concentrations exist they are generally orientated perpendicular to the layering (Fig. 3). Therefore the position of any initial $c$-axis concentration is governed by the angle the layering makes with the edges of the experimental block.

In most of the columnar ice samples a distinction on grain size was generally not made and all grains were plotted together (e.g. Fig. 3d–f all grains are represented as solid-dots). However, it

**Fig. 3.** Structural changes and fabrics associated with 10% shortening of the anisotropic columnar ice in the experiment 1-13. Shortening direction is vertical. (**a**) Thick $XZ$ section in plane polarised light showing the boundaries between the initial layers and deformation of the initially circular strain makers that are locally deformed into ellipses. (**b**) Thin-section in crossed nicols showing areas preserving original grain structures (areas A & B) areas of significant recrystallization occur in regions of higher strain (areas C–E). (**c**) Microstructural change associated with sample after annealing for 30 days at −1°C. (**d–f**) The solid dots represent the distribution of $c$-axis orientations; no attempt was made in this specimen to separate the fine from the coarse grains in the undeformed sample, and the fine recrystallized grains in the deformed sample were too fine-grained to measure successfully optically. However, the concentrations of $c$-axes relates to the elongate grains observed in both the undeformed and deformed ice as indicated by the dashed line in (f).

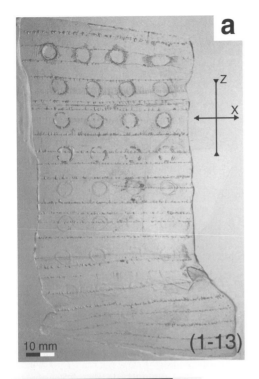

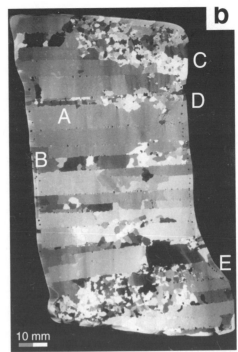

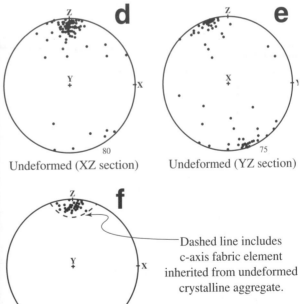

Undeformed (XZ section)

Undeformed (YZ section)

Dashed line includes
c-axis fabric element
inherited from undeformed
crystalline aggregate.

Deformed (XZ section)

**Table 1.** *Plane strain-compression experiments (Series-1) undertaken at* −1°C

| Expt no. | Type of starting material and load | Shortening (%) | Orientation of layering | Initial and (final) axial stress (MPa) |
|---|---|---|---|---|
| 1-12 | Columnar 0.83 (1.5) kN | 9.2 | Perpendicular to $Z$ and // to $XY$ plane | With load of 1.5 kN 0.63 (0.59) |
| 1-13 | Columnar 0.83 (1.5) kN | 9.1 | 15° to $Y$ in $YZ$ plane | 0.64 (0.52) |
| 1-14 | Columnar 0.83 kN | 9.1 | 30° to $Y$ in $YZ$ plane | 0.38 (0.30) |
| 1-15 | Columnar 1.5 kN | 11.0 | 45° to $Y$ in $YZ$ plane | 0.67 (0.66) |
| 1-16 | Columnar 0.83 kN | 8.5 | 45° to $Y$ in $YZ$ plane | 0.37 (0.33) |
| 1-17 | Columnar 0.83 kN | 10.1 | Parallel to $Z$ in $YZ$ plane | 0.37 (0.33) |
| 1-18 | Columnar 0.83 kN | 30.2 | Parallel to $Z$ in $YZ$ plane | 0.37 (0.26) |
| 1-19 | Columnar 0.83 kN | 29.3 | 45° to $Z$ in $XZ$ plane | 0.38 (0.28 top 0.34 bottom surface) |
| 1-20 | Columnar 0.83 kN | 26.9 | 45° oblique to $X$, $Y$ and $Z$ | 0.36 (0.3) |

Each experiment was conducted at a constant load comprising lead weights with initial loads of either 0.83 kN or 1.5 kN. Because there is elongation in the $X$ direction ($Y$ remains constant length), there is a change in the initial and final stresses (value in brackets) on the $XY$ surface of the specimen. In experiments 1-19 where buckling of the layering elongates the top surface there is a significant stress variation across the sample.

was noted that the initial undeformed $c$-axis distribution is a random distribution (in the fine grains) combined with a strong single maxima (in the coarse elongate grains). In the deformed samples (Figs 3–6) the distinction between the fine versus coarse grains is not made as grain boundary modification during the deformation precludes the identification of a pre-existing grain population.

By using the method of Wilson & Russell-Head (1982) for preparing the polycrystalline ice it was possible to produce an equiaxed polygonal ice grain aggregate with a random distribution of $c$-axes. The blocks of polycrystalline ice used in experiments 2-07, 2-09 and 2-19) were cut from blocks containing equiaxed grains with such randomly oriented $c$-axes and a grain size of <1.2 mm.

**Table 2.** *Compressive stress* ($\sigma$) *and octahedral shear stress* ($\tau$) *at* −5°C *in various combinations of polycrystalline and anisotropic ice aggregates and minimum strain rates in compression* ($\varepsilon_m$)

| | Experiment no. | Temperature (°C) | $\sigma$ (MPa) | $\tau$ (MPa) | $\varepsilon$ (%) | $\gamma$ | $\varepsilon_m$ |
|---|---|---|---|---|---|---|---|
| Polycrystalline | 2-07 | −5 | 0 | 0.4 | 0 | n/a | $1.9 \times 10^{-8}$ |
| | 2-09 | −5 | 0.22 | 0.4 | 2.4 | 0.08? | $1.2 \times 10^{-8}$ |
| | 2-19 | −5 | 0.22 | 0.4 | 5.7 | 0.25 | $1.6 \times 10^{-8}$ |
| Columnar parallel to $XY$ plane | 2-10 | −5 | 0 | 0.4 | 0 | 0.2 | $1.7 \times 10^{-7}$ |
| | 2-14 | −5 | 0 | 0.4 | 0 | 0.5 | $2.0 \times 10^{-7}$ |
| Columnar parallel to $XZ$ plane | 2-16 | −5 | 0.22 | 0.4 | 8.5 | 0.5 | $1.4 \times 10^{-8}$ |
| Columnar parallel to $YZ$ plane | 2-13 | −5 | 0.22 | 0.4 | 3.2 | 0.5 | $5.0 \times 10^{-8}$ |
| | 2-15 | −5 | 0.22 | 0.4 | 5.3 | n/a | $6.8 \times 10^{-8}$ |
| | 2-18 | −2 | 0.22 | 0.4 | 6.9 | 0.6 | $2.4 \times 10^{-7}$ |

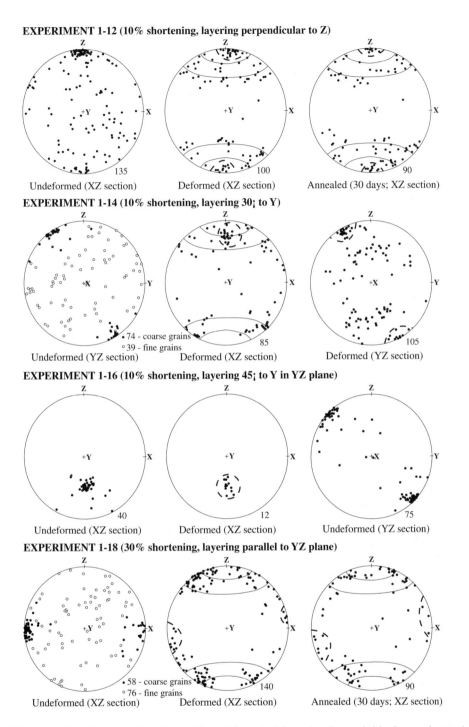

**Fig. 4.** Patterns of *c*-axis orientations observed in undeformed, deformed and annealed ice in experiments 1-12, 1-14, 1-16 and 1-18. Plotted as scatter diagrams on the lower hemisphere equal area stereographic net. Where grain types have been differentiated open-dots represent fine grains and filled-dots represent coarse grains (experiments 1-14 and 1-18). In all other diagrams grain types have not been differentiated (represented as solid dots). However, where there is a clustering of similar grain orientations this can usually be related to groups of relic grains inherited from the initial ice aggregate, these are outlined by the dashed lines. The superimposed small-circles are at 25° and 45° to the shortening axis *Z*. The number of *c*-axes is shown on the bottom right of each projection and the number of days any sample was annealed is shown in brackets.

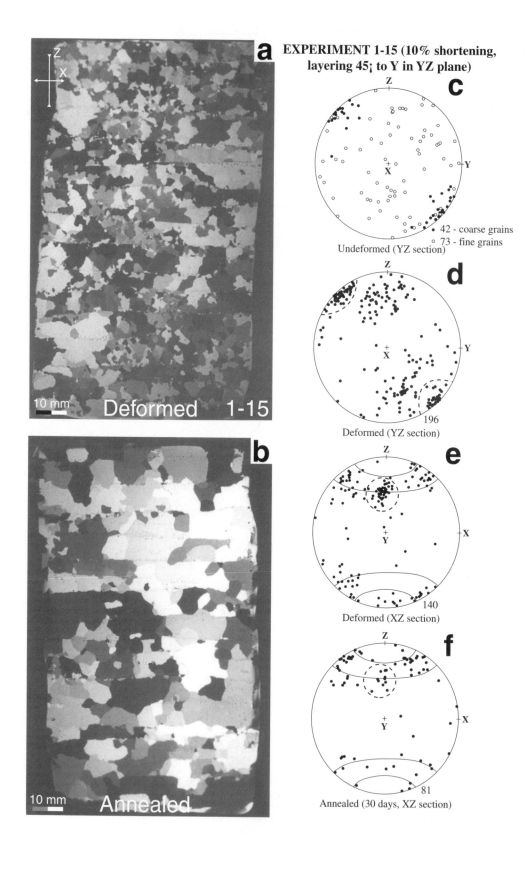

**a**

**b**

Deformed 1-15

Annealed

10 mm

10 mm

**EXPERIMENT 1-15 (10% shortening, layering 45¡ to Y in YZ plane)**

**c**

Z

Y

X

• 42 - coarse grains
○ 73 - fine grains

Undeformed (YZ section)

**d**

Z

Y

X

196

Deformed (YZ section)

**e**

Z

X

Y

140

Deformed (XZ section)

**f**

Z

X

Y

81

Annealed (30 days, XZ section)

## Results from plane strain-compression experiments

These experiments were intended to replicate the effect of a pure shear component that may be recognized on a small scale in variously oriented anisotropic ice samples within an ice stream (Fig. 1). Four sets of experiments were performed (Fig. 2). One with an initial anisotropy perpendicular to the shortening direction (experiment 1-12), experiments with layering inclined to the shortening direction (experiments 1-13, 1-14, 1-15, 1-16), experiments where the layering is parallel to the shortening direction (experiments 1-17, 1-18) and where the layering is inclined at 45° to the shortening axis (experiments 1-19, 1-20).

Where the layering was perpendicular to $Z$ (experiment 1-12) or tilted at less than 15° to the horizontal (experiment 1-13), there is layer-parallel simple shear (Fig. 3). Such a situation would be identified in a glacier where the layering sub-parallels the ice surface (Fig. 1). In these experiments there was only localized grain boundary migration within some of the finer grains of the columnar ice. Deformation was only induced by increasing the load from 0.83 to 1.5 MPa (Table 1; experiment 1-12) with deformation localized to the ends of the sample (e.g, areas C and E Fig. 3b) and in select zones (e.g. area D Fig. 2b) between coarse columnar ice. Strain rate in experiments 1-12 was initially constant and then decreased (Fig. 2) during the initiation of localized zones of recrystallization. Therefore at low stresses, the majority of grains are in hard-glide orientations, and the ice locally deforms accompanied by recrystallization. At higher stresses, subsequent cycles of recrystallization begin before the previous ones are finished, and the strain rate decelerates.

When the layering was inclined at 45° (experiments 1-15 and 1-16), as in the situation where the ice is folded (Fig. 1), there was an acceleration of strain rates during the initiation of numerous more zones of recrystallization (Fig. 5a). These areas represent regions in which extensive dynamic recrystallization occurred (the nucleation and growth of new strain free grains), between the larger elongate grains, that define the anisotropy, and resulted in the production of two grain populations. The initial acceleration of strain rate (Fig. 2) probably suggests that the shape of the flow curve depends on the activation of easy-glide orientations in the crystals. Again there was a constant deceleration of strain rate with time, which probably reflects a hardening of deformation during a process of ongoing recrystallization and new fabric development.

With layering parallel to the shortening axis (experiments 1-17 and 1-18) the strain markers are all elongate in a plane perpendicular to $Z$ with extension parallel to $X$ (in $XY$ plane). The folded layer boundaries contain numerous small, microscopic, parasitic flexures superimposed on a broad half-wavelength warping. The columnar ice structure disappears after 10% shortening, with the aggregate developing a duplex grain structure (illustrated by Wilson 1981) characterized by large deformed grains versus a smaller grain population of small, equant and polygonal recrystallized grains. The strain rate is initially decreased but as the aggregate becomes more recrystallized the rate increases. This suggests that once the easy-glide systems are rotated they are then activated, but the strain rate curve (Fig. 2) then depends on the proportion of recrystallized to starting grains.

In columnar ice samples skewed to the $XZ$ plane (Fig. 2; experiments 1-19, 1-20) the $c$-axis fabric within the layering is essentially in an easy glide orientation. In these samples there was always buckling of the layering (Fig. 6a) accompanied by extensive recrystallization (Fig. 6b) with the development of a duplex grain structure. This duplex grain structure was characterized by equiaxed grains, with high-angle boundaries replacing the old elongate grains (e.g. section parallel to the $XY$ plane; Fig. 6c). After an initial decrease in strain rates there were large increases followed by rapid deceleration of the type that may accompany the propagation of brittle fractures. However, no trace of brittle fractures were obvious in the final grain microstructure.

## Preferred orientation of deformed material

The fine grains related to the initial undeformed $c$-axis distribution is a random distribution (e.g.

---

**Fig. 5.** Structural changes associated with deformation and annealing of the anisotropic columnar ice in the experiment 1-15. Shortening direction is vertical and the superimposed small-circles are at 25° and 45° to the shortening axis $Z$. The dashed lines include grain populations that were inherited from the undeformed ice. (**a**) $XZ$ section in crossed nicols with recrystallized grains overgrowing the original coarse elongate grain fabric. (**b**) $XZ$ section in crossed nicols through sample after 30 days annealing. (**c–f**) Patterns of $c$-axis orientations observed in undeformed, deformed and annealed samples.

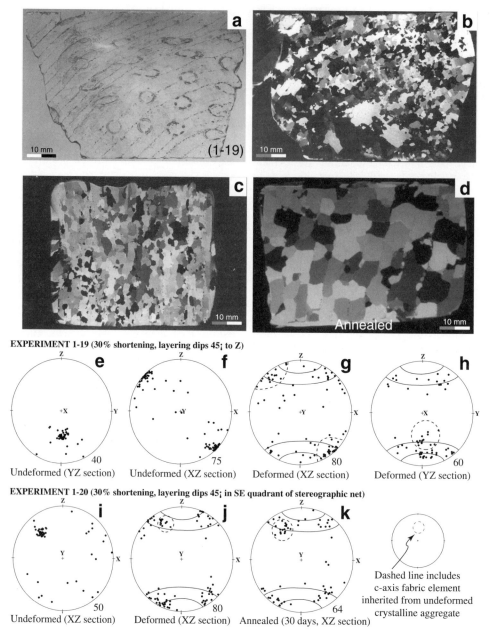

**Fig. 6.** Structural changes associated with 30% shortening and annealing of the anisotropic columnar ice in experiment 1–19 and fabric changes in experiments 1-19 and 1-20. Shortening direction is vertical and the superimposed small-circles are at 25° and 45° to the shortening axis Z. The dashed lines include grain populations that were inherited from the undeformed ice. (**a**) Thick $XZ$ section in plane polarized light showing buckled layering and distribution of strain as identified by the deformation of a set of initially circular strain makers. (**b**) $XZ$ section in crossed nicols with substantial recrystallization occurring in regions of higher strain; identified by transformation of circles into ellipses in (**a**). (**c**) Horizontal $YZ$ section in crossed nicols through the centre of experiment 1-19. (**d**) Microstructural change, in $XZ$ section, associated with sample after annealing for 30 days at $-1°C$. (**e–h**) Patterns of $c$-axis orientations observed in undeformed and deformed samples of experiment 1-19. In the deformed sections there is a small-circle distribution of $c$-axes that can be attributed to the new recrystallized grains and the preservation of the old host orientation pattern. (**i–k**) Patterns of $c$-axis orientations observed in undeformed, deformed and annealed samples of experiment 1-20. Note that in the deformed and annealed sections there is a small-circle distribution of $c$-axes that can be attributed to the new recrystallized grains with a small concentration that is probably an inherited host $c$-axis orientation pattern.

Figs 4–6) and this is combined with a strong single maximum in the elongate coarse grains, the latter grains define the layering. In these experiments this layering has been variably oriented with respect to the compression axis $Z$ (Fig. 2). In the situation where the strong maxima is parallel to $Z$ the ice is initially ductile, but with time becomes harder to compress (experiment 1-12 in Fig. 2) then there is a change in the $c$-axis distribution that accompanies the dynamic recrystallization. Similarly at low strains (10% shortening, experiment 1-12, 1-14 and 1-16 in Fig. 4) there is a strong inherited preferred orientation, from the initial maxima. This concentration exists with a small-circle $c$-axis distribution, with symmetry axis for the small-circle parallel to the compression direction; with a median $c$-axis opening angle of approximately 40. The small-circle pattern is symmetrically disposed with respect to the main foliation and the new grain structure. At higher strains (30% shortening, experiment 1-18) the skeletal outline of the pre-existing fabric, that is single point maxima with respect to the main foliation, is still preserved (see rotated concentration adjacent to $X$-axis). However, the symmetrical small-circle pattern represents a regime of coaxial flow that has been superimposed on the pre-existing fabric.

In experiments 1-18 to 1-20 there is flexural flow to produce the folds, accompanied by bulk grain rotation, induced by vertical compression. This produces a strong small-circle distribution of $c$-axes and dynamic recrystallization that has just about removed any sign of the earlier strong $c$-axis fabric (Fig. 6). As pointed out with the computer simulation of Zhang & Wilson (1997) there is very little difference in the final crystallographic fabric where the grain undergoes rotation in a pure shear or simple shear environment. Indeed the slip planes oriented in a domain of about 25° to the shear plane can rotate both towards and away from the shear plane in simple shear. It is also why the slip planes with initial orientations parallel to the bulk extension and shortening axes can still rotate in pure shear (Zhang & Wilson 1997). Therefore in a folded polycrystalline aggregate, grain interactions can significantly modify lattice rotation, as recrystallization accommodates incompatible deformation between neighbouring grains.

Therefore on the stereographic projection it can be seen that certain orientations are populated with increasing concentrations of $c$-axes as deformation proceeds. Specific areas can be delineated as the source areas, as these provide the initial orientations for developing the maxima and the small-circle distributions.

Since the relative extent of the source area determines the strength of the maximum, preferred orientation is affected by initial orientation distribution. Peculiar symmetry effects will result in the final fabric as a consequence of non-random starting populations. The inherited maxima still form in the same orientations, only the intensity distributed over the fabric skeleton is different.

## Modification of preferred orientation by annealing

Dynamic recrystallization affects both the grain-size distribution and the $c$-axis fabrics in the experimentally deformed ice. There was the development of a small-circle distribution about the shortening axis, together with a remnant concentration that can be attributed to the pre-existing fabric anisotropy. Therefore what are the effects of static annealing on the experimentally deformed ice aggregates?

Using the techniques described by Wilson (1982b) the starting materials were slabs of the deformed samples that were annealed in a silicon oil bath at $-1°C$ for periods of time up to 30 days. The deformed sample had an average recrystallized grain size of 0.9 mm within variously shaped grains of columnar ice (see Figs 3b, 5a and 6d). The annealing of the ice deformed at $-1°C$ produced irregular increases in grain size and grain shape (Figs 3c, 5b and 6d) with an obvious duplex grain structure, which in part reflects preservation of some host grains, produced in the less deformed samples (e.g. Fig. 3c). Where there has been extensive recrystallization an equiaxed texture is produced (Figs 5 and 6).

Scatter diagrams showing the pre-annealed orientations and annealed $c$-axis orientations are shown in Fig. 5. The number of grains decreases, particularly the single maximum pattern, but the texture is not altered under prolonged isothermal conditions without any imposed load. The shape of the deformed and annealed grains with the retention of the $c$-axis patterns, provide reason to believe that nuclei for the growth of grains came from the deformed state. Attributing large crystals with multi-maxima fabrics in glaciers purely to annealing recrystallization phenomena related to the preferred growth of nuclei in certain crystallographic orientations (Matsuda & Wakahama 1978) is not supported by this study. Instead such textures might reflect incomplete fabric development and the preservation of fabrics in a lower strained portion of the ice mass.

## Results from combined simple shear-compression experiments

In the laboratory prepared polycrystalline ice (experiments 2-07, 2-09 and 2-19) the initial grain size ranged from 1.1 to 1.6 mm, but was substantially reduced to a mean grain-size of 0.9 mm by recrystallization and nucleation of new grains during deformation. The creep curves for these polycrystalline ice tests are shown as plots of octahedral strain rate as a function of octahedral strain (Fig. 7 and Table 2). These demonstrate that, under these test conditions, the isotropic octahedral strain rates (i.e. the strain rates down to minimum) are very similar in all polycrystal-

line ice samples and comparable to the results of Li et al. (1996) and Li & Jacka (1998). During this process the ice sample shows a marked transition on loading, during which the creep rate decreases to a minimum, then increased with strain to a steady-state tertiary value factor of about approximately 10 greater than the minimum. The crystal orientation fabric was similar to the plane strain-compression experiments being a small-circle with symmetry axis parallel to the compression direction and opening angle of approximately 40°. This compares favourably with the results of Li et al. (1996) and Li & Jacka (1998) who have undertaken similar experiments on the same type of apparatus.

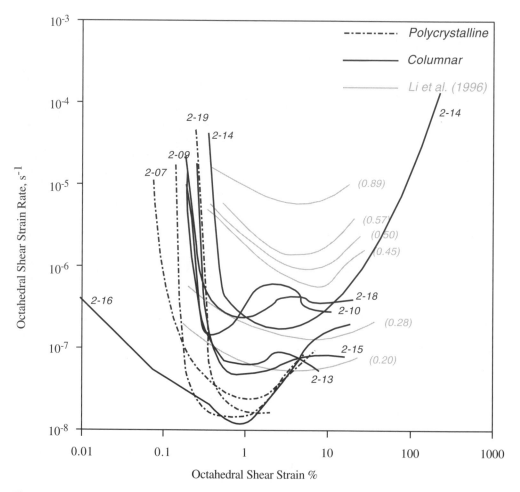

**Fig. 7.** Creep curves for the Series-2 experiments. Log–log plots of octahedral strain rate as a function of octahedral strain for the deformation tests on layered polycrystalline ice, columnar ice and the isotropic ice. All experiments were undertaken at −5°C except for experiment 2-18 that was undertaken at −2°C. The isotropic ice tests are those of Li et al. (1996) undertaken at −2°C and at various octahedral shear stress (MPa), indicated by the number beside each creep curve.

The strain rate response of the horizontally layered anisotropic ice (experiment 2-10) decreased to a minimum value with a factor c. 12 times greater than the minimum isotropic strain rate. This was then followed by a further decrease in strain rate as deformation was localized at the base of the sample in a region that was accompanied by a localized grain-size decrease, parallel to the horizontal layering (Fig. 8a). The adjacent anisotropic grains remained in an easy-glide orientation throughout the deformation, with the development of intergranular deformation features such as undulose extinction (Fig. 8a). This certainly supports the idea that it is easier to localise the shear horizontally in a glacier. In experiment 2-14 there was also

an initial decrease in strain rate, under conditions where there is only a shear strain and no compression (Table 2). In this situation recrystallization is localised within the horizontal layering. Further strain is accompanied by a rapid acceleration of strain rate (Fig. 7) and the development of folds in the layering (Fig. 8b) together with brittle fractures that eventually develop into thrusts, parallel to the shear zone boundary (Fig. 8b).

In the samples with the layering parallel to the XZ plane (Fig. 4; experiment 2-16) there is a hardening process (Fig. 9) that occurs during the initial deformation with a slow decrease to a minimum strain rate (Fig. 7). This hardening process in the early stages of the deformation is

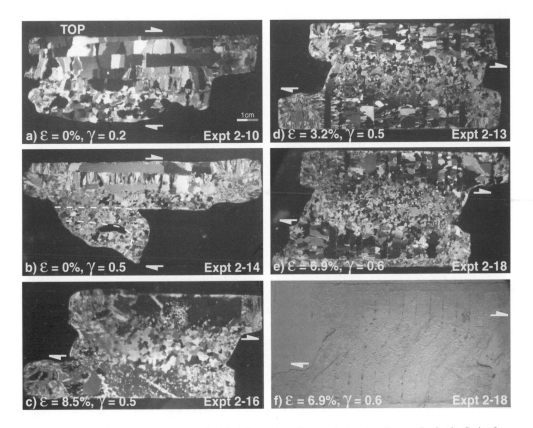

**Fig. 8.** Microstructural changes associated with deformation of the anisotropic columnar ice in the Series-2 experiments in sections parallel to XZ. (**a**) Horizontal layering with compression perpendicular and shearing parallel to layering (in crossed nicols). Bar scale is 1 cm and same scale for other photographs. (**b**) Top-half of sample preserved after failing in a brittle manner. Strain was initially localized in the centre of the sample and accompanied by extensive recrystallization. This zone then became folded and thrusted during the localization of the shear strain in the horizontal layering. The broken white lines indicate the thrust boundary and the location of the folded zone of recrystallized grains. (**c**) Vertical layering initially parallel to XZ (in crossed nicols). (**d & e**) Vertical layering initially parallel to YZ (in crossed nicols). (**f**) Vertical layering initially parallel to YZ in plane polarized light showing rotation of layering.

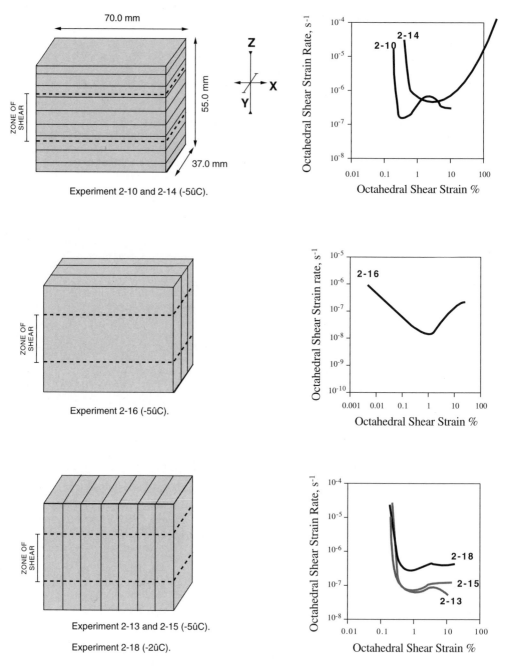

**Fig. 9.** Schematic representation of Series-2 layered samples with respect to the bulk deformation axes $X$, $Y$ and $Z$, together with representative creep curves. The compression (applied to the upper surface) and simple shear are applied simultaneously with the shear strain is confined to a $c$. 15 mm wide zone, bounded by an undeformed region where the sample is frozen between two support boxes. The size dimensions shown in experiments 2-10 and 2-14 is applicable for all samples.

probably a reflection of the hard glide orientation of the ice crystals in this anisotropic sample. However, the strain rate increase after the minimum by a factor of about 10 probably reflects the complete recrystallization of the anisotropic ice aggregate within the zone of shearing Pettit

The strain curves where the vertical layering initially parallels *YZ* (Fig. 9; experiments 2-13 & 2-15 at $-5°C$) all show a marked decrease in the creep to a minimum, then an increase with strain followed by a further decrease or a steady state (experiment 2-15). This probably reflects the amplification of the buckles (Figs 8d and e) in the layering that is occurring as dynamic recrystallization is accelerated by decrease in grain size and the development of grain orientations that are at a maximum critical resolved shear stress. In these experiments the final decrease in the creep rate, as the sample undergoes further shear, is invariably followed by brittle failure with the propagation of a thrust oriented 30° to the shear zone boundary.

The effect of increased temperature on the mechanical behaviour is markedly reflected in comparing experiment 2-15 (at 5°C) with 2-18 (at $-2°C$) where strain rates decrease to a minimum value that is a factor of *c*. 12 greater than the minimum isotropic strain rate. However, there is little notable difference in microstructure and grain-size between the $-5°C$ and $-2°C$ experiments (Fig. 8d and e).

## Discussion

Single crystals of ice have a strong mechanical anisotropy, as glide on the basal plane is at least two orders of magnitude easier than other crystallographic systems (Duval *et al.* 1983; Wilson & Zhang 1994). Therefore with prolonged deformation the individual crystals, in an aggregate, tend to rotate or recrystallize through a process of strain localization (Wilson & Zhang 1994), to produce an orientation more appropriate for basal glide. During this process it is noticeable that *c*-axes rotate either towards the principal compressive stress or away from the principal tensile stress to form small-circle distribution about the shortening axis.

If there is a localization of strain on a macroscopic scale there may well be the development of alternating domains of high-strain versus low-strain. With sites of recrystallization in the high-strain regions and a preservation of the pre-existing *c*-axis fabric in the low-strain regions as described by Tison *et al.* (1994) and Alley *et al.* (1997). The effects of such deformation and the role of the *c*-axis fabric on the

mechanical behaviour of a polar ice mass and vice versa is uncertain. Although fabric reorientation is towards forming a *c*-axis small-circle distribution the effect of annealing and accompanying grain growth is to preserve the pre-existing *c*-axis preferred orientation.

The stress and strain rate conditions used in the two series of deformation tests are different, but both demonstrate the marked effect of anisotropy in comparison to isotropic ice. In the isotropic polycrystalline samples, with randomly oriented crystals, the mechanical behaviour of individual crystals negates each other, and the aggregate deforms in a manner that follows Glen's Law (Li *et al.* 1996). However, the Glen Law cannot predict the onset or development of anisotropic instabilities at small scales, nor can it accurately characterise the ice flow pattern where a uniform but strong bulk anisotropic fabric develops. The polycrystalline ice (Fig. 7) shows well-developed minimum strain rates and the curves follow the form and the results presented by Li *et al.* (1996). The variously oriented columnar ice samples show similar strain rates in the early stages of deformation (except experiment 2-16) and reach a minimum at shear strains of <1, but the point of inflection occurs sooner where the anisotropy in the ice allows easy glide (experiments 2-10 and 2-14). In hard glide orientations (experiment 2-16) the minima are reached very much later, but where there is buckling (experiments 2-13, 2-15, and 2-18) significant variations in behaviour is obtained.

In the majority of the combined compression and simple shear experiments graphs of strain rate versus time (total strain) reach minimum creep rates (Fig. 7), which are reached after a strain of about 10%. In isotropic ice strain rate goes through a minimum and increases steadily regardless of stress levels and temperature (Li *et al.* 1996). The horizontally layered ice is considered to be an analogue of what is happening at the base of a glacier, that is with ice crystals that have an orientation favouring deformation in shear parallel to the base. Then in contrast to the polycrystalline tests (experiments 2-07; 2-09; 2-19) and most other laboratory tests (Budd & Jacka 1989; Li *et al.* 1996) the extent of the strain rate minimum, if zones of strain localization are initiated, is shorter in terms of the total strain.

In the plane strain experiments the strain rate variations change with shortening and are a function of layer dip. The results are probably more realistic than the more conventional Series-2 experiments as the rapid changes in strain rate simulate the localization of deformation and 'stick slip'-type displacements. During which strain is localized into new sites and would be equivalent to the ice-quakes that are often recorded in glaciers. It can also be interpreted that movement was achieved via flexural slip along many basal planes within individual anisotropic layers, as well as a certain amount of flexural flow that produced the buckles and folds. In applying these observations to a polar ice sheet it can be seen that the relative strain between layers is not constant. It is related to a complex deformation history, where the strain rate varies with the change in geometry of the anisotropic layering.

Comparing the microstructural development between the two series of experiments there are notable differences. The Series-1 experiments did not have a major component of shear strain, with the microstructural development primarily controlled by recrystallization occurring parallel or within the layering (experiments 1-12 to 1-16) or related to the buckling of the layering (experiments 1-17 to 1-19). In all cases the contribution of inherited grains to both the microstructure and final fabric is significant. In the Series-2 experiments shear strain was localized to a zone of recrystallization (Fig. 8) that were bounded by folded and buckled regions preserving the original layered ice.

## Conclusions

The presence of a strong pre-existing $c$-axis fabric in an ice mass has a major influence and profound effect on strain rates, the nature of the creep curve and the nucleation of small scale instablities. In particularly the dominance of grains with an easy-glide versus hard-glide orientation pattern will either decrease or increase the strain rate. Similarly the creep curve is further complicated if there is the development of folds or thrusts during the localization of deformation. The final strength of the $c$-axis fabric also depends on contributions from the pre-existing fabric particularly where strain becomes localized in recrystallized regions. Such high-strain zones may alternate with domains preserving pre-existing grain structures and $c$-axis fabrics. Therefore secondary $c$-axis maximum, in a bulk sample, may continue to exist at high-strains and may not be reflecting textures generated during the deformation.

Although folding and thrusting have long been reported on a macroscale (e.g. Hudleston 1983), their occurrence on a microscopic scale in natural ice masses is generally unknown except for an example described by Alley et al. (1997)

and Tison et al. (1998). In such cases, irregularities and local $c$-axis fabric variations may suggest their existence. If the ice undergoes flexural slip and folding then there is a rotation of $c$-axes towards the normal to the shear plane (the vertical in polar glaciers), which further strengthens the fabric and so increases the deformation rate. Folding facilitates the development of the fabric with recrystallization accommodating incompatible deformation between neighbouring grains and by eliminating grains unfavourably oriented for basal slip in the shear direction.

Annealing to produce large crystals preserves pre-existing recrystallized fabrics, but may weaken the inherited fabric from the older deformed grains. Therefore in natural polar ice (e.g. Thorsteinsson et al. 1997) local variations in $c$-axis patterns from the generally observed small-circle or single maximum patterns may reflect the existence of a pre-annealing fabric. Such variations may be attributable to local disturbances in the layering, but little evidence of this is preserved in the grain microstructure.

H. Sim is thanked for his enthusiastic help in undertaking the Series-2 experiments and preparing the figures. The manuscript has benefited from thoughtful reviews by J.-L. Tison and E. Pettit. Financial support from Australian Research Grant A39601139 is gratefully acknowledged.

## References

ALLEY, R. B., GOW, A. J., MEESE, D. A., FITZPATRICK, J. J., WADDINGTON, E. D. & BOLZAN, J. F. 1997. Grain-scale processes, folding, and stratigraphic disturbances in the GISP2 ice core. *Journal of Geophysical Research*, **102**, 26 819–26 830.

AZUMA, N. & GOTO-AZUMA, K. 1996. An anisotropic flow law for ice-sheet ice and its implications. *Annals of Glaciology*, **23**, 202–208.

BOUCHEZ, J. L. & DUVAL, P. 1982. The fabric of polycrystalline ice deformed in simple shear: Experiments in torsion, natural deformation and geometrical interpretation. *Textures and Microstructures*, **5**, 171–190.

BUDD, W. F. & JACKA, T. H. 1989. A review of ice rheology for ice sheet modelling. *Cold Regions Science and Technology*, **16**, 107–144.

CASTELNAU, O., SHOJI, H., MANGENEY, A., MILSCH, H., DUVAL, P., MIYAMOTO, A., KAWADA, K. & WATANABE. O. 1998. Anisotropic behaviour of GRIP ices and flow in central Greenland. *Earth and Planetary Science Letters*, **154**, 307–322.

DAHL-JENSEN, D., THORSTEINSSON, T., ALLEY, R. & SHOJI, H. 1997. Flow properties of the ice from the Greenland Ice Core Project ice core: The reason for folds. *Journal of Geophysical Research*, **102**, 26 831–26 840.

DUVAL, P., ASHBY, M. F. & ANDERMAN, I. 1983. Rate-controlling processes in the creep of polycrystalline ice. *Journal Physics and Chemistry*, **87**, 4066–4074.

GLEN, J. W. 1955. The creep of polycrystalline ice. *Proceedings of the Royal Society*, **A228**, 513–538.

GOLD, L. W. 1963. Deformation mechanisms in ice. *In*: KINGERY,W. D. (ed.) *Ice and snow*. M.I.T. Press, Cambridge, Mass., 8–27.

GOW, A. J. & WILLIAMSON, T. 1976. Rheological implications of the internal structure and crystal fabrics of the West Antarctic ice sheet as revealed by deep core drilling at Byrd Station. *Geological Society of America Bulletin*, **87**, 1665–1677.

HUDLESTON, P. J. 1983. Strain patterns in an ice cap and implications for strain variation in shear zones. *Journal of Structural Geology*, **5**, 455–463.

HUGHES, T. H. 1998. *Ice Sheets*. Oxford University Press, New York.

KAMB, B. 1972. Experimental recrystallisation of ice under stress. *In*: HEARD, H. C., BORG, I. Y., CARTER, N. L. & RALEIGH, C. B. (eds) *Flow and Fracture of Rocks*. AGU Geophysical Monograph Series, **16**, 211–241.

LI, J. & JACKA, T. H. 1998. Horizontal shear rate of ice initially exhibiting vertical compression fabrics. *Journal of Glaciology*, **44**, 670–672.

——, —— & BUDD, W. F. 1996. Deformation rates in combined compression and shear for ice which is initially isotropic and after the development of strong anisotropy. *Annals of Glaciology*, **23**, 247–252.

MATSUDA, M. & WAKAHAMA, G. 1978. Crystallographic structure of polycrystalline ice. *Journal of Glaciology*, **21**, 607–620.

MARMO, B. A. & WILSON, C. J. L. 1999. A verification procedure for the use of FLAC to study glacier dynamics and the implementation of an anisotropic flow law. *In*: DETOURNAY, C. & HART, R. (eds) *FLAC and Numerical Modeling in Geomechanics*. Balkema, Rotterdam, 183–190.

POIRIER, J. P. 1980. Shear localisation and shear instablity in materials in the ductile field. *Journal of Structural Geology*, **2**, 135–142.

THORSTEINSSON, T. J., KIPFSTUHL, J. & MILLER, H. 1997. Textures and fabrics in the GRIP ice core. *Journal of Geophysical Research*, **102**, 26 583–26 599.

THWAITES, R. J., WILSON, C. J. L. & McCRAY, A. P. 1984. Relationship between bore hole closure and crystal fabrics in Antarctic ice core from Cape Folger. *Journal of Glaciology*, **30**, 171–179.

TISON, J.-L., LORRAIN, R. D., BOUZETTE, A., BONDESAN, A & STIÉVENARD, M. 1998. Linking landfast sea ice variablity to marine ice accretion at Hells Gate Ice Shelf, Ross Sea. *In*: *Antarctic Sea Ice: Physical processes, interactions and variablity*. AGU Antarctic Research Series, **74**, 375–407.

——, THORSTEINSSON, T., LORRAIN, R. D. & KIPFSTHUL, S. 1994. Origin and development of textures and fabrics in basal ice at Summit, Central Greenland, *Earth and Planetary Science Letters*, **125**, 421–437.

WILSON, C. J. L. 1981. Experimental folding and fabric development in multilayered ice. *Tectonophysics*, **78**, 139–159.

——1982a. Fabrics in polycrystalline ice deformed experimentally at −10°C. *Cold Regions Science and Technology*, **6**, 149–161.

——1982b. Texture and grain growth during the annealing of ice. *Textures and Microstructures*, **5**, 19–31.

—— & RUSSELL-HEAD, D. S. 1982. Steady-state preferred orientation of ice deformed in plane stain at −1°C. *Journal of Glaciology*, **28**, 145–160.

—— & ZHANG, Y. 1994. Comparison between experiments and computer modelling of plane strain simple shear ice deformation. *Journal of Glaciology*, **40**, 46–55.

ZHANG, Y. & WILSON, C. J. L. 1997. Lattice rotation in polycrystalline aggregates and single crystals with one slip system: a numerical and experimental approach. *Journal of Structural. Geology*, **19**, 875–885.

# The stress distribution related to the boudinage of a visco-elastic material: examples from a polar outlet glacier

BRETT A. MARMO & CHRISTOPHER J. L. WILSON

*School of Earth Sciences, The University of Melbourne, Parkville, Victoria 3052, Australia*
*(e-mail: bmarmo@mail.usyd.edu.au; cjlw@myriad.its.unimelb.edu.au)*

**Abstract:** Quantitative stress measurements related to the development of mesoscale structures in rock are difficult, if not impossible. A method for determining the stress distribution and history during the development of mesoscale boudinage structures in ice is introduced here. Boudinage structures in fracture trace ice have been observed in the outlets glaciers of the Framnes Mountains, east Antarctica. Fracture traces are preserved when crevasses fill with surface melt water, which freezes to form coarse-grained columnar ice. A 4.0 km flow-parallel traverse across the Central Ice Stream of the Framnes Mountains is presented to illustrate the boudinage of fracture traces with a mean width of 0.30 m. Field-based measurement of the geometric evolution of boudinage structures has been combined with surface flow rate measurements to quantitatively determine the strain rate at which the boudinage structures formed. The strain rate measurements provide boundary constraints on several two-dimensional finite difference models that have been used to analyse the stress distribution related to the formation of boudinage structures in a visco-elastic solid. The results reveal the development of pressure-shadows during the boudinage of layered rocks, and demonstrate the degree of refraction of stress across rheological boundaries, which have important ramifications for the analysis of planar and linear fabrics in rocks.

The magnitude, orientation and history of stress fields associated with the development of mesoscale and macroscale structures in the Earth are not well understood. When a deviatoric stress is applied to a layered rock mass, the local stress distribution is complicated by the refraction of the stress field by the rheological differences between layers. Variations in the geometry of the layered rock associated with the development of structures, such as folds, shear zones and boudins, leads to localized temporal variations of the magnitude and orientation of the stress field. The extremely complex histories of the stress fields during deformation is of great interest to geologist as it controls the development of the linear and planar fabrics used in structural measurements, and produces local pressure variations that control fluid flow.

Glacier ice and rock at high metamorphic grade deform according to the same non-linear flow laws. Ice is therefore an ideal analogue for the study of crustal deformation. Folds, faults, boudinage structures and shear zones can be observed in glaciers at both the mesoscale and macroscale. It is also possible to measure directly the strain rate associated with the development of structures in glaciers. The integration of strain rate measurements with constitutive flow laws and material parameter allows one to determine the distribution and history of the stress field. This method is demonstrated in this paper with finite difference models of boudinage structures observed in the Framnes Mountains outlet glaciers.

Boudinage is an example of the role played by rheological contrasts in the development of structures in layered materials. Boudinage occurs when deformation becomes localized within a more competent layer. It leads to an acceleration of the strain rate, and the progressive reduction in the width of the layer to form a pinch-and-swell structure and eventually to the break-up of the layer into individual boudins. Problems related to strain localisation in a layered medium are generally considered in terms of the relative viscosity between the layers. However, the deformation of layered ice and rocks at temperatures close to their respective melting points is best described by a non-linear power law of the form $\dot{\varepsilon} = A\sigma^n$, where $A$ and $n$ are constants, so that the relative viscosity is dependent on the strain rate (Treagus 1993). Thus, the relative viscosity between layers is not constant, but is related to the complex deformation history where the strain rate varies with the progressive change in geometry of the boudinage structures. The non-

*From*: MALTMAN, A. J., HUBBARD, B. & HAMBREY, M. J. (eds) *Deformation of Glacial Materials*. Geological Society, London, Special Publications, **176**, 115–134. 0305-8719/00/$15.00 © The Geological Society of London 2000.

constant viscosity ratio makes an analytical approach to the study of the development of boudins in a non-linear material impossible, and a numerical method must be used.

Boudinage studies have in general followed either a geometric approach, where mesoscale boudins in crustal rocks are related to the bulk strain (Flinn 1961; Ramsay 1961), or an analogue approach where soft materials such as wax, plasticine and cheese (Ramberg 1955; Neurath & Smith 1982), or rocks (Paterson & Weiss 1968) are deformed at known strain rates and the stress field is inferred. A third approach is to embed rigid objects in a photo-elastic material so that the stress field can be observed and analysed (Strömgård 1973). While these approaches have been extremely valuable, it is difficult to relate them to natural deformation due to differences in the geometry, rheology and/or the time scale. This paper studies the development of boudinage structures in competent fracture traces that are hosted by relatively incompetent bubbly blue glacier ice. Fracture traces form when open crevasses fill with surface water and/or snow

which freezes to form coarse-grained columnar ice. When fracture traces are elongated subparallel to their length, metre-scale boudins form that look the same as those observed in high-grade metamorphic rocks. A 4 km traverse parallel to a flow line in the Framnes Mountains, east Antarctica (Fig. 1), reveals the way that boudinage structures develop with respect to time. This presents us with a unique opportunity to study the boudinage of a crystalline non-linear material as the rheology and strain rate are well constrained, and the flow line represents an unbroken sequence of structural development.

## The nature of glacier ice and fracture trace ice

The bubbly blue glacier ice is exposed in the ablation zones of the outlet glaciers. A well-defined foliation has developed in the ice in high shear strain areas. The foliation, $S_1$, is composed of interleaved sub-vertical compositional layers of dark and light blue ice. The layers

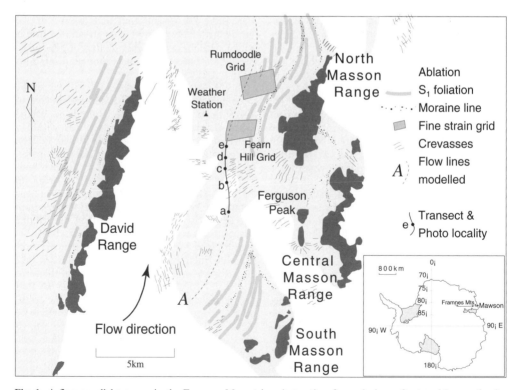

**Fig. 1.** A flow-parallel traverse in the Framnes Mountains, Antarctica, from virgin, unfractured ice south of a large crevasse field to the south west corner of the Fearn Hill strain grid where well-developed boudinage structures are exposed. The lower case letters show the location of the photographs in Fig. 2, that illustrate the progressive boudinage of crevasse traces. Inset: The location of the Framnes Mountains.

are hundreds of millimetres to tens of metres wide and are generally parallel to flow. In some areas, the foliation is isoclinally folded with an axial surface that is parallel to the flow direction. However, in the area under discussion, the $S_1$ foliation parallels the flow direction, and has been used as a guide for the longitudinal traverse.

The compositional layers that compose the $S_1$ foliation are defined by a variation in the grain size and bubble content of the bubbly blue ice. Generally, the grains in the bubbly blue glacial ice are 5–10 mm in diameter and are equigranular, euhedral with a hexagonal habit. The equigranular texture is expected to continue to significant depth below the surface as processes such as recrystallization and polygonization tend to produce a uniform grain size (Budd & Jacka 1989). Bubbles in the light layers of the $S_1$ foliation are $c$. 1 mm in diameter and evenly distributed with $c$. 5 mm between each bubble. The darker bands have noticeably fewer bubbles which are randomly dispersed through the ice (separated by 1–20 mm) and the bubble size varies greatly from 1 mm to 10 mm.

Fracture traces form when open crevasses fill with snow or, more commonly, surface melt water, which re-freezes to form coarse-grained translucent columnar ice. Crystals grow from each fracture wall and join in the centre, often trapping air bubbles to form a suture plane parallel to the fracture wall. Some bubbles are also trapped between neighbouring grains as they grow away from the fracture walls, to produce elongate air bubbles perpendicular to the walls of the trace. These textures are identical to those in quartz or carbonate syntectonic veins in rocks (Durney & Ramsay). The coarse columnar trace ice appears dark blue, and strikingly different to the light, bubbly blue glacial ice that hosts them. The relative bubble content, intercrystalline geometry and the bulk crystallography of fracture traces, also give the trace ice a very different rheology to that of the bubbly blue host ice.

New ice crystals begin to nucleate on the existing crystals in the fracture walls once a crevasse becomes filled with water. As the crystals grow on the walls, they encounter neighbouring crystals and their growth parallel to the wall is impeded. The growth rate of crystals is many orders of magnitude higher that the local strain rate ($10^{-9}$ s$^{-1}$), effectively producing static conditions so that crystal growth is dependent only on intercrystalline surface tensions. Wilson (1994) demonstrated that static conditions allow face-free growth of crystals away from the fracture walls to produce elongated crystals that have well-defined dihedral angles at the terminus of each crystal. Ketcham &

Hobbs (1967) demonstrated that surface tension effects between neighbouring grains growing in face-free growth conditions favour crystals with $c$-axes perpendicular to the growth direction. Thus, the great majority of grains in a fracture trace are oriented with $c$-axes parallel to the fracture walls. This is referred to as the crystal growth fabric of the fracture trace.

If a stereographic projection of $c$-axes is considered, with the pole to the fracture wall at the zenith, a large population of $c$-axes plots on a small-circle on the periphery of the projection, with an angle of $c$. 90° to the zenith, and a smaller population of $c$-axes will plot close to the normal to the fracture wall (Fig. 3). This distribution of $c$-axes is very similar to the pervasive fabric that forms in polycrystalline ice under uniaxial tension, with the unique stress direction orthogonal to the fracture wall.

## The development of boudinage structures along a flow-parallel traverse

The numerical models used to study the development of boudinage structures are based on structural observations from a 4 km traverse parallel to flow, beginning 4.5 km west of Ferguson Peak, in the Central Masson Range, and ending at the south west corner of the Fearn Hill strain grid (Fig. 1). A particle on the surface would take $c$. 180 years to travel the length of the traverse, based on observed surface velocities (Marmo & Dawson 1996; Marmo & Wilson 1998). During this period the ice moves through a crevasse field, where fracturing is predominantly orthogonal to flow, then fracture traces are formed, and are subsequently deformed to produce boudinage structures. The average ablation rate, measured in the nearby Fearn Hill grid, is 0.1 m a$^{-1}$ (Marmo & Dawson 1996), so $c$. 18 m of ice is exhumed along the length of the traverse.

The longitudinal traverse began in an area where the surface was composed of virgin unfractured, glacier ice (Fig. 2a). The ice had either not been fractured previously, or all the fracture traces that formed up-flow had annealed. As the strain rate used in the numerical models is constrained by the geometry of fracture traces, and the distance that the traces travelled, it is important to know where the traces were formed. All fracture traces that were observed down-flow must have formed within the crevasse field, as no fracture traces were passively transported into the crevasse field.

The crevasse field is approximately 1.5 km long parallel to flow, and is composed of fractures that

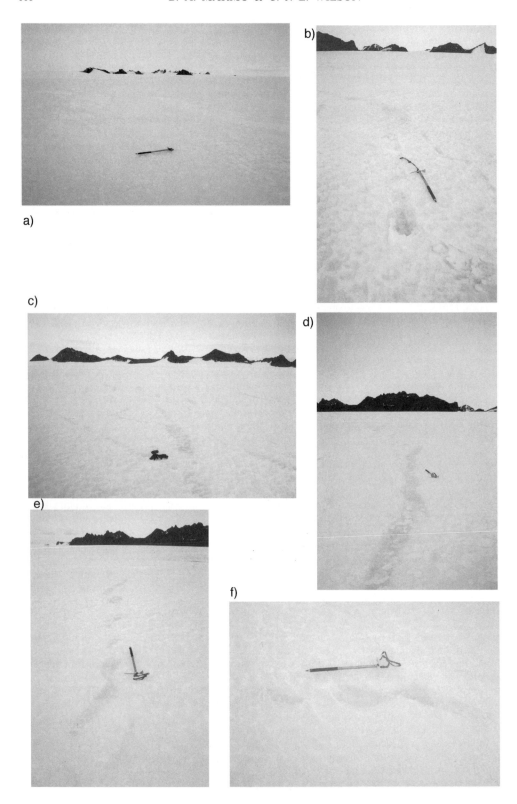

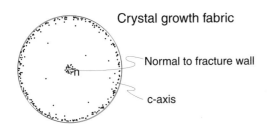

**Fig. 3.** A stereographic projection of a growth fabric where the zenith is parallel to the normal to the fracture wall. Surface tension effects favours the growth of crystals with *c*-axes either perpendicular to the fracture wall, or parallel to the fracture wall, over the growth of crystals with *c*-axes oblique to the wall. This leads to the development of a pervasive fabric in fracture trace ice.

are 0.2–0.4 m wide and can be over 50 m long. The crevasses tend to be linear at the surface with some meter scale jogs at *c*. 15° which have sharp, well defined corners. The uppermost meter of any crevasse becomes filled with snow in the days and weeks after the fractures open (Fig. 2b). The percolation of water from surface melt appears to fill the crevasses at greater depths.

In the area directly down stream from the last observed open crevasse, the crevasse traces had an identical habit to open fractures within the crevasse field (Fig. 2c). The mean trend of the traces was 080°. The traces directly north of the crevasse field tend to have a cloudy appearance, with no well-defined sutures or crystal structure. This is attributed to the uppermost part of the fracture consisting of a different ice type that is produced by the coalescence of snow and percolated water. Therefore, the crystals do not grow from each of the fracture walls to produced an interlocking aggregate as observed deeper in fracture traces.

Five hundred meters further north of the crevasse field, all the traces had become slightly attenuated and had a general trend of 070°. The corners between linear sections of the fractures and small jogs had become rounded and were

no longer well defined (Fig. 2d). Ablation had removed the section of trace formed by snow coalescence, and the traces were clear with well-defined crystal structures, sutures and elongate bubbles that extended from the wall of the fracture to the suture.

Just south of the Fearn Hill grid, and 950 m down-flow from the crevasse field, well-defined boudins were recognized (Fig. 2e). Individual boudins were approximately a metre long and had asymmetric shapes with thick centres and attenuated tips. Some traces also had well-defined pinch-and-swell structures (Fig. 2e). The necking regions of pinch-and-swell structures tended to be cloudy, which suggests that the crystals had been re-organized and the well defined bubble structure had also been disturbed.

## Geometric constraints to the model

The analysis of the stress history related to boudinage formation can be determined using numerical methods if the strain rate can be determined and the constitutive equations that relate the strain rate to stress are known. Thus, the stress history can be studied if the strain rate history is determined using field-based observations. The geometric development of boudinage structures from the northern extent of the crevasse field to the end of the traverse has been used to provide the geometric constraints. The numerical models are based on the deformation that occurs along the last kilometre of the traverse.

The surface velocity is constrained by the Fearn Hill grid which lies at the northern extent of the traverse and moves $21 \pm 0.3\,\mathrm{m\,a^{-1}}$ (Marmo & Wilson 1998). The surface velocity of the Fearn Hill area was measured using differential GPS methods in the 1994–95, 1995–96 and 1997–98 austral summers. A seasonal flow rate variation was observed with an increase during summer in the order of $0.2\,\mathrm{m\,a^{-1}}$. Marmo- (1999) demonstrated with a series of numerical models

**Fig. 2.** The progressive development of fracture trace boudins. The locality of each photograph is shown in Fig. 1. (**a**) Virgin, unfractured ice. Any fracture trace ice formed in crevasse fields up-flow have annealed, leaving clean bubbly blue ice. The view is due south towards the Southern Masson Range (visible on the horizon) with a meter long ice axe in the foreground. (**b**) A snow-filled crevasse in the centre of the crevasse field. The snow has become crusty and has begun to coalesce to form a crevasse trace. The ice axe is 1 m long. (**c**) Undeformed crevasse traces immediately down-stream from the crevasse field. The view is west towards the David Range with gloves in the foreground. (**d**) Pull-apart areas have begun to form in the jogs of the fracture trace. The view is due east with the Northern Masson Range in the background and a meter long ice axe in the middleground. (**e**) Boudinaged crevasse trace. The relative transportation of boudins indicates a sinistral sense of movement with extension parallel to the boudins. The view is east towards the North Masson Range and the ice axe is 1 m long. (**f**) Close up of the pull-apart structure in the foreground of (e).

**Table 1.** *Properties of ice used to model ice sheets adapted from Frost & Ashby (1982)*

| Bulk modulus | $K$ | $9.31 \times 10^9$ Pa |
|---|---|---|
| Shear modulus | $G$ | $3.57 \times 10^9$ Pa |
| Density | $\rho$ | $917\,\text{kg}\,\text{m}^3$ |

*Power-law creep parameters for glacial ice*

| Flow parameter | $A$ | $7.5 \times 10^{-25}\,\text{s}^{-1}\,\text{Pa}^{-3}$ |
|---|---|---|
| Flow parameter | $n$ | 3 |
| Activation energy | $Q$ | $1200\,\text{cal}\,\text{mol}^{-1}$ |
| Gas constant | $R$ | $8.31\,\text{J}\,\text{K}^{-1}\,\text{mol}^{-1}$ |

*Power-law creep parameters for trace ice*

| Flow parameter | $A$ | – |
|---|---|---|
| Flow parameter | $n$ | 3 |
| Activation energy | $Q$ | $1200\,\text{cal}\,\text{mol}^{-1}$ |
| Gas constant | $R$ | $8.31\,\text{J}\,\text{K}^{-1}\,\text{mol}^{-1}$ |

that there in a slight reduction of surface flow rate up-stream. To simplify the numerical models the temporal and spatial variation of flow rate along the flow line are ignored and the velocity is assumed to be a constant $21.0\,\text{m}\,\text{a}^{-1}$.

The bulk extensional strain perpendicular to the flow line, $\varepsilon_{yy}$, and the shear strain in the horizontal plane, $\delta_{xy}$, have been determined by the geometric analysis of boudinage structures (Fig. 4). The shear strain is given by the angle between the $S_1$ foliation and the crevasse traces or the general trend of boudins. The effects of bulk rotation have been removed by relating all structures to the $S_1$ foliation. Directly down-stream from the crevasse field, the undeformed crevasses trend $80°$ and the $S_1$ foliation trends $354°$, making an angle of $86°$ (Fig. 4a). Pinch-and-swell structures can be observed $510\,\text{m}$ down-flow from the crevasse field and the general trend of crevasse traces was $070°$ while the $S_1$ foliation had a trend of $001°$ and make an angle of $69°$ (Fig. 4b). Thus, the angle between the crevasse traces and the $S_1$ foliation had been reduced by $17°$ and $\delta_{xy} = 0.296$ radians. The segments between pull-apart regions had a trend of $086°$ and made an angle of $85°$ with $S_1$, so that they had only rotated about $1°$ from their original orientation.

Individual boudins were recognized $970\,\text{m}$ down-stream from the crevasse field, where they had a general trend of $060°$ and the $S_1$ trend was $005°$, making an angle of $55°$. Thus, the shear strain from the undeformed crevasse trace locality to the boudinage locality was $\delta_{xy} = 0.541$ radians and from the pinch-and-swell locality it was $\delta_{xy} = 0.244$ radians (Fig. 4c).

An estimate of the extensional strain perpendicular to flow, $\varepsilon_{yy}$, was determined by measuring the sum lengths of individual boudins that had formed from one crevasse trace and com-

paring it to the length across which the boudins were distributed:

$$\varepsilon_{yy} = \frac{l_1 - l_0}{l_0}$$

where $l_1$ is the length that the boudins were distributed across and $l_0 = \sum_{i=1}^{n} l_i$ where $l_i$ is the length of each boudin. By using this method, the extensional strain between the undeformed crevasse traces and the boudins is $\varepsilon_{yy} = 1.6$. This estimate of bulk strain is a slight underestimate, as the internal deformation of the fracture trace is not considered.

## An anisotropic flow law for polycrystalline ice

Single crystals of ice have an extremely strong mechanical anisotropy as glide is around two orders of magnitude easier on the basal plane, than in non-basal glide systems. If a polycrystalline aggregate of ice has a random orientation of crystals then the anisotropy of individual crystals negates each other, and the aggregate deforms in an isotropic manner to any applied stress. Isotropic ice is generally modelled using Glen's Law:

$$\dot{\varepsilon}_{ij} = A\sigma_{ij}^n \exp\left(\frac{-Q}{kT}\right) \qquad (1)$$

where $\dot{\varepsilon}_{ij}$ is the strain tensor, $\sigma_{ij}$ is the deviatoric stress tensor and $A$ and $n$ are flow parameters and $n = 3$ for polycrystalline ice. Glen's Law is only applicable for isotropic ice as the relation is independent of the orientation of the stress regime.

However, if the aggregate has a preferred crystallographic fabric, Glen's Law breaks down as the fabric produces a strong mechanical anisotropy and the strain rate is no longer independent of the orientation of the applied stress. A geometric term based on the mean orientation of $c$-axes in the polycrystalline aggregate must be introduced to equation 1 so that the strain rate varies with the orientation of any applied stress. To simplify the approach it is common to assume that single crystals deform by basal glide only (Azuma 1994; Azuma & Goto-Azuma 1996). This method will be followed in the anisotropic model presented in this paper.

For the case when glide is assumed to occur on basal systems only, the resolved shear stress on the basal plane of any crystal under uniaxial compression or tension, can be related by a

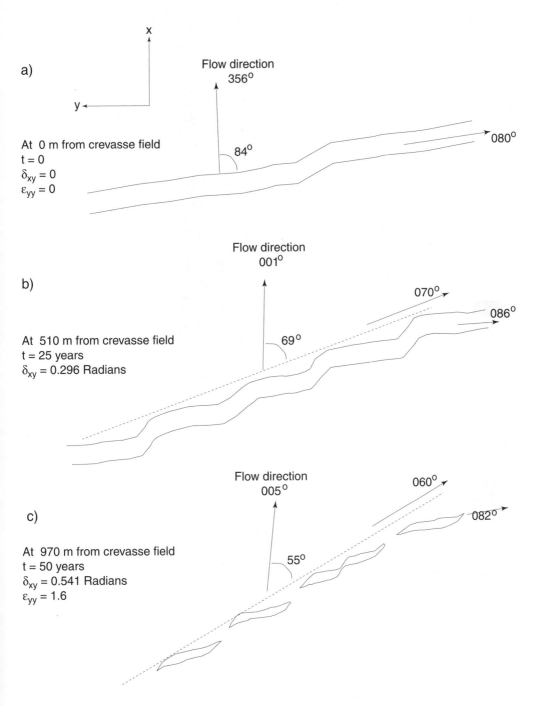

**Fig. 4.** The geometric development of boudins used to determine strain. (**a**) An undeformed crevasse trace immediately down-flow from the crevasse field. (**b**) The development of pinch-and-swell structures 510 m down-flow from (**a**). (**c**) Individual boudins 970 m down-flow from (**a**).

geometric function referred to as the Schmid Factor, $S_g$ (Azuma & Higashi 1985):

$$S_g = \cos \phi_0 \sin \phi_0 \qquad (2)$$

where $\phi_0$ is the angle between the c-axis and the unique stress axis. The Schmid factor for an aggregate as a whole, $\bar{S}$, is given by the mean distribution of c-axes relative to the unique stress axis:

$$\bar{S} = \frac{1}{N_T} \sum_{g=1}^{N_T} S_g \qquad (3)$$

When the c-axes are randomly oriented $\bar{S} = \frac{1}{3}$, for pure shear and $\bar{S} = \frac{3}{10}$ for simple shear, and when a fabric is completely developed such that all the c-axes are inclined at $45°$ to the compressive stress axis, $\bar{S} = \frac{1}{2}$.

$$\dot{\varepsilon} = B_0 \bar{S}^{(n+1)} \sigma^n \exp\left(\frac{-Q}{kT}\right) \qquad (4)$$

The geometric factor introduced to Glen's Law acts as a scalar quality on the flow parameter $A (A = B_0 \bar{S}^{(n+1)})$. As $n = 3$ for polycrystalline ice, the strain rate varies with the fourth power of the Schmid factor. A fully developed fabric $(\bar{S} = \frac{1}{2})$ therefore enhances the strain rate by about five times under uniaxial compression, compared to an isotropic aggregate $(\bar{S} = \frac{1}{3})$.

Azuma & Goto-Azuma (1996) have generalized this anisotropic flow law to deal with triaxial stress fields, by introducing a geometric tensor term, $\mathbf{G}$, that acts on the stress tensor in equation 1:

$$\dot{\varepsilon} = B_0 \mathbf{G}^{sym} (\mathbf{G} : \sigma)^n \exp\left(\frac{-Q}{kT}\right) \qquad (5)$$

As in the uniaxial stress relationship, the strain rate varies with the geometric term to the fourth power (for $n = 3$). The geometric tensor, $\mathbf{G}$ in equation 5, which describes a crystallographic fabric where the c-axes lie in a great circle on a stereographic projection, such as is the case with the crystal growth fabric in a fracture trace is:

$$\mathbf{G} = \begin{bmatrix} \frac{1}{2} & 0 & 0 \\ 0 & -\frac{1}{6} & 0 \\ 0 & 0 & -\frac{1}{6} \end{bmatrix} \qquad (6)$$

where the x-axis is perpendicular to the fracture wall. This is very similar to a fabric that would form after prolonged uniaxial tension in the x direction. Azuma (1994) estimates that a fabric with a zenith angle of $90°$ would form after extension of 82%, where the zenith angle is

measured from the zenith to the small circle that contains the c-axes. This fabric produces a very strong mechanical anisotropy where the flow is enhanced if tension acts perpendicular to the fracture wall, while the fabric is very resistant to tension and shear stresses that act parallel to the fracture wall. Azuma & Goto-Azuma (1996) report that the flow enhancement goes towards zero when a shear stress acts on a small circle fabric with a zenith angle of $90°$, and the shear stress is applied to the same plane as that which contains the c-axes. This observation based on numerical simulation of the fabric is in good agreement with the experimental observations of Budd & Jacka (1989).

## The finite difference model

The bulk strain rate determined from field observation has been imposed on a finite difference mesh that deforms according to a non-linear visco-elastic flow law to represent deformation by glide-controlled creep in polycrystalline ice. A fracture trace is represented in the finite difference mesh by several layers of zones that have different rheological properties to the surrounding bubbly blue glacier ice. A flow law in the form of equation 1 is used to model deformation in the bubbly blue glacier ice and an anisotropic flow law in the form of equation 5 is used to model deformation in the fracture trace.

The surface layer of the glacier along the traverse is assumed to have a random crystallographic fabric, as grain boundary migration processes would have remove any pre-existing fabric in the surface layer produced by deformation in the accumulation zone (Budd & Jacka 1989). Marmo & Wilson (1999) showed that the surface ice along the traverse is being transported in a passive manner, with little deformation in the ablation zone, and has therefore retained the random bulk crystallographic fabric. The rheology of the fracture trace was determined by comparing the results of different numerical models with observed structures. The effects of the crystallographic fabric have been incorporated into the numerical models, and follows the approach of Azuma (1994) and Azuma & Goto-Azuma (1996). Once the nature of the rheological contrast between the trace ice and the glacial ice is determined, the models are used to study the nature of strain localization and the stress history related to the development of boudinage structures.

The stress and strain histories related to the development of both pinch-and-swell structures, and boudins are examined using FLAC (fast Lagrangian analysis of continua). FLAC (Detournay & Hart 1999) is a continuum code so it is not possible to rupture the finite difference mesh to represent the transition from a pinch-and-swell structure to boudins without introducing unreasonable numerical errors. The models do not include this transitional phase of the deformation, and have instead been separated into two parts. The first model studies the development of pinch-and-swell structures in a fracture trace, and the second model begins with individual boudins and analyses the deformation history related to their progressive separation.

## The starting geometry for ice models

A $50 \times 26$ finite difference mesh was configured to represent the geometry of fracture traces down-flow from the crevasse field (Fig. 5). The numerical problem has a horizontal centre of symmetry, so only the eastern side (right half in plan) is modelled to reduce computational time. An associative flow law is used to model deformation so the initial geometry of the fracture trace must have a perturbation in it, otherwise the trace will deform uniformly and no necking region will develop. In nature the perturbation may be a rheological heterogeneity produced by a difference in bubble content or crystallographic geometry, or it may be a jog in the initial crevasse. The numerical models include a jog in the undeformed fracture trace, similar to those observed in the field. The geometry of the trace is represented in the numerical model by six rows of grid elements that extend from the left boundary to the centre of the mesh. The fracture trace is kept a large distance from any of the active boundaries so edge effects do not affect the analysis of the deformation. Fracture traces observed in the field terminate at a single point, so the end of the trace in the numerical model is tapered (Fig. 5).

## Bulk strain rate and boundary conditions for models

The bulk strain rate determined from the field-based observations is imposed on the model by assigning velocities to the nodes on the boundaries of the mesh. The imposed boundary velocity is fixed for each calculation cycle so that it is independent of internal deformation. The imposed boundary velocity sets up a velocity field within the mesh that represents the local strain rate field.

The passive transportation of the fracture traces due to glacial flow (in the x-direction)

has been removed from the model so that all velocities are relative to the point where the centre of the fracture trace meets the left boundary (Fig. 5). All the imposed boundary velocities are graduated along each of the boundaries so that no erroneous strain occurs in the corners of the model (Fig. 5). The shear strain rate, $\dot{\delta}_{xy}$, is represented by imposing a velocity in the x-direction on the right boundary of the model and by graduating it along the upper and lower boundaries such that the nodes at the upper left and lower left corners have no shear contribution. The extension perpendicular to flow, $\dot{\varepsilon}_{yy}$, is represented by imposing a velocity in the negative y direction on the right boundary.

The surface area of the ice is assumed to be constant because no fractures were observed north of the crevasse field. The extension in the y-direction must be balanced by compression in the x-direction. The compression parallel to flow, $\dot{\varepsilon}_{xx}$, varies with time as the shape of the grid changes, so it must be re-calculated with each time step according to:

$$u_t = \frac{-vA}{(l + vt)^2} \qquad (7)$$

where $u_t$ is the velocity at any time in the x-direction, $v$ is the velocity in the y-direction which is imposed on the model, $l$ is the initial length of the lower boundary and $t$ is time. The velocity, $u$ is divided in two and applied to the upper and lower boundaries (Fig. 5). The temperature regime in polar outlet glaciers is extremely complicated with seasonal and diurnal variations and contributions from shear heating. To simplify the models the temperature fluctuations are ignored and the average temperature for the area, 264 K is assumed to be constant.

## Material parameters for ice used in models

The development of boudins in ice is dependent on the rheological contrast between bubbly blue glacier ice and fracture trace ice. The material parameters used to describe the rheology of bubbly blue ice were determined by Marmo & Wilson (1999) to be $A = 7.5 \times 10^{-25}\,\mathrm{s^{-1}\,Pa^{-3}}$ and $n = 3$. However, the rheology of the fracture trace ice is not well known, and it is strongly dependent on the anisotropic growth fabric. The same elastic properties are used for both the glacier and the trace ice, and the flow parameter $n$ is assumed to equal 3 for two types of ice. Thus the flow parameter $A$ controls the rheological difference between glacial and trace ice (Table 1). The final geometry of boudins produced by several numerical models with different

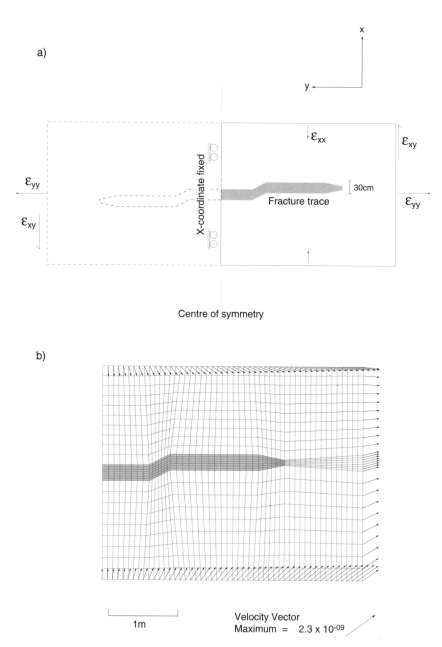

**Fig. 5.** The initial geometry of the finite difference mesh with the boundary conditions illustrated. (**a**) The initial geometry with the centre of symmetry and the strain rates that are applied to the boudaries illustrated. (**b**) The strain rate is converted into velocities that are applied to the nodes on the boundary of the model.

values of $A$ is compared to the geometry of pinch-and-swell structures observed in the field (Fig. 2) to determine an appropriate value for the flow parameter. The flow parameter $A$ for the trace ice must be significantly less than $A = 7.5 \times 10^{-25}\,\mathrm{s}^{-1}\,\mathrm{Pa}^{-3}$ otherwise the trace ice and glacier ice would deform at the same strain rate, and boudinage structures would not develop. This provides the upper limit to the possible value of $A$ for the fracture trace ice.

## Crystal growth fabric in the fracture trace ice

The majority of the grains that form the columnar ice in fracture traces have their $c$-axes aligned parallel to the fracture wall, due to surface tension effects during the growth of crystals from each wall. The mechanical anisotropy due to the growth fabric has been included in the numerical models by following the approach of Azuma & Goto-Azuma (1996) (eqn 5). The anisotropic flow law is incorporated into the boudinage models by assuming that the elongate nature of ice crystals perpendicular to fracture walls restricts the rotation of individual crystals so that the mechanical anisotropy is fixed relative to the fracture walls. Thus, the orientation of the local stress relative to the crystalline fabric is dependent on the relationship between the fracture wall and the stress field. The mesh has been configured such that rows of grid nodes are initially parallel to the fracture wall. Deformation of the fracture trace will result in the rotation of the crystallographic fabric. The local rotation of any section of the crystallographic fabric within the fracture trace is represented by the rotation and deformation of individual grid zones. It is, therefore, possible to determine the effect of the crystallographic fabric on flow by monitoring the angle between each grid zone and the maximum tensile stress.

The geometric tensor, $G$ (eqn 6) in equation 5, which describes the crystallographic fabric of a fracture trace can be enacted in the finite difference code by writing it in terms of an ellipse and then relating the direction of the maximum tensile stress to the orientation of each grid zone. The ellipse is considered to lie with its long axes (which equal $\frac{1}{2}$) perpendicular to the fracture wall, and the small axis ($\frac{1}{6}$) parallel to the fracture wall. If $\alpha$ is the angle between the maximum tensile stress and any mesh zone (small axes of the ellipse) then the enhancement of flow will be given by:

$$S = \frac{1}{2}\sqrt{\left(\frac{1 + \tan^2 \alpha}{9 + \tan^2 \alpha}\right)}. \tag{8}$$

The geometric factor is used as a scalar which acts on the flow parameter $A$ following equation 4, such that $A = B_0 S^4$, where $B_0$ is a constant. Thus, if the tensile stress is parallel to the grid zone and in the plane of the $c$-axis fabric, then the resistance to flow is a maximum and $A = B_0(\frac{1}{6})^4$, while if the tensile stress is perpendicular to the grid zone then the resis-

tance is a minimum and $A = B_0(\frac{1}{2})^4$. Writing the geometric tensor in terms of an ellipse allows it not to act on the deviatoric stress tensor, which greatly improves the efficiency of the numerical code. The angle between the maximum tensile stress and each grid zone in the fracture trace is calculated every time step and the flow parameter $A$ is re-adjusted to emulate the effect of the anisotropic fabric.

## Rheology of fracture trace ice

The value of the flow parameter $B_0$ (eqn 4) for the fracture trace was varied between $2.5 \times 10^{-23}\,\mathrm{s}^{-1}\,\mathrm{Pa}^{-3}$ ($A = 1.9 \times 10^{-25}$ when $S = \frac{1}{6}$) and $5.0 \times 10^{-24}\,\mathrm{s}^{-1}\,\mathrm{Pa}^{-3}$ ($A = 3.8 \times 10^{-25}$ when $S = \frac{1}{6}$) and geometry of the pull-apart structure, after ten years, was compared to those observed in the field. As no brittle fractures were observed in the area between the crevasse field and the Fearn Hill grid, the models are invalid if the tensile stress exceeds the tensile strength of ice, 1.0 MPa (Marmo 1999).

The variation in deformation style with an increase in the rheological contrast between the trace and host is shown in Fig. 6. As the value of $B_0$ for the fracture trace increases, strain is partitioned into the host bubbly blue ice to a greater extent. The pull-apart region for each of the models has a similar length after 10 years. The pull-apart region in the traces with lower values of $B_0$ also has better developed pinch-and-swell structures, which are highly sheared as observed in the field (Fig. 6c and d). The value for $B_0$ in the columnar trace ice is therefore less than $B_0 = 5.0 \times 10^{-23}\,\mathrm{s}^{-1}\,\mathrm{Pa}^{-3}$ (where $A = 3.86 \times 10^{-26}\,\mathrm{s}^{-1}\,\mathrm{Pa}^{-3}$ when $S = \frac{1}{6}$). Models that contain trace ice where $B_0$ was less than $5.0 \times 10^{-24}\,\mathrm{s}^{-1}\,\mathrm{Pa}^{-3}$ ($A = 3.86 \times 10^{-27}\,\mathrm{s}^{-1}\,\mathrm{Pa}^{-3}$ when $S = \frac{1}{6}$) developed tensile stresses greater than 1.0 MPa in the pull-apart region. This stress would be high enough to nucleate a brittle fracture that would propagate through both the fracture trace and the host ice. As no brittle fractures were observed down flow from the crevasse field, the value for the flow parameter $B_0$ must be greater than $5.0 \times 10^{-24}\,\mathrm{s}^{-1}\,\mathrm{Pa}^{-3}$. Thus, the value of $B_0 = 2.5 \times 10^{-23}\,\mathrm{s}^{-1}\,\mathrm{Pa}^{-3}$ best approximates the rheology of fracture trace ice, as models with this value produce a pinch-and-swell geometry that best matches those observed in the field, while the tensile stresses are not great enough to produce brittle fractures. For the value $B_0 = 2.5 \times 10^{-23}\,\mathrm{s}^{-1}\,\mathrm{Pa}^{-3}$, the flow parameter $A$ in equation 1 has a range of value that vary between $A =$

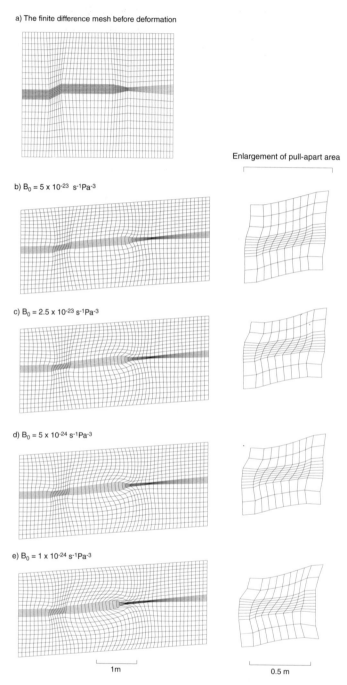

**Fig. 6.** Deformation of four grids with different rheologies to determine the most suitable value for the flow parameter $A$ for fracture trace ice. Ten years of deformation are shown. The selection of the most appropriate rheology for the fracture trace is based on the geometry of the pull-apart area which is shown enlarged on the right. (**a**) The starting grid for all the models. Boundary conditions are shown in Fig. 5. (**b**) to (**e**) The finite difference mesh after ten years of deformation, where the flow parameter $B_0$ for the fracture trace is; (**b**) $5 \times 10^{-23}\,\text{s}^{-1}\,\text{Pa}^{-3}$, (**c**) $2.5 \times 10^{-23}\,\text{s}^{-1}\,\text{Pa}^{-3}$, (**d**) $5 \times 10^{-24}\,\text{s}^{-1}\,\text{Pa}^{-3}$ and (**e**) $1 \times 10^{-24}\,\text{s}^{-1}\,\text{Pa}^{-3}$.

$1.93 \times 10^{-26} \, s^{-1} \, Pa^{-3}$ when the principal tensile stress is parallel to the fracture trace $(S = \frac{1}{6})$ and $A = 1.56 \times 10^{-24} \, s^{-1} \, Pa^{-3}$ when the principal tensile stress is oriented perpendicular the fracture trace $(S = \frac{1}{2})$.

## The geometric development of the boudinage structures

The geometric development of the pinch-and-swell and boudinage structures in the numerical

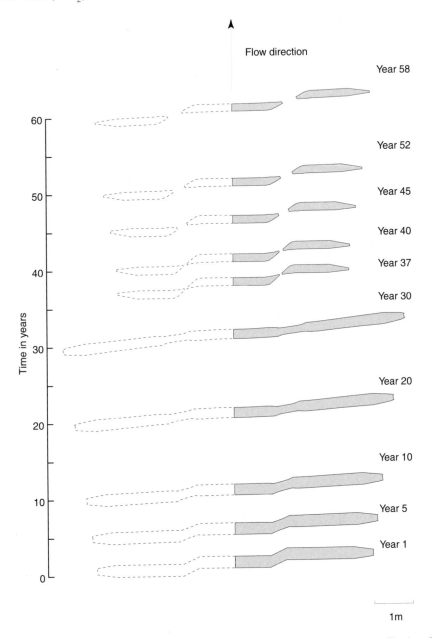

**Fig. 7.** The geometric development of boudins in numerical models, where $B_0 = 2.5 \times 10^{-23} \, s^{-1} \, Pa^{-3}$ for the fracture trace. The change in the geometry of the fracture trace section of the finite difference mesh is shown, with its reflection (dashed) that has not been modelled due to the plane of symmetry in the numerical problem. Years 1 to 30 show the progressive development of the geometry in a single model. The model was stopped at Year 37 and 40 due to mesh problems in the pull-apart area between the boudins.

models is shown in Fig. 7. The sequence from zero to 30 years shows the development of a pinch-and-swell structure in one continuous numerical model. The numerical representation of the separation of individual boudins has been divided into three separate models, as the initial separation requires a very fine mesh that is not capable of large strains before the zones become too attenuated and the mixed descretisation process breaks down. To avoid the geometric breakdown of the grid, a model with 70 mm between boudins was run to represent two years of deformation before being stopped. Several new grid zones were added to the pull-apart area and the model was started again with a distance of 100 mm between boudins. The second model was stopped after a further two years of numerical time, and more finite difference zones added to the pull-apart area. The final model was then run to represent 12 years of deformation with an initial distance 150 mm between boudins. Each geometric result in Fig. 7 has a year assigned to it which represents the period of numerical deformation, where year zero shows the undeformed fracture trace.

The small jog in the fracture trace provides a perurbation in the layering of the ice and acts as a site for the initiation of a pull-apart structure. After ten years the centre of the jog is visibly narrower than the surrounding fracture trace (Fig. 7). Extension parallel to the fracture trace has straightened the jog area after 20 years, but the area continues to narrow further and deformation is partitioned into the host ice. After 30 years the pull-apart region is approximately three quarters the width of the rest of the fracture trace, which is significantly greater than the width of necking regions observed in the field that have deformed for the same period of time. The elongation of the pull-apart area increased linearly at the rate of $0.0169 \, s^{-1}$, which was approximately two and half times faster than the remainder of the trace, which extended at $6.46 \times 10^{-3} \, s^{-1}$.

Once the individual boudins separate they are passively transported away from each other and slightly elongated, though the geometry does not change significantly (Fig. 7). The elongation rate of the fracture trace increases slightly to $7.12 \times 10^{-3} \, s^{-1}$ and remains constant. The elongation of the pull-apart area increases by almost three times to $0.0475 \, s^{-1}$ when the boudins first separate, then declines to $0.0352 \, s^{-1}$ by the fortyfifth year and becomes constant.

## Stress distribution

When the fracture trace was transported passively away from the crevasse field, into an environment with extension sub-parallel to the length of the trace, the stress was localized within the trace close to the jog. A maximum differential stress of 0.82 MPa occurred at the inside inflection points of the jog, while the opposite inflection point had a relatively low differential stress of 0.26 MPa (Fig. 8a). The flank regions of the trace and the centre of the jog had a stress of $c$. 0.6 MPa, while stress was less than 0.4 MPa at the tips of the fracture trace. Two relatively high stress areas occurred in the corners of the tip. However, these appear to be due to the mesh geometry not tapering out to a single point as was observed in the field. The differential stress in the host, bubbly blue ice was two to three time less than that for the majority of the fracture trace. A region of relatively high stress in the host propagated across the jog in the fracture trace, and was joined to another high stress zone that formed a butterfly pattern at the tip of the fracture trace (Fig. 8a).

The zones of relatively high stress became less localized after ten years and the maximum differential stress declined further to 0.76 MPa, the adjacent low stress area increased to 0.36 MPa and the zone of relatively high stress in the host had begun to disperse (Fig. 8b). By the twentieth year of deformation the traces became elongated, so that the jog no longer existed and the stress distribution changed dramatically (Figs 7 & 8c). The two zones of localized high stress were replaced by a broad region of relatively high stress ($>0.7$ MPa) that propagated through the pull-apart region (Fig. 8c). The zone of relatively high stress in the host, that crossed the pull-apart region disappeared, and the high stress zone at the tip of the trace became more localized. After 30 years the maximum stress in the trace was 0.71 MPa and was localized to the centre of the pull-apart region. The relatively high stress zone at the tip became more localized (Fig. 8d).

Once the necking region separated into individual boudins, the stress in the host increased by $c$. 0.1 MPa, and was localized in the region between the tips of the boudins (Fig. 8e). Initially the host accommodated up to 0.32 MPa in the necking region, while areas of relatively low stress ($<0.1$ MPa) occurred close to the flanks of the boudins (Fig. 8e). Zones of low stress also occurred up-stream and down-stream of the necking region in the bubbly blue ice. The maximum differential stress in the boudinaged traces occurred at the tips and was 0.78 MPa and the minimum differential stress of 0.30 MPa occurred on the outer edge of the trace and close to the necking area (Fig. 8e). The area of minimum stress in the boudinaged trace

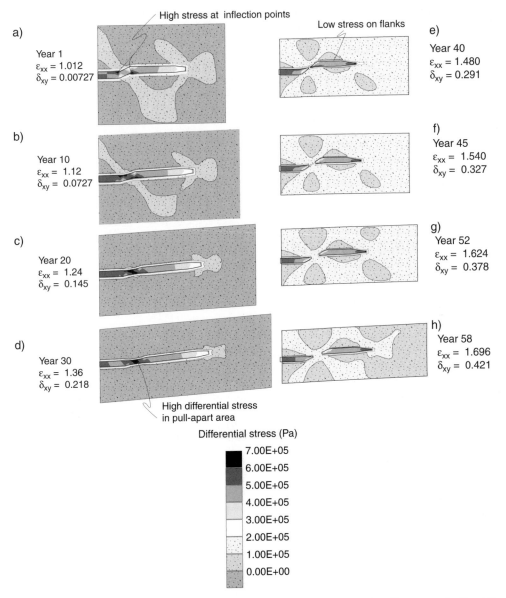

**Fig. 8.** Differential stress distribution associated with the development of pinch-and-swell structures (**a–d**) and the separation of boudins (**e–f**).

correlates with the minimum stress area in the pinch-and-swell models that occurred on the outside inflection point of the jog (Fig. 8a and e). The stress distribution in the trace ice generally remained unchanged as the boudins progressively separated, though the maximum stress at the tips of each boudin reduced slightly (Fig. 8f & g). However, after the fiftyeighth year, the stress distribution in the host ice did become pro-

gressively more localized into two broad bands that crossed in the necking region (Fig. 8h).

## Rheological control on the orientation the stress field

The bulk extensional strain produced a stress field that was oblique to the fracture trace in areas of the host ice that were some distance from the fracture trace. In the area surrounding the trace, and within the trace itself, the orientation of the stress field was controlled by the geometry of the rheological contrast between the fracture trace ice and the bubbly blue ice. The principal tensile stress in the undeformed fracture trace tended to parallel the boundary of the fracture trace in the areas some distance from the jog, and curved through the centre of the jog itself (Fig. 9a). Initially, the abrupt change in the orientation of the trace at the jog resulted in a stress field that was oblique to the jog in areas close to the trace wall. However, with increased elongation the jog was removed and the stress field was then close to parallel to

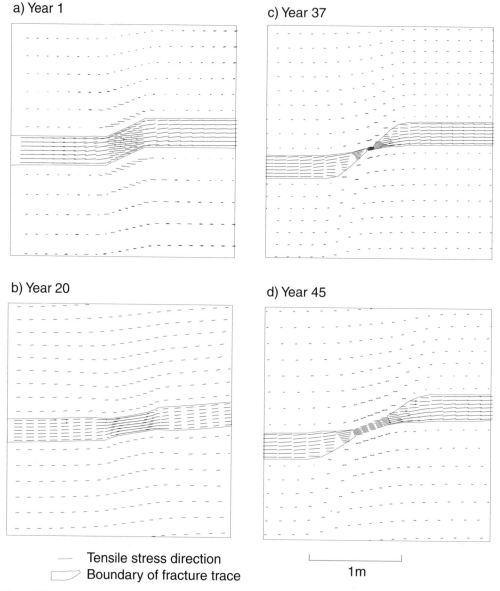

a) Year 1

c) Year 37

b) Year 20

d) Year 45

— Tensile stress direction
▱ Boundary of fracture trace

1m

**Fig. 9.** The tensile stress distribution and the effect of a rheological contrast. The orientation of the stress field is controlled by the geometry of the rheological contrast between the fracture trace and the bubbly blue ice. The orientation of the principal tensile stress is shown for the pull-apart models after (**a**) one year of deformation, (**b**) 20 years, and the boudinage models after (**c**) 37 years and (**d**) 45 years.

the fracture wall throughout the trace, including the pull-apart area (Fig. 9b).

The orientation of the stress field continued to be controlled by the rheological contrast after the trace separated into individual boudins. Within the trace, the principal tensile stress was parallel to the walls of the trace in the centre of the boudin, but tended to splay out in the tips of the boudins (Fig. 9c & d). The stress orientation in the host ice tended to parallel the stretching direction, except in the areas close to the necking region where the orientation of the principal compressive stress was inclined toward the space between the two boudins (Fig. 9c & d).

## Pressure distribution

The mean stress of each grid zone was monitored throughout the deformation of the numerical models. The localisation of tensile stresses in the fracture traces produced strong negative mean stresses when the trace was transported into an environment where extension was sub-parallel to the orientation of the fracture. The mean stress had a minimum of $-0.36$ MPa at the inner inflection points of the jog, which correlates with the maximum differential stress areas (Fig. 10). The jog region, and the area in the centre of the trace had a mean stress of $c. -0.25$ MPa, while the areas on the flanks of the trace had a mean stress of $c. -0.15$ MPa. The bubbly blue host is dominated by compressive stresses so the mean stress was significantly higher ($c. 0.05$ MPa), except in areas around the perimeter of the fracture trace, where the mean stress was also $c. -0.15$ MPa. As the deformation proceeded, the distribution of mean stress remained constant with low mean stress concentrated in the pull-apart region of the fracture trace, though the minimum mean stress at the inflection points reduced to $0.31$ MPa.

The mean stress distribution changed dramatically when the fracture trace separated into individual boudins, as the mean stress in the host ice reduced to $-0.13$ MPa in the area between the boudins at year 37, and a large area of mean stress lower than $-0.05$ MPa extended radially about the pull-apart area (Fig. 10). The mean stress through the centre of the fracture trace remained below $-0.2$ MPa, while in the flank areas of the boudins it was $c. -0.1$ MPa, and a minimum mean stress of $-0.31$ MPa occurred at the tips of the boudins. The mean stress at the centre of the pull-apart area began to rise above $-0.05$ MPa once the boudins had separated to $0.25$ m after 37 years, while the areas of the host ice abutting the tips of the boudins had a

mean stress of $-0.16$ MPa (Fig. 10). The distribution of the mean stress in the boudinaged fracture trace remained unchanged, though it was slightly higher with the minimum mean stress of $-0.20$ MPa at the tips of boudins (Fig. 10).

## Discussion

### Deformation processes not incorporated in the numerical models

The finite difference models provide an excellent representation of the progressive boudinage sub-parallel to the length of a fracture trace under extension. However, the final geometry of pinch-and-swell structures does not match those observed in the field. After a period of 30 years the pull-apart region of the pinch-and-swell structure is at least twice the thickness of those observed in the field. The numerical models only consider deformation by glide-controlled creep, and the effect of the anisotropy in the fracture trace. Clearly, other processes are also important and contribute to the final break-up of the fracture trace into individual boudins.

An important process not incorporated in the numerical models is recrystallization. The models assumed that fracture traces have a strong mechanical anisotropy which is fixed with relation to the fracture walls due to the elongate, interlocking nature of the individual crystals that comprise the columnar ice. However, after significant shortening across the width of the fracture trace, the highly-strained areas would begin to recrystallize and break-up the interlocking fabric. It has been noted that the highly-deformed pull-apart areas and the tips of boudins are no longer translucent like the rest of the fracture trace, but have a milky appearance. This is due to a re-distribution of the air bubbles that were trapped during the formation of the fracture trace and is an indication that the crystallographic fabric has been re-organised during the deformation of the pull-apart area. Wilson & Russell-head (1982) report that recrystallization is initiated in polycrystalline ice after 10% shortening in plane strain. The centres of the pull-apart region are shortened across their width by this amount after seven years. Recrystallization in the pull-apart area results in their break-down of the fixed mechanical anisotropy assumption. Recrystallization would replace crystals in an unsuitable orientations for glide with crystals that are in the optimal orientation for easy glide. This process leads to strain-softening that would be localized in the pull-apart region. The absence of recrystallization in

a) Year 1

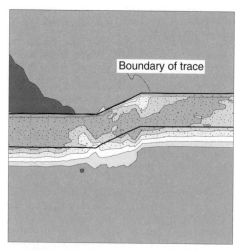

c) Year 37

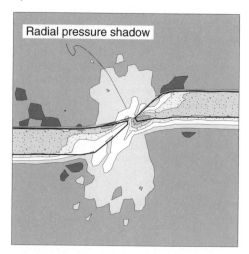

b) Year 10

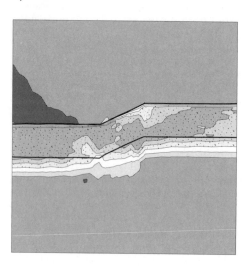

d) Year 45

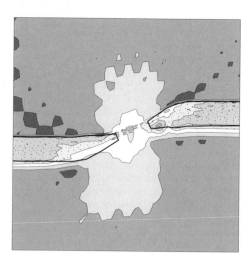

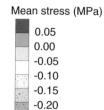

0.4 m

Mean stress (MPa)

| | |
|---|---|
| ■ | 0.05 |
| | 0.00 |
| | -0.05 |
| | -0.10 |
| | -0.15 |
| | -0.20 |

**Fig. 10.** The distribution of the mean stress in the pull-apart and boudinage models. The mean stress distribution is analogous to the pressure distribution about a deforming rock mass. Negative mean stress indicates that tensile stress dominates. The mean stress decreases towards the negative end of the scale. The low mean stresses are concentrated in the pull-apart region while the fracture trace is deformed to produce a pinch-and-swell structure. Once the individual boudins form, the mean stress in the host drops dramatically in the area surrounding the pull-apart area to form a large pressure shadow. By the forty-fifth year of deformation, the minimum mean stress in the host occurs in the area abutting the tips of the boudins.

the numerical models may explain why the finite difference models did not replicate the rapid break-up of the fracture traces into individual boudins.

## Implications for boudins in high grade metamorphic rocks

Crustal rocks at high metamorphic grade deform according to non-linear flow laws that have the same form as equation 1, where the stress exponent, $n$, has a value of either $c.3$ or $c.5$ (Kirby & Kronenberg 1987; Tsenn & Carten 1987). The deformation history and stress distribution around boudinage structures in ice, can therefore be related to the formation of pinch-and-swell and boudinage structures in amphibolite and granulite grade rocks, as similar geometric features form, and the deformation is governed by the same flow laws. The flow line traverse detailed in this paper provides a unique opportunity to study the boudinage of a non-linear material as an unbroken sequence of structures. This has fundamental implications for the development of textures in rocks, such as the presence of different mineral assemblages, veins or partial melt in the pull-apart region between boudins, and the variation in the orientation of lineation or planar fabrics due to the refraction of stress by rheological contrasts.

A common observation in boudinaged rocks is the presence of small areas where the mineralogical composition has changed during the deformation of the rock mass. Such structures are referred to as *pressure-shadows* (Passchier & Trouw 1996), and generally consist of veins, fibrous growths, partial melts or metasomatic alteration in the pull-apart region between boudins. These textures are dependent on the movement of fluids, or the diffusion of elements, through the mass of the rock, which in turn is controlled by pressure gradients (Ramberg 1952; Strömgård 1973). The pressure in a deforming rock mass is analogous to the mean stress modelled about the boudinage structure in ice, with the addition of lithostatic pressure due to the height of the rock column above the deforming rock mass. The numerical models have demonstrated that the mean stress is initially 0.4 MPa lower in the pull-apart region of the more competent fracture trace, than it is in the host ice. In a rock mass the relative low pressure in the pull-apart area would drive diffusion of fluids toward the necking region. The inundation of the pull-apart area by fluids would then be dependent on the permeability of the rheological boundary. If the boundary is permeable, then the presence of fluid may weaken the pull-apart region via metasomatic reactions and increase the local strain rate. Once the more competent layer is divided into individual boudins, the pressure in the host is dramatically reduced in the region between the boudins, and a pressure gradient forms that extends radially from the pull-apart region into the host (Fig. 10). A pressure gradient is also observed along the flanks of the boudins, so that any fluids that are present in the host, will tend to move towards the pull-apart area and along the boundaries of the boudins. As the boudins separate, the stress field changes leading to a small increase in pressure at the centres of the pull-apart area, while the lowest pressure is observed in the host ice that abuts the nose of the boudins. Thus, any fluid-related textures that formed at the centre of the boudins when the competent layer first broke-up will tend to be attenuated by the ongoing deformation, while the textures close to the nose of the boudins will tend to be protected from subsequent deformation.

The finite difference models also have implications for the development of structural elements that are used to interpret finite strain, such as foliations and lineations. The local orientation of a foliation or lineation is dependent on the local stress field, however, rheological boundaries tend to refract the orientation of the stress field. Treagus (1993) discussed in detail the refraction of stress fields in layered materials that deform according to linear and non-linear flow laws.

The degree of the refraction of the stress field is dependent on the geometry of the more competent layer with respect to the bulk extension or compression direction, and on the ratio of viscosity between the layers. The refraction of stress in a non-linear visco-elastic material is complex as the ratio of viscosity between layers is dependent on the magnitude of the stress. It is made more complex in a material with a strong mechanical anisotropy, such as a fracture trace, as the viscosity ratio also changes with the orientation of the stress field. Thus, it is only possible to solve problems that involve stress refraction in an anisotropic, non-linear material using numerical methods. The models have demonstrated how the stress orientation varies as the geometry of the rheological boundary changes with time, and has implications for the orientation of foliations and lineations in and around boudins in rocks.

## Conclusions

(1) It is possible to derive quantitative data about the spatial distribution and temporal

nature of stress fields associated with the development of boudinage structures by combining field based structural glaciology data with finite difference numerical methods.

(2) Numerical models that have been constrained by strain and strain-rate measurements from a polar outlet glacier demonstrate that regions with significantly lower mean-stress develop in the pull-apart area between boudins. In crustal rocks, such a mean stress distribution would drive the movement of fluids toward the pull-apart area and result in the formation of pressure-shadow textures, such as metasomatic alteration zones, fibrous growths, the accumulation of partial melt or veins.

(3) The numerical models of boudinage in ice show that the orientation of stress in a deforming, layered non-linear system has a complex history. Geologists who study crustal rocks use foliations and lineations to interpret the macro-scale evolution of a terrain. However, linear and planar element in rocks form in response to the local stress orientation that may not be representative of the far field stress regime.

This project would not be possible without the GPS survey data obtained in conjunction with J. Dawson and G. Butcher of the University of Melbourne, Department of Geomatics. We would like to extend thanks to all officers and expeditioners of the Australian National Antarctic Research Expeditions (ANARE) for assistance and co-operation during the 1993–94, 1994–95, 1995–96 and 1997–98 field. M. Seigert and M. Sharp are thanked for their review comments on this manuscript and the financial support from ASAC grant number 599 is gratefully acknowledged.

# References

AZUMA, N. 1994. A flow law for anisotropic ice and its implication to ice sheets. *Earth and Planetary Science Letters*, **128**, 601–614.

—— & GOTO-AZUMA, K. 1996. An anisotropic flow law for ice-sheet ice and its implications. *Annals of Glaciology*, **23**, 202–208.

—— & HIGASHI, A. 1985. Formation processes of ice fabric patterns in ice sheets. *Annals of Glaciology*, **6**, 130–134.

BUDD, W. F. & JACKA, T. H. 1989. A review of ice rheology for ice sheet modelling. *Cold Regions Science and Technology*, **16**, 107–144.

DETOURNAY, C. & HART, R. 1999. *FLAC and Numerical Modeling in Geomechanics*. A. A. Balkema, Rotterdam.

DURNEY, D. W. & RAMSAY, J. G. 1973. Incremental strain measured by syntectonic crystal growth. *In*: DE JONG, K. A. & SCHOLTEN, R. (eds) *Gravity and Tectonics*. Wiley, New York, 67–96.

FLINN, D. 1961. Deformation at thrust planes in Shetland and the Jotunhiem area of Norway. *Geological Magazine*, **8**, 245–258.

FROST, H. J. & ASHBY, M. F. 1982. *Deformation mechanisms maps. The plasticity and creep of metal and ceramics*. Pergamon Press, New York.

KETCHAM, W. M. & HOBBS, P. B. 1967. The preferred orientation in the growth of ice from the melt. *Journal of Crystal Growth*, **1**, 263–270.

KIRBY, S. H. & KRONENBERG, A. K. 1987. Rheology of the lithosphere: selected topics. *Reviews in Geophysics*, **25**, 1219–1244.

MARMO, B. A. 1999. *Deformation processes in polar outlet glaciers, Framnes Mountains, east Antarctica*. PhD Thesis, The University of Melbourne.

—— & DAWSON, J. 1996. Movement and structural features observed in ice masses, Framnes Mountains, MacRobertson Land, east Antarctica. *Annals of Glaciology*, **23**, 388–395.

—— & WILSON, C. J. L. 1998. Strain localisation and incremental deformation within ice masses, Framnes Mountains, east Antarctica. *Journal of Structural Geology*, **20**, 149–162.

—— & ——1999. A verification procedure for the use of FLAC to study glacial dynamics and the implementation of an anisotropic flow law. *In*: *FLAC and Numerical Modeling in Geomechanics*. A. A. Balkema, Rotterdam, 183–190.

NEURATH, C. & SMITH, R. B. 1982. The effect of material properties on the growth rates of folding and boudinage, experiments with wax models. *Journal of Structural Geology*, **4**, 215–229.

PASSCHIER, C. W. & TROUW, R. A. J. 1996. *Microtectonics*. Springer-Verlag, Berlin.

PATERSON, M. S. & WEISS, L. E. 1968. Folding and boudinage of a quartz-rich layer in experimentally deformed phyllite. *Geological Society of America Bulletin*, **79**, 795–812.

RAMBERG, H. 1952. *The origin of metamorphic and metasomatic rocks*. University of Chicago Press.

——1955. Natural and experimental boudinage and pinch and swell structures, *Journal of Geology*, **63**, 512–526.

RAMSAY, J. G. 1961. *Folding and fracturing of rock*. McGraw-Hill, New York.

STRÖMGÅRD K. E. 1973. Stress distribution during formation of boudinage and pressure shadows. *Tectonophysics*, **16**, 215–248.

TREAGUS, S. H. 1993. Flow variations in power-law, multilayers: implications for competence contrast in rocks. *Journal of Structural Geology*, **15**, 423–434.

TSENN, M. C. & CARTEN, N. L. 1987. Upper limits of power law creep in rocks. *Tectonophysics*, **136**, 1–24.

WILSON, C. J. L. 1994. Crystal growth during a single-stage opening event and its implications for syntectonic veins. *Journal of Structural Geology*, **16**, 1283–1296.

—— & RUSSELL-HEAD, D. S. 1982. Steady-state preferred orientation of ice deformed in plane strain at $-1°C$. *Journal of Glaciology*, **28**, 145–159.

# The potential contribution of high-resolution glacier flow modelling to structural glaciology

ALUN HUBBARD[1] & BRYN HUBBARD[2]

[1] Department of Geography, University of Canterbury, Private Bag 4800, Christchurch, New Zealand (e-mail: ahubbard@geog.canterbury.ac.nz)
[2] Centre for Glaciology, Institute of Geography and Earth Sciences, University of Wales, Aberystwyth, Ceredigion SY23 3DB, Wales, UK (e-mail: byh@aber.ac.uk)

**Abstract:** The three-dimensional stress and strain fields derived through high-resolution flow modelling of Haut Glacier d'Arolla, Switzerland are used to predict the generation, passage and surface expression of a variety of structural forms at the glacier. Flow vectors and strain ellipses are computed and illustrated in plan-form and long section. The model is used to predict the formation and orientation of surface crevasses, and, once healed, the downglacier evolution of their traces. Similarly, the evolution of primary stratification where it crops out at the glacier surface is predicted. The resulting stratification pattern compares well with that revealed in aerial photographs of the glacier. Finally, the three-dimensional strain field is used to track the (accumulation area) burial, (englacial) transport and (ablation area) exposure of ice deposited within a pre-defined elevation range. This *deposition–transport–exposure* tracking allows the location of ice that was initially deposited at any defined location on the glacier to be identified. Such information is of significance in interpreting, for example, the distribution of ash and isotopic horizons within a glacier. We conclude that high-resolution three-dimensional flow modelling has the potential to provide a powerful tool for investigating the genesis and evolution of valley glacier structures.

Recent developments in the techniques for solving of the equations governing ice flow have enabled the influence of longitudinal stress gradients to be included in high-resolution glacier models (e.g. Blatter 1995; Hanson 1995; Hubbard *et al.* 1998; Gudmundsson 1999). Consequently, such models have the capacity to recreate fully the three-dimensional stress and strain field in relatively small, confined valley glaciers. The boundary conditions required for such models are glacier geometry, ice viscosity and the basal sliding distribution, the latter two of which may be approximated in the light of more general glaciological data and parameterisations. The resulting modelled stress and strain fields can be used to predict a variety of structural glaciological information, from flow vector fields, through three-dimensional strain ellipse patterns, to the locations and orientation of surface crevasse fields, and, once formed, their full spatial expression as healed crevasse traces. The implications of such an approach are significant, bearing on interpretations of observed glacier flow and structural patterns, as well as of secondary factors such as the passage of ash layers through an ice mass and the isotopic composition of ice exposed in the glacier's ablation area. In the latter case, model predictions could be used to direct surface sampling regimes where ice of a specific origin is sought.

In this paper we illustrate, through a variety of examples, the potential contribution that high-resolution ice-flow modelling can make to structural glaciological investigations and interpretations. These include:

- the three-dimensional stress and strain fields of the glacier;
- the planform location and orientation of surface crevasses that open where a tensile-stress failure-criterion is exceeded;
- the evolution of the surface, planform expression of crevasse traces as they pass through the glacier;
- the evolution of the vertical, long-section expression of deformation profiles as they pass through the glacier;
- the evolution of the surface, planform expression of primary stratification as it flows through the glacier;
- the three-dimensional, spatial distribution of ice that was initially deposited within a specified elevation range, termed *deposition–transport–exposure* modelling.

*From:* MALTMAN, A. J., HUBBARD, B. & HAMBREY, M. J. (eds) *Deformation of Glacial Materials.* Geological Society, London, Special Publications, **176**, 135–146. 0305-8719/00/$15.00 © The Geological Society of London 2000.

## Field site and methods

Examples are drawn from Haut Glacier d'Arolla, Switzerland (Fig. 1), which has been intensively studied through a series of glaciological projects over the past ten years, providing a large amount of supporting and background data (e.g. Sharp *et al.* 1993; Hubbard *et al.* 1995; Richards *et al.* 1996). The glacier is predominantly warm-based, and extends within a well-defined valley from *c.* 3500 m a.s.l. at its headwall to *c.* 2560 m a.s.l. at its snout, covering an area of *c.* 6.3 km$^2$. The glacier is geometrically simple with a maximum thickness of *c.* 140 m and contains no ice falls, limited crevasse fields, and only one significant tributary basin, which enters the main body of the glacier from the SW (Fig. 1).

The three-dimensional model used to derive the stress and strain fields operates at a spatial resolution of 70 m in the horizontal and 2.5% of the local ice thickness, equating to 40 vertical layers, and its application to Haut Glacier d'Arolla is fully is described in Hubbard *et al.* (1998). The model is steady-state (i.e. time-independent) and is based on a finite-difference first-order solution of the ice-flow equations

(Muller 1991; Blatter 1995) that includes the effect of longitudinal or normal deviatoric stresses. These stresses are particularly important in valley-glacier modelling, where significant compressive and tensile stresses are induced by local changes in glacier surface and bed slopes and basal and lateral friction (Kamb & Echelmeyer 1986). To replicate the annual flow regime, Hubbard *et al.* (1998) conducted three independent modelling experiments, each constrained by a contrasting basal velocity boundary scenario defined by measured surface displacements made at 34 markers between 1994 and 1995:

- winter base flow – assuming zero basal motion and providing a means by which the value of the flow law rate factor $A$ of 0.063 a$^{-1}$ bar$^{-3}$ could be determined by tuning model output to winter surface velocities observed between 3 February and 8 February 1995;
- summer flow – a basal motion distribution representative of mean melt-season conditions, defined by the difference between observed summer (between 30 June and 3 September 1994) and winter surface velocities;

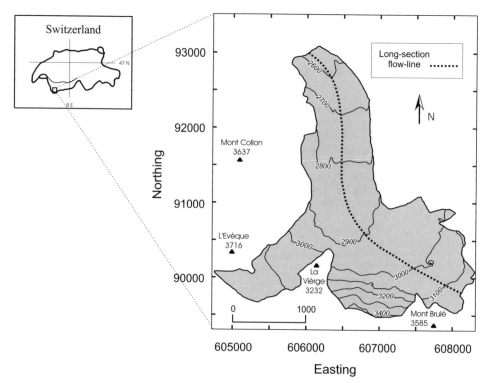

**Fig. 1.** Haut Glacier d'Arolla, Switzerland. The dotted line defines the long-section centreline illustrated in Figs 3, 6 and 8.

- enhanced spring flow – an enhanced distribution of basal motion representing conditions at the bed during a major glacier-wide speed up event which lasted 10 days, defined by the difference between observed spring event (measured between 19 May and 29 May 1994) and winter surface velocities.

Each scenario represents a characteristic, steady-state snap-shot of the distinctive flow scenarios operating over the course of a year at Haut Glacier d'Arolla and which have been repeatedly observed over successive years (e.g. Hubbard & Nienow 1997). The careful bracketing of these representative flow periods means minimal departure between time-averaged and time-integrated flow conditions, and the three independent regimes are subsequently time-weight averaged to yield the composite annual flow regime. The modelled composite was successfully verified against further observations at Haut Glacier d'Arolla reproducing both the annual pattern of surface velocity and the distribution of downglacier velocity in cross-section derived from repeat, down-borehole inclinometry measured over the same period (Harbor et al. 1997; Hubbard et al. 1998). This three-dimensional modelled velocity composite provides the basis for the present study, which further utilizes a Lagrangian co-ordinate tracking algorithm to advect structural assemblages downglacier.

## Results

Although the analysis is in three-dimensions, results are presented in orthogonal (two-dimensional) surfaces: planform and long-section. The latter follows the glacier centreline, as indicated on Fig. 1. Results may be compared with actual structural features at the glacier as revealed on an annotated aerial photograph (Fig. 2).

### Ice velocity and strain rate

Ice velocity and strain fields are presented in planform and long section in Fig. 3. Surface ice velocity (Fig. 3a) is generally greatest where the glacier is thickest: the maximum surface velocity ($18 \, \mathrm{m \, a^{-1}}$) is predicted where the glacier is both thick and steep, towards the base of the tributary glacier between the peaks of La Vierge and L'Evêque (Fig. 1). Large velocity gradients near the lateral margins of the glacier are reflected in the ice surface strain field (Fig. 3b), where initially circular strain ellipses elongate strongly

parallel to the flow direction (indicating longitudinal extension). Conversely, strain ellipses elongate strongly orthogonal to the flow direction (indicating longitudinal compression) near the glacier terminus. The surface strain field also reveals gradients between the two major flow units downglacier from the base of La Vierge. Indeed, the boundary between these two flow units may be clearly identified from the surface strain field by the rotation and downglacier elongation of the strain ellipses along this axis.

In long-section, the ice flow field (Fig. 3c) reveals, as expected, that ice buried at the head of the glacier follows a deep transport trajectory to the glacier terminus, while that buried near the model's equilibrium line (located at $c. 2875 \, \mathrm{m}$) emerges at the ice surface a short distance down-flow, at the top of the glacier's ablation area. The maximum long-section velocity of $c. 16 \, \mathrm{m \, a^{-1}}$ is located at the ice surface above the deepest ice in the central area of the glacier (this velocity is lower than that recorded in planform (above) since the centre-line long-section does not extend up the tributary glacier; fig. 1). The modelled velocity field decreases vertically from the ice surface to the glacier bed, where basal velocities are prescribed. Long-section strain ellipses (Fig. 3d) correspondingly flatten dramatically as they approach the glacier bed, where shear strain rate is highest. These ellipses elongate further parallel to the glacier bed as they progress downglacier, such that they may be considered to form a bed-parallel basal foliation by the time they reach the glacier terminus. This prediction agrees with previous, qualitative assessments of the formation of basal, flow-parallel foliation from initially obliquely-orientated features (e.g. Hooke & Hudleston 1978). In contrast, strain rate decreases markedly at higher elevations within the glacier, barely deforming high-level strain ellipses, until near the glacier's frontal margin. Here, longitudinal compression causes ellipses to rotate away from the glacier bed and to flatten at a high angle to the glacier surface (Fig. 3d).

### Crevassing and crevasse traces

Model validation against field data at Haut Glacier d'Arolla (Hubbard et al. 1998) included comparison of predicted crevasse fields with those observed at the glacier surface. Crevasse prediction was based on adoption of the von Mise's criterion of a threshold octahedral stress (computed from the surface-parallel principal stresses) for ice failure following Vaughan (1993). The threshold failure criterion was defined as the

**Fig. 2.** Annotated aerial photograph of Haut Glacier d'Arolla. Crevassed areas are outlined, and crevasse orientations are indicated by (parallel) internal hatching. The ice stratification exposure traced downglacier in Fig. 7 is indicated by the thick dashed line, with specific features referred to in the text labelled (**1**) and (**2**). The modelled glacier outline (Figs 1, 3, 4, 5, 7 & 8) is superimposed (dotted line).

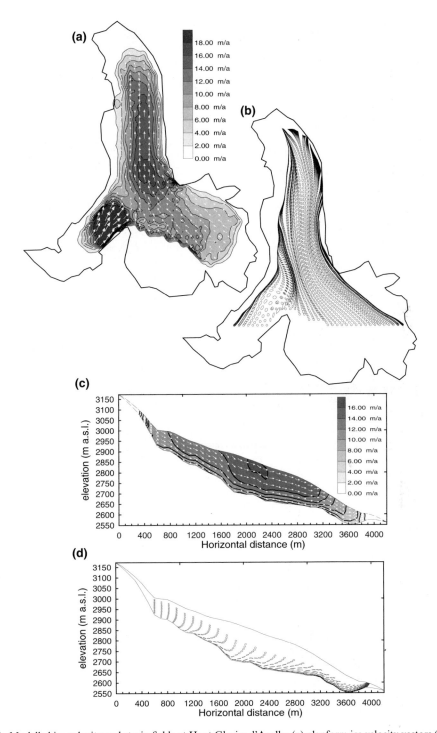

**Fig. 3.** Modelled ice velocity and strain fields at Haut Glacier d'Arolla: (**a**) planform ice velocity vectors (arrows) and magnitude (contoured and shaded); (**b**) planform strain field, illustrated as the deformation of initially circular strain ellipses; (**c**) long-section ice velocity vectors (arrows) and magnitude (contoured and shaded); (**d**) long-section strain ellipses. The long-section follows the glacier's centreline, as indicated on Fig. 1.

maximum 10% of calculated values, yielding a good match between computed zones of maximum predicted failure (Fig. 4) and actual surface crevasse location (Fig. 2). Furthermore, within the zones of predicted failure, there is also high correspondence between actual crevasse orientations and the directions of the principal tensile stress. Here, we extend this approach by modelling the down-flow evolution of the traces that remain following the closure of these crevasses. In reality, such traces take the form of distinctive ice-crystallographic layers from which bubbles are commonly excluded, producing sub-vertical 'blue' ice layers, commonly some decimetres across and some metres to tens of metres long (e.g. Hambrey et al. 1980). Several examples of the predicted planform surface expression of the downglacier evolution of crevasse traces at Haut Glacier d'Arolla are presented in Fig. 5. The surface strain field dictates that most crevasses open up in response to excess longitudinal tension, and are correspondingly initially orientated orthogonal to the general flow-line direc-

tion. The subsequent imposition of lateral shear on these crevasse traces (in accordance with the surface strain field depicted in Fig. 3b) commonly rotates these traces into a foliation that is orientated approximately longitudinally. This effect is particularly marked near the glacier margins (e.g. Fig. 5d) and near the common margin of the two major flow units comprising the glacier (e.g. Fig. 5b and c). In the latter cases, two sets of crevasse traces, each initiated within a separate flow unit, rotate and merge to form a relatively uniform foliation centred on the resulting La Vièrge medial moraine (compare Fig. 5b and c with Fig. 2). In contrast, crevasse traces that pass downglacier more centrally within the major flow unit are only minimally rotated (e.g. Fig. 5e). In all cases, the orientation of crevasse traces inherited from upglacier may contrast sharply with that of crevasses formed locally. Indeed, structural glaciology field studies frequently encounter several generations of crevasse traces that cut across one another, causing a high degree of complexity (e.g. Hambrey & Milnes

**Fig. 4.** The magnitude and direction of modelled planform principal stresses at Haut Glacier d'Arolla. Shaded areas represent zones of maximum computed tensile stresses, indicating regions most likely to experience crevassing (after Hubbard et al. 1998). These areas, and the orientations of the tensile stresses within them, compare well with the actual patterns of crevassing at the glacier, as indicated on Fig. 2.

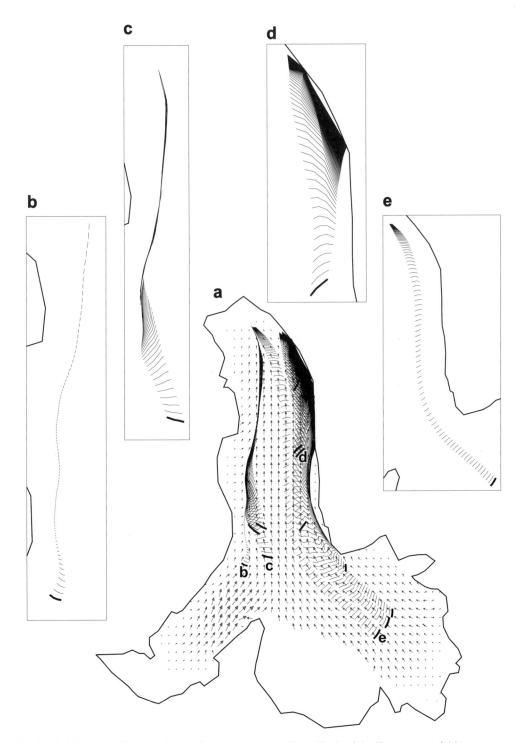

**Fig. 5.** Modelled downglacier evolution of crevasse traces at Haut Glacier d'Arolla, represented (**a**) as a composite of several selected crevasses (surface velocity vectors are represented as arrows), and (**b**) to (**e**) as a selection of individual crevasse traces.

**(a)**

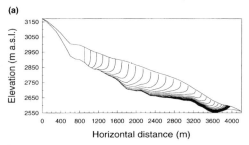

**(b)**

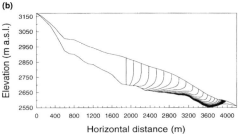

**Fig. 6.** Modelled long-section deformation of two, initially vertical strain profiles located (**a**) *c.* 900 m and (**b**) *c.* 1900 m from the head of the glacier. The long-section follows the glacier's centreline, as indicated on Fig. 1.

1977; Hambrey & Müller 1978). Crevasse-trace spacing also varies spatially in the model output (Fig. 5), indicating an associated variability in longitudinal tension and compression. Crevasse-trace spacing particularly decreases within some hundreds of metres of the glacier terminus, again indicating the dominance of longitudinal compression in this zone.

Model output also allows crevasse-trace deformation, as well as that of moulins and boreholes, to be predicted in the vertical plane. We illustrate this capacity by entering two initially vertical profiles into the model (these may be stipulated as lines, mimicking moulins or boreholes, or surfaces, mimicking crevasses) and observing their deformation in long-section as they flow downglacier (Fig. 6). These features are initially located 900 m (Fig. 6a) and 1900 m (Fig. 6b) from the head of the glacier. In accordance with the long-section strain-rate field (Fig. 3d), the resulting deformation patterns for both profiles indicate significant flow-parallel shear strain close to the glacier bed. This pattern is supplemented by the influence of strong longitudinal compression near the glacier terminus, which decreases the spacing between profiles and forces them to deviate towards the glacier surface, similar to the long-section strain field analysis presented above (Fig. 3d).

## Primary stratification and deposition-transport-exposure modelling

Primary sedimentary stratification is laid down as surface-parallel layers that survive within the accumulation area of the glacier. Such layers are progressively buried by subsequent surface accumulation, forming alternating bubble-rich and bubble-poor ice layers that correspond to seasonal variations in firnification processes. These layers flow downglacier, often rotating and eventually emerge as longitudinal foliation at the ice surface in the ablation area of the glacier (e.g. Hambrey 1975, 1976; Hooke & Hudleston 1978). Where such layers are exposed by the surface erosion of fold hinges, their expression may be traced downglacier in accordance with the velocity field (Fig. 3). We model such an exposure, as revealed on the aerial photograph of Haut Glacier d'Arolla (Fig. 2). This example, however, exemplifies one of the principal shortcomings of steady-state modelling, since the glacier's present day equilibrium line altitude is over 200 m higher than that required to maintain equilibrium in the model. This discrepancy implies that the glacier's current geometry is markedly out of balance with its climate, an imbalance that is also indicated by the glacier's recent rapid retreat and thinning (e.g. Willis *et al.* 1998). Under conditions of steady-state equilibrium, indicated by the modelled long-section velocity field (Fig. 3c), this high-elevation primary stratification would continue to be buried within an extended accumulation zone downglacier. In this particular case, we surmount the problem by advecting the primary stratification downglacier with the two-dimensional surface-strain field which yields a satisfactory prediction as revealed from comparison with aerial photography. However, this discrepancy between the equilibrium strain field and the actual one is significant, indicating that such a steady-state approach would not be sufficient for long-term modelling of glacier structure, particularly in situations where glacier geometry may be responding rapidly to climatic change. The predicted pattern (Fig. 7), however, still compares well with that of previous stratification layers exposed at the actual glacier surface down-flow of the modelled layer on Fig. 2. For example, the evolution of individual fold hinges, such as those marked (**1**) and (**2**) on Figs 2 and 7 may be traced downglacier in both the modelled and the actual fields. Inspection of these figures reveal that in both cases (**1**) gradually tightens parallel to flow and (**2**) opens over *c.* 2 km then tightens as it approaches the glacier terminus.

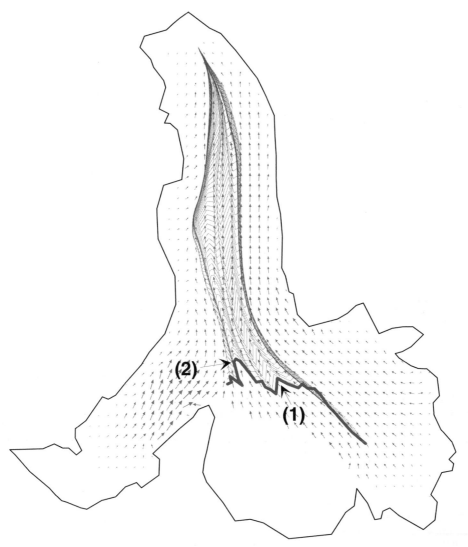

**Fig. 7.** Modelled planform downglacier evolution in the surface expression of an ice stratification layer exposed by erosion at the ice surface (surface velocity vectors are represented as arrows). The feature and its downglacier evolution, in particular the elements labelled (**1**) and (**2**) (see text) may be compared with aerial photographic evidence presented in Fig. 2.

Analysis of the downglacier evolution of primary stratification as it crops out at the glacier surface may be extended to reconstruct the glacier-wide extent of ice layers initially deposited within pre-defined elevation ranges; a procedure we term *deposition–transport–exposure* modelling. The utility of this approach lies in its ability to locate ice, either exposed at the glacier surface or within cores, that was initially deposited within a stipulated area or elevation range. In order to illustrate the potential of this method we select the elevation band 2990 to

3000 m a.s.l. as our initial input and trace its subsequent passage through the glacier (Fig. 8). This elevation range incorporates two glacier surface bands, one on the main glacier and the other on the base of the tributary glacier. Both are located within the glacier's steady state accumulation area, resulting in burial of the material at the site of deposition. The ice then flows via an englacial transport pathway to the ablation area, where it is exposed in accordance with the glacier's three-dimensional velocity field. The plan-form pattern of surface re-exposure

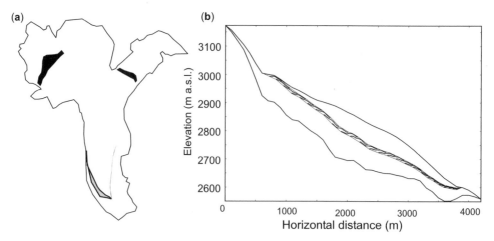

**Fig. 8.** Modelled englacial transport and exposure (light grey) of ice initially deposited (dark grey) between the altitudes of 2990 and 3000 m a.s.l. in (a) planform (inverted for comparison) and (b) long section.

of the ice in the glacier's ablation area (Fig. 8a) again reflects variations in the three-dimensional flow field of the glacier, with both bands being rotated from their initial flow-orthogonal orientation to approach parallelism with the general direction of ice flow upon re-exposure. In this case, both ice bands have flowed more rapidly on their in-glacier edges (where ice is thickest), resulting in a downglacier-pointing, V-shaped planform expression. In centreline long-section (Fig. 8b), the englacial passage of this ice indicates burial to a depth of c. 75 m before re-emergence at the ice surface in accordance with the velocity field in this plane (Fig. 3c).

## Discussion and conclusions

High-resolution, three-dimensional glacier flow modelling has the capacity to make a major contribution to structural glaciological investigations. This capacity centres on the ability of such models to generate full, three-dimensional flow vectors for any point (or modelled cell) within the glacier. Such cells can then be traced individually or as groups to define ice surfaces or volumes approximating the geometries of actual glaciological structures. Critically, the inclusion of longitudinal stresses within such models means that they may now be applied with confidence to relatively small valley glaciers. The validity of this approach, however, is subject to a number of limitations that should be considered in any study that proposes to adopt a modelling-based component.

First, the spatial resolution of this method is dictated by the cell size of the model used. In the present case this was 70 m × 70 m in the horizontal, and 2.5% of the local ice thickness in the vertical.

Second, the model assumes steady-state geometry. It is important to recognise that the modelled flow field, and thus the model output, is based on the equilibrium flow field that maintains the steady-state geometry, rather than that actually present at a given time at the glacier. Thus, at Haut Glacier d'Arolla for example, modelling places the equilibrium line altitude at c. 2875 m a.s.l., at least 200 m lower than that observed at the glacier for the past 10 years. Over shorter, multi-annual time scales this equilibrium flow field will approximate the actual glacier flow regime averaged. However, over longer time-scales, when the glacier geometry (and hence ice flow) is responding to large shifts in climate, then there will be an increasing discrepancy between the modelled equilibrium flow field and that actually present at the glacier. Under these conditions, it would be necessary to integrate the three-dimensional flow model through time using a mass continuity solution coupled to a climatic reconstruction. This computationally intensive task (which requires integration of a three-dimensional time-dependent glacier flow model) may soon be possible, particularly given recent developments in the numerical techniques for solving the ice continuity equation (e.g. Hindmarsh & Payne 1996) and concurrent hardware developments.

Third, the model is sensitive to the boundary conditions used to constrain it. Thus, high-resolution digital elevation models of both the glacier surface and the glacier bed are required. The flow law parameters used in the model should also, where possible, be tuned by empirical velocity data. At Haut Glacier d'Arolla, this exercise resulted in a 50% reduction to the value of the ice hardness parameter (*A*) commonly assumed in the application of Glen's flow law (Glen 1955) to temperate glaciers (Hubbard *et al.* 1998). Further, where possible, the basal motion (sliding and bed deformation) field should be prescribed in order to avoid over-estimation of englacial ice deformation velocities. At Haut Glacier d'Arolla, the summer sliding field was approximated as the difference between measured summer and winter ice surface velocities (under an assumption of no winter sliding). Basal sliding at the glacier accordingly accounts for roughly 30% of the summer ice surface velocity, although this contribution varies spatially from 20 to 55%.

Despite these limitations, we believe that ice-flow modelling has a major role to play in structural glaciological investigations and interpretations at glaciers and ice-sheets. This information may, in turn, be used to direct ice-surface and ice-core sampling programmes. For accurate results, however, empirical data are required to constrain the boundary conditions employed by the model and for the present study, this parameterisation required extensive, field-based research. However, with the increased coverage and sophistication of satellite- and aeroplane-based remote sensing techniques, much of this information may now be obtained remotely. Thus, high-resolution airborne laser ranging (e.g. Sapiano *et al.* 1998) and radio-echo sounding (e.g. Kennett *et al.* 1993) may be used to produce digital elevation models of the ice surface and bed, and to indicate the internal thermal structure of the ice mass in question. Satellite-based synthetic aperture radar interferometry (InSAR), is available at increasingly high temporal and spatial resolutions (e.g. Forster *et al.* 1999) and may be used to monitor temporal variations in the ice surface motion field, allowing approximation of the ice hardness parameter and basal motion field required by the model. In this way, full three-dimensional ice flow and structural glaciological information may be generated for a glacier or ice-sheet based only on remotely sensed information.

We thank H. Blatter for assistance in model development, J.-M. Bonvin of GDSA for providing aerial photographs of the glacier, and T. Payne and H. Gudmundsson for helpful comments on the manuscript. A.H. gratefully acknowledges receipt of funding from the Leverhulme Trust. Field investigations at Haut Glacier d'Arolla were supported by Natural Environment Research Council Grants GR9/02530 and GR3/11216.

# References

BLATTER, H. 1995. Velocity and stress fields in grounded glaciers: a simple algorithm for including deviatoric stress gradients. *Journal of Glaciology*, **41**, 333–343.

FORSTER, R. R., Rignot, E., ISACKS, B. L. & JEZEK, K. C. 1999. Interferometric radar observations of Glaciares Europa and Penguin, Hielo Patagónico Sur, Chile. *Journal of Glaciology*, **45**, 325–337.

GLEN, J. W. 1955. The creep of polycrystalline ice. *Proceedings of the Royal Society of London*, **A228**, 519–538.

GUDMUNDSSON, G. H. 1999. A three-dimensional numerical model of the confluence area of Unteraargletscher, Bernese Alps, Switzerland. *Journal of Glaciology*, **45**, 219–230.

HAMBREY, M. J. 1975. The origin of foliation in glaciers: evidence from some Norwegian examples. *Journal of Glaciology*, **14**, 181–185.

——1976. Structure of the Glacier Charles Rabots Bre, Norway. *Geological Society of America Bulletin*, **87**, 1629–1637.

—— & MILNES, A. G. 1977. Structural geology of an Alpine glacier (Griesgletscher, Valais, Switzerland). *Eclogae Geologicae Helvetiae*, **70**, 667–684.

—— & MÜLLER, F. 1978. Ice deformation and structures in the White Glacier, Axel Heiberg Island, Northwest Territories, Canada. *Journal of Glaciology*, **20**, 41–66.

——, MILNES, A. G. & SIEGENTHALER, H. 1980. Dynamics and structure of Griesgletscher, Switzerland. *Journal of Glaciology*, **25**, 215–228.

HANSON, B. 1995. A fully three-dimensional finite-element model applied to velocities on Storglaciaren, Sweden. *Journal of Glaciology*, **41**, 91–102.

HARBOR, J., SHARP, M., COPLAND, L., HUBBARD, B., NIENOW, P. & MAIR, D. 1997. Influence of subglacial drainage conditions on the velocity distrubution within a glacier cross section. *Geology*, **25**, 739–742.

HINDMARSH, R. C. A. & PAYNE, A. J. 1996. Time-step limits for stable solutions of the ice-sheet equation. *Annals of Glaciology*, **23**, 74–85.

HOOKE, R.LeB. & HUDLESTON, P. J. 1978. Origin of foliation in glaciers. *Journal of Glaciology*, **20**, 285–299.

HUBBARD, A., BLATTER, H., NIENOW, P., MAIR, D. & HUBBARD, B. 1998. Comparison of a three-dimensional model for glacier flow with field data at Haut Glacier d'Arolla, Switzerland. *Journal of Glaciology*, **44**, 368–378.

HUBBARD, B. & NIENOW, P. 1997. Alpine subglacial hydrology. *Quaternary Science Reviews*, **16**, 939–955.

——, SHARP, M., WILLIS, I. C., NIELSEN, M. & SMART, C. C. 1995. Borehole water-level variations and the structure of the subglacial hydrological system of Haut Glacier d'Arolla, Switzerland, *Journal of Glaciology*, **41**, 572–583.

KAMB, B. & ECHELMEYER, K. A. 1986. Stress-gradient coupling in glacier flow: IV. Effects of the 'T' term. *Journal of Glaciology*, **32**, 342–349.

KENNETT, M. I., LAUMANN, T & LUND, C. 1993. Helicopter-borne radio-echo sounding of Svartisen, Norway. *Annals of Glaciology*, **17**, 23–26.

MULLER, H. C. 1991. *Une méthode iterative simple pour résoudre les équations de mouvement d'un glacier.* Mémoire de Diplôme en Mathematique, Université de Genève.

RICHARDS, K. S., SHARP, M., ARNOLD, N., GURNELL, A. M., CLARK, M., TRANTER, M., NIENOW, P., BROWN, G. H., WILLIS, I. C. & LAWSON, W. 1996. An integrated approach to modelling hydrology and water quality in glacierised catchments. *Hydrological Processes*, **10**, 479–508.

SAPIANO, J. J., HARRISON, W. D. & ECHELMEYER, K. A. 1998. Elevation, volume and terminus changes of nine glaciers in North America. *Journal of Glaciology*, **44**, 119–135.

SHARP, M. J., RICHARDS, K. S., WILLIS, I. C., NIENOW, P., LAWSON, W. & TISON, J-L. 1993. Geometry, bed topography and drainage system structure of the Haut Glacier d'Arolla, Switzerland. *Earth Surface Processes and Landforms*, **18**, 557–72.

VAUGHAN, D. G. 1993. Relating the occurrence of crevasses to surface strain rates. *Journal of Glaciology*, **39**, 255–266.

WILLIS, I., ARNOLD, N., SHARP, M., BONVIN, J.-M. & HUBBARD, B. 1998. Mass balance and flow variations of the Haut Glacier d'Arolla, Switzerland calculated using digital terrain modelling techniques. *In*: LANE, S. N., CHANDLER, J. H. & RICHARDS, K. S. (eds) *Landform Monitoring, Modelling and Analysis*. Wiley, Chichester, 343–361.

# Folding in the Johnsons Glacier, Livingston Island, Antarctica

L. XIMENIS, J. CALVET, D. GARCIA, J. M. CASAS & F. SÀBAT

*Departament de Geodinàmica i Geofísica, Universitat de Barcelona, Martí i Franquès s/n, Barcelona 08028, Spain (e-mail: lximenis@geo.ub.es)*

**Abstract:** An active fold system revealed by interbedded tephra layers is visible on the ablation surface of Johnsons Glacier (Livingston Island, South Shetland Islands, Antarctica). Johnsons Glacier is a cirque-shaped glacier of $5 \, km^2$ area located in a temperate ice cap. Converging flow-lines as a consequence of the reducing channel section extend from ice divides and terminate in a calving ice-cliff. Recent tephra layers from the volcano on Deception Island constitute excellent markers of the internal structure of the glacier and, when dated, provide valuable information about deformation kinematics.

Detailed field mapping of the clearly visible tephra markers revealed several folded layers that define a set of folds with sub-horizontal and fan-distributed axes being sub-parallel to the converging flow-lines. Folds become tighter towards the centre of the channel, where a cylindrical anticline is clearly exposed. Related deformational structures, comprising minor folds, foliations, thrust faults and crevasses, are observed in the ablation area. Strain-rates reflect extensional flow and higher deformation values where maximum confluence occurs.

We examine how the occurrence of this structure results from the transverse shortening in response to the reducing channel section and the consequent confluence of ice masses, increase in differential flow-rates between the centre of the glacier and the margins, and the development of passive folding processes.

An active large-scale fold system revealed by interbedded tephra layers is the most striking feature visible on the ablation surface of Johnsons Glacier. The purpose of this paper is to describe the geometry of the folds and to examine their relationship with other deformational structures, with the aim of understanding the internal structure and relate it with previous studies about the kinematics of the glacier (Calvet *et al.* 1998; Furdada *et al.* in press; Ximenis *et al.* in press). A further attempt is made to provide a hypothesis on deformational processes involved in the generation of such structures.

Folds of varying scale and orientation are commonly observed in glaciers. Vertical plunging folds several kilometres in amplitude have been recognised both within piedmont glaciers (Sharp 1958) and in surging glaciers as a response to fluctuations in discharge at the junction with a non-surging glacier (Post 1966). Recumbent folds with horizontal hinges perpendicular to flow have been modelled and interpreted as being propagated by irregularities in bedrock topography against a background of changes in the flow regime (Hamilton & Hayes 1961; Hudleston 1976; Sharp *et al.* 1988; Chinn 1989). Folds of metric size with axes normally dipping gently up-glacier, parallel to flow-lines are described in

Allen *et al.* (1960), Ragan (1969) and Hambrey *et al.* (1999). Large-scale folds of this type form the main focus of this paper.

Complete structural analyses of glaciers are presented by Meier (1960), Allen *et al.* (1960), Hambrey & Milnes (1977), Hambrey & Müller (1978), Hambrey *et al.* (1980), and by Sharp *et al.* (1988) and Lawson *et al.* (1994) in surging glaciers. Modes of debris entrainment and subsequent transfer in the context of the structural evolution of a glacier as the ice deforms during flow are presented by Hambrey *et al.* (1999).

## Glaciological setting

Johnsons Glacier is a small $(5 \, km^2)$ cirque-shaped glacier, located in a temperate ice cap in Livingston Island, South Shetland Islands, Antarctica (Figs 1, 2). Flow extends from smooth ice divides and terminates over the sea in a 50 m high calving ice-cliff, where the margins are steep, grounded ice-ramps. The glacier surface is convex around the ice divides and concave downstream, with maximum slopes of 20% at the NE sector. The upper part of the glacier occupies a steep semi-circular sector from where flow converges to a narrow outlet due to the reduc-

*From*: MALTMAN, A. J., HUBBARD, B. & HAMBREY, M. J. (eds) *Deformation of Glacial Materials*. Geological Society, London, Special Publications, **176**, 147–157. 0305-8719/00/$15.00 © The Geological Society of London 2000.

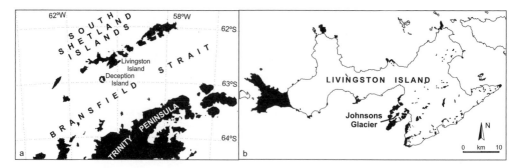

**Fig. 1.** Regional setting of the study area. (**a**) South Shetland Islands; (**b**) Johnsons Glacier on Livingston Island. Rock outcrops represented in black.

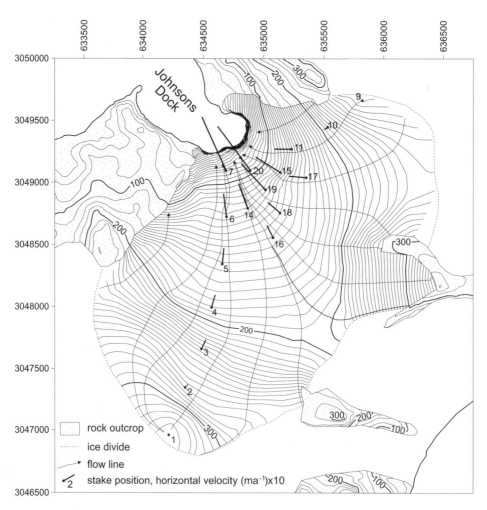

**Fig. 2.** Johnsons Glacier map showing flow-lines, stake positions and horizontal velocities. Contour interval is 5 m at the glacier and 25 m at rock outcrops. Data were obtained by the authors during 1998–99 austral summer by theodolite surveying.

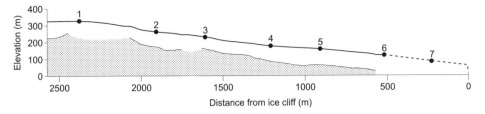

**Fig. 3.** Bedrock and surface topography profile along stakes 1–7 based on seismic results. Discontinuities in bedrock representation correspond to segments where seismic reflections were poorly defined. Stake positions are also plotted.

ing channel section. Estimates of the transverse shortening reveal values higher than 80%. The spacing of the plotted flow-lines, obtained by considering the direction parallel to maximum slopes, becomes tighter to the centre of the channel.

A maximum ice thickness around 100 m was obtained from reflection and refraction seismic data (Benjumea *et al.* 1998). The bedrock topography is undulated with a marked basin in the upper part of the SW sector (Fig. 3).

Horizontal velocities, obtained by stake surveying, increased downstream to reach $50\,\text{m}\,\text{a}^{-1}$ near the snout (stake 7) during 1998. Maximum values are observed to occur at the centre of the channel. Annual mass balance determinations showed maximums of 0.45 m w.e. (water equivalent) accumulation (stake 2) and 3.8 m w.e. ablation (stake 7) (Ximenis *et al.* in press). The equilibrium line altitude ranges from 250 m to 150 m from SW to NE in response to drifted snow accumulation in the NE sector (Furdada *et al.* in press). A rise in the equilibrium line altitude of about 100 m during the last 30 years (Furdada *et al.* in press; Ximenis *et al.* in press) together with an ice cap thinning of 15 m on average from 1956 onwards around the island (Calvet *et al.* 1998) indicate that Johnsons Glacier is losing in volume and hence, it is not in equilibrium.

## Methodology: tephra layer markers

Tephra layers constitute excellent markers of the internal structure and stratigraphy of the ice and, when dated, provide valuable information about ice age and mass balance (Knox 1993). Snow-free glacier surfaces and ice-cliffs around Livingston Island reveal several tephra layers interstratified in the glacial ice (Calvet *et al.* 1993). Sampling was made at several points around the island with the aim of establishing a tephra stratigraphy. Geochemical correlation with nearby Deception Island series, resulted in

the identification of three main groups of tephra layers, corresponding to 1970, 1829–39 (dating process in progress) and previous, undated eruptions (Calvet *et al.* 1993; Smellie pers. comm. 1996; Casas *et al.* 1998).

Livingston tephra stratigraphy can be briefly summarized as follows (Fig. 3). The uppermost group is constituted by a thin single layer with a thickness ranging between a few millimetres and two centimetres. It is fine grained with a predominant grain size of 0.25 mm. This layer corresponds to the 1970 volcanoclastic eruption of Deception which covered the eastern part of Livingston Island (Baker & McReath 1971). Radioactivity analysis of ice cores including this layer confirms this age (Furdada *et al.* in press).

The intermediate group corresponds to a pair of layers that are 0.5 cm thick. They present the finest grain size, with values around 0.125 mm. They are also from Deception Island eruptions and estimates date them to be between 1829 and 1839 (?) (Smellie pers. comm. 1996).

The lowest group is formed by at least five different layers of unknown age, separated by 80–100 cm of ice, with the thickness of each individual layer ranging between 1 and 1.5 cm. It is the most heterogeneous layer in terms of grain size, with values ranging between 2 and 0.125 mm.

This pattern is readily identifiable at all points around the island, suggesting that this sequence follows a homogeneous distribution. It therefore constitutes a useful means of dating the ice rapidly when tephra ages are well constrained.

Detailed mapping of the most clearly visible tephra markers on the snow-free glacier surface was conducted from both theodolite surveying from fixed points on bedrock and terrestrial photogrametry. Bed attitudes were measured by means of a compass clinometer where tephra layers crop out on the glacier surface and sometimes in crevasse walls.

Data were collected during 1995–1998 austral summers.

## Deformational structures

### Types of structures

The structures of a glacier are classified into two groups: those of primary and those of secondary origin (Hambrey 1976). The primary structure that is most visible on the surface of Johnsons Glacier is sedimentary stratification marked by tephra deposits. Flow and related deformation processes give rise to secondary structures, including folds, foliation, thrust faults and crevasses. In Johnsons Glacier, sedimentary stratification is not obliterated by secondary structures and thus any relationship between them can be established.

### Folded sedimentary stratification

Several folded tephra layers crop out on the snow-free glacier surface and in the ice-cliff section perpendicular to the flow direction (Figs 4 & 5). When tephra is not present, it is difficult to discern the sedimentary layers, delineated only by different ice textures. Tephra layers are very good markers of the sedimentary stratification and can unequivocally be interpreted as isochrons. In Johnsons Glacier, the Livingston tephra stratigraphy described above was clearly recognized thus making local dating of ice possible.

The 1970 tephra layer corresponds to the most recent volcanoclastic fall. It is the least deformed tephra layer and its gross structure defines an open anticline, with the NE limb dipping more steeply than that of the SW. The axis strike is 104° and it is horizontal. Ice coring at stake 1 near the ice divide, revealed 1970 tephra at 6.7 m w.e. (Furdada *et al.* in press).

The 1829–39 (?) tephra layers give rise to the most striking surface expression of folding. Mapped traces define a cylindrical anticline, strongly asymmetric, at the centre of the channel (Fig. 6a) with syncline-anticline structures at their limbs. The whole set of fold axes is sub-horizontal and fan-distributed, the central cylindrical anticline striking at 137°. At the ice-cliff, the NE limb drops vertically into the sea and continues in an anticline exposed a few meters above the sea level. Both layers in this group are locally unconformable in response to irregularities on the surface topography when sedimentation occurred, and appear on the surface as either a double or a single layer.

The oldest group of tephra layers comprises a set of 5 or more layers. Mapping them on the ablation ramp is difficult due to the quantity of debris they spread over the surface and to the large number of outcrops corresponding

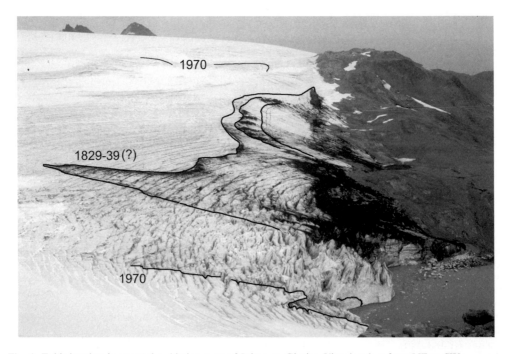

**Fig. 4.** Folded tephra layers at the ablation zone of Johnsons Glacier. View is taken from NE to SW.

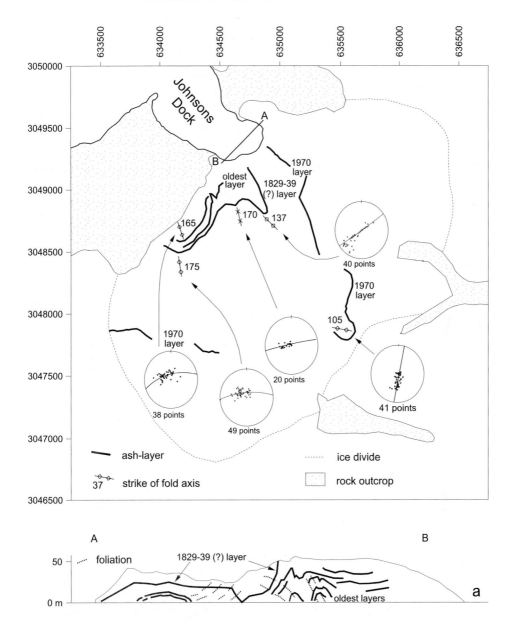

**Fig. 5.** (**a**) Mapped tephra layers with fold axes and poles to tephra layers planes plotted in lower hemisphere projections. (**b**) Ice-cliff section showing position of tephra layers and foliation.

to original layers and traces multiplied by strong deformation. Nevertheless, two mapped tephra layers define a structure similar to that of the upstream 1829–39 (?) folded traces, both on the surface and at the lowest position in the ice-cliff.

If we look at the structure as a whole, the tephra layers in the NE sector (mainly cropping out at the ice-cliff) lie deeper than they do in the SW sector (where they are present on the glacier

surface). This indicates transverse differences in ice age on the surface, presumably due to mass balance differences (Furdada *et al.* in press, Ximenis *et al.* in press).

*Minor folds*

Intense minor sub-isoclinal folding affects the 1829–39 (?) tephra layers (Fig. 7). The two

**Fig. 6.** Ice structures. (**a**) Core of the central anticline defined by 1829–39 tephra layers. Thrust fault affects right limb. (**b**) Minor folds at 1829–39 tephra layer. (**c**) Debris rich ridges along foliation. (**d**) Bluish lenses defining foliation at the ice-cliff, also shown in Fig. 5b.

layers that constitute this level form folds of different geometry. Wide folds of metric wavelength and a few decimetres in amplitude mainly affect the upper layer whereas folds of decimetric wavelength and centimetric amplitude (Fig. 6b) tightly fold the lower layer. Minor fold axes are parallel to the large-scale folds and plunge slightly up- and down-glacier at the SW and NE limbs of the gross anticline, respectively.

As tightly folded tephra layers emerge, hinge lines develop into linear ice-cored ridges of debris that widen downstream (Hambrey *et al.* 1999). The ridges typically extend from several metres to a few tens of metres across the ice surface, progressively losing their linearity and becoming a chaotic field of ice-cored cones of debris.

*Foliation*

Several structures define foliation in the zone around the central anticline defined by the 1829–39 (?) tephra layer (Fig. 7). The very dense alternation of millimetric ice layers of different texture defines the foliation on the glacier surface, commonly revealed by parallel layers of fine debris. Where observed, this foliation shows a

strike around 135° and is sub-vertical. It is not homogeneously visible due to the rugose ice surface caused by the spreading of tephra particles. However, local observations seem to indicate that it becomes more intense towards the core of the anticline. Small-scale folding of foliation occurs locally in crevasse-disturbed areas.

Secondly, decimetre-spaced penetrative fractures were observed to broadly affect the NE limb and the core of the central anticline. They are sub-vertical and the strike changes downstream from 155° to 115°. The extrusion of englacial tephra along these fractures generates longitudinal ridges tens of metres in length (Fig. 6c) as documented in Hambrey *et al.* (1999).

A third type of foliation is revealed in the NE sector of the ice-cliff. Observation of the structure in two dimensions in the ice-cliff wall, reveals thick discontinuous bodies in the form of bluish lenses of non-bubbly ice (Fig. 6d). They take steep dips to NE and SW, with increasing values towards the centre of the channel, becoming parallel to the axial surface of the central anticline (Fig. 5a). This kind of foliation is thought to correspond to rotated crevasse traces as described in Hambrey *et al.* (1980).

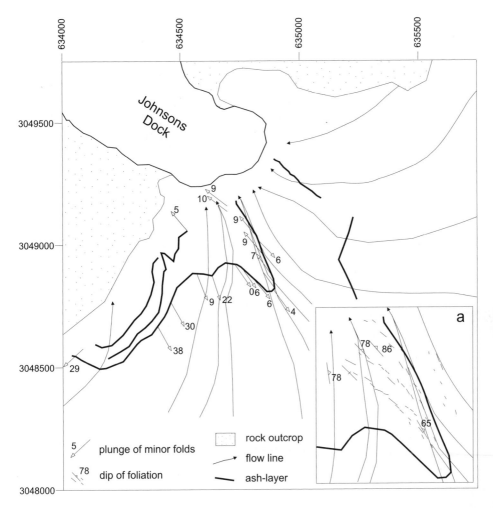

**Fig. 7.** Detail of the confluence area showing minor folds. Inset: Enlarged area of the main anticline showing the foliation.

## Thrust faults

Arcuate planar surfaces striking normal to the flow direction and dipping up-glacier at angles of 20–30° are revealed when overlying stratification truncates underlying stratification (Fig. 8). Such structures could result from both sedimentary accumulation or thrust faulting. If the latter, tephra layers lying at the sliding plane would facilitate motion while the underlying ice slows down.

Smaller-scale planar structures were observed in crevasse walls within the main anticline defined by the 1829–39 (?) tephra layer. They comprise near-vertical planes dipping towards the margins and striking parallel to the axis of the large fold. In the NE limb of the anticline, this structure results in the duplication of tephra layer outcrops and, at the deeper layers of its core, near stake 7 small-scale fault-related folds have developed at the hanging wall (Fig. 6a). They would seem to represent thrust faults related to transverse shortening.

## Crevasses

Widespread arcuate tensional fractures form a regular system visible at the lower part of the glacier (Fig. 8). Crevasses are concave downstream, and they range from nearly closed to tens of meters open near the snout, where a slip

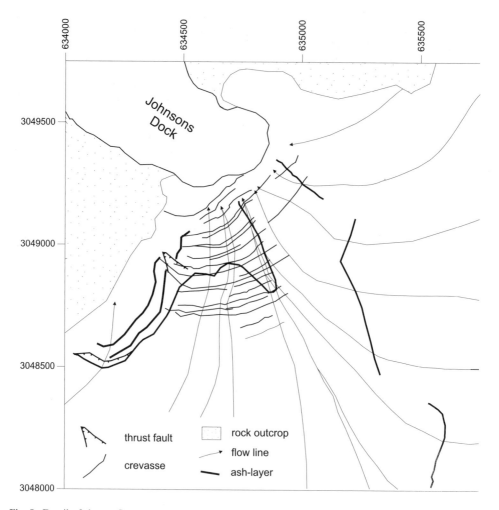

**Fig. 8.** Detail of the confluence area showing main crevasses and thrusts faults.

component exists. Their geometry changes progressively down-glacier, becoming more curved and sometimes S-shaped.

En echelon crevasse systems occur at the core of the anticline and near its NE limb, resulting in dextral strike-slip rotation which could be quantified by measuring the distance between the two segments of a broken longitudinal structure. Values obtained at stake 7 reach 30° clockwise rotation.

## Surface strain-rates at the confluence area

In detailed strain analyses over the glacier surface, our interest has centred on the orientation and magnitude of the surface strain-rates and on their relation to glacial structures (Hambrey &

Müller 1978). These values were established in an attempt to determine how the structural pattern changes in relation to progressive deformation in a mass of ice, as it moves downstream. Nevertheless, in terms of the development of structures such as foliation and folds, it is important to consider the total sum of these increments, i.e. the cumulative strains (Hambrey *et al.* 1980). Incremental strain takes into account deformational processes during a period of time which could be fortuitous with regard to structural development, whereas cumulative strains include the whole deformational history, considering the originally non-deformed body in relation to the present deformed state.

Deformation of triangular arrays with sides of about 100 m derived from stakes surveying, was considered to determine principal strain-rates in Johnsons Glacier. The assumption of homogeneous strain within the strain nets has to be made, thus considering the results to be an average of deformation within triangles. The area studied was limited to the low ablation zone (Fig. 9).

Horizontal co-ordinates of stakes were considered and adjusted to periods of exactly one year. Strain triangles were constructed, resulting in strain-rates that represent the accumulated deformation during annual periods. Plane strain conditions were assumed when considering only horizontal co-ordinates, i.e. velocity independent of depth. Any other errors are likely to have arisen from surveying inconsistencies.

Deformation is expressed by principal elongation rates and their orientations relative to the deformed triangles, obtained from the Mohr circle construction (Ragan 1973, pp. 47–48). Deduced principal logarithmic strain-rates for annual periods and angular shear strain-rates are presented in Fig. 9.

Strain-rates became higher downstream at the centre of the channel. Maximum values appeared to be $0.9\,a^{-1}$ extension and $0.6\,a^{-1}$ compression. Most rates reflect longitudinal extensional flow, extension being sub-parallel to flow-lines. The orientation of extension axes rotates progressively with flow, converging around $135°$ at the centre of the channel. Shear strain-rates obtained from the principal strain-rate data are mapped in Fig. 9 (inset). Maximum values reaching $0.05\,a^{-1}$ are observed at the centre of the channel.

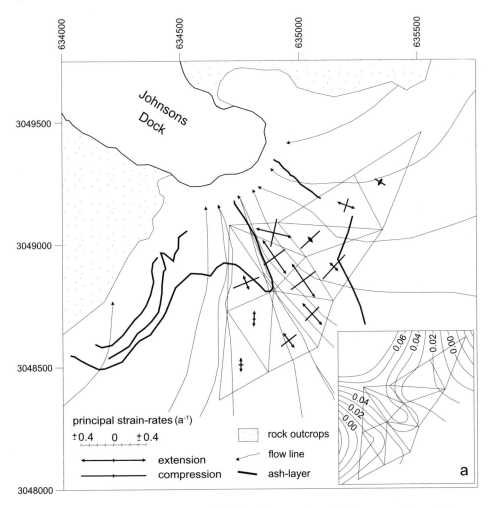

**Fig. 9.** Detail of the confluence area showing principal strain-rates. Inset: Contour map of shear-strain distribution.

## Discussion

### *Relation of motion and strain to structures*

The distribution of deformational structures within the ablation zone is far from homogeneous. The large-scale folds become tighter towards the centre of the channel, where the main cylindrical anticline is exposed. Longitudinal foliation and sub-vertical thrust faults dipping to the margins are well developed within the core and around its steeply dipping NE limb. Such structural association reflects an area where strong deformation occurs.

The orientation of large-scale and minor fold axes is slightly fan-distributed, converging towards the snout. At the centre of the channel, these deformational structures have a consistent strike around 135°.

The relation of flow direction to structures denotes a clear association between major confluence and the strongly deformed zone where most structures are developed, including the tightest expression of folding.

A striking 30° divergence downstream the main anticline defined by the 1829–39(?) tephra layer responds to a sudden change in flow direction from that point onwards. This area is broadly affected by dextral strike-slip faulting, generating a rotation of such values and in the same sense. Thus, the lateral displacement of ice blocks by this process is thought to be responsible for this anomaly.

In the ablation zone, the highest horizontal velocities are registered within the main anticline. Nevertheless, more detailed data along a transverse profile would be needed to determine the exact location of maximum velocities.

In relation to vertical velocities, the strong asymmetry of large-scale folds in terms of higher dips in the NE limb can be related to the low equilibrium line altitude at this sector, where a decrease of 100 m occurs due to accumulated drifted snow.

The strain-rates pattern accounts for the distribution and the nature of deformational structures quite well. Again, maximum strain-rates are obtained at the so-called strongly deformed zone where major confluence occurs. The orientations of principal axes of strain are coherent with structures in that large-scale and minor folds, foliation and sub-vertical thrusts are nearly perpendicular to the direction of com-pression whereas crevasses and arcuate thrusts lay perpendicular to the direction of extension.

### *Deformational process*

Strongly folded ice typically is of similar type (Hudleston 1977). Construction of dip isogons demonstrates that the large-scale folds of Johnsons Glacier are sub-similar (P. Santanach pers. comm. 1992). A true similar fold geometry implies a lack of contrast in effective viscosity between the layers. This type of folding has been termed 'passive' because the layering merely reflects the geometry of the deformation and has no mechanical significance, thus the ice behaves homogeneously at the fold-scale (Donath & Parker 1964). Hudleston (1977) examines how differential simple shear is able to account for the formation of recumbent, sub-similar folds striking parallel to the ice-cap margins and suggests very large shear strains under the influence of gravity and no contrast in effective viscosity between layers. The folds studied here present a different disposition with a sub-vertical axial plane sub-parallel to flow, but they seem to respond to the same deformational process with different stress orientations. In this case, the reducing channel section may generate large shear strains.

## Conclusions

Fold axes and foliation directions together with extensional strain-rates in the confluence area show a consistent strike around 135°. This direction shows a 30° divergence with flow-lines and velocity-vector directions downstream the 1829–39(?) tephra layer outcrop. Dextral strike-slip faulting is suggested as being responsible for this anomaly.

Where higher strain-rates occur, tight fold geometry and widespread deformational structures are developed. This corresponds to the area of maximum velocity and maximum confluence, both found in the reducing channel section. Hambrey *et al.* (1999) show that large-scale folding may be associated with glaciers of a particular morphology, that is, those containing multiple accumulation basins, feeding into a narrow tongue with no icefall, explaining in this way, the strongly converging flow and derived transverse compression.

In the NE sector, tephra layers do not appear on the surface but rather in the ice-cliff itself, as a consequence of transverse differences in vertical flow due to higher snow accumulations in the NE sector.

Tephra layers of varying age, considered as isochrons, represent different steps in the evolution of the large-scale structure of the glacier, thus showing different stages of deformation. Then, the upstream folds in each train are expected to mature in time to forms similar to those seen downstream in the same train. Steady-state conditions during the deformation process are needed in order to assess this.

The cirque-shaped geometry of the whole glacier and the rapid narrowing downstream result in the folding of the sedimentary stratification. We propose that this structure results from the transverse shortening in response to the reducing channel section and the consequent confluence of ice masses, the increase in differential flow-rates between the centre of the glacier and its margins, and the development of passive folding processes.

This study was supported by the Projects CICYT ANT-93/0852 and CICYT ANT-96/0734.

# References

ALLEN, C. R., KAMB, W. B., MEIER, M. F. & SHARP, R. P. 1960. Structure of the lower Blue Glacier, Washington. *Journal of Geology*, **68**, 601–625.

BAKER, P.E & McREATH, I. 1971. 1970 volcanic eruption at Deception Island. *Nature* (Physical Science), **231**, 5–9.

BENJUMEA, B., TEIXIDÓ, T., CASAS, J. M., XIMENIS, L. & FERICHE, M. 1998. 2D seismic survey on Johnsons Glacier, Livingston Island (Antarctica). *In*: CASAS, A. (ed.) *IV Meeting of the Environmental and Engineering Geophysical Society*. Instituto Geográfico Nacional, Madrid, Proceedings, 591–594.

CALVET, J., CASAS, J. M., CORBERA, J., ENRIQUE, J., FERNANDEZ DE GAMBOA, C., FURDADA, G., PALÀ, V., PALLÀS, R., VILAPLANA, J. M. & XIMENIS, L. 1998. Disminución del casquete glaciar de la isla Livingston. *Geogaceta*, **24**, 71–74.

——, PALLÀS, R., SÀBAT, F. & VILAPLANA, J. M. 1993. Los niveles de cenizas de los glaciares de Livingston, Criterios para su datación. *In*: SERRAT, D. (ed.) *V Simposio Español de Estudios Antárticos*. Comisión Interministerial de Ciencia y Tecnología, Madrid, Actas 195–208.

CASAS, J. M., SÀBAT, F., VILAPLANA, J. M., PARÉS, J. M. & POMEROY, D. M. 1998. A new portable ice-core drilling machine: application to tephra studies. *Journal of Glaciology*, **44**, 179–181.

CHINN, T. J. H. 1989. Single folds at the margins of dry-based glaciers as indicators of a glacial advance. *Journal of Glaciology*, **12**, 23–30.

DONATH, F. A. & PARKER, R. B. 1964. Folds and folding. *Geological Society of America Bulletin*, **75**, 45–62.

FURDADA, G., POURCHET, M. & VILAPLANA, J. M. in press. Characterization of Johnsons Glacier (Livingston Island, Antarctica) by means of shallow ice cores and their tephra and [137]Cs content. *Acta Geológica Hispánica*.

HAMBREY, M. J. 1976. Structure of the glacier Charles Rabots Bre, Norway. *Geological Society of America Bulletin*, **87**, 1629–1637.

—— & MILNES, A. G. 1977. Structural geology of an Alpine glacier (Griesgletcher, Valais, Switzerland) *Eclogae Geologicae Helvetiae*, **70**, 667–684.

—— & MÜLLER, F. 1978. Structures and ice deformation in the White Glacier, Axel Heiberg Island, Northwest Territories, Canada. *Journal of Glaciology*, **20**, 41–66.

——, BENNETT, M. R., DOWDESWELL, J. A., GLASSER, N. F. & HUDDART, D. 1999. Debris entrainment and transfer in polythermal valley glaciers. *Journal of Glaciology*, **45**, 69–86.

——, MILNES, A. J. & SIEGENTHALER, H. 1980. Dynamics and Structure of Griesgletscher, Switzerland. *Journal of Glaciology*, **25**, 215–228.

HAMILTON, W. & HAYES, P. T. 1961. Structure of lower Taylor Glacier, South Victoria land, Antarctica. *In*: *Short papers in the Geologic and Hidiologic Sciences*. US Geological Survey, Professional Papers, 424-C, C206–C209.

HUDLESTON, P. J. 1976. Recumbent folding in the base of the Barnes Ice Cap, Baffin Island, Northwest Territories, Canada. *Geological Society of America Bulletin*, **87**, 1684–1692.

——1977. Similar folds, recumbent folds, and gravity tectonics in ice and rocks. *Journal of Geology, Geological Notes*, **85**, 113–122.

KNOX, R. W. O'B. 1993. Tephra layers as precise chronostratigraphical markers. *In*: HAILWOOA, E. A. & KIDD, R. B. (eds) *High Resolution Stratigraphy*. Geological Society, London, Special Publications, **70**, 169–186.

LAWSON, W. J., SHARP, M. J. & HAMBREY, M. J. 1994. The structural geology of a surge-type glacier. *Journal of Structural Geology*, **16**, 1447–1462.

MEIER, M. F. 1960. *Mode of flow of Saskatchewan Glacier, Alberta, Canada*. US Geological Survey, Professional Papers, **351**.

POST, A. S. 1966. The recent surge of Walsh Glacier, Yukon and Alaska. *Journal of Glaciology*, **6**, 375–381.

RAGAN, D. M. 1969. Structures at the base of an ice fall. *Journal of Geology*, **77**, 647–667.

——1973. *Structural Geology*. New York, John Wiley & Sons Inc., 208.

SHARP, M., LAWSON, W. & ANDERSON, R. S. 1988. Tectonic processes in a surge-type glacier. *Journal of Structural Geology*, **10**, 499–515.

SHARP, R. P. 1958. Malaspina Glacier, Alaska. *Geological Society of America Bulletin*, **69**, 617–646.

XIMENIS, L., CALVET, J., ENRIQUE, J., CORBERA, J. & FURDADA, G. in press. The measurement of ice velocity, mass balance and thinning-rate on Johnsons Glacier, Livingston Island, South Shetland Islands, Antarctica. *Acta Geológica Hispáncia*.

# The Pasterze glacier, Austria: an analogue of an extensional allochthon

PAUL HERBST & FRANZ NEUBAUER

*Institute of Geology and Palaeontology, University of Salzburg, Hellbrunnerstr. 34,*
*A-5020 Salzburg, Austria (e-mail: paul.herbst@sbg.ac.at; franz.neubauer@sbg.ac.at)*

**Abstract:** Structures of the Pasterze glacier (Austria) have been studied in detail and interpreted as representing a natural model of an extensional allochthon formed on top of an orogenic wedge, and also a model for raft tectonics at passive continental margins. Structures of the lower, ablation-controlled portion of the glacier, including ductile structures at the base, are forming close to the melting point of ice, with predominantly brittle structures close to the surface of the glacier. Formation of these structures results from self-weight, gravity-driven spreading, similar to extensional allochthons.

S-planes are common and clearly related to three individual flow units. They show a typical, spoon-like arrangement. These three flow units are bordered by centimetre- to decimetre-wide shear zones which also include shear folds. Brittle structures include ice-mineralized tension gashes, thrust and normal faults, and hybrid, shear-extensional fractures. Together, these structures show that glacial flow is more rapid in middle to upper sectors of the glacier than along the lateral, and lower margins. The coherent upper sheet is behaving in a brittle manner and is elongating slightly along the flow direction by tensional deformation.

The distribution of structures allows three structural domains within the lower, ablation-controlled sector of the glacier to be distinguished: (1) an upper sector with predominant extensional structures due to rapid flow; (2) a lower sector with ductile and brittle thrust faults, penetrating from the ground and dipping strictly opposite to the local flow direction; (3) a few normal faults at the terminus that developed by rapid melting along the steep lower frontal margin of the glacier. These three structural domains are also found within extensional allochthons as exemplified by the Neogene Alpine–Carpathian system where a huge allochthon, partly driven by gravity, extruded from the Eastern Alps towards the Carpathian arc. Three similar structural domains are also found in recent analogue models and field examples of passive continental margins.

The structural geology of glaciers has been researched in temperate (Harper & Humphrey 1995) and surge-type-glaciers (Sharp *et al.* 1988; Pfeffer 1992; Lawson *et al.* 1994; Lawson 1996) in Alaska and Canada (Hambrey & Müller 1978) and on glaciers in the European arctic and subarctic region (Croot 1987; Hodgkins & Dowdeswell 1994; Glasser *et al.* 1998; Soerrensen *et al.* 1998). Only limited structural work has been performed on Alpine glaciers (e.g. Schwarzacher & Untersteiner 1953; Hambrey & Milnes 1977). This paper presents new structural data on the Alpine Pasterze valley-glacier and compares the structures with analogues in geological units that are similarly driven by lateral downward motion, that is, in the extensional allochthons and raft tectonics at passive continental margins.

An extensional allochthon represents a near-surface, gravity-driven tectonic body with a basal low-angle normal fault. The faults form at the base of tilted blocks at passive continental margins (Lister *et al.* 1986) or at the base of extensional allochthons formed during late orogenic stages (e.g. Platt 1986; Malavieille 1993; Wang *et al.* 1999). An extensional allochthon can be often found to represent upper, coherent brittle crust where a normal fault forms at the transition with lower ductile crust. Consequently, a common feature is the rheological control of the location of the basal normal fault. This can be either a specific, weak lithology or a 'weak' mineral forming a major rock constituent or temperature (with normal downwards increase of temperature). The structure of extensional allochthons is very similar to raft tectonics found at passive continental margins (Mauduit *et al.* 1997; Mauduit & Brun 1998). Such structures form in a slightly tilted system when a mechanically strong, brittle, upper layer floats on a weak lower layer (e.g. lower crust). The types of structures are then entirely controlled by the strain-rate influenced behaviour of the lower ductile layer (Mauduit *et al.* 1997).

*From*: MALTMAN, A. J., HUBBARD, B. & HAMBREY, M. J. (eds) *Deformation of Glacial Materials*. Geological Society, London, Special Publications, **176**, 159–168. 0305-8719/00/$15.00 © The Geological Society of London 2000.

We compare the structures of the Pasterze glacier with the large extensional allochthon that extends from the central Eastern Alps (eastern margin of the Tauern window) via the Pannonian basin to the frontal Carpathian mountain belt (Ratschbacher *et al.* 1989, 1991*b*; Peresson & Decker 1997; Neubauer *et al.* 2000, and references therein). This extensional allochthon initiated by Neogene shortening in the Eastern Alps and is laterally confined by approximately orogen-parallel wrench corridors. The extensional allochthon floats on metamorphic lower crustal rocks which were exhumed within metamorphic core complexes during the main stage of eastward motion, due to east-west extension within the allochthon. The main force after initiation of the eastward motion is believed to be gravity, due to a strong topographic gradient from the high eastern Alps to the Pannonian basin, an idea supported by analogue modelling (Ratschbacher *et al.* 1991*a*).

## Glaciology of the Pasterze

The Pasterze glacier is situated in Carinthia, Austria, in the Hohe Tauern National Park (Fig. 1 insert), just beyond the highest mountain of Austria, the Grossglockner (3797 m a.s.l.). With a glaciated area of more than 20 km² in 1993 it represents the biggest glacier in Austria.

The glacier itself is divided morphologically into three parts. (1) The accumulation area with a broad lateral extension up to 5 km. It covers an area of almost 15 km², shows gentle slopes (22° on average) and an average surface velocity of 4 m a$^{-1}$ (Patzelt 1998). (2) The so-called Hufeisenbruch, which is a steep, heavily crevassed ice fall, where the whole ice volume from the wide accumulation area must flow through a pipe only 800 m wide. The icefall shows slopes up to 50°. (3) The lowest part is the long glacier tongue which falls very gentle to the terminus (5°) over a length of 4250 m.

The glacier ends (in 1998) at 2120 m above sea level. The whole glacier is regarded as temperate, by analogy with other glaciers in similar locations latitude in the Alps (Haeberli 1976). The maximum thickness of 177 m was measured by ground penetrating radar (Span pers. comm. 1997), 2.5 km down the ice-fall.

Like most Alpine glaciers this one is in retreat (Patzelt 1999), with ablation of up to 6.4 m a$^{-1}$ (Tintor 1997). Since the glacier's maximum at the end of the 'Little Ice Age' in the 1850s it lost 130 m of its thickness (Wakonigg & Lieb 1996). Through the loss of so much volume, some rock windows opened in the ice-fall, which separate the glacier into three individual flow units with good visible, till-covered margins, different velocities and different structural features.

The velocities have been measured almost continuously since 1899 on several cross-profiles. They are highest just beyond the icefall and reach almost zero at the terminus (Patzelt 1998, 1999).

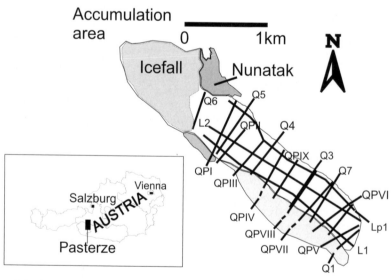

**Fig. 1.** Map of the Pasterze Glacier, Austria showing the area of investigation and lines of measurements. Insert displays location of the Pasterze glacier in Austria.

**Fig. 2.** Structures of the Pasterze glacier. (**a**) Foliation of the ice marked by varying debris content of the ice and variable grain size of ice crystals. (**b**) Shear-folds with steeply plunging fold axis and subvertical axial surface on the boundary between two icestreams. (**c**) Open fold with subvertical axial plane and subhorizontal fold axis in the compressive milieu just beyond the icefall. Flow from right to left. (**d**) Till transportation to the surface at a thrust fault. Flow from left to right. (**e**) Tension gash perpendicular to flow; the ends are already closed by new epitaxially grown ice crystals. (**f**) Thrust fault near the terminus. Flow from right to left. (**g**) Overview of the Pasterze glacier. Note the icefall in the background. Tangential thrust faults are visible at the terminus, and sigmoidal crevasses.

## Structures

### Foliation

The term *foliation* is used here in the sense of a planar, metamorphic-like structure (Ragan 1967). In the usage of Passchier & Trouw (1996), it is a continuous secondary foliation, ranging through varying grain-sizes from cleavage to schistosity. The foliation therefore, represents a good marker for the local strain and stress regime within the glacier.

The foliation is the oldest feature within the whole glacier. Contents of dust and debris and varying contents of air bubbles make the foliation visible (Fig. 2a). These inhomogenities, which are in part already built up during sedimentation in the accumulation area, are brought by the stress regime in the glacier to a position which is more or less rectangular to the local existing direction of shortening (Hooke & Hudleston 1978). The orientation of the foliation was measured along several profiles (for locations of profiles, see Fig. 1) covering the entire area of the investigations. A summary of the measurements can be seen on Fig. 3a. The foliation on the glacier shows a range of appearances, varying in spacing from few centimetres to decimetres, grain-size (mm to cm) and debris content. The orientation of the foliation follows strictly the flow lines, and is therefore a good marker for separating the three flow units. On each flow unit (termed 'NE', 'middle' and 'SW') the foliation shows the same spoon-like orientation, with high angles on the sides of the ice-streams, and smaller angles in the centre and also at the terminus of those streams (Fig. 3a). Lateral sectors of individual flow units are characterized by steep dips towards the shallow-dipping central sectors. Several generations of foliation can be interpreted from overprint criteria (see ductile Shear Zones).

### Folds

Due to the well-developed foliation, folds are visible with many details. They are found in two settings on the glacier: (1) along the shear zones that mark the boundaries between the three flow units; (2) below the ice fall in the northwestern sector of the area of investigation.

Along the shear zones the fold-axes plunge very steeply (angles between 50 and 80°) and plot along a great-circle (see Fig. 3b). They are generally shear-folds (Fig. 2b) with thickened layers in the fold hinges, and appear with limb-lengths between a few centimetres to some

metres. The interlimb angle can reach 50° but most of the folds can be regarded as isoclinal. The axial planes are orientated nearly parallel to the local direction of strike-slip-movement.

The folds below the ice fall show almost vertical axial planes with an orientation perpendicular to the flow direction which is also the direction of the principal stress $\sigma 1$. These folds are open folds with interlimb angles around 100° (Fig. 2c). Sometimes they show the appearance of 'mushroom folds' (e.g. Ragan 1969), thought to have formed in a compressional regime.

### Ductile shear zones

Ductile shear zones mark the borders between the three individual ice streams in association with the shear-folds (see above) and multiple generations of foliation. They are typically decimetres to metres wide, show strongly developed, regular planar foliations that cut earlier ones, and recrystallized ice grains that are smaller than outside the shear zones. Zones can be traced over several metres along strike, and have an en échelon arrangement.

### Normal faults

Normal faults are found along the lateral margins of the glacier, especially on the northeastern margin. They often appear as cracks with widths of up to 4 m (Fig. 4a); their planes dip steeply towards the margin. These faults are a sign of the relaxation along the margin due to loss of ice-mass and therefore loss of constraint from the surrounding bedrock. Normal faults also appear both at the terminus of the glacier, again as a result of relaxation and loss of confinement, and in the middle part of the glacier due to high surface velocities (Fig. 4a).

### Thrust faults

Thrust faults occur in the lowermost part of the glacier. They appear over the whole glacier, affecting all the flow units (Fig. 3g). Several major thrust faults can be followed across the whole glacier near the terminus (Fig. 4a). They seem to develop as a result of the push from the back which cannot be transported through the glacier since the velocity at the terminus is zero. The surfaces dip at right angles to the direction of flow, with dip angles between 25° and more than 90° so that even overturned reverse thrusts can be seen (Fig. 4a). The thrust faults are in contact

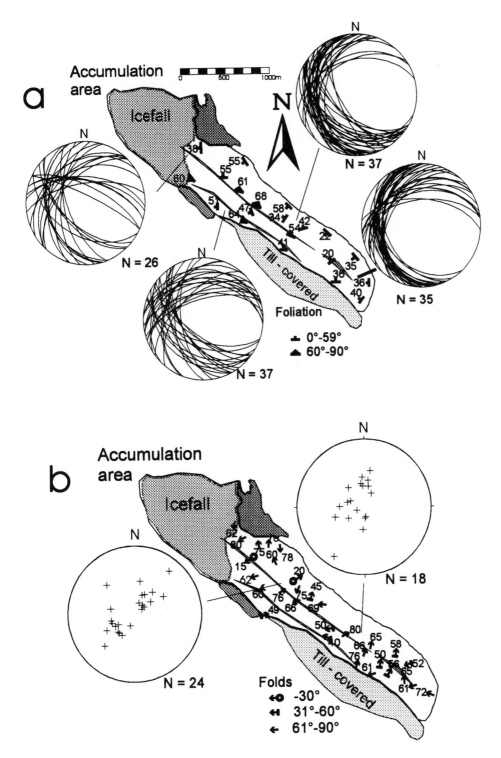

**Fig. 3.** (a) Map displaying the distribution and orientation of foliation. (b) Map displaying the distribution and orientation of fold axes. Diagrams are equal-area projection, lower hemisphere.

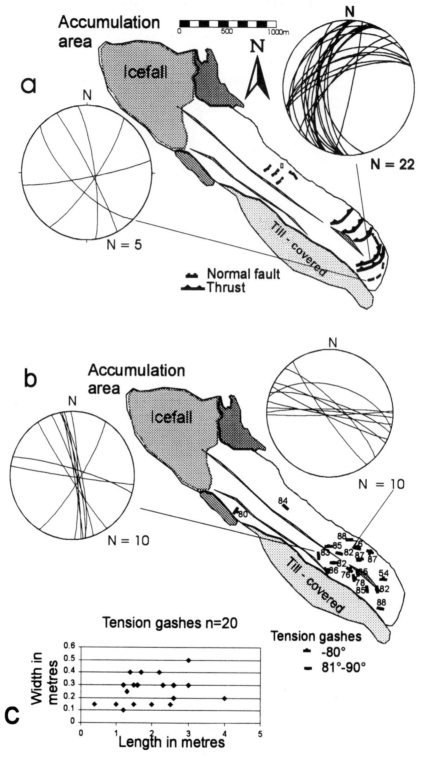

**Fig. 4.** (a) Map displaying the distribution and orientation of thrust faults and normal faults. (b) Map displaying the distribution and orientation of tension gashes. Diagrams are equal-area projection, lower hemisphere. (c) Diagram displaying the length v. width of tension gashes.

with the bed, since there is till transportation to the glacial surface at some locations (Fig. 2d). Both brittle (Fig. 2f) and ductile thrust faults (Fig. 2g) occur close to the terminus.

## Tension gashes

Subvertical tension gashes occur typically in the middle of the flow units, showing an orientation at right angles to flow (Fig. 4b). They are commonly arranged in an en échelon pattern. Their length and depth vary from 1 to 5 m; the width from 10 to 50 cm (Fig. 4c). They are filled with water and show epitaxial mineralization beginning with dendritic growth of small, several millimetres thick, ice-crystals which grow to a length of more than 10 centimetres. The space between those crystals then fills with similar, larger crystals until the whole gash is filled and closed (Fig. 2e). These gashes are typical for

an extensional regime with a lateral velocity-decrease perpendicular to the flow direction.

## Shear-extensional fractures

Generally shear-extension fractures only occur near the lateral margins of the glacier. They are in two principal orientations: (1) at a small angle to the local flow-direction; (2) parallel to the side-margins of the glacier. These are interpreted to represent Riedel shear fractures, which were later transformed in part into tensional fractures by rotation (Fig. 2g).

## Discussion

### Glacial structures

The structures reported above allow the division of the ablation area of the glacier into several

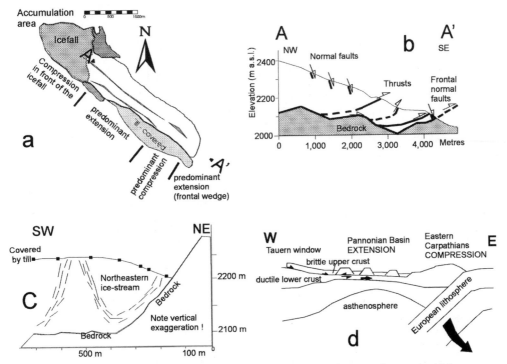

**Fig. 5.** Interpretation of the Pasterze glacier as an extensional allochthon in comparison to the Alpine–Carpathian region. (**a**) Variation of style of deformation along the glacier. (**b**) Longitudinal section, based on results of Ground Penetrating Radar measurements (Span pers. comm. 1997; orientation A–A′ is located in a). (**c**) Cross-section of the glacier (c. QP3 in Fig. 1b). (**d**) Schematic model of an extensional allochthon formed during late-stage orogenesis; the example of the Alpine–Pannonian–Carpathian system. The section is approximately perpendicular to the orogenic front. Note the topographic gradient from high mountains (Tauern window area) in the west and the low Pannonian basin. The upper brittle crust is extended by normal faults resulting from the eastward motion. The East Carpathians represent the compressive domain due to collisional blocking of the system.

structural domains (Fig. 5a). These include (1) an upper, very long area with tensional structures and three individual, well-developed flow units, (2) a lower area with thrusts and (3) a frontal wedge with normal faulting where ablation prevails (Fig. 5b). The upper tensional sector represents an area where the individual flow units flow in a near steady state manner, mainly governed by the flow along a basal normal fault surface. The lateral shear zones may illustrate what this basal surface of motion looks like. Some water may enhance the ductile behaviour at the basal shear zone. At the surface, the glacier shows transitions to brittle behaviour due to the high tensional stresses there (due to the low load) and more rapid flow than along the basal shear zone.

A model of the deep structure of one flow unit is shown in Fig. 5c. Note the locations of lateral shear zones which are thought to match the basal shear zone.

The sector of thrusting shows that velocity at the basal shear zone decreases in relation to the upper sectors and that the glacier may thin out there, individual thrust surfaces splay from the basal shear zone. Similar structures are generally observed in frontal sectors of gliding rock masses, where they accumulate due to velocity decrease.

The formation of the extensional frontal wedge is driven by superposition of several effects including: ablation by erosional melting of the glacier, oversteepening of the topography and its destruction due to basal erosion by melt water.

## Extensional allochthon

The Pasterze glacier is divided into several structural domains which vary from the upper part of the flow unit to the terminus (Fig. 5a, b). These domains are: (1) areas with predominant extension in direction of overall low, in upper portions of the glacier; (2) areas with predominant compression in direction of motion in lower parts of the glacier; (3) a small area with extensional deformation in frontal sectors of the terminus of the glacier.

These domains represent prominent features of the glacier and may be explained in the following way: the upper extensional area is an area with rapid flow, controlled by the strain rate along the basal, active shear zone. The tension forces induced in the upper brittle sectors of the glacier are high and cannot be accommodated by the plastic flow that is likely from the lower temperature and larger grain sizes of the ice crystals.

The compressional sector accumulates the rapidly flowing material from the extensional sector. The physical expression is the thrust faulting which also forms a steeper surface topography than that present in the upper extensional sector. Consequently, the frontal extensional sector of the glacier represents an extensional wedge along which the oversteepened frontal terminus of the glacier decays by the following processes: normal faulting due to melting at the base, gravity driven ice sliding, and formation of block breccia due to loss of contacts.

This situation of this valley glacier is similar to the extensional allochthons that form in late orogenic stages within mountain belts. We use the Neogene to Recent Alpine–Carpathian system as an example (Ratschbacher et al. 1989, 1991b; Peresson & Decker 1997; Neubauer et al. 2000; Fig. 5d).

The extensional allochthon is evident from the lower margin, ductile low-angle normal fault, exposed at the eastern margin of the Tauern window. The Tauern window forms a topographic high from which the extensional allochthon is moving away, in part driven by gravity, towards the east (Ratschbacher et al. 1991b; Neubauer et al. 2000; and references therein). From there and towards the east, the extensional allochthon is extended in an east–west direction along high- and low-angle normal faults. The frontal terminus of the extruding wedge is formed by the East Carpathian orogen (Fig. 5d). The extruding wedge is laterally confined by wrench corridors, such as the sinistral northern and dextral southern corridor. In analogy with the glacier, the extruding wedge forms an upper brittle sector. Furthermore structural domains are similar, including the main extensional sector, here expressed by normal faults. The East Carpathian orogen forms the accumulation area with compressive structures, mainly outwards verging thrusts. The external side of the East Carpathian orogen can be interpreted to represent the frontal erosional wedge with extensional structures.

Two structural domains also equivalent to glaciers can be found at passive continental margins (Mauduit et al. 1997): (1) the upper extensional area, with tilted blocks and roll-over structures; (2) a lower, oceanward domain with some compressional structures. The analogy between glacier and passive continental margins suggests that similar processes may be responsible for the formation of structures. Mauduit et al. (1997) argued that the governing process is the strain-rate control of deformation within the lower ductile layer, which controls the distribution and type of structures within such systems.

In summary, valley glaciers represent a good natural analogue to both extensional allochthons within convergent orogenic systems and raft tectonics at passive continental margins. The formation of both bodies is guided by similar processes and similar relative, external factors, such as temperature and rheological behaviour.

We acknowledge discussions with H. Slupetzky. The final version of the manuscript was influenced by fruitful discussions with J.-P. Brun (Rennes). We thank N. Span, Innsbruck University, for radar data from the Pasterze glacier. We also thank the National Park Institute of Salzburg, Haus der Natur, for the logistic support. We acknowledge careful reviews and suggestions by N. Glasser, B. Marmo and A. Maltman. These helped to clarify ideas and their presentation. P.H. gratefully acknowledges support by the Herbst Family & Co. Foundation for continuous support during the course of the study.

# References

CROOT, D. G. 1987. Glacio-tectonic structures: a mesoscale model of thin-skinned thrust sheets? *Journal of Structural Geology*, **9**, 797–808.

GLASSER, N. F., HAMBREY, M. J., CRAWFORD, K. R., BENNETT, M. R. & HUDDART, D. 1998. The structural glaciology of Kongsvegen, Svalbard, and its role in landform genesis. *Journal of Glaciology*, **44**, 136–148.

HAEBERLI, W. 1976. Eistemperaturen in den Alpen. *Zeitschrift für Gletscherkunde und Glazialgeologie*, **11**, 203–220.

HAMBREY, M. J. & MILNES, A. G. 1977. Structural geology of an Alpine glacier (Griesgletscher, Valais, Switzerland). *Eclogae geologicae Helvetiae*, **70**, 667–684.

—— & MÜLLER, F. 1978. Structures and ice deformation in the white glacier, Axel Heiberg Island, Northwest territories, Canada. *Journal of Glaciology*, **20**, 41–66.

HARPER, J. T. & HUMPHREY, N. F. 1995. Borehole video analysis of a temperate glacier's englacial and subglacial structure: Implications for glacier flow models. *Geology*, **23**, 901–904.

HODGKINS, R. & DOWDESWELL, J. A. 1994. Tectonic processes in Svalbard tide-water glacier surges: evidence from structural geology. *Journal of Glaciology*, **40**, 553–560.

HOOKE, R. L. B. & HUDLESTON, P. J. 1978. Origin of foliation in glaciers. *Journal of Glaciology*, **20**, 285–299.

LAWSON, W. 1996. Structural evolution of Variegated glacier, ALASKA, USA, since 1948. *Journal of Glaciology*, **42**, 261–270.

——, SHARP, M. J. & HAMBREY, M. J. 1994. The structural geology of a surge-type glacier. *Journal of Structural Geology*, **16**, 1447–1462.

LISTER, G. S., ETHERIDGE, M. A. & SYMONDS, P. A. 1986. Detachment faulting and the evolution of passive continental margins. *Geology*, **14**, 246–250.

MALAVIEILLE, J. 1993. Late orogenic extension in mountain belts: insights from the Basin and Range and the late Paleozoic Variscan belt. *Tectonics*, **12**, 1115–1130.

MAUDUIT, T. & BRUN, J.-P. 1998. Growth fault/rollover systems: Birth, growth, and decay. *Journal of Geophysical Research*, **103**, 18 119–18 136.

——, GUERIN, G., BRUN, J.-P. & LECANU, H. 1997. Raft tectonics: the effects of basal slope angle and sedimentation rate on progressive extension. *Journal of Structural Geology*, **19**, 1219–1230.

NEUBAUER, F., FRITZ, H., GENSER, J., KURZ, W., NEMES, F., WALLBRECHER, E., WANG, X. & WILLINGSHOFER, E. 2000. Structural evolution within an extruding wedge: model and application to the Alpine-Pannonian system. *In*: LEHNER, F. K. & URAI, J. L. (eds) *Aspects of tectonic faulting*. Springer, Berlin, Heidelberg, New York, 141–153.

PASSCHIER, C. W. & TROUW, R. A. J. 1996. *Microtectonics*. Springer, Berlin.

PATZELT, G. 1996. Gletscherbericht 1994/95. *Mitteilungen des österreichischen Alpenvereins*, **51**, 20–27.

——1998. Gletscherbericht 1996/97. *Mitteilungen des österreichischen Alpenvereins*, **53**, 6–12.

——1999. Gletscherbericht 1997/98. *Mitteilungen des österreichischen Alpenvereins*, **54**, 6–12.

PFEFFER, W. T. 1992. Stress-induced foliation in the terminus of Variegated glacier, ALASKA, USA, formed during the 1982–1983 surge. *Journal of Glaciology*, **38**, 213–222.

PLATT, J. P. 1986. Dynamics of orogenic wedges and the uplift of high-pressure metamorphic rocks. *Geological Society of America Bulletin*, **97**, 1037–1053.

PERESSON, H. & DECKER, K. 1997. Far-field effects of Late Miocene subduction in the Eastern Carpathians: E–W compression and inversion of structures in the Alpine–Carpathian–Pannonian region. *Tectonics*, **16**, 38–56.

RAGAN, D. M. 1967. Planar and layered structures in glacial ice. *Journal of Glaciology*, **6**, 565–567.

——1969. Structures at the base of an ice fall. *Journal of Geology*, **77**, 647–667.

RATSCHBACHER, L., FRISCH, W., LINZER, H.-G. & MERLE, O. 1991*a*. Lateral extrusion in the Eastern Alps, part II: structural analysis. *Tectonics*, **10**, 257–271.

——, ——, NEUBAUER, F., SCHMID, S. M. NEUGEBAUER, J. 1989. Extension in compressional orogenic belts: the eastern Alps. *Geology*, **17**, 404–407.

——, MERLE, O., DAVY, P. & COBBOLD, P. 1991*b*. Lateral extrusion in the Eastern Alps, part I: boundary conditions and experiments scaled for gravity. *Tectonics*, **10**, 245–256.

SCHWARZACHER, W. & UNTERSTEINER, N. 1953. Zum Problem der Bänderung des Gletschereises. *Sitzungsberichte der Österreichischen Akademie der Wissenschaften, mathematisch-naturwissenschaftliche Klasse Abteilung IIa*, **162**, 111–145.

SHARP, M., LAWSON, W. & ANDERSON, R. S. 1988. Tectonic processes in a surge-type glacier. *Journal of Glaciology*, **10**, 499–515.

SOERRENSEN, A. M., WALMANN, T., JAMTVEIT, B., FEDER, J. & JOESSANG, T. 1998. Modeling and characterization of fracture patterns in the Vatnajökull glacier. *Geology*, **26**, 931–934.

TINTOR, W. 1997. Witterung und ablation an der pasterze von 1990 bis 1994. *Carinthia II*, **187**, 583–590.

WAKONIGG, H. & LIEB, G. K. 1996. Die Pasterze und ihre Erforschung im Rahmen der Gletschermes-sungen. *Kärntner Nationalparkschriften – Wissenschaft im NP Hohe Tauern, Kärnten*, **8**, 99–115.

WANG, X., NEUBAUER, F., GENSER, J. & YANG, W. 1999. The Dabie UHP unit, Central China: a Cretaceous extensional allochthon superposed on a Triassic orogen. *Terra Nova*, **10**, 260–267.

# Glaciotectonic Studies

(Overleaf) Intense deformation in the 30 m-high advancing frontal cliff of Thompson Glacier, Axel Heiberg Island, Canadian Arctic. At the base of this polythermal glacier, sediment is subjected to intense deformation, and much of it is incorporated into the body of the glacier along thrust-faults, reaching a high level in the face. Some dirty layers are almost pure debris. As the cliff advances, ice-blocks fall from the face and are incorporated as an ice-breccia in the base of the glacier. To the left of the waterfall, in mid-cliff is evidence of ductile deformation in the form of isoclinal folding. (Photo: M. J. Hambrey).

# Continuity comes first: recent progress in understanding subglacial deformation

R. B. ALLEY

*Environment Institute and Department of Geosciences, The Pennsylvania State University,
Deike Building, University Park, PA 16802, USA (e-mail: ralley@essc.psu.edu)*

**Abstract:** Subglacial till deformation is glaciologically and geologically important, but is difficult to model owing to numerous uncertainties related to till generation and the flow law for till deformation. I review recent results with the hope of at least focusing the uncertainties. Fine-grained subglacial tills often are sufficiently soft that continuity issues are more important than the 'flow law' of the till in affecting glacier behavior. Non-steady forcing, coarse clasts and an irregular ice-till contact favour till deformation to significant depths and thus rapid subglacial transport of till. Over long times, site history is important in till continuity and thus in glacier behaviour.

More than 20 years have passed since Engelhardt *et al.* (1978) and Boulton (1979) demonstrated the importance of deforming beds to glacial processes. Nearly 15 years have passed since Boulton (1986) suggested that growing evidence of till deformation, including the results of Blankenship *et al.* (1986) from Ice Stream B, constituted 'A Paradigm Shift in Glaciology?' The intervening years have produced a wealth of new information (e.g. Murray 1997), with the primary effect of removing the question mark from Boulton's (1986) title.

In this short review, I provide my interpretation of what we have learned and where we are going. I do this by listing a series of hypotheses based on the available data, and then discussing these hypotheses.

(1) Subglacial tills deform.
(2) Till deformation produces significant sediment fluxes through layers of order 0.3 m thick or thicker.
(3) Although the 'flow law' for tills appears to be approximated well by a Coulomb-plastic model, the nonsteady subglacial conditions that distribute deformation over tens of centimetres of till thickness also may cause the depth- and time-averaged shear-strain rate to increase with a low power of the shear stress averaged in the same way.
(4) But, subglacial tills containing abundant fine-grained materials often are sufficiently weak that till continuity rather than the till flow-law is the major issue in modelling ice flow.

(5) And, an increase in the thickness or lateral extent of subglacial till slows further erosion and generation of more till.
(6) Thus, the history of ice occupation and previous sedimentation often is the most important factor controlling the distribution of deforming beds and of 'streaming' ice flow.

## Hypothesis 1: subglacial tills deform

While a student, I learned that widespread ground moraine or basal till was produced by lodgment, the clast-by-clast smearing of sediment from moving ice onto its bed to form a dense, compacted layer, which thereafter sat in place. Deformation was not so much forbidden as ignored; despite well-known evidence of features indicating deformation to strains of order one, the idea of extensive subglacial deformation to large strains was not prominent in important writings in the field (e.g. Sugden & John 1976, pp. 232–234).

The direct verification of subglacial till deformation by Engelhardt *et al.* (1978) and Boulton (1979) began to change this view. Boulton (1979) showed that the deforming bed beneath Breiða-mérkurjökull in Iceland was producing a deposit that a glacial geologist likely would classify as a lodgment till. This implied that either: (a) lodgment tills are polygenetic or (b) all lodgment tills were produced by deforming-bed processes.

I suspect that basal tills are polygenetic, primarily because no single answer ever suffices in

*From:* MALTMAN, A. J., HUBBARD, B. & HAMBREY, M. J. (eds) *Deformation of Glacial Materials.* Geological Society, London, Special Publications, **176**, 171–179. 0305-8719/00/$15.00 © The Geological Society of London 2000.

**Table 1.** *Glaciers with deforming subglacial tills; thickness of deformation where measured*

---

Black Rapids Glacier (Truffer *et al.* in press); deformation >2 m deep

Blue Glacier (Engelhardt *et al.* 1978)

Breiðamérkurjökull (Boulton 1979; Boulton & Hindmarsh 1987); order of 30 cm

Columbia Glacier (Humphrey *et al.* 1993); order of 30 cm

Ice Stream B (Engelhardt & Kamb 1998); most order of 3 cm, some deeper

Ice Stream C (Anandakrishnan & Alley 1997)

Ice Stream D (Engelhardt 1999, pers. comm. 2000); more than order of 30 cm

Storglaciären (Iverson *et al.* 1995); order of 30 cm

Trapridge Glacier (Blake *et al.* 1992, 1994); order of 30 cm

Variegated Glacier (Harrison *et al.* 1986)

---

a system as complicated as a glacier. Somewhere, and possibly in many places, till is being plastered onto the upglacier sides of roche moutonees or is otherwise being deposited as a rigid layer.

The limited but growing evidence from beneath glaciers paints a different picture, however: subglacial tills typically are deforming through at least part of their thickness. Table 1 lists glaciers with known deformation; I am not aware of any situations in which widespread subglacial tills are motionless through their entire thickness. Certainly, Engelhardt & Kamb (1998) argued that most (but not all) basal velocity is concentrated near the base of ice stream B, Porter *et al.* (1997) calculated that limited deformation as well as extensive deformation exist beneath different regions of Bakaninbreen and Truffer *et al.* (in press) provide evidence for nearly rigid behavior of the upper 2 m of a till as much as 7 m thick beneath Black Rapids Glacier, but with much of the ice velocity arising from processes deeper than 2 m in the till. Despite their great variability, all of these involve till deformation, or possibly sliding of till over substrate in the case of Black Rapids Glacier, rather than motionless and rigid 'lodgment' tills.

The realization that subglacial tills typically are moving through at least part of their thickness is a fundamental result in the study of glacier-bed processes. It frames a number of additional lines of study, reviewed next.

## Hypothesis 2: till deformation produces significant sediment fluxes

Subglacial tills can allow rapid ice motion by pervasive deformation to significant depths (Boulton & Hindmarsh 1987), by ploughing of ice-contact clasts through largely undeforming sediment (Brown *et al.* 1987), and by rather deep deformation or failure along a surface or surfaces (Alley 1989; Truffer *et al.* in press). These probably grade into each other. Pervasive deformation through layers of order 30 cm thick or more is common, based on theoretical understanding and on available data, as discussed next. Such deformation will produce significant sediment fluxes (see Hooke & Elverhoi 1996; Alley *et al.* 1997).

Glaciers tend to couple tightly to the sediment beneath them unless the ice is floated off that sediment by high-pressure water (Engelhardt *et al.* 1978; Alley 1989; Iverson *et al.* 1998; Tulaczyk 1999*a*). Brown *et al.* (1987) argued that the tendency for basal shear stress of ice on substrate to be concentrated on obstacles of certain sizes (Weertman 1957) will favour 'ploughing' of certain sizes of clasts through till beds. Ploughing of a clast through till may cause deformation about three times deeper than the clast (Tulaczyk 1999*a*), producing a deforming layer ranging from centimetres or less to tens of centimetres or more in thickness depending on the grain-size distribution and other factors. Tulaczyk (1999*b*) extended this model by suggesting that an irregular ice base also can plough through sediments, producing deeper deformation than by clast ploughing if the roughness scale of the ice base is larger than the clasts, as seems likely in at least some cases.

The depth of pervasive deformation might reach a few tens of centimetres even without ploughing owing to the effects of water-pressure variations at the ice–till interface. Higher water pressure softens till, so deformation is expected to tend to localize where water pressures are highest (Iverson *et al.* 1998; Tulaczyk 1999*a*). If variations in the water pressure at the top of the till (from diurnal variation in input of surface melt water, from tidal processes, or for other reasons) can be approximated by a sine wave, the water pressures in the till at any depth will also have sinusoidal variation, with the amplitude decreasing exponentially with increasing depth, and with the phase increasingly lagged with increasing depth (ignoring the complications of dilatancy discussed below). When the surface water pressure is dropping, water pressure will be higher at some depth in the till than at the surface. If the deformation focuses where water pressure is highest, the depth of deformation would propagate downward until the surface water pressure rises above the highest subsurface value, at which time the deformation would tend to jump back to the surface (Iverson *et al.* 1998;

Tulaczyk 1999a). The depth to which the signal propagates would depend on the hydraulic properties of the till, but may be of order 30 cm (Iverson *et al.* 1998; Tulacyzk 1999a).

At least partial profiles of velocity with depth in till are available from beneath seven glaciers (Table 1). Of these, all show significant till fluxes. As discussed next, one (Ice Stream B) has velocity concentrated near the base of the ice (Engelhardt & Kamb 1998). Four (Breiðamér-kurjökull, Boulton & Hindmarsh 1987; Storgla-ciären, Iverson *et al.* 1995; Trapridge Glacier, Blake *et al.* 1992, 1994; Columbia Glacier, Humphrey *et al.* 1993) show significant deformation to a depth of a few tens of centimetres. The final two of these deforming-bed glaciers (ice stream D Engelhardt 1999, pers comm. 2000; Black Rapids Glacier, Truffer *et al.* in press) probably have deeper deformation, which may or may not be localized. These data indicate that ploughing typically is subsidiary to pervasive deformation, although water-pressure increase over a deform-ing bed has been observed to decrease ice-bed coupling and bed deformation while likely allowing continued ploughing (Engelhardt *et al.* 1978; Iverson *et al.* 1995).

The contrast between Ice Streams B and D on the Siple Coast of West Antarctica is quite interesting (Engelhardt & Kamb 1998; Engelhardt 1999, pers. comm. 2000). Differential motion was measured between instruments placed near the bottoms of holes melted through Ice Streams B and D and stakes tethered to the instruments and inserted into subglacial sedi-ments. The differential motion is achieved by some combination of true sliding and till deformation between the base of the borehole and the 'effective depth' of the tethered stake.

Measurements over 28 days beneath Ice Stream B indicated that true sliding and defor-mation within the top 3–25 cm of the subglacial till (depending on the stake orientation, with the authors favouring the smaller depth), accounted for approximately 70–80% of ice-surface motion. Surface motion of Ice Stream B is calculated to arise almost entirely from basal motion, leaving deeper till deformation signifi-cant but subsidiary to some combination of true sliding, ploughing, and very shallow deforma-tion. For Ice Stream D an experiment over a longer time interval indicates that 80–90% of the basal motion of the ice stream arises from processes occurring deeper than 30–60 cm below the base of the ice.

Seismic results are not available for Ice Stream D but seismic experiments on Ice Stream B near the site of the Engelhardt & Kamb (1998) tethered-stake experiment (Blankenship *et al.*

1986, 1987; Rooney *et al.* 1987) showed several metres of high-porosity, high-water-pressure material, which has been interpreted as a deforming till (Alley *et al.* 1986, 1987). It is slightly puzzling that the interpretation of the seismic experiments on Ice Stream B is more consistent with the tethered-stake results from Ice Stream D than from Ice Stream B.

Many interpretations of this puzzle are possible. Alley *et al.* (1986) may have used the seismic data incorrectly in building models; Rooney *et al.* (1987) did observe internal reflectors in the Ice Stream B till, which may have dynamic significance. The seismic data may have been complicated by non-steady effects – the shutdown of Ice Stream C just over a century ago has been linked to diversion of basal meltwater down Ice Stream B (Alley *et al.* 1994; Anandakrishnan & Alley 1997), which would favour true sliding over bed deformation beneath Ice Stream B perhaps with insufficient time for the tills to have adjusted fully to these changed conditions. Perhaps the seismically detected high porosity in comparison to lodg-ment tills can be maintained by the rather slow rates of deformation that are consistent with the Engelhardt & Kamb (1998) experiment, which does indicate some deformation deeper than a few centimetres producing some till flux. There also is a slight possibility that the Engelhardt & Kamb (1998) experiment was not representative of widespread natural conditions beneath Ice Stream B. The water-pressure record from that experiment shows two 10 bar water-pressure fluctuations, much larger than values typically found beneath the ice stream (Engelhardt & Kamb 1997). In addition, one of the several boreholes studied by Engelhardt & Kamb (1997) showed slight overpressurization for at least a week during borehole freeze-in, which might have promoted sliding rather than bed deforma-tion if a similar overpressurization happened in association with the tethered-stake experiment (Engelhardt *et al.* 1978). Personal communica-tions from H. Engelhardt (1999, 2000) provide strong support for the published interpretations of Engelhardt & Kamb (1997, 1998), but additional field data would certainly be valuable.

The data of Truffer *et al.* (in press) from Black Rapids Glacier, Alaska provide a counterpoint to the results of Engelhardt & Kamb from the Upstream B camp on ice stream B. Most of the basal velocity beneath Black Rapids Glacier appears to arise from processes deeper than 2 m into till that is locally 7 m thick beneath the ice, with no information on the localized or dis-tributed nature of that deeper deformation. Exceptionally low summertime basal water pressures compared to the ice-overburden pressure may contribute to consolidation of the upper till layers and to causing the locus of deformation to be deep in the till (Truffer et al. in press).

Taken together, these data indicate that subglacial deformation generally is spatially and temporally variable, responding to changes in water pressure and supply, thermal conditions, clast and ice ploughing, and other factors (Iverson et al. 1995, 1998; Tulaczyk 1999a, b; Tulaczyk et al. 2000a, b). Typically, deformation is significant to a depth of order 30 cm, although the thickness of till experiencing significant velocities can drop to near zero or expand to many metres (also see MacClintock & Dreimanis 1964; Alley 1991). Associated sediment fluxes can be important or dominant in glacial sediment budgets (Alley et al. 1997).

## Hypothesis 3: low effective stress exponent

This hypothesis is more speculative than the others. However, several results could be reconciled if it proved to be accurate, and there is some evidence in its favour.

The only source of classical viscosity in a deforming till is likely to be the water itself, with viscosity so low as to be incapable of significantly restraining ice flow. Clast–clast interactions should follow some frictional law. The flow law for till thus would reduce to Coulomb-plastic behaviour in which strain rate is independent of stress above a yield strength, which depends on material properties and on the difference between the overburden and water pressures (e.g. Iverson et al. 1998). This can also be modelled as strain rate increasing with the shear stress raised to a power in the limit as the power becomes very large.

Nonetheless, much early work on till deformation assumed more nearly linear-viscous behavior, in which the strain rate increases with a power of the shear stress of order one (e.g. Boulton & Hindmarsh 1987; Alley et al. 1987). In large part, this was based on publication of a low-power flow law derived from subglacial observations near the terminus of Breiðamér-kurjökull (Boulton & Hindmarsh 1987). The complexity of the stress state in terminus regions of glaciers suggests caution.

Linear-viscous flow was also favoured in early work because of the observed deformation to significant depths in till; in a homogeneous, fine-grained Coulomb-plastic material subjected to constant forcing, deformation typically collapses to a surface. As summarized above, however, subsequent work has demonstrated that distributed subglacial deformation is not just possible but often should be expected beneath glaciers even if till is a Coulomb-plastic material. Thus, the main arguments originally used in favour of a low-power flow law are either uncertain or wrong. In addition, several experiments have now shown that steady deformation of tills produces a Coulomb-plastic or nearly Coulomb-plastic behavior (e.g. Kamb 1991; Iverson et al. 1998) (although one set of laboratory experiments for small deformation probably not to steady state has been interpreted to indicate a low-power flow law (Jenson et al. 1996)).

In contrast, one set of observations from Ice Stream C indicates that the till can be described with a more nearly linear-viscous model (Anandakrishnan & Alley 1997). The ice stream is restrained largely by a combination of 'sticky spots' and distributed basal drag over till between sticky spots. When a sticky spot breaks in a microearthquake, nearby sticky spots preferentially break thereafter, following a delay that increases with distance. The model invoked is that the stress supported on the first sticky spot is propagated to the others after the first microearthquake, with the delay caused by time-dependent deformation of the till. That in turn is most easily modelled as low-power deformation of the till between the sticky spots. Linear-viscous behaviour is fully consistent with the observations, but dependence of the strain rate in the till on a high power of the shear stress is not consistent with the data in the model used by Anandakrishnan & Alley (1997).

The evidence for rapidly changing conditions beneath Ice Stream C (Anandakrishnan & Alley 1997) suggests that reconciliation of the observations there with Coulomb-plastic till behaviour may involve nonsteady effects. Here, dilatancy may be important. Iverson et al. (1998) argued that subglacial tills typically should be dilatant at the onset of their deformation. Fluctuating water pressures during times without local till deformation should cause tills to become more consolidated than in the critical state of deformation (e.g. Clarke 1987). Onset of deformation would begin to dilate such a till, dropping the water pressure in the dilating region. Higher water pressure external to the dilating region then would restrict further dilation and deformation. However, higher water pressure external to the dilating region also would cause water flow through pore spaces to the dilating region, allowing further dilation with a time-delay linked to the rate of water flow. The water-inflow rate would increase with the water-pressure difference, and thus with the applied shear stress causing the deformation. Hence, the deformation rate would increase with shear stress during the transient period when dilation is occurring. The water loss by adjacent regions would cause some consolidation in them producing porosities lower than in the critical state, so subsequent migration of the locus of deformation under non-steady forcing would continue to exhibit stress-dependence because of the effects of dilatancy.

If dilatant-strengthening behaviour such as postulated by Iverson *et al.* (1998) occurs subglacially, then empirical estimation of constants in a power-law flow model using data averaged over many cycles of the water-pressure variability is likely to yield a dependence of strain rate on a low power of the stress. If steady-state deformation is achieved and the deforming band is fully dilated, then the Coulomb-plastic steady-state flow-law applies (Iverson *et al.* 1998). Iverson *et al.* (1998) also calculated that the time for achieving steady state depends on permeability of the sediments (longer times for lower permeabilities associated with finer-grained tills), but typically is hours to days. Hence, the diurnal tidal cycle and faster changes induced by microearthquakes on sticky spots beneath Ice Stream C (Anandakrishnan & Alley 1997) may induce a 'pseudo-viscous behaviour' when modelled over times much longer than the perturbation interval.

Engelhardt & Kamb (1997) observed that water pressures beneath ice stream B are highly variable, often including a diurnal component. Modulation and timing of the fluctuations do not obviously indicate a tidal signal (as is observed for the motion of neighboring Ice Stream C; Anandakrishnan & Alley 1997), but no other clear explanation is available. Regardless of the explanation, the observation points to important nonsteady effects on the till deformation beneath ice stream B.

If a diurnal cycle on Ice Stream B forces the failure surface to move up and down through the tills (Tulaczyk *et al.* 2000*a*), steady state may never be reached. The situation on Ice Stream C with its numerous microearthquakes, strongly time-varying local basal shear stress, and tidal signals, may be even less steady. Most small glaciers exhibit diurnal forcing at least through the summer. Steady subglacial conditions thus may be rare. However, reduced winter variability of water pressure on small glaciers suggests that Coulomb-plastic tills could approach steady state, lose the extra strength imparted by their dilatant nature, and allow increased flow velocity perhaps including wintertime triggering of surging.

Thus, subglacial tills may be dilatant Coulomb-plastic materials. Strain rates averaged over the significantly deforming thickness and over many days or longer may increase with a low power of the stress averaged in the same way, with possible implications for the results of Anandakrishnan & Alley (1997) and Boulton & Hindmarsh (1987), among others.

An important qualification relates to initial deformation of sediments with porosity higher than for critical-state deformation (Clarke 1987; N. Iverson pers. comm. 1999). Deformation of such sediments will cause water expulsion and till consolidation rather than water inflow and till dilation. Dilatant strengthening would not occur during water expulsion and consolidation. Eventually, non-steady conditions plus the expulsion should consolidate such sediments sufficiently to allow dilatant strengthening as suggested by Iverson *et al.* (1998); however, the non-steady behaviour reaching that state may be quite interesting. Spatially variable processes also may be important (Hindmarsh 1997).

## Hypothesis 4: till continuity comes first

A good understanding of the till flow law is certainly important, but new results from laboratory and field indicate that the flow law may not be the most important till characteristic affecting glacier flow. Rather, till continuity may exercise the dominant control of ice motion.

Rapid ice flow can be achieved in many ways, including by internal deformation where basal shear stresses are exceptionally high (Iken *et al.* 1993), or by enhanced basal lubrication. Basal lubrication includes sliding over thick layers of fresh or salt water in lakes or marginal seas (e.g. Thomas *et al.* 1988), or sliding over water thickened by failure of channelized basal drainage systems (Kamb *et al.* 1985). However, basal water drainage systems show a strong tendency to collapse to efficient channels (Walder 1982), and subglacial lakes seem to occur preferentially in regions where frozen margins slow basal water drainage but also slow ice motion (Siegert & Dowdeswell 1996). Effective basal lubrication by thick water layers producing high ice velocities over large areas and long times is not likely to be common except for marginal ice shelves of cold ice sheets.

In contrast, the till-lubricated fast flow of Ice Streams B and D Rutford ice stream (Smith 1997), the lobes of the southern margins of the former Laurentide and Fennoscandian ice sheets (e.g. Alley 1991; Clark 1992, 1997; Boulton 1996), and possibly of the Hudson Strait ice stream feeding Heinrich Event surges (MacAyeal

1993; Alley & MacAyeal 1994), argue that till deformation can produce extensive regions of well-lubricated glacier beds persisting for centuries to millennia or longer, although typically with rather non-steady behaviour.

Sufficiently coarse-grained tills, such as those beneath some mountain glaciers and perhaps beneath portions of large ice sheets, can have rather high strengths that figure significantly in the stress balance of the ice even with basal water pressures within one bar of the level needed to float the glacier (Iverson et al. 1995; Hooke et al. 1997). However, when basal water pressures are high, sufficiently fine-grained tills, including those found beneath Ice Stream B and some along the southern margin of the Laurentide ice sheet, have strengths an order of magnitude or more lower than typical basal shear stresses of glaciers (Kamb 1991; Iverson et al. 1998).

Non-steady effects may strengthen tills, as noted above, but indications are that this still would leave many subglacial tills weak compared to typical basal shear stresses of glaciers. Data from the Siple Coast ice streams indicate that much of the shear stress for ice-stream flow is not supported on their beds but is transmitted through the ice-stream margins and supported on the beds of interstream ridges (e.g. Whillans & Van der Veen 1997), which likely are frozen in the key areas (Engelhardt & Kamb 1997). Additional stress may be supported on 'sticky spots' in the ice streams (MacAyeal 1992; Alley 1993; Anandakrishnan & Alley 1997), features of uncertain cause but that probably represent discontinuities in the lubricating till linked to geological inhomogeneities beneath (Rooney et al. 1987). In upstream reaches, where till generation may occur, the margin of one ice-stream tributary is closely tied to a geological boundary, with the ice stream sitting over materials more likely to yield till easily (Anandakrishnan et al. 1998).

Where soft tills occur, their lateral extent and continuity likely are the critical factors in ice behaviour. Thermal boundaries suggest some interesting feedbacks (see Payne 1995; Tulaczyk et al. 2000b), in which efficient basal lubrication keeps the ice thin enough to allow efficient heat transfer from the bed to a cold surface, thus allowing frozen regions to exist and prevent even more efficient flow and further thinning.

## Hypothesis 5: till slows its own replacement

Sticky spots also suggest feedbacks. From the work of Cuffey & Alley (1996), deforming tills do not abrade their beds efficiently. Further, the plucking/quarrying models of Iverson (1991) and Hallet (1996) indicate that extensive tills would slow or stop erosion by these mechanisms.

Although inwash or rockfalls may contribute significantly to the sediment budgets of some mountain glaciers, most ice sheets and many mountain glaciers must generate most of their sediment from their beds (Hallet et al. 1996; Alley et al. 1999). If a glacier with no subglacial till began generating sediment faster than it was removed by subglacial streams or entrainment to the ice, then till would begin to increase beneath the glacier. Sediment flux would increase because of deformation in the till layer, and sediment generation would decrease because of the effects of the till layer. With less sediment generation and more sediment removal, till accumulation would become more difficult until a steady state was reached. For weak, fine-grained tills, this steady state likely would leave some sticky spots and possibly some interstream ridges supporting the basal shear stress of the ice. I believe it would be quite difficult to build a plausible model in which a glacier generates a continuous, extensive soft till from hard bedrock.

Tulaczyk (1999b) suggested that large irregularities (roughness elements or 'ice keels' on the base of the ice) can plough through tills, reaching sediments beneath and perhaps generating tills from them, much like icebergs ploughing through and disturbing submarine sediments. This mechanism would allow till generation to persist beneath tills provided the subjacent materials were sufficiently 'soft' compared to ice, as they may be beneath Ice Stream B and some other sites (Rooney et al. 1991). Increased till thickness nevertheless would decrease the rate of till generation by this mechanism as well as by other possible mechanisms.

## Hypothesis 6: history matters

Till generation can be quite easy in certain situations dictated by history. A glacier that advances over unconsolidated sediments in lake or ocean basins, or that advances over thick regolith produced by weathering over long times, does not so much generate tills as mobilize them – the hard work of disaggregating the bedrock has already been done. In this respect, it is notable that many tills of the southern margin of the Laurentide ice sheet were smeared out of lake basins during rapid oscillations of the ice margin that alternately opened the lakes and then readvanced into them and their sediments (e.g. Alley 1991). The ice streams of West

Antarctica typically are related to sedimentary basins containing poorly consolidated materials, especially in upglacier regions where till generation may be more important than till transport from farther upglacier (Rooney *et al.* 1991; Anandakrishnan *et al.* 1998), although ice-stream and sedimentary-basin margins are not in exact correspondence especially in down-glacier regions. Available data indicate that the sediments in the basins are poorly consolidated and could be mobilized easily by moving ice (Rooney *et al.* 1991; Tulaczyk *et al.* 1998).

If ice manages to persist over soft sediment for long enough times, the easily eroded sediments are likely to be exhausted and the soft till to disappear. The ice then may evolve from a West Antarctic form with thin, fast flow to an East Antarctic pattern with thicker, steeper surface. The coastal ice flux and even the ice velocity ultimately may end up being similar, but the pattern of flow and its stability will be quite different. The inherent instability of thin, fast-moving marginal ice-sheet regions entering lakes, oceans, or simply warm climates favours deforming tills, because the ice can be removed easily to allow sedimentation or weathering to generate more raw materials for till deformation.

An interesting application of till continuity was provided by Clark & Pollard (1998). Glacial-geological evidence shows that the southern Laurentide ice sheet was more extensive during earlier glaciations, when isotopic ratios of marine foraminiferal shells indicate that total ice volume on Earth was smaller, than during later glaciations when the Laurentide ice sheet covered less area on its southern margin but is inferred to have had larger volume. Clark & Pollard (1998) noted that the onset of glaciation would have occurred on a regolith-covered continent (soils), which would have favoured laterally extensive but thin ice owing to till deformation. But, this till deformation would have transported sediments more rapidly than additional loose materials were generated, eventually reducing the extent of deforming beds and thus of the ice sheets.

## Summary

Deforming glacier beds clearly are important in ice flow. Limited data indicate that the beds of well-lubricated, grounded glaciers are usually deforming tills. Subglacial tills are widespread, and subglacial tills usually deform. This deformation can localize near the base of the ice or metres deep in the till, but more typically occurs through at least the upper tens of centimetres,

giving geologically significant till fluxes. The deformation rate may show stress-dependence because of strong non-steady effects despite the likely Coulomb-plastic nature of the tills. Fine-grained subglacial tills often are sufficiently soft compared to typical glaciogenic stresses that the details of the flow law are unimportant in modeling the ice flow; most of the driving stress for ice flow is supported somewhere other than on the till. Thermal or geological boundaries may allow lateral stress transmission to inter-stream ridges, and geological inhomogeneities or other causes may allow sticky spots within till. Till generation beneath a deforming bed is slow unless the materials there are weak or unconsolidated. Hence, geological history is important or dominant in determining the distribution of deforming beds, well-lubricated ice flow, and thin, unstable regions of ice sheets.

I thank N. Iverson, B. Kamb, S. Tulaczyk and H. Engelhardt for helpful discussions, H. Engelhardt for permission to cite unpublished data, M. Truffer, W. Harrison and K. Echelmeyer for the preprint of their work, and the US National Science Foundation for support.

## References

ALLEY, R. B. 1989. Water-pressure coupling of sliding and bed deformation. II. Velocity-depth profiles. *Journal of Glaciology*, **35**, 119–129.
——1991. Deforming-bed origin for southern Laurentide till sheets? *Journal of Glaciology*, **37**, 67–76.
——1993. In search of ice-stream sticky spots. *Journal of Glaciology*, **39**, 447–454.
—— & MACAYEAL, D. R. 1994. Ice-rafted debris associated with binge/purge oscillations of the Laurentide Ice Sheet. *Paleoceanography*, **9**, 503–511.
——, ANANDAKRISHNAN, S., BENTLEY, C. R. & LORD, N. 1994. A water-piracy hypothesis for the stagnation of ice stream C. *Annals of Glaciology*, **20**, 187–194.
——, BLANKENSHIP, D. D., BENTLEY, C. R. & ROONEY, S. T. 1986. Deformation of till beneath ice stream B West Antarctica. *Nature*, **322**, 57–59.
——, ——, ROONEY, S. T. & BENTLEY, C. R. 1987. Till beneath ice stream B. 4. A coupled ice-till flow model. *Journal of Geophysical Research*, **92B**, 8931–8940.
——, CUFFEY, K. M., EVENSON, E. B., STRASSER, J. C., LAWSON, D. E. & LARSON, G. J. 1997. How glaciers entrain and transport basal sediment: Physical constraints. *Quaternary Science Reviews*, **16**, 1017–1038.
——, STRASSER, J. C., LAWSON, D. E., EVENSON, E. B. & LARSON, G. J. 1999. Glaciological and geological implications of basal-ice accretion in over-deepenings. *In*: MICKELSON, D. M. & ATTIG, J. W. (eds) *Glacial Processes Past and Present*,

Geological Society of America Special Paper, Boulder, Colorado, USA, 1–9.

ANANDAKRISHNAN, S. & ALLEY, R. B. 1997. Tidal forcing of basal seismicity of ice stream C West Antarctica, observed far inland. *Journal of Geophysical Research*, **102B**, 15 183–15 196.

——, BLANKENSHIP, D. D, ALLEY, R. B. & STOFFA, P. L. 1998. Influence of subglacial geology on the position of a West Antarctic ice stream from seismic observations. *Nature*, **394**, 62–65.

BLAKE, E., CLARKE, G. K. C. & GERIN, M. C. 1992. Tools for examining subglacial bed deformation. *Journal of Glaciology*, **38**, 388–396.

——, FISCHER, U. H. & CLARKE, G. K. C. 1994. Direct measurement of sliding at the glacier bed. *Journal of Glaciology*, **40**, 595–599.

BLANKENSHIP, D. D., BENTLEY, C. R., ROONEY, S. T. & ALLEY, R. B. 1986. Seismic measurements reveal a saturated porous layer beneath an active Antarctic ice stream. *Nature*, **322**, 54–57.

——, ——, —— & ——1987. Till beneath ice stream B. 1. Properties derived from seismic travel times. *Journal of Geophysical Research*, **92B**, 8903–8911.

BOULTON, G. S. 1979. Processes of glacier erosion on different substrates. *Journal of Glaciology*, **23**, 15–38.

——1986. A paradigm shift in glaciology? *Nature*, **322**, 18.

——1996. Theory of glacial erosion, transport and deposition as a consequence of subglacial sediment deformation. *Journal of Glaciology*, **42**, 43–62.

—— & HINDMARSH, R. C. A. 1987. Sediment deformation beneath glaciers: rheology and geological consequences. *Journal of Geophysical Research*, **92B**, 9059–9082.

BROWN, N. E., HALLET, B. & BOOTH, D. B. 1987. Rapid soft bed sliding of the Puget glacial lobe. *Journal of Geophysical Research*, **92B**, 8985–8997.

CLARK, P. U. 1992. Surface form of the southern Laurentide ice sheet and its implications to ice-sheet dynamics. *Geological Society of America Bulletin*, **104**, 595–605.

——1997. Sediment deformation beneath the Laurentide ice sheet. *In*: MARTINI, I. P. (ed.) *Late glacial and postglacial environmental changes; Quaternary, Carboniferous-Permian, and Proterozoic*. Oxford University Press, Oxford, 81–97.

—— & POLLARD, D. 1998. Origin of the middle Pleistocene transition by ice sheet erosion of regolith. *Paleoceanography*, **13**, 1–9.

CLARKE, G. K. C. 1987. Subglacial till: a physical framework for its properties and processes. *Journal of Geophysical Research*, **92B**, 9023–9036.

CUFFEY, K. M. & ALLEY, R. B. 1996. Erosion by deforming subglacial sediments: Is it significant? (Toward till continuity). *Annals of Glaciology*, **22**, 17–24.

ENGELHARDT, H. 1999. Sliding and other observations on Ice Stream D (Abstract). *Sixth Annual West Antarctic Ice Sheet Workshop, September 1999*, http://igloo.gsfc.nasa.gov/wais/Agenda99.html. Also see http://skua.gps.caltech.edu/hermann/upd/upd.htm.

—— & KAMB, B. 1997. Basal hydraulic system of a West Antarctic ice stream: constraints from borehole observations. *Journal of Glaciology*, **43**, 207–230.

—— & —— 1998. Basal sliding of Ice Stream B West Antarctica. *Journal of Glaciology*, **44**, 223–230.

——, HARRISON, W. D. & KAMB, B. 1978. Basal sliding and conditions at the glacier bed as revealed by bore-hole photography. *Journal of Glaciology*, **20**, 469–508.

HALLET, B. 1996. Glacial quarrying: a simple theoretical model. *Annals of Glaciology*, **22**, 1–8.

——, HUNTER, L. & BOGEN, J. 1996. Rates of erosion and sediment evacuation by glaciers: A review of field data and their implications. *Global and Planetary Change*, **12**, 213–235.

HARRISON, W. D., KAMB, B. & ENGELHARDT, H. 1986. Morphology and motion at the bed of a surge-type glacier (Abstract). *Mitteilungen der Versuchsanstalt fur Wasserbau, Hydrologie und Glaziologie*, **90**, 55–56.

HINDMARSH, R. 1997. Deforming beds; viscous and plastic scales of deformation. *Quaternary Science Reviews*, **16**, 1039–1056.

HOOKE, R. LeB. & ELVERHOI, E. 1996. Sediment flux from a fjord during glacial periods, Isfjorden, Spitsbergen. *Global and Planetary Change*, **12**, 237–249.

——, HANSON, B., IVERSON, N. R., JANSSON, P. & FISCHER, U. H. 1997. Rheology of till beneath Storglaciaren, Sweden. *Journal of Glaciology*, **43**, 172–179.

HUMPHREY, N., KAMB, B., FAHNESTOCK, M. & ENGELHARDT, H. 1993. Characteristics of the bed of the lower Columbia Glacier, Alaska. *Journal of Geophysical Research*, **98B**, 837–846.

IKEN, A., ECHELMEYER, K., HARRISON, W. & FUNK, M. 1993. Mechanisms of fast flow in Jakobshavns Isbrae, West Greenland: Part I. Measurements of temperature and water level in deep boreholes. *Journal of Glaciology*, **39**, 15–25.

IVERSON, N. R. 1991. Potential effects of subglacial water-pressure fluctuations on quarrying. *Journal of Glaciology*, **27**, 27–36.

——, HANSON, B., HOOKE, R. LeB. & JANSSON, P. 1995. Flow mechanism of glaciers on soft beds. *Science*, **267**, 80–81.

——, HOOYER, T. S. & BAKER, R. W. 1998. Ring-shear studies of till deformation: Coulomb-plastic behavior and distributed strain in glacier beds. *Journal of Glaciology*, **44**, 634–642.

JENSON, J. W., MACAYEAL, D. R., CLARK, P. U., HO, C. L. & VELA, J. C. 1996. Numerical modeling of subglacial sediment deformation: implications for the behavior of the Lake Michigan Lobe, Laurentide ice sheet. *Journal of Geophysical Research*, **101B**, 8717–8728.

KAMB, B. 1991. Rheological nonlinearity and flow instability in the deforming-bed mechanism of ice-stream motion. *Journal of Geophysical Research*, **96B**, 16 585–16 595.

——, RAYMOND, C. F., HARRISON, W. D., ENGELHARDT, H., ECHELMEYER, K. A., HUMPHREY, N.,

BRUGMAN, M. M. & PFEFFER, T. 1985. Glacier surge mechanism: 1982–1983 surge of Variegated Glacier, Alaska. *Science*, **227**, 469–479.

MACAYEAL, D. R. 1992. The basal stress distribution of Ice Stream E Antarctica, inferred by control methods, *Journal of Geophysical Research*, **97B**, 595–603.

——1993. Binge/purge oscillations of the Laurentide Ice Sheet as a cause of the North Atlantic's Heinrich events. *Paleoceanography*, **8**, 775–784.

MACCLINTOCK, P. & DREIMANIS, A. 1964. Reorientation of till fabric by overriding glacier in the St. Lawrence valley. *American Journal of Science*, **262**, 133–142.

MURRAY, T. 1997. Assessing the paradigm shift; deformable glacier beds. *Quaternary Science Reviews*, **16**, 995–1016.

PAYNE, A. J. 1995. Limit cycles in the basal thermal regime of ice sheets. *Journal of Geophysical Research*, **100B**, 4249–4263.

PORTER, P. R., MURRAY, T. & DOWDESWELL, J. A. 1997. Sediment deformation and basal dynamics beneath a glacier surge front: Bakaninbreen, Svalbard. *Annals of Glaciology*, **24**, 21–26.

ROONEY, S. T., BLANKENSHIP, D. D, ALLEY, R. B. & BENTLEY, C. R. 1987. Till beneath ice stream B. 2. Structure and continuity. *Journal of Geophysical Research*, **92B**, 8913–8920.

——, ——, ——, & ——1991. Seismic reflection profiling of a sediment-filled graben beneath ice stream B West Antarctica. *In*: THOMSON, M. R. A., CRAME, J. A. & THOMPSON, J. W. (eds) *Geological Evolution of Antarctica*. Cambridge University Press, Cambridge, 261–265.

SIEGERT, M. J. & DOWDESWELL, J. A. 1996. Spatial variations in heat at the base of the Antarctic ice sheet from analysis of the thermal regime above subglacial lakes. *Journal of Glaciology*, **42**, 501–509.

SMITH, A. M. 1997. Basal conditions on Rutford Ice Stream, West Antarctica, from seismic observations. *Journal of Geophysical Research*, **102B**, 543–552.

SUGDEN, D. E. & JOHN, B. S. 1976. *Glaciers and Landscape: A Geomorphological Approach*. Edward Arnold, London.

THOMAS, R. H., STEPHENSON, S. N., BINDSCHADLER, R. A., SHABTAIE, S. & BENTLY, C. R. 1988. Thinning and grounding-line retreat on Ross Ice Shelf, Antarctica. *Annals of Glaciology*, **11**, 165–172.

TRUFFER, M., HARRISON, W. D. & ECHELMEYER, K. A. In press. Glacier motion dominated by processes deep in underlying till. *Journal of Glaciology*.

TULACZYK, S. 1999a. Ice sliding over weak, fine-grained tills: dependence of ice-till interactions on till granulometry. *In*: MICKELSON, D. M. & ATTIG, J. W. (eds) *Glacial Processes Past and Present*, Geological Society of America Special Papers, **137**, 159–177.

——1999b. Slippery when wet: Sub-ice-stream tills as mechanical and hydrological ... (Abstract). Sixth Annual West Antarctic Ice Sheet Workshop, September 1999, http://igloo.gsfc.nasa.gov/wais/Agenda99.html.

——, KAMB, B. & ENGELHARDT, H. 2000a. Basal mechanics of Ice Stream B. I. Till mechanics. *Journal of Geophysical Research*, **105**, 463–481.

——, —— & —— 2000b. Basal mechanics of Ice Stream B. II. Plastic-undrained-bed model. *Journal of Geophysical Research*, **105**, 483–494.

——, ——, SCHERER, R. & ENGELHARDT, H. F. 1998. Sedimentary processes at the base of a West Antarctic ice stream: Constraints from textural and compositional properties of subglacial debris. *Journal of Sedimentary Research*, **68**, 487–496.

WALDER, J. S. 1982. Stability of sheet flow of water beneath temperate glaciers and implications for glacier surging. *Journal of Glaciology*, **28**, 273–293.

WEERTMAN, J. 1957. On the sliding of glaciers. *Journal of Glaciology*, **3**, 33–38.

WHILLANS, I. M. & VAN DER VEEN, C. J. 1997. The role of lateral drag in the dynamics of ice stream B Antarctica. *Journal of Glaciology*, **43**, 231–237.

# Behaviour of subglacial sediment and basal ice in a cold glacier

SEAN J. FITZSIMONS[1], REGI D. LORRAIN[2] &
MARCUS J. VANDERGOES[1]

[1] *Department of Geography, University of Otago, Dunedin, New Zealand*
*(e-mail: sjf@perth.otago.ac.nz)*
[2] *Département des Sciences de la Terre et de l'Environnement,*
*Université libre de Bruxelles, Belgium*

**Abstract:** Tunnels in glaciers offer unique opportunities for examining basal processes. At Suess Glacier in the Taylor Valley, Antarctica, a 25 m tunnel excavated into the bed of the glacier provides access to a 3.2 m thick basal zone and the ice–substrate contact. Measurements of ice velocity over two years together with glaciotectonic structures show that there are distinct strain concentrations, a sliding interface and thin shear zones or shear planes within the basal ice. Comparison of ice composition, debris concentrations and the shear strength of basal ice samples suggest that strength is controlled by ice chemistry and debris concentration. The highest strain rates occur in fine-grained amber ice with solute concentrations higher than adjacent ice. Sliding occurs at the base of the ice that experiences the highest strain rates. The substrate and blocks of the substrate within basal ice are characterized by brittle and slow ductile deformation whereas ice with low debris concentrations behaves in a ductile manner. The range of structures observed in the basal ice suggests that deformation occurs in a self-enhancing system. As debris begins to deform, debris and ice are mixed resulting in decreased debris concentrations. Subsequent deformation becomes more rapid and increasingly ductile as the debris and sedimentary structures within the debris are attenuated by glacier flow. The structural complexity and thickness of the resulting basal ice are considerably greater than previous descriptions of cold glaciers and demonstrate that the glacier is or was closely coupled to its bed.

Although cold glaciers are widespread today and were common during the Pleistocene, the behaviour of ice and sediment at the base of cold glaciers has not attracted a great deal of attention. There are only two comprehensive published accounts of the behaviour of basal ice and substrate in cold glaciers. The first of these was a study of Meserve Glacier, which is a small alpine glacier in the Wright Valley, Antarctica (Holdsworth & Bull 1970). Meserve Glacier has a basal temperature of −18°C and rests on a substrate of ice-cemented rock debris ranging in size from clay to boulders up to 2 m in diameter. The basal ice is characterized by a slight discolouration and solute concentrations about twice the values for overlying englacial ice. Holdsworth (1974) called this ice facies 'amber ice' and it can be recognized in most of the small alpine glaciers in the McMurdo dry valleys. Measurements of basal ice deformation demonstrated that the glacier has a parabolic velocity profile and that deformation is concentrated in the amber ice. These observations led Holdsworth to conclude that the amber ice deforms much more readily

than the clean glacier ice. A more recent examination of basal ice in Meserve Glacier has suggested that sliding at −17°C could be attributed to the presence of interfacial water in basal ice (Cuffey *et al.* 1999). The other published account of the behaviour of a cold glacier is a paper by Echelmeyer & Wang (1987) which describes observations of basal ice characteristics and measurements of ice deformation at the bed of Urumqi No. 1 Glacier. The glacier substrate consists of ice laden drift with ice concentrations between 21 and 39% by weight and an average density of 3.25 Mg m$^{-3}$ (46–68% by volume). Measurements made by Echelmeyer & Wang show that sliding occurred at a temperature of −4.6 and deformation of the ice-laden drift at a temperature of −2°C. Deformation of the ice-laden drift together with movement across shear bands and basal sliding account for 60–80% of the overall glacier motion. Echelmeyer & Wang argued that a significant proportion of surface motion was provided by deformation of a thin layer of 'basal drift' about 35 cm in thickness. Subsequently

*From:* MALTMAN, A. J., HUBBARD, B. & HAMBREY, M. J. (eds) *Deformation of Glacial Materials.* Geological Society, London, Special Publications, **176**, 181–190. 0305-8719/00/$15.00 © The Geological Society of London 2000.

this research has been cited as evidence for substrate deformation at subfreezing temperatures. However, because the deformation that they describe occurs in material with high ice concentrations, it could be argued that they have described deformation of debris-laden ice rather than ice-laden drift. The distinction is more than semantic because once the ice concentration exceeds saturation the frictional strength of the material decreases dramatically and rheology is increasingly defined by the strength of the ice. One could argue that Echelmeyer and Wang describe deformation of basal ice that has accreted to the base of the glacier. The accounts of basal processes in Merserve and Urumqi No. 1 glaciers present views of the behaviour of ice below pressure melting point that are not entirely compatible.

The research programme that this paper derives from aims to increase our understanding of how cold glaciers interact with landscapes. In this paper we focus on the behaviour of basal ice and ask how does it deform close to the bed and what effect does the presence of the debris have on glacier behaviour?

## Methods

Suess Glacier is a 5 km long glacier that descends from 1350 m on the Asgard Range to about 50 m on the floor of the Taylor Valley. A 2 m × 1 m × 25 m tunnel was excavated at the bed of the glacier in 1996 using chainsaws and a demolition hammer. The tunnel was extended in 1997 and a vertical shaft 4.5 m high was cut at the end of the tunnel to expose the entire debris zone. At the end of the tunnel thermocouples and alcohol thermometers showed that the basal ice was at a temperature of $-17°C$.

The velocity of the glacier at the bed was measured using linear variable displacement transducers (LVDTs) and precision dial gauges. Strain markers that consisted of 300 mm rods of wood cut into 20 mm segments were inserted into holes drilled in the substrate. The LVDTs were bolted to the glacier bed and onto the tops of boulders in basal ice using rock bolts and attached to moving ice using thin wires and wooden pegs frozen into holes drilled in the ice. Plumblines as described by Holdsworth (1974) were used to determine the velocity through the basal ice. The plumblines dropped onto brass targets bolted to the bed and displacements were measured from a line of wooden markers drilled and frozen into ice behind the line. Initial results of these measurements were reported by Fitzsimons *et al.* (1999) and in this paper we describe the results of over two years of data from the plumblines.

## Results

The structure of the basal zone consists of a frozen sand and gravelly sand substrate, 2.5 m of stratified basal ice within which we identify four distinct ice facies (Table 1), a 1 m thick plate of frozen debris, an amber ice layer about 0.8 m thick, and clean englacial ice that is about 20 m thick. The basal zone also contains occasional boulders up to 1.2 m in diameter and numerous gas-filled cavities. The cavities occur in five associations: at the upstream and down stream ends of boulders, within and adjacent to broken blocks of frozen debris, at the contact between the layer of solid debris and the amber ice, within relatively clean basal ice, and at the basal ice-substrate contact (Figs 1 & 2).

Displacement transducers anchored to the bed recorded no sliding over 60 days before a battery developed a fault and the data logger stopped recording. Dial gauges anchored to the bed and attached to ice 5 mm above the

**Table 1.** *Characteristics of ice facies*

| Ice facies | Shear strength (MPa) | Debris content (%) | Ice crystal size[†] (mm) |
|---|---|---|---|
| Englacial diffused | 1.39 | 0.06 | 2.96 |
| Basal amber | 0.90 | 0.87 | 0.50 |
| Basal solid | 2.53 | 96.00 | – |
| Basal diffused* | 1.35 | 0.30 | 8.65 |
| Basal laminated* | | | |
|    debris-bearing | 1.81 | 73.99 | 1.31 |
|    clean | 1.30 | 13.26 | 1.37 |
| Basal clear* | – | 3.03 | – |

* Collectively called the basal stratified facies in the text.
† Measured from the optical axis.

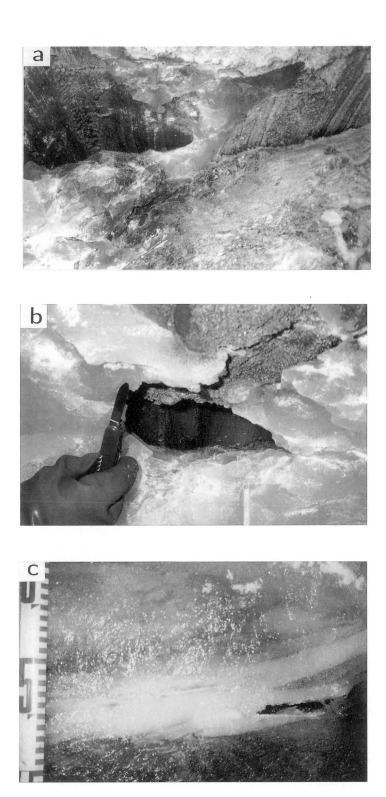

**Fig. 1.** Gas-filled cavities in basal ice at the downstream side of a granite boulder (**a**) (field of view is 250 mm), at the contact between the solid facies and amber ice (**b**), and within relatively clean basal ice (**c**).

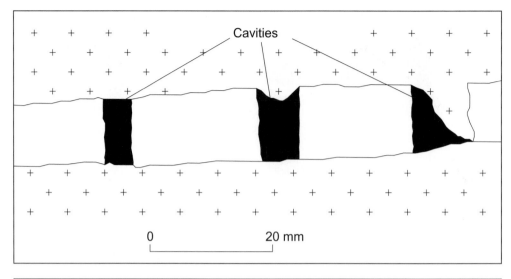

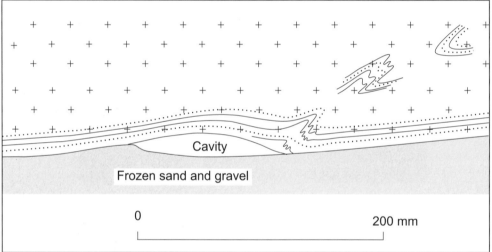

**Fig. 2.** Sketches of boudinage structures with gas filled cavities and a cavity formed at the basal ice–substrate interface with a fold at the cavity closure. Flow is from left to right.

ice-substrate interface in different locations measured 0.93, 0.4, 5.65 and 4.35 mm a$^{-1}$ measured over 348 days.

Figure 3 shows a velocity profile through 4 m of basal ice together with a column that summarizes ice stratigraphy and a velocity profile from Meserve Glacier. Three patterns can be identified in the velocity profile. In the lower part of the profile (0–2 m) there is a near linear increase in velocity, between 2 m and 2.8 m there is no apparent change in velocity and above 3 m there is a logarithmic increase. Comparison of the velocity profile with the basal ice stratigraphic column shows that the velocity varia-

tions can be directly associated with changes in the physical characteristics of the ice and the occurrence of debris-rich ice. The near linear increase in velocity occurs in the basal stratified facies, the area of no increase in velocity occurs within the basal solid facies and the section of logarithmic increase in velocity occurs within the amber facies. As the tunnel was excavated we encountered cavities along the contact of the solid facies and the overlying ice. As we excavated ice surrounding the lower part of these cavities we exposed a sharp contact between overlying ice and the frozen debris. Viewed from below, the ice flowing over the frozen debris was

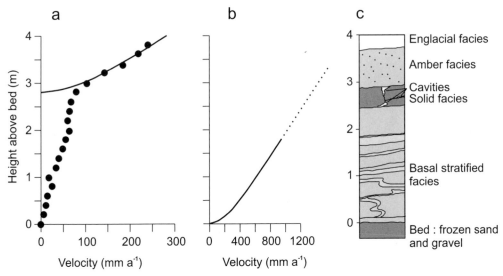

**Fig. 3.** Velocity profile 24 m into the tunnel through 4 m of the basal zone (**a**), velocity profile from Meserve Glacier (Holdsworth 1974) (**b**), and column describing the characteristics of the basal zone adjacent to the site where the velocity profile was measured (**c**). The top of the cavity shown by (**c**) consists of slickenslides on the ice surface and the flow is from the bottom to the top of the photograph.

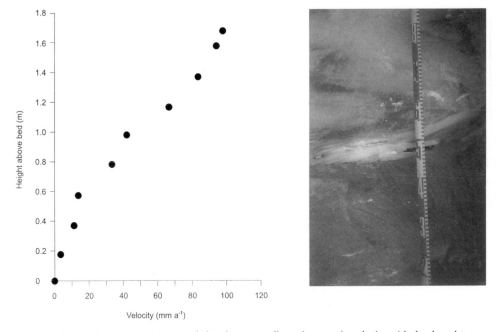

**Fig. 4.** Velocity profile 10 m into the tunnel showing a near linear increase in velocity with depth and two apparent displacements at 0.6 and 1.0 m. The photograph at right shows an unconformity in the ice at 0.6 m.

marked by slickenslides which reflect the form roughness of the top of the debris layer (Fig. 1c). Although the strain markers used to measure the velocity profile are too widely spaced to convincingly demonstrate sliding, the profile suggests an offset at 2.8 m. (Fig. 3). The slickenslides, together with displacements measured with a dial gauge in the ice layer within 5 mm of the contact, demonstrate that the ice is sliding on the debris layer. Within the debris layer the near vertical

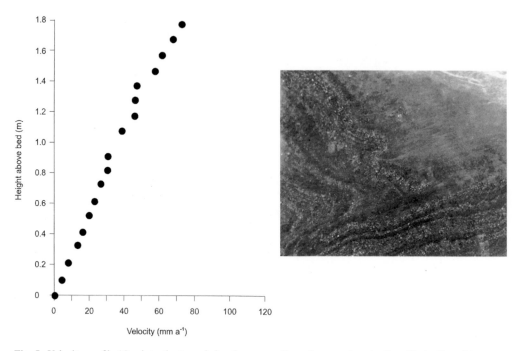

**Fig. 5.** Velocity profile 15 m into the tunnel showing a near linear increase in velocity with depth and two apparent displacements at 0.9 and 1.4 m. The photograph at left shows an unconformity in laminated ice that appears to be a sheared fold.

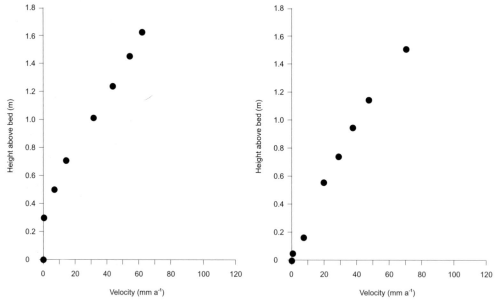

**Fig. 6.** Velocity profiles in relatively clean basal ice 14 and 16 m into the tunnel.

nature of the velocity profile indicates that a section of frozen debris within the basal ice is either not creeping or creeping very slowly compared to the adjacent ice.

A velocity profile measured below the solid facies layer in laminated basal ice is shown by Figure 4. The profile records a near linear increase in velocity similar to the lower 3 m of the velocity profile shown by Fig. 3. Two small offsets occur in the profile at 0.6 m and 1.0 m. At both these locations obvious unconformities were present in the ice which was highly deformed into tight, sheared folds (Fig. 3). The photograph shows ice with a near vertical foliation truncated by a low angle shear that dips in an up glacier direction at 4°. The bubble-rich shear zone was about 80 mm thick and contained a flat cavity that was 100 mm long, 20 mm high and at least 120 mm deep. The roof of the cavity was marked by slickenslides parallel to the ice flow direction which suggested that there was a sliding interface that was not detected by the plumbline measurements. When punctured, gas rushed from the cavity which suggested that the gas was above atmospheric pressure. Above the shear zone the ice is folded and consists of layers of strongly stretched bubbles and relative bubble-free ice which appears dark on Fig. 4. The top of the fold shown in the photograph in Fig. 4 was truncated by another low angle shear zone.

Figure 5 shows another velocity profile in laminated basal ice. The profile contains offsets at 0.9 and 1.4 m where unconformities in ice structure were observed. One of these shears is shown by the photograph in Fig. 5 where debris-bearing and clean laminae of the basal laminated facies are perpendicular and separated by a thin layer (<10 mm) of relatively clean ice with low gas content.

Two velocity profiles measured in relatively clean ice below the main debris layer show little evidence of shear zones either in the profile or from observations of the physical characteristics of the ice (Fig. 6).

## Implications for glacier behaviour

### Strain concentrations

The most important implication of the velocity measurements and observations is that deformation is highly concentrated in the basal ice. Three forms of deformation concentration that can be recognised are high strain in the amber facies, sliding at the interface between the amber ice and the solid facies, and thin shear zones or shear planes in the stratified basal ice. Taken together the uneven distribution of strain means that a simple application of Glen's Law is a poor representation of the behaviour of the basal ice (Fig. 3).

The amber ice facies is clearly the weakest material that we have tested in direct shear tests (Table 1 and Fitzsimons *et al.* 1999). The mean peak shear strength of the amber facies is less than 65% of the overlying englacial facies and 35% of the underlying solid facies. Consequently, the amber facies will deform at a greater rate that the adjacent ice. This data suggests that the amber facies in Suess Glacier behaves in a very similar manner to the amber ice originally described from Meserve Glacier by Holdsworth (1974). The main characteristics of the amber ice in Suess Glacier include very low debris concentrations, concentrations of solutes that are more than five times that of the adjacent glacier ice (unpublished data) and mean ice crystal size of 0.5 mm (Table 1). Although the controls of the rheology of debris-bearing ice are governed by complex and inter-related physical and chemical properties it seems likely that the solute content of the ice plays an important role in allowing it to deform relatively easily at low temperatures (Jones & Glen 1969; Gilpin 1979; Cuffey *et al.* 1999).

Slickensides and cavities at the interface between the amber and solid facies together with the velocity profile (Fig. 3) suggest that the amber ice is sliding over the solid facies at −17°C. The conventional view of basal motion by sliding is that it requires liquid water for the regelation process which limits sliding to ice that is close to the pressure melting point. However, Shreeve (1984) and Fowler (1986) have suggested that very slow sliding can occur at a few degrees below the pressure melting point. There is also limited empirical evidence of sliding at subfreezing temperatures. For example Hallet *et al.* (1986) measured intermittent movement of ice over a lock ledge at −1°C and Echelmeyer & Wang (1987) observed sliding of clean ice over a large boulder in basal ice at −4 to −6°C at several times the velocity predicted by Shreeve. More recently another investigation of amber ice in Meserve Glacier used Gilpin's (1979) interfacial film theory to argue that sliding and ice segregation are best explained as manifestations of unfrozen water in films at ice-rock interfaces (Cuffey *et al.* 1999).

The presence of cavities and slickensides at all of the sites where sliding was measured suggests a close association between sliding and cavitation. The importance of cavities in sliding was first recognised by Lliboutry (1968). Subsequent examination of cavities has focused on their formation in ice close to pressure melting point where pressure in a water film become equal to the triple point pressure where ice, water and water vapour co-exist. Fowler (1979) offered a different explanation and argued that cavities could be the natural equivalent to vortices formed at a boundary layer separation.

Although we cannot make an unambiguous interpretation Fowler's suggestion has some appeal because all the cavities we observed occurred at distinct flow separations. A less complex statement about the manner of cavity formation is that they form when the tensile stresses due to ice flow exceeds the cryostatic pressure. Drewry (1986) summarized the case of cavity formation for a bed protuberance as

$$. \tau_b \left( \frac{\lambda_b}{l} \right)^2 s > \rho_i g h$$

where $\tau_b$ is the basal shear stress, $\lambda_b$ is the spacing between bed bumps, $l$ is the length of the side of the bed bump, $s$ is a parameter related to the dimensions of the bed hummock, $\rho_i$ is ice density, $g$ is acceleration due to gravity and $h$ is ice thickness. Although this relationship may be used to explain cavities at the bed and around boulders and blocks within basal ice, it does not offer an explanation of cavities that occur at planar contacts between ice and the bed or where frozen sediment bands have been pulled apart (Fig. 2). In these situations tensional stresses appear to be generated by differences in rheology of the materials which is reflected in the different strengths of materials (Table 1 and Fitzsimons *et al.* 1999). The effect of a cavity is to produce part of the glacier that is not in contact with the substrate thereby reducing contact friction and increasing the basal shear stress at the points that remain in contact. The points that remain in contact may therefore experience enhanced creep and higher sliding velocities. Although the observations and data that we have do not allow us to conclude whether cavities are the cause or an outcome of sliding it seems likely that their presence results in localised increases in sliding velocities.

Previous descriptions of cavities close to the bed of cold glaciers have suggested that they have formed as the glacier advanced over a subaerial landscape trapping air between the glacier and the substrate (Holdsworth 1974). However, at least some of the cavities we have observed have formed within the basal zone well above the substrate (e.g. Fig. 2b). Nevertheless, Holdsworth's interpretation indirectly raises the important question of what is the source of the gas in the cavity if they form within basal ice. We do not yet have a explanation of the formation of cavities although we speculate that gas bubbles may be squeezed from the ice. Cavities that were above atmospheric pressure are likely to have experienced partial closure after they have formed.

The offsets that we have measured in the velocity profiles within the basal stratified facies (Figs 4 and 5) have been produced by either narrow shear zones or shear planes. Although the spatial resolution of the pegs that were used to measure the velocity profiles was not sufficient to determine whether the offsets were caused by thin zones of high strain or sliding along planes, the occurrence of slickenslides in the roof of one cavity suggests the development of a sliding plane within the ice at some time. Our experiments on the strength of ice suggest that there is no difference in the strength of ice above and below the shear zones although the direct shear apparatus provides estimates of the bulk strength of the ice and would not be sensitive to changes in ice fabric or small-scale variations in debris concentration. We have, however, observed numerous instances of small-scale variations in debris concentrations, which is a first-order control of rheology, associated with flow perturbations including brittle and ductile boudinage, isoclinal folds and recumbent folds.

## Uncoupling within the basal ice

High strain in the amber ice facies together with sliding at the interface of the amber and solid facies produce a major flow discontinuity in the velocity profile 3 m above the bed. At this location in the basal ice the glacier is partly uncoupled from the basal stratified facies that rests below the interface. If there was a complete uncoupling of flow there would be no movement below the sliding interface. We interpret this discontinuity in the velocity profile as a major flow separation within the basal ice that appears similar to patterns of strain proposed for glaciers that experience subglacial bed deformation (e.g. Boulton 1996).

Three consequences of the partial uncoupling within the basal ice are that the bed may be protected, the debris-bearing ice moves at a fraction of the velocity of the glacier, and that a major temporal unconformity exists near the bed of the glacier. The bed may be protected because the majority of strain occurs 3 m above the basal ice–frozen sediment interface potentially resulting in isolation of the bed and suppression of glacier-substrate interactions. Bed material that is deformed and entrained (see Fitzsimons *et al.* 1999) is transported from the site of entrainment relatively slowly because the basal stratified ice flows at a significantly lower velocity than the overlying ice. The age unconformity close to the bed has important implications for palaeoclimatological reconstructions based on such an ice sequence because the record is discontinuous.

## Comparison with other cold glaciers

Comparison of the velocity measurements and observations of structures in the basal ice with two published accounts of the basal conditions of cold glaciers highlights some interesting contrasts in the composition and behaviour of cold ice. Meserve Glacier is located on the southern side of the Wright Valley about 6 km from Suess Glacier. It has a basal temperature of around −17°C and rests on a bouldery substrate compared to the unconsolidated sand and gravel substrate of Suess Glacier. Both surface and bed slopes are considerably greater than Suess Glacier (3° v. 13° at the terminus) which explains why the velocity is considerably higher (Fig. 3). The base of Meserve Glacier consists of 1.8 m of amber ice resting directly on the bouldery substrate. In contrast, the basal ice sequence of Suess Glacier is 3 m thicker, structurally more complex and contains significantly more debris. In previous publications we have attributed the thickened basal sequence to bed deformation and entrainment (Fitzsimons *et al.* 1999) and to accretion of lake water beneath the glacier (Lorrain *et al.* 1999). The structural complexity of the Suess Glacier together with the variability in ice composition, are matched by complex patterns of strain relating to differences in ice rheology.

A second published account of the behaviour of a cold glacier is a study of the Urumqi No. 1 Glacier in China which describes observations of basal ice characteristics and measurements of ice deformation (Echelmeyer and Wang 1987). Echelmeyer & Wang (1987) described the substrate of the glacier as 'ice laden drift' with ice concentrations between 21 and 39% by weight and an average density of 3.25 Mg m$^{-3}$ (46–68% by volume). Measurements showed that sliding and deformation of the ice-laden drift occurred at a temperature around −4°C. Deformation of the ice-laden drift together with movement across shear bands and basal sliding account for 60–80% of the overall glacier motion and Echelmeyer & Wang argued a significant proportion of surface motion was provided by deformation of a thin layer of 'basal drift' about 35 cm in thickness. Although there is a large difference between the basal ice temperatures of the two glaciers there appear to be strong similarities between the debris concentration and structure of the 'substrate' of Urumqi

No. 1 Glacier and the solid and stratified basal ice facies of Suess Glacier. One important difference between the two studies is what we describe as basal ice Echelmeyer and Wang describe as the glacier substrate. While it could be argued that the difference is semantic it raises the question of what is the bed of a glacier. It could be argued for example that the flow separation at interface between the solid and amber facies is the effective bed of the glacier and that a very small amount of strain is propagated into the bed (cf. Echelmeyer & Wang 1987). Alternatively, the bed is to be defined as the horizon below which there is no strain. The latter definition does not recognise that the position of the bed is likely to be variable in time and space.

## Conclusions

(1) In Suess Glacier strain is concentrated in the amber ice facies, which is considerably weaker than adjacent ice facies, at a sliding contact between the amber and solid facies and in thin shear zones or planes in stratified basal ice. Together the high strain within the amber ice and sliding behaviour at its base have resulted in a major flow separation within the basal zone. The flow separation, which occurs 3 m above the glacier bed, shows that the glacier is partly uncoupled from the slower moving underlying basal ice.

(2) At least some cavities form within the basal ice and at the bed, which results in localized variations in basal shear stress and velocity, which may result in complex patterns of velocity at the amber–solid facies contact and at the interface between the stratified facies and the glacier substrate. It is not clear whether the cavities are the product or cause of the flow separations or both. Even though we have direct observations of ice structure and measurements of deformation made over two years it is very difficult to separate cause from effect.

(3) The displacement data together with glaciotectonic structures observed in basal ice show that shear zones and folding are two mechanical means by which debris can be elevated from the bed and mixed into basal ice.

(4) Comparison of data from the Meserve, Urumqi No. 1 and Suess glaciers suggest that the rheology of frozen debris is more complex than previously realised at not yet well understood.

We would like to thank the University of Otago Research Committee, the Marsden Fund of the Royal Society of New Zealand and the Belgian Antarctic Programme Science Policy Office for financial support and Antarctica New Zealand for logistical support.

# References

BOULTON, G. S. 1996. Theory of glacial erosion, transport and deposition as a consequence of subglacial sediment deformation. *Journal of Glaciology*, **42**, 43–62.

CUFFEY, K. M., CONWAY, H., HALLET, B., GADES, A. M. & RAYMOND, C. F. 1999. Interfacial water in polar glaciers and glacier sliding at −17°C. *Geophysical Research Letters*, **26**, 751–754.

DREWRY, D. 1986. Glacial geologic processes. Edward Arnold, London.

ECHELMEYER, K. & WANG, Z. 1987. Direct observation of basal sliding and deformation of basal drift at sub-freezing temperatures. *Journal of Glaciology*, **33**, 83–98.

FITZSIMONS, S. J., LORRAIN, R. & McMANUS, K. J. 1999. Structure and strength of basal ice and substrate of a dry-based glacier: evidence for substrate deformation at sub-freezing temperatures. *Annals of Glaciology*, **28**, 236–240.

FOWLER, A. C. 1979. A mathematical approach to the theory of glacier sliding. *Journal of Glaciology*, **23**, 131–141.

——1986. Sub-temperate basal sliding. *Journal of Glaciology*, **32**, 3–5.

GILPIN, R. R. 1979. A model of the 'liquid-like' layer between ice and a substrate with applications to wire regelation and particle migration. *Journal of Colloid Interfacial Science*, **68**, 235–251.

HALLET, B., GREGORY, C. E., STUBBS, C. W. & ANDERSON, R. S. 1986. Measurements of ice motion over bedrock at subfreezing temperatures. *Mitteilungen der Versuchsanstalt fur Wasserbrau, Hydrologie und Glaziologie an der, E.T.H., Zurich*, **90**, 53–54.

HOLDSWORTH, G. 1974. *Meserve Glacier, Wright Valley, Antarctica: Part I. Basal processes.* Ohio State University Institute of Polar Studies Report, 37.

—— & BULL, C. 1970. The flow law of cold ice: investigations on the Meserve Glacier, Antarctica. *In: Symposium at Hanover, New Hampshire 1968 – Antarctic Glaciological Exploration (ISAGE)*, International Association of Scientific Hydrology Publications, **86**, 204–216.

JONES, S. J. & GLEN, J. W. 1969. The effect of dissolved impurities on the mechanical properties of ice crystals. *Philosophical Magazine*, Ser. 8, **19**, 13–24.

LORRAIN, R. D. FITZSIMONS, S. J., VANDERGOES, M. J. & STIEVENARD, M. 1999. Ice composition evidence for the formation of basal ice from lake water beneath a cold-based Antarctic glacier. *Annals of Glaciology*, **28**, 277–281.

LLIBOUTRY, L. A. 1968. General theory of subglacial cavitation and sliding of temperate glaciers. *Journal of Glaciology*, **7**, 21–58.

SHREEVE, R. L. 1984. Glacier sliding at subfreezing temperatures. *Journal of Glaciology*, **30**, 341–347.

# Use of a viscous model of till rheology to describe gravitational loading instabilities in glacial sediments

RICHARD C. A. HINDMARSH[1] & KENNETH F. RIJSDIJK[2]

[1] *British Antarctic Survey, Natural Environment Research Council, High Cross,*
*Madingley Rd, Cambridge CB3 0ET, UK (e-mail: rcah@bas.ac.uk)*
[2] *Geo-Marine and Coast Department, National Geological Survey, NITG-TNO,*
*PO Box 80015, 3508 TA Utrecht, The Netherlands*

**Abstract:** This paper models the operation of loading (Rayleigh–Taylor) instabilities in sediments using an effective-pressure-dependent viscosity such has been used to model the deformation of sediment beneath glaciers. A particular feature is a strong increase of viscosity with depth, resulting from the fact that the effective pressure increases with depth.

Observations suggest that more than one wavelength is generally present (e.g. diapirism and loadcasting) which requires at least three layers with uniform properties to be present. Three layers permit wavelength growth maxima at two distinct wavelengths. We investigate whether an effective-pressure dependent rheology is consistent with RT instabilities, and whether the non-uniformity it produces is able to increase the number of growth-rate maxima.

The investigation starts from the point where sediment in an underlying layer is less dense than the overlying sediment, and the Rayleigh–Taylor instability starts to operate. The mechanics of two layers of finite thickness but infinite extent are modelled by the Stokes equations. The equation set is linearized, and the Fourier transform taken in order to describe the periodic horizontal variation of flow fields at a specified wavelength.

The influences of layer thickness and viscosity ratio on the flow fields are considered. It is found that, for a given wavelength, layer thickness has a far stronger influence on flow fields than does viscosity ratio. For all configurations inspected, the dependence of growth rate on wavenumber exhibited one maximum, meaning that a variable viscosity model does not produce multiple wavelengths. Maximum growth rates occur at wavelengths corresponding to the layer thicknesses.

We infer that loading instabilities occurring at wavelengths around the layer thicknesses are consistent with the effective-pressure-dependent viscous model.

The hypothesis that the large-scale flow of sediments underneath glaciers and ice sheets can best be described by a viscous rheological model (Boulton & Hindmarsh 1987) has remained controversial. Some recent papers have reviewed this controversy (Hindmarsh 1997; Boulton & Dobbie 1998) and have concluded that on the large scale (>100 m horizontal) the idea of a viscous flow remains tenable, notwithstanding a series of observations in the 1990s that suggest that on the small scale, subglacial sediment deformation is event-driven and by inference, plastic (Iverson *et al.* 1995; Murray & Clarke 1995). In particular, the viscous theory has had recent success in describing some aspects of drumlin formation (Hindmarsh 1997, 1998*a, b*).

This paper seeks to determine whether the viscous theory of sediment deformation, with the hypothesized effective-pressure-dependent visc-osity, can explain or is at least consistent with observations of loading (Rayleigh–Taylor) instabilities in glaciogenic sediments. A gravitationally unstable density gradient exist when a higher density sediment overlies a lower density sediment, comparable to a Rayleigh–Taylor instability in fluids (Ramberg 1968, 1972; Anketell *et al.* 1970; Allen 1984; Rönnlund 1989). These reverse-density systems strive to exchange the density layering, resulting in deformation at the sediment layer interfaces. An interesting feature of the theoretical studies lies in the fact that the dominant wavelength can be related to the viscosity ratio of the two layers, as well as other factors, notably their thickness (Ramberg 1968; Rönnlund 1989). This forms one of the principal motivations for the study.

This paper is not concerned with modelling how the denser layer comes to overlie a lighter layer; it takes it as the initial condition,

*From*: MALTMAN, A. J., HUBBARD, B. & HAMBREY, M. J. (eds) *Deformation of Glacial Materials.* Geological Society, London, Special Publications, **176**, 191–201. 0305-8719/00/$15.00 © The Geological Society of London 2000.

hypothesizes that subsequent sedimentary deformation is viscous, and models this situation in order to determine the outcome. Lower layers of less dense sediment can for example arise when there is rapid deposition onto an unconsolidated layer leading to liquefaction or when fluidization is triggered by earthquakes (Owen 1987).

Rayleigh–Taylor (RT) instabilities have been discussed extensively in the geological literature (Ramberg 1972; Rönnlund 1989). RT instabilities lead to deformation at and motion of the interface of the sediment layers, which progressively develop to form (i) loadcasts of the higher density sediments in the lower density sediment and (ii) diapirs of the lower density sediments in the higher density sediments (Butrym et al. 1964; Anketell et al. 1970; Eisman 1981; Allen 1984; Rijsdijk 1999a). This leads eventually to a stable density gradient in the sediments with the higher density material ultimately lying at the base and vice versa. In intermediate stages of deformation, teardrop shaped perturbations which detach as rafts of higher density material may develop in the lower density material (e.g. Kuenen 1958; Rijsdijk 1999a). Deformation continues until the system stabilizes; in sediments this does not necessarily result in a complete reversal of density, and in practise deformation halts at an intermediate stage.

Examples of these density driven deformation structures are reported from a wide range of sedimentary settings including marine, periglacial and glacial (Butrym et al. 1964; Anketell et al. 1970; Eisman 1981; Allen 1984). In fossil glacial settings large-scale density driven deformational structures are described in gravels and sands overlying glacilacustrine muds or low viscosity (flow-) tills. Deformation structures include synforms up to tens of metres wide or rafts of overlying gravels and sands in fine-grained diamict coinciding with up to 10 m high diapirs in underlying fine grained diamicts (Schlüchter & Knecht 1979; Schwan et al. 1980; Brodzikowski & van Loon 1980; Aber et al. 1989; Brodzikowski et al. 1987; Eyles et al. 1987; Eyles & McCabe 1989; Allen 1990; Paul & Eyles 1990; Ehlers et al. 1991; Rijsdijk 1999a). In pro-glacial settings RT instabilities usually form as a result of sudden rapid deposition of sand and gravels on top of lacustrine clays or (flow-) tills, which fail plastically and form diapirs. Observations also include glacitectonized diapirs in subglacial tills, demonstrating that RT instabilities may develop in tills in subglacial conditions (Aber 1989; Rijsdijk 1999b).

The field examples of Aber (1984, figs 7–5 & 7–11) are particularly interesting: diapirism is argued to be triggered by an advancing ice-wedge (similar examples exist in Killiney, Ireland, Rijsdijk 1999b). In these cases it seems that the advance raised pore-water pressures as well as the glacitectonic stresses in the tills, inducing RT instabilities. The associated RT deformation structures (diapirs and loadcasts) were ultimately glaciotectonised by the over-riding ice.

The novel feature of the present study is the application of a rheological relationship

$$e_{ij} = \frac{1}{2B} \frac{\tau_{ij}}{(c + p_e)^b} \tag{1}$$

where $e_{ij}$ is the deformation rate, $\tau_{ij}$ is the shear stress, $B$ is a viscous coefficient, $p_e$ is the effective pressure, $c$ is the cohesion and $b$ is an index. This rheological relationship is analogous to a form suggested by Boulton & Hindmarsh (1987) $e_{ij} = (\tau^{a-1}/2B)(\tau_{ij}/p_e^b)$, $a \neq 1$, $\tau^2 = \tau_{ij}^2$. This rheological relationship has been shown to be useful in modelling a variety of glacial geological situations (Hindmarsh 1997), in particular drumlin formation (Hindmarsh 1996, 1998a, b). The rheological relationship (1) used in this paper contains only a linear dependence of deformation rate upon stress and includes a form of effective-pressure-dependence proposed explicitly by Fowler & Walder (1993); these are both included in order to avoid certain mathematical difficulties. The index $a$ is sufficiently poorly constrained by empirical evidence to permit the simplification of taking it to be unity. This does mean of course that we cannot consider plastic rheologies in this paper.

In this paper, we use rheological relationship (1) together with RT theory to improve our understanding of the issue of whether these viscous models can explain loading instabilities in glaciogenic instabilities. The rheological model omits several important features of observed sediment rheology, for example shear-thinning (P. Harrison pers. comm.) and is not intended at all to be a complete description of sediment rheology, but rather an exploration of whether effective-pressure-dependent viscosities are consistent with the operation of RT instabilities. The issues of whether an instability can occur can be dealt with by linear stability analysis, which can also give a guide to the wavelength of an instability as this is likely to be the fastest growing wavelength. Linear stability analysis ceases to be reliable once the amplitude of the perturbation becomes greater than about 20% of the thickness of the layer.

An important feature of the problem is the behaviour of the effective pressure with depth. In this paper we assume that the effective pressure $p_e = p - p_w$ increases with depth according to static balances, i.e.

$$\frac{\partial p}{\partial z} = -\rho g, \tag{2a}$$

$$\rho \equiv (1 - \phi)\rho_s + \phi\rho_w, \tag{2b}$$

$$\frac{\partial p_w}{\partial z} = -\rho_w g, \tag{2c}$$

$$\frac{\partial p_e}{\partial z} = -(1 - \phi)(\rho_s - \rho_w)g, \tag{2d}$$

where $(x, z)$ defines a coordinate system with $z$ pointing up, $\phi$ is the porosity of the sediment, $\rho_s, \rho_w$ are the densities of sediment and water respectively (taken to be $2.7 \, \text{kg m}^{-3}$ and $1 \, \text{kg m}^{-3}$ respectively) and $g$ is the acceleration due to gravity ($9.81 \, \text{m s}^{-2}$). In this relationship $p$ is the bulk soil pressure and $p_w$ is the water pressure. An alternative way of looking at this equation is to regard it as a statement of 'solid matrix support', i.e. some stresses are transmitted through intergranular contacts, and the material is not fluidized. (If it were fluidized, one would not expect dependence of the viscosity on the effective pressure.) It can be seen from relationship (2a) and relationship (1) that deformation rates are expected to decrease rapidly with depth. This can inhibit flow at depth and potentially inhibit the development of the RT instability. Acting against this is the fact that the viscous coefficient $\eta$ can change at a layer boundary. In particular, if $B$ were to decrease in the lower layer this could result in greater deformation rates in the lower layer. Finally, we do not know *a priori* what the effective pressure is at the surface of the soil. If the soil is completely saturated, then the effective pressure at the (by construction unloaded) surface is zero, since the bulk pressure must equal the water pressure at the point. This is true whether the soil is under water or not. It may be that the free water surface is below the soil surface, in which case, in the dry zone, the density of water should be replaced by the density of air in equation (2).

The exploration of these effects is carried out theoretically. The momentum balance equations for slow creeping flow are combined with the rheological relationship (1) to describe the plane flow of two infinitely broad layers of different density and rheology abutting each other. The resulting equations are linearized (since we are dealing primarily with stability issues) and the Fourier transform then taken (which reduces the dimension of the system by exploiting symmetry in the horizontal dimensions). The resulting system of equations can be solved for

the flow and stress fields within the layers, and the stability of the system examined (i.e. will a perturbation in the thickness of the layers grow as a result of loading instabilities?). This procedure is conceptually equivalent to that of Ramberg, but we consider more general configurations not tractable by Ramberg's methods.

The quantities $B$, $c$ and $b$ are very poorly constrained empirical quantities. Since we are not dealing with observations of the instability proceeding and do not in consequence have observations of rates, the actual value of the viscosity parameter is not relevant. RT theory shows that the viscosity ratio of the two layers affects the wavelength of the instability, and we shall explore the effect of the viscosity and density ratios on the operation of the system.

Loading instabilities in glacigenic instabilities have been recently studied by Rijsdijk (1999$a$, $b$), who considered several cases with two layers. There are several forms of load structures, which occur at different wavelengths; in particular,

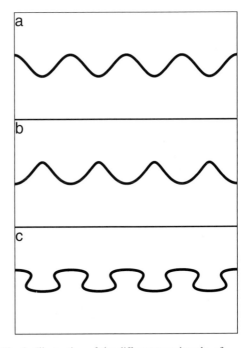

**Fig. 1.** Illustration of the different wavelengths of Rayleigh–Taylor instability. The diagram illustrates the geometry of the interface between an upper, denser layer and a lower more porous layer. (**a**) One wavelength case; (**b**) two wavelengths associated with loadcasts (downwards operating) and diapirs (upwards operating); (**c**) Two wavelengths associated with teardrop formation.

within the same section, more than one wavelength manifests itself (Fig. 1). For example, the concave, downwards-pointing manifestation of the instability is often a longer-wavelength loadcast, while the convex, upper-pointing part is a shorter-wavelength diapir. Sometimes, the loadcast breaks into teardrop forms. This has been considered empirically and qualitatively by Anketell *et al.* (1970), who suggest that the detailed structure of diapirs and loadcasts depends upon the viscosity ratio of the two layers.

For this to be explainable by the linear theory, multiple maximum growth rates must manifest themselves. Ramberg (1968) showed that this can happen if there are more than two layers. By multiple maximum growth rates we mean that the growth rate, when considered as a function of wavelength, must exhibit multiple maxima. Load structures are expected to develop at the wavelengths corresponding to the maximum growth rate. From a mathematical point of view, layers with non-uniform properties can be regarded as a very large number of layers. Thus, the situation modelled in this paper, where the viscosity is a function of depth, might produce multiple growth rate maxima. If not, we will certainly need to search for additional physical mechanisms to produce multiple maximum growth rates. This would mean that the formation of diapirs is a non-linear phenomenon that occurs once the amplitude of the disturbance becomes large, rather than a consequence of linear behaviour.

## Physical model

Physically, we solve the equations for momentum balance (the Stokes' equations). We ignore dilation in the till, and thus assume it to be incompressible. This is represented by the equation

$$\frac{\partial u}{\partial x} + \frac{\partial w}{\partial z} = 0, \tag{3}$$

where $(u, w)$ are the velocities in the $(x, z)$ directions. We are not seeking to argue that dilation is negligible under all circumstances, but we proceed by taking the porosities as given, initial states. These may have arisen as a consequence of till dilation. The effects of the evolution of the instability on till dilation and porosity are likely to be significant but are not considered in this paper for the sake of tractability.

Most generally, we could consider $N$ layers of sediment with different properties overlying each other, although in this paper we restrict consideration to two layers of sediment with different densities (in all cases) and viscous coefficients $B$ (in some of the cases). In plane flow the Stokes' equations and the boundary conditions are (Morland 1984)

$$\partial_z \tau_{xz} + \partial_x \tau_{xx} = \partial_x p, \tag{4a}$$

$$\partial_x \tau_{xz} - \partial_z \tau_{xx} = \partial_z p + \rho_i g, \tag{4b}$$

$$\mathbf{T}^s = \theta . \tau \nu, \tag{4c}$$

$$i \in (1, N) \tag{4d}$$

where $\tau_{xx}, \tau_{xz}$ are the deviator stresses, $p$ is the pressure, $\rho_i$ is the density of the $i$th layer, $g$ is the acceleration due to gravity $\nu$ and $\theta$ are the normal and tangent vectors of the surface and $\mathbf{T}^s$ is the vector of tractions acting on the layer surfaces/interfaces. Velocities are regarded as continuous across layer boundaries (no décollements). This has the consequence that in general, with layers of different viscosity, stresses need not be continuous across the layer boundaries, although tractions must be continuous boundaries in order to respect Newton's Laws. On the upper surface, we prescribe a traction-free surface that is also free to move, while on the lower surface we fix both velocity components to be zero. This configuration is analogous to, but not exactly the same, as Rönnlund's (1989) 'free-slip, no slip' configuration. These equations are for creeping flow and do not apply if inertial terms become important during, for example, a liquefied flow.

For solution purposes we combine the two field equation in a diff-integro form (see e.g. Colinge & Blatter 1998). The equations are solved on a transformed domain, with normalized co-ordinate $\zeta_i$ in each layer defined by

$$\zeta_i = (z - b_i)/H_i \tag{5}$$

where $b_i$ is the base of the $i$th layer and $H_i$ is the thickness of this layer. This means that the equations must be transformed using the appropriate relationships (e.g. Hindmarsh & Hutter 1988).

The constitutive relationship is the effective-pressure dependent relationship (1) defining the viscosity to be

$$\eta = B(p_e + c)^b, \tag{6}$$

we obtain the usual relationships

$$\tau_{xz} = 2\eta e_{xz}, \tag{7a}$$

$$\tau_{xx} = 2\eta e_{xx}, \tag{7b}$$

where $e_{xx}, e_{xz}$ are the strain-rate components.

Now we perform a perturbation expansion about a base rest state characterized by flat layer interfaces. By perturbing the layer interfaces slightly we can see whether the Rayleigh–Taylor instability will operate. For each layer we write

$$H_i = H_{i0} + \mu H_{i1}, \tag{8a}$$

$$(\tau_{ij}, e_{ij}) = \mu(\tau_{ij1}, e_{ij1}), \tag{8b}$$

$$(p, p_w, p_e) = (p_0 + \mu p_1, p_{w0} + \mu p_{w1}, p_{e0} + \mu p_{e1}), \tag{8c}$$

$$(u, w) = \mu(u_1, w_1), \tag{8d}$$

where $\mu$ is a small parameter. We have used the fact that the zeroth-order (base state) is a rest state and thus $(\tau_{ij0}, e_{ij0}) = 0$. We substitute these relationships into the Stokes' equations to obtain equations describing the evolution of the perturbation (see e.g. Ramberg 1972). These are known as the first-order Stokes' equations. Computation of these perturbations is a standard procedure and will not be discussed further here. The no-deformation zeroth-order state is the technical reason as to why we have to restrict consideration to a linear rheology $a = 1$ since the viscosity is singular in the unperturbed state $\tau = 0$. This simplification is unlikely to be crucial as viscosity variation is likely to be due primarily to effective pressure variations, but it does mean that we cannot explore the ability of plastic rheologies to produce loading instabilities.

One also requires a relationship between the first order stress fields and the first-order velocity fields, and in particular one needs to consider the effect of perturbations on the viscosity of the fluid. The zeroth-order pressures at $\zeta_i$ in layer $i$ are given by

$$p_0 = p_0^{s_i} = \rho g H_{i0}(1 - \zeta_i) \tag{9}$$

$$p_{w0} = p_{w0}^{s_i} + \rho_w g H_{i0}(1 - \zeta_i), \tag{10}$$

$$p_{e0} = p_{e0}^{s_i} + (\rho_s - \rho_w)(1 - \phi_i) g H_{i0}(1 - \zeta_i). \tag{11}$$

(see equation 2) The superscript $s_i$ indicates evaluation at the upper surface of layer $i$. We are ignoring Darcian flow effects, so we can write

$$p_{w1} = p_{w1}^{s_i} + \rho_w g H_{1i}(1 - \zeta_i). \tag{5}$$

The perturbed constitutive relationships are found to be

$$\tau_{xz1} = B(p_{e0} + c)^b(\partial_z u_1 + \partial_x w_1), \tag{13}$$

$$\tau_{xx1} = B\frac{(p_{e0} + c)^b}{2}\partial_x u_1, \tag{14}$$

meaning that perturbations to the viscosity induced by the water pressure variations do not enter into the first-order problem problem. (This is a subtlety and a consequence of the fact that the zeroth-order state is a rest-state, so there is no coupling between zeroth-order stresses and first-order viscosity which depends upon the first-order effective pressures.) To first-order the viscosity is therefore given by

$$\eta_0 = B(p_{e0} + c)^b. \tag{15}$$

Finally, since we are dealing with infinitely long planes, we can take Fourier transforms (where the Fourier coefficients are represented by carets) of the relevant field quantities

$$\tau_{xz1} = \hat{\tau}_{xz1}\exp(-jkx), \qquad \tau_{xx1} = \hat{\tau}_{xx1}\exp(-jkx), \tag{16}$$

$$u_1 = \hat{u}_1\exp(-jkx), \qquad w_1 = \hat{w}_1\exp(-jkx), \tag{17}$$

where $k$ is the wavenumber of the perturbation and $j^2 \equiv -1$. These relationships express the sinusoidal variation of the quantities with wavelength $2\pi/k$. The hatted variables are complex numbers such that $|\hat{\tau}_{xx1}|$ for example represents the amplitude of $\tau_{xx1}$ and $\arg(\hat{\tau}_{xx1})$ provides information about the phasing of the variables. We can take similar transformations of the Stokes equations, incompressibility condition and constitutive relationships. This is a standard technique for solving partial differential equations, and, as in related problems, this yields coupled ordinary differential equations with independent variable $\zeta_i$ for the Fourier transformations, in particular for the vertical velocity field $\hat{w}$. These equations are solved numerically using pseudo-spectral methods (Fornberg 1996), with 'Chebyshev clustering' of points. Momentum balance across layers is ensured by equating tractions. This does not assume the stress-field is continuous, as this will not occur where there are jumps in the density or viscosity fields. Pseudo-spectral methods work optimally where fields are continuously differentiable, which is the case within any given layer under the assumptions of this model. Numerical methods are needed to deal with the cases where the material properties vary within a layer. The solution was checked by comparing it with two analytical solutions, one for $k \ll 1$ (the lubrication theory approximation), the other for $k \gg 1$. Excellent agreement was found, and in particular the pseudo-spectral method was found to be good at capturing boundary layers.

The vertical velocity field at the surface of any layer determines whether it is thickening or thinning. Of significance is the fact that we can manipulate algebraically the Fourier transformed system to compute a matrix $\mathbf{D}$ such that

$$\dot{\mathbf{H}}_1 = \mathbf{D}\mathbf{H}_1 \tag{18}$$

where $\mathbf{H}_1$ is a vector of layer thickness perturbations $H_{i1}$. The eigenvalues of the matrix $\mathbf{D}$ determine the stability of the system, and if it exhibits the Rayleigh–Taylor instability, one of the eigenvalues must be positive for the two layer system.

## Examples of solutions

In this section we consider examples of solutions, and discuss the associated flow and stress fields. Velocity fields depend significantly on the assumed layer geometry, not only upon the thicknesses of the two layers, but also upon the size of the perturbations to the layer thicknesses. Different combinations of layer thicknesses give rise to different velocity and stress fields in the two layers. This last observation is crucial to the way we display the results. The evolution of the layer thicknesses is governed by relationship (18) and the matrix $\mathbf{D}$ is a matrix of order $N \times N$, where $N$ is the number of layers. Linear systems theory can be used that to show that this

implies the linearized physical problem has $N$ timescales, and that the dynamic response will be exponential decay or growth. In most cases, configurations of the layer geometry will evolve with $N$ characteristic timescales, but there are $N$ configurations which evolve with only one time constant, the so-called 'normal modes'. The normal modes are a result of the calculation. In the cases considered here it turns out that they correspond to particularly simple configurations; one mode corresponds to a lower layer of nearly uniform thickness, and an upper layer with undulating thickness. In this configuration, the system relaxes, meaning that the thickness undulation of the upper layer decay. This configuration always relaxes faster than the configuration corresponding to the other normal mode, essentially because the density difference between air and sediment is very much greater than the density difference between the two layers.

The second normal mode corresponds to an upper surface of nearly constant elevation, and an undulating interface between the two

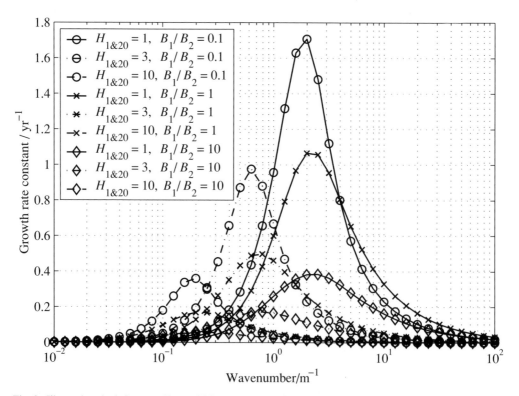

**Fig. 2.** Illustrating the influence of layer thickness, wavenumber and viscosity ratio on the growth rate constant. Folding is expected to develop around the wavelength of fastest growth. Layer thickness has a strong influence on wavelength of maximum growth rate, while viscous coefficient ratio has a scarcely discernible influence. Layers 1 and 2 are of the same thickness in each case.

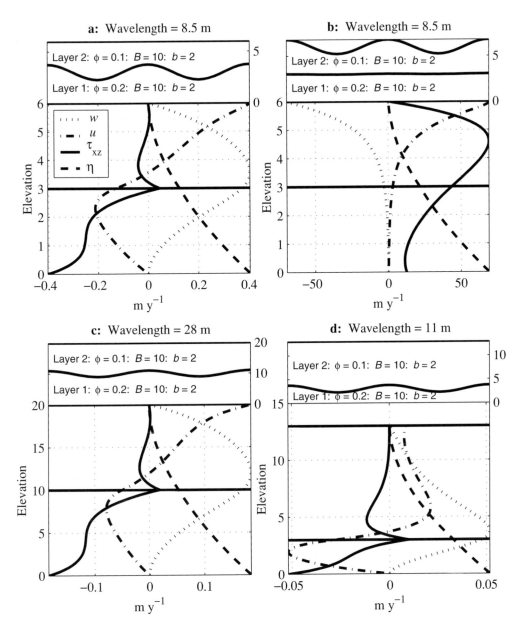

**Fig. 3.** Illustrating the influence of layer thickness geometry on the flow fields generated by undulations at the layer interface (**a, c, d**) and at the upper surface (**b**). Shown are the Fourier transforms of the velocity fields ($\hat{u}, \hat{w}$) and shear stress $\hat{\tau}_{xz}$. These correspond to the vertical velocity component *in phase* with the thickness perturbation and the horizontal velocity and shear stress components *out of phase* with the thickness perturbations. Pseudo-spectral methods were used; 41 points per layer spaced using Chebyshev clustering were used. All diagrams are shown at the wavelength where the RT instability grows most quickly.

sediment layers. The amplitude of this interface grows (the RT instability) and there is also a smaller increase in relief of the upper surface. (Note that the net potential energy of the system is decreasing, as lighter material is moving up while denser material is moving down.)

A parameter search was done over the direct product of the sets

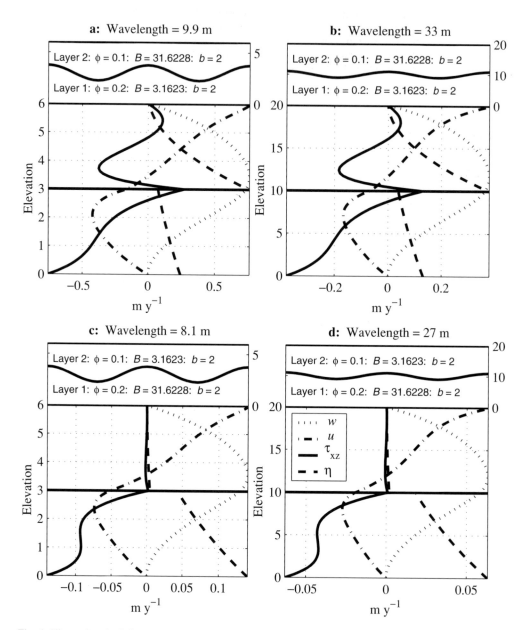

**Fig. 4.** Illustrating the influence of viscosity ratio and layer thickness geometry on the flow fields generated by undulations at the layer interface. See caption of Fig. 3 for a more detailed explanation of the quantities. Note how velocity fields are independent of viscosity ratio while stress fields strongly depend upon it.

- $H_{10}/m \in \{0.1, 0.3, 1, 3, 10, 30\}$
- $H_{20}/m \in \{0.1, 0.3, 1, 3, 10, 30\}$
- $\phi_1 \in \{0.15, 0.2, 0.25, 0.3\}$
- $B_1/Pa^{1-b}s \in \{0.1, 0.3, 1, 3, 10\}$
- $k/m^{-1} \in \{0.1, 0.2, 0.5, 1, 2, 5, 10, 20, 50, 100\}$

with the other parameters being $\phi_2 = 0.1$, $B_2 = 1\,Pa^{1-b}s$. This means that the stability of the system was computed for all 7200 combinations of the parameters listed above. There were 104 cases where the RT instability did not operate, and eight cases where the upper surface was unstable. The growth rates for all these anomalous cases was very small, indicating that they were more or less in neutral equilibrium, and that where instabilities existed they grew too slowly to be of practical consequence.

Firstly, we investigate in detail the influence of wavenumber on the growth constant for lower layer interface. Figure 2 shows plots of growth rate against wavenumber for various layer thicknesses and viscous coefficient ratios. Of interest is the wavenumber corresponding to the growth-rate maximum, since this is expected to be the wavelength at which the RT instability will develop. This notion needs to be tested by solution of the non-linear equations. Layer thickness has the determining effect on the wavelength of the mode with maximum growth rate, while the viscosity ratio has a scarcely discernible influence. In particular, there are no manifestations of multiple maximum growth rates, indicating that observations of multiple dominant wavelengths cannot be explained by the present model, and more complicated rheological descriptions may be needed as well as solution of the non-linear equations.

In the remaining two figures examples are shown of stress and flow fields at the wavelength of maximum growth rate of the unstable mode. These plots are of the Fourier coefficients, but in this case they can equally be regarded also as plots of the vertical velocity at the surface maxima and of horizontal velocity and shear stress at the surface *slope* maxima. In other words, they can be regarded as vertical sections at locations where the plotted quantity exhibit maximum, and therefore relevant, quantities. Figure 3 shows examples of computed first-order velocity and stress fields for the two normal mode configurations, with two layers of various thickness (all combinations of 1, 3 and 10 m), the lower layer with porosity 0.2 and the upper layer with porosity 0.1. The viscous coefficient $B$ was the same in both layers (10 $Pa^{1-b}s$). Figure 3b shows the stable decay of surface undulations. These calculations show velocity distributions broadly comparable with the calculations of velocity fields under the hydrostatic thin-till approximation, which assumes that the shear stress is constant in the till (Alley 1989; Hindmarsh 1998a). Note that the scale is for the vertical velocity, and that all other fields are normalized by their maximum value, and that for these quantities the axis range is $[-1, 1]$. The sediment is at its weakest near the top, and this causes large vertical gradients in the horizontal velocity and the formation of a shear layer at the upper surface. This happens whatever the distribution of shear stress. Vertical velocities are negative, and increase in magnitude upwards

as expected. The values depend upon the viscous parameters chosen and the amplitude of the surface perturbation. In our present state of knowledge, the viscous parameter could be changed substantially and the computed velocities thereby changed.

Figures 3a, c and d show the computed stress fields for the unstable normal mode configuration, where there are undulations at the interface between the two layers. Computed vertical velocities are positive over significant depths centred on the layer interface, much smaller in magnitude than those computed for the decay of surface undulations, and exhibit markedly different characteristics depending on the layer geometry. Where the upper layer is thicker than the lower layer, vertical velocities are positive at the interface but negative at the upper surface. Where the layers are of roughly the same thickness, the areas where the vertical velocities are significant are centred on the interface, but decay to small amounts at the upper surface. In this case the layer geometry can evolve without there being a significant surface expression. In computations not plotted in this paper, it is found that where the lower layer is thicker than the upper layer, vertical velocities remain positive throughout both layers.

Figure 4 shows the influence of layer thickness and viscosity ratio on the computed stress and velocity fields. The situations considered have layers of equal thickness, and 10:1, 1:1 and 1:10 viscosity ratios. Velocity field distribution seems largely unaltered by the viscosity ratio, although magnitudes are substantially altered. Stress fields distributions are necessarily significantly altered.

## Discussion and conclusions

The principal result of this paper is that the RT instability is predicted to operate for effective-pressure-dependent rheologies similar to those hypothesized to apply to sub-glacial deformation of sediments. This should not be taken as direct evidence of the validity of the viscous law, but simply as an indication that viscous behaviour can explain some of the observed phenomena. This model has ignored some severe difficulties, for example the physical processes which create a lower layer of high porosity, and how the co-evolution of layer geometry and water pressure might affect the outcome. Nevertheless, it does shed some light on the viscous processes which are hypothesized to operate in the deformation of sedimentary layers.

As expected, layer thickness geometry has a very strong influence on the wavelength corresponding to the maximum growth rate, while viscosity ratio has a scarcely discernible effect. These results are not really in contradiction with previous theory; previous theory has considered viscosity ratios of several orders of magnitudes, while it is difficult to conceive of this being possible between fairly similar sediments. Having a strongly varying viscosity within layers does not affect the situation. It is conceivable that situations where the lower layer has fluidized (with a very low viscosity) could be differentiated from situations where it has crept.

The linearized theory in this paper predicts only one maximum growth rate as a function of wavelength, which means that the linearized theory does not explain such multiple-wavelength phenomena such as associated loadcasting and diarpirism. Thus, even though having a varying viscosity corresponds to many layers, which can produce multiple growth rate maxima as a function of wavelength (Ramberg 1968), the situations modelled in this paper do not produce multiple growth rate maxima. This could perhaps be remedied by inclusion of dynamic evolution of pore-water pressures and porosity with till deformation which have consequent effects upon the density and viscosity (e.g. Boulton & Hindmarsh 1987). Further modelling utilizing both these approaches is necessary to understand the development of RT instabilities in sediments and how the evolving interface geometry is determined by the rheological properties of the contributing layers. There have been suggestions (Anketell *et al.* 1970) that in the non-linear regime multiple wavelength phenomena *do* inform about viscosity ratios, and the development of these phenomena needs to be described using physically-based models. Solution of the problem will significantly illuminate our understanding of sediment rheology.

# References

ABER, J. S., CROOT, D. F. & FENTON, M. M. 1989. *Glacitectonic landforms and structures*. Kluwer, London.

ALLEN, J. R. L. 1984. *Sedimentary structures: their character and physical basis*. Developments in Sedimentology, **30**. Elsevier, Amsterdam.

ALLEN, P. 1990. Deformation structures in British Pleistocene sediments. *In*: EHLERS, J., GIBBARD, P. L. & ROSE, J. (eds) *Glacial deposits in Great Britain and Ireland*. Balkema, Rotterdam, 455–469.

ALLEY, R. B. 1989. Water-pressure coupling of sliding and bed deformation: II. Velocity-depth profiles. *Journal of Glaciology*, **35**, 119–129.

ANKETELL, J. M., CEGLA, J. & DZULYNSKI, S. 1970. On the deformational structures in systems with reversed density gradients. *Annales de la société géologique de Pologne*, **40**, 3–30.

BRODZIKOWSKI, K. & VAN LOON, A. 1985. Inventory of deformational structures as a tool for unraveling the Quaternary geology of glaciated areas. *Boreas*, **14**, 175–188.

——, GOTOWALA, R., HALUSZCZAK, A., KRZYSZKOWSKI, D. & VAN LOON, A. J. 1987. Soft-sediment deformations from glaciodeltaic, glaciolacustrine and fluviolacustrine sediments in the Kleszczow Graben (central Poland). *In*: JONES, M E. & PRESTON, R. M. F (eds) *Deformation of Sediments and Sedimentary Rocks*. Geological Society, London, Special Publications, **29**, 255–267.

BOULTON, G. S. & DOBBIE, K. E. 1998. Slow flow of granular aggregates: the deformation of sediments beneath glaciers. *Philosophical Transactions of the Royal Society of London*, **A356**, 2713–274.

—— & HINDMARSH, R. C. A. 1987. Sediment deformation beneath glaciers: rheology and geological consequences. *Journal of Geophysical Research*, **92**, 9059–9082.

—— & BLATTER, H. 1998. Stress and velocity fields in glaciers: Part 1. Finite difference schemes for higher-order glacier models. *Journal of Glaciology*, **44**, 448–456.

BUTRYM, J., CEGLA, J., DZULINSKI, S. & NAKONIECZNY, S. 1964. New interpretation of periglacial structures. *Folia Quaternaria*, **17**, 1–41.

EHLERS, J., GIBBARD, P. & WHITEMAN, C. A. 1991. The glacial deposits of northwestern Norfolk, *In*: EHLERS, J., GIBBARD, P. & ROSE, J. (eds) *Glacial deposits in Britain and Ireland*. Balkema, Rotterdam, 223–232.

EISMAN, L. 1981. *Periglaziäre Prozesse und Permafroststrukturen aus sechs Kaltzeiten des Quartärs: Ein Beitrag zur Periglazialgeologie aus der Sicht des Saale-Elbe-Gebietes*. Altenburger Naturwissenschaftliche Forschungen, Heft 1, Altenburg.

EYLES, N., CLARK, B. M. & CLAGUE, J. J. 1987. Coarse grained sediment gravity flow facies in a large supra-glacial lake. *Sedimentology*, **34**, 193–216.

—— & McCABE, A. M. 1989. The Late Devensian Irish Sea basin: The sedimentary record of a collapsed ice-sheet margin. *Quaternary Science Reviews*, **8**, 307–351.

FORNBERG, B. 1996. *A practical guide to pseudospectral methods*. Cambridge University Press.

FOWLER, A. C. & WALDER, J. R. 1993. Creep closure of channels in deforming subglacial till. *Proceedings of the Royal Society London*, **A441**, 17–31.

HINDMARSH, R. C. A. 1996. Sliding of till over bedrock: scratching, polishing, comminution and kinematic wave theory. *Annals of Glaciology*, **22**, 41–47.

——1997. Deforming beds: viscous and plastic scales of deformation of subglacial sediment. *Quaternary Science Reviews*, **16**, 1039–1056.

——1998a. Drumlinization and Drumlin-Forming Instabilities: Viscous Till Mechanisms. *Journal of Glaciology*, **44**, 293–314.

——1998b. The Stability of a Viscous Till Sheet Coupled with Ice Flow, Considered at Wavelengths

Less than the Ice Thickness. *Journal of Glaciology*, **44**, 285–292.

—— & HUTTER, K. 1988. Numerical fixed domain mapping solution of free surface flows coupled with an evolving interior field. *International Journal for Numerical and Analytic Methods in Geomechanics*, **12**, 437–459.

IVERSON, N. R., HANSON, B., HOOKE, R. L. & JANSSON, P. 1995. Flow Mechanism of Glaciers on Soft Beds. *Science*, **267**, 80–81.

KUENEN, P. H. 1958. Experiments in Geology. *Transactions of the Geological Society of Glasgow*, **23**, 1–28.

MORLAND, L. W. 1984. Thermo-mechanical balances of ice sheet flows. *Geophysical and Astrophysical Fluid Dynamics*, **29**, 237–266.

MURRAY, T. & CLARKE, G. K. C. 1995. Black-box modelling of the sub-glacial water system. *Journal of Geophysical Research*, **100B**, 10 231–10 245.

OWEN, G. 1987. Deformation processes in unconsolidated sands. *In*: JONES, M. E. & PRESTON, R. M. F. (eds) *Deformation of sediments and sedimentary rocks*. Geological Society, London, Special Publications, **29**, 11–24.

PAUL, M. A. & EYLES, N. 1990. Constraints on the preservation of diamict facies (melt-out tills) at the margins of stagnant glaciers. *Quaternary Science Reviews*, **9**, 51–69.

RAMBERG, H. 1968. Instability of layered systems in the field of gravity, I. *Physics of the Earth and Planetary Interiors*, **1**, 427–447.

——1972. Theorectical models of density stratification and diapirism in the Earth. *Journal of Geophysical Research*, **77**, 877–889.

RIJSDIJK, K. F. 1999*a*. Density-driven deformation structures in consolidated glacial diamicts: Examples from Traeth y Mwnt, Cardiganshire, Wales, Great Britain. *Journal of Sedimentary Research*, **129**, 111–126.

——1999*b*. *Reconstructing the Late Devensian (26 ka– 13 ka BP) deglaciation history of the southern Irish Sea basin – Testing of competing hypotheses*. PhD thesis, University of Wales, Swansea.

RÖNNLUND, P. 1989. Viscosity ratio estimates from natural Rayleigh–Taylor instabilities. *Terra Nova*, **1**, 344–348.

SCHLÜCHTER, C. H. & KNECHT, U. 1979. Intrastratal contortions in a glacio-lacustrine sediment sequence in the eastern Swiss Plain. *In*: SCHLÜCHTER, C. H. (ed.) *Moraines & Varves*. Balkema, Rotterdam, 433–441.

SCHWAN, J., VAN LOON, A. J., STEENBEEK, R. & VAN DER GAUW, P. 1980. Intraformational clay diapirism and extrusion in Weichselian sediments at Ormehoj (Funen, Denmark). *Geologie en Mijnbouw*, **59**, 241–250.

# Evidence against pervasive bed deformation during the surge of an Icelandic glacier

SARAH FULLER & TAVI MURRAY

*School of Geography, University of Leeds, Leeds, LS2 9JT*
*(e-mail: s.fuller@geog.leeds.ac.uk)*

**Abstract:** Hagafellsjökull Vestari, Iceland, surged in 1980 and advanced about 700 m. Subsequent retreat of the ice front has exposed a suite of streamlined bedforms and former subglacial sediments. Investigations were undertaken on the recently deglaciated forefield to explore the degree of ice–bed coupling during the surge. The morphometry of flutes at this site suggest that the ice was relatively well coupled to its bed. Widespread till wedges show that ploughing was a dominant process. These observations suggest that pervasive deformation of the sedimentary bed at high water pressures in surge was a strong possibility. However, evidence from the micromorphology of drumlin sediments shows laterally continuous clay bands preserved between till layers within a drumlin. Thus pervasive deformation of the whole sediment thickness did not occur. There is evidence only of a thin layer of deformation (<16 cm). We propose that at low effective pressures there was a reduction in the strength of the ice–bed interface rather than a reduction in sediment strength. Motion was concentrated at the sole with localized and shallow deformation where clasts ploughed through the substrate.

Glacier surging remains an incompletely understood cyclic flow phenomenon. It is commonly accepted that fast flow during surge is a result of extreme subglacial water pressures. However, it is unknown whether fast flow is promoted and sustained by sliding at the ice/bed interface (e.g. Kamb 1987) or by deformation of weakened, saturated, subglacial sediments (Clarke *et al.* 1984). The former implies decoupling at low effective pressures with rigid bed conditions while the latter requires a coupling between the ice and bed. Since Boulton and Jones (1979) showed that when a glacier overlies a soft unlithified bed coupling between the ice and the sediment may occur, sediment deformation has become intrinsically linked with fast flow in the literature (e.g., Clarke *et al.* 1984; Alley *et al.* 1987). The implications of bed deformation are thought to be potentially far reaching in terms of understanding glacier dynamics and stability of ice sheets. There are a growing number of sites from modern glaciers where bed deformation has been measured (eg., Breiðamérkurjökull – Boulton & Hindmarsh 1987; Trapridge Glacier – Blake *et al.* 1992; Storglaciären – Iverson *et al.* 1995; Bakaninbreen–Porter *et al.* 1997). The fast flow of Ice Stream B has been attributed to the presence of a deforming substrate (Alley *et al.* 1987). Furthermore, bed deformation is argued to have been an intrinsic component to the movement of surging glaciers (e.g. Clarke *et al.* 1984), which have been shown to preferentially overlie geologies that produce deformable beds in Svalbard (Jiskoot *et al.* 1998). The Quaternary ice sheets are also known to have advanced over large areas of soft sediment (e.g. Beget 1986; Hart *et al.* 1996).

A glacier may overlie soft sediments but need not cause pervasive deformation of them (e.g. Kamb 1991); in order for a glacier to deform its bed it must couple to and interact with the underlying sediments. Evidence of this coupling should be implicit in the sediments themselves. Most authors argue that elevated pore-water pressures are required to weaken the sediments below a critical point beyond which motion would occur in a dilatant deforming horizon (e.g. Boulton & Hindmarsh 1987). However, basal sliding, at low effective pressures, over unconsolidated sediment has also been observed (Englehardt & Kamb 1998; Iverson *et al.* 1995, 1999) and inferred from sedimentary deposits (Piotrowski & Kraus 1997). High water pressures at Storglaciären were seen to enhance decoupling of the ice and bed rather than promoting deformation (Iverson *et al.* 1995). The motion of Ice Stream B originally considered to result from pervasive deformation (Alley *et al.*

*From*: MALTMAN, A. J., HUBBARD, B. & HAMBREY, M. J. (eds) *Deformation of Glacial Materials*. Geological Society, London, Special Publications, **176**, 203–216. 0305-8719/00/$15.00 © The Geological Society of London 2000.

1987), has also been inferred to occur within a thin layer at the glacier sole (Englehardt & Kamb 1998; Tulaczyk *et al.* 1998). Piotrowski & Kraus (1997) and Piotrowski & Tulaczyk (1999) show that the advance of the Weischelian ice sheet in Germany probably did not lead to pervasive deformation of the bed as had been previously inferred (e.g. Hart *et al.* 1996), but resulted from enhanced basal sliding. Furthermore, Brown *et al.* (1987) suggested that movement of the Puget lobe of the North American ice sheet was by sliding and localised deformation associated with ploughing.

A glacier surge is often accompanied by frontal advance, which is followed by downwasting and retreat exposing subglacial sediments and bedforms. Drumlins, flutes and ploughed boulders are common subglacial landforms of both modern and Pleistocene deglaciated forefields. Such features form at the ice-bed interface and thus also potentially provide information regarding basal boundary conditions, allowing testing of hypotheses on the degree of ice-bed coupling during surge advance. In this paper we examine the morphometry, sedimentology and micromorphology of drumlins, flutes and ploughed boulders from the recently deglaciated forefield of Hagafellsjökull Vestari, Iceland exposed since a surge in 1980. Evidence is used to differentiate between pervasive deformation and sliding (reduction in strength of the ice–bed interface) as mechanisms for fast flow during periods of high pore water pressure.

## Field site and methodology

Hagafellsjökull Vestari together with its neighbour, Hagafellsjökull Eystri, drain the southern parts of the ice cap Langjökull in central Iceland (Fig. 1). Both outlets surged in late winter/early summer of 1979/1980 and Hagafelsjökull Vestari advanced a total of 718 m (Sigursson 1998). After the surge Hagafellsjökull Vestari entered a stagnant phase which was followed by a period of steady retreat which continues to the present day. The glacier is approximately 5 km wide at its snout and has a low surface slope and a relatively steep ice margin, where a few high discharge streams were emerging from subglacial tunnels. Evidence from the recently deglaciated forefield suggests that the snout is underlain by till and alluvium, which overlies bedrock of very young Holocene basalt lavas. Aside from the 1979/80 surge push moraine and a small outer moraine marking the Little Ice Age maximum, which occurred around 1890 in Iceland (Thorarinsson 1969), other proglacial features have not been dated.

Hart (1995) divided the forefield of Hagafellsjökull Vestari into four distinct geomorphic zones. These zones comprised an outer tumuli zone of pahoehoe lava, a lake bed zone, and an outer and inner drumlin zone of streamlined bedforms. The inner drumlin zone was overridden in the 1980 surge advance, and is delineated by a prominent push moraine (Fig. 1). At the margin the ice is thin and steadily retreating, which is exposing streamlined

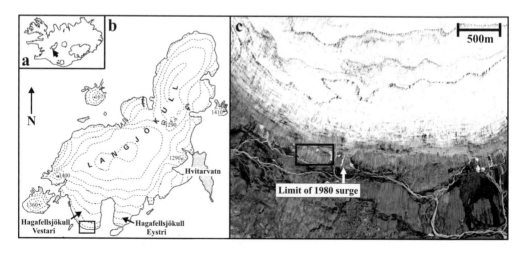

**Fig. 1.** Field site: (**a**) location within Iceland; (**b**) location of Hagafellsjökull Vestari and field site in relation to Langjökull; (**c**) the glacier tongue and forefield area, limit of last surge and field site is indicated (black rectangle). Part of Aerial photograph No. 9243 (taken 11/8/91) (Reproduced with permission of Landmælingar Islands).

bedforms and a saturated upper till layer, which represent former subglacial sediments from the surge advance.

To examine bedform morphometry 50 flutes were measured from within a $10\,m^2$ area, which was situated approximately 100 m from the ice margin. Twenty drumlins from within the surge zone were also surveyed (Fig. 1). Measurements of length, width, height, orientation, and for flutes only, initiating boulder size and boulder burial depth were made. In order to examine the internal structure of these bedforms, a total of 12 sample blocks were collected from a flute and drumlin and thin sectioned for micromorphological analysis (Figs 5 and 7). The drumlin selected for analysis was exposed by a longitudinal streamcut, samples were extracted after a further thickness of >50 cm of surface material was removed from the bedform exposure. Aluminium 'Kubiëna' tins were pressed into the sediment face and the sample block carefully removed with a knife. Samples were first air dried, and then impregnated, cut and mounted using the method described by van der Meer (1987). Three orientated thin-sections were produced from each sample block. The sample block is cut so that each oriented thin-section is progressively smaller than the previous one. This precludes the examination of individual micro structures in 3D, but allows definition of the overall 3D sedimentary structure.

Thin sections were viewed under a petrographic microscope (Olympus B201) in unpolarized, plane-polarized, cross-polarized and cross-polarized light with a $\frac{1}{4}\lambda$ plate. The terminology we use to describe the sections is that developed by pedologists (see van der Meer 1987, 1993), which describes the patterns and orientations of clay sized material (plasma) within the sediment body. Domains containing particles that are aligned show as areas of higher 'birefringence', which change colour synchronously in cross-polarized light when the microscope stage is rotated. The same magnifications and illumination were used for each section in order to make valid sample comparisons. Magnifications between 4 and 10 were commonly used.

## Results

### Bedform morphometry

The length of subglacial bedforms, such as flutes or drumlins, has been shown to depend on the magnitude of stress enhancement around an obstacle (Boulton 1976). The theory suggests that if an obstacle moves only slightly slower than the ice then the flute length is less than for an obstacle that is considerably retarded with respect to the ice. Although the process of lodgement remains a contentious issue, the development of a till wedge by clast ploughing may provide a mechanism to decelerate the clast. Alternatively the clast may bridge across the deforming layer to a more stable substratum (Benn 1994). It follows that the resultant morphology of subglacial bedforms can be used to make inferences regarding basal sliding and ice-bed coupling. Morphometry has been summarised using the length:width ratio (e.g. Rose 1987), which describes the elongation of a bedform. The dimensions of the initiating obstacle must also be a factor in controlling the size of a flute, since a larger boulder would be expected to produce a larger pressure shadow in its lee. Therefore the size and burial depth of each boulder/clast is also considered.

The largest bedforms were drumlinised lava blisters (nomenclature after Hart 1995), which consisted of a basalt rock core that was smoothed and striated with a short stoss side and long leeside accumulation of till. From here on we refer to these features as drumlins. The drumlins varied little in morphometry, their mean elongation ratio was 6 and they were generally between 50 and 100 m in length (Fig. 2).

Flutes at Hagafellsjökull are meso-scale features, which consisted of a cobble- or boulder-sized clast with a relatively long lee side accumulation of till (e.g. Fig 3a). In some cases the flutes had no relief and were defined only by a sorting of surface clasts, the flute tail appearing

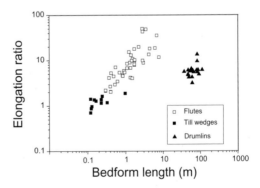

**Fig. 2.** Graph illustrating the basic morphometry of till wedges (solid squares), flutes (hollow squares) and drumlins (solid triangles). Flutes are generally short and tapering with a wide range of elongation ratios. Drumlins are similar to one and other in terms of their morphometry.

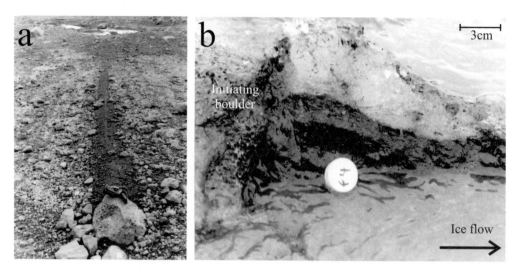

**Fig. 3.** (a) A 6 m long flute at Hagafellsjökull. (b) Cross-section of flute emerging from the ice margin frozen and associated with a casing of ice, which contains bubbles stretched in the direction of flow. The flute form is preserved and consists of a concave till wedge which rapidly decreases in height leading to a flute of constant height.

as a clast-free zone. Flutes were present within the inner drumlin zone only, being ubiquitous close to the ice margin but showing no occurrences beyond about 100 m from the ice margin. It is unknown whether this was related to their formation or whether it was a consequence of post-depositional subaerial modification of the till surface. Flutes varied from a few centimetres to 8 m in length and often occurred superimposed on the backs of drumlins. Figure 2 shows the flutes were generally short and tapering with a range of elongation ratios from 2 to 50 with a mean of 9.8. Flute elongation ratio increased with bedform length, which was not seen for the drumlin population (Fig. 2). Hart & Smith (1997) interpreted such a pattern as evidence of flutes at difference stages of their evolution.

There is no clear relationship between flute length and the depth to which the initiating obstacle was embedded in the substrate (Fig. 4a). Flutes were associated with both lightly embedded and more deeply buried boulders (Fig. 4a). However, there was a relationship between the size of obstacle and the length of flute if the boulder was more than 60% embedded within the substrate (Fig. 4b); there was no relationship between flute length and obstacle size for more lightly embedded boulders (Fig. 4b). This supports the hypothesis that the relative velocity of the ice (velocity of ice minus velocity of boulder) controls bedform length. One significant factor for retardation would be an increase in the burial

depth within the substrate. This would effectively increase the contact area between the obstacle and the till and/or allow contact with a more stable sub-horizon to aid in lodgement. Thus, boulder size exerts a control on bedform size only when the obstacle is sufficiently slowed with respect to the ice.

Boulders and clasts were also commonly associated with till wedges (Boulton 1976), which are related to the ploughing process (Boulton 1976; Clark & Hansel 1989). These bedforms consisted of a concave lee-side wedge of till, which quickly decreased in height away from the obstacle. There was often no associated flute development. These features are shown on Fig. 2 and are distinguishable from flutes by an elongation ratio of <2. Till wedges were associated only with boulders that had shallow burial depths (Fig. 4). No till wedges or flutes with low elongation ratios were associated with boulders that were more than 30% embedded within the substrate. Based on a large number of observations where both the till wedge and flute are preserved in the lee of an obstacle, we hypothesize that flutes develop from till wedges. Figure 3b illustrates a till wedge that has been preserved in the lee of a boulder together with flute of constant height, which emerges from it. Both flutes and till wedges were observed emerging from the ice margin frozen. The frozen flutes were associated with a casing of regelation ice covering the lee-side till accumulation. This casing contained bubbles

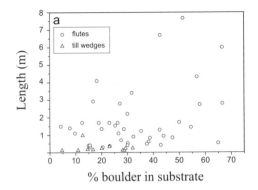

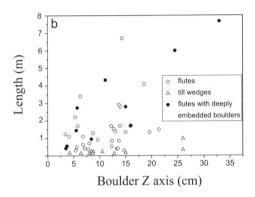

**Fig. 4. (a)** Graph showing relationship between flute and till wedge length and the depth of burial of the initiating boulder. Till wedges (hollow circles) are associated with boulders that rest lightly on the substrate. Flutes (solid squares) can be associated with both lightly and deeply embedded boulders.
**(b)** Relationship between flute and till wedge length and the size of the initiating boulder (given as vertical height of boulder (z-axis)). Boulder size does not control flute length (open circles) or till wedge length (open triangles), except for those bedforms whose initiating boulders are buried more than 50% in the substrate (solid circles).

that were elongated in the direction of ice flow. Till wedges without flutes were not associated with an ice casing. These observations confirm the relative velocity differences between the ice and the obstacles forming till wedges and flutes. It is envisaged that as a boulder melts out it is moving faster than the till and slightly slower than the ice and thus a till wedge can form. As the boulder becomes more deeply buried within the till it begins to move much slower relative to the ice and consequently a flute can form, with associated development of regelation ice as the ice slides past the boulder.

## Drumlin sedimentology

Figure 5a and b shows locations and sedimentary logs for three excavated sections within the drumlin tail. There were no obvious macroscale deformation structures within the drumlin and no obvious intra-clasts. The stratigraphy within the drumlin was essentially simple; a massive, grey-coloured till overlay a fissile, orange coloured till throughout the bedform. The upper grey till had two principle sub-units (Fig. 5b). The uppermost sub-unit was massive, matrix supported and water saturated to a depth of a clay band, which was present between the two units in stratigraphic sections one and two only. The lower grey sub-unit had similar textural properties (Fig. 5c) and was also massive and matrix supported. The lower till layer (Fig. 5b) was orange in colour, massive, matrix supported and had no distinct sub-units. Under the scanning electron microscope (SEM) grains from the lower till showed an abundance of surface textures indicative of extensive *in situ* chemical alteration (86% of 30 grains) and few of recent brittle fracture or mechanical abrasion (25%). Based on these observations we tentatively suggest that the lower till was not affected by most recent surge advance. The upper saturated grey unit in contrast is hypothesized to relate to the 1979/80 surge. SEM examination showed few grains from the upper till with textures of chemical alteration (23% of 30 grains) but many with recent fracture surfaces or mechanical abrasion textures (87%). In all three sections the upper grey till and lower orange till was separated by a 2mm thick clay band similar in appearance to the band between the upper grey till sub-units. Several sand lenses complicated the picture, particularly in distal parts of the drumlin.

## Grain size

Particle size characteristics of a sample are an important factor in determining how a sediment will deform. The grain-size distribution can also give valuable information concerning processes acting within the subglacial environment to produce a certain distribution. A full discussion of the particle size characteristics of sediments from this site is beyond the scope of this paper; however, a brief consideration is relevant since particle size has important implications for the development of a plasmic fabric and the interpretation of microstructures. There is little difference between the textural characteristics of the till throughout the drumlin. Particle size distributions are given for stratigraphic

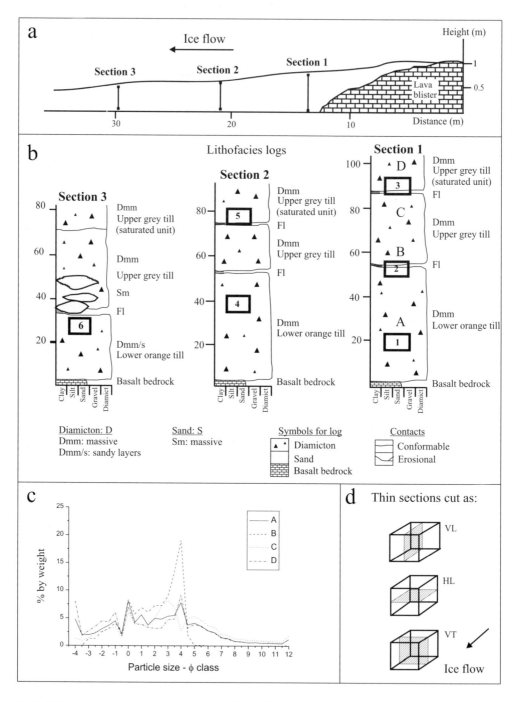

**Fig. 5.** (a) Location of excavated stratigraphic sections within drumlin long profile. (b) Sedimentary logs showing thin-section sampling locations (1–6) and bulk sample locations (A–D). (c) Particle size distributions for samples A–D in section 1 given as percentage weight per phi class. (d) Orientation of thin sections relative to ice flow direction: VL, vertical and longitudinal; HL, horizontal and longitudinal; VT, vertical and transverse.

Section 1 (Fig. 5c). The distributions were all similar with modes in the fine sand class. The lowermost till also had a distinct peak within the medium gravel size range. All samples were deficient in clay-sized material, which comprised <2% by weight for all samples (Fig. 5c).

## Drumlin micromorphology

The location of thin section samples within the drumlin tail and vertical section are given in Fig. 5. Figure 5d shows how the sample blocks were cut to produce three thin sections to define microstructures in 3D; these sections are referred to relative to ice flow directions. Sample D1, D4 and D6 are all from the lowest orange-coloured till unit at proximal, middle and distal points respectively. Sample D2 was taken through a clay band at the top of the lower orange till. Samples D3 and D5 also cut across a laterally continuous clay band, which can be traced between the two log locations below the upper saturated grey till layer. All sections were lacking in a plasmic fabric which may indicate either that (1) applied stress was insufficient to cause plasma re-orientation or (2) clay content was insufficient for birefringence to show in thin section (van der Meer 1987).

*D1.* This sample was taken from the lower till layer close to the initiating lava blister, 20 cm above the bedrock. This layer was capped by a laminated red clay band. The thin sections showed poorly sorted sands with a high porosity and variable proportion of interstitial silts. In silt-rich areas, weak, high angled discrete shears were apparent. Sections VL and VT exhibited a clear subhorizontal fissility of largely continuous fissures in strings. Isolated small grains occurred within the fissures, and thus, it may be that the fissures acted as passageways for water. Several larger grains had deep embayments, which were commonly filled with a finer-grained silt-sized material (Fig. 6a). Infills generally had a sharp contact with surrounding till and were sometimes delineated by voids. Small areas of moderately sorted sand were also apparent in all three oriented sections, which had sharp boundaries with surrounding material.

*D2.* This sample was taken from the upper part of the lower till layer 25 cm above D1. The sections sampled the clay layer that separates the orange till from the upper grey till. In section VL the clays had been broken into blocks by a combination of reverse and normal faulting (Fig. 6b). Small thrust faults were also evident where clay blocks were displaced (Fig. 6b). Silt

layers were well-sorted, discontinuous and separated by individual red clay bands. Individual fault blocks largely retained internal laminations, however, in places mixing with silt layers had occurred (Fig. 6b). The material immediately underlying the clays showed some evidence of ductile deformation, which had resulted in some layer mixing and small open folding. The kinematics of this sample are highly complex, yet it is possible to say that the sample has undergone low strain since the contact between clays and the overlying sands was generally sharp and only limited mixing of the units had occurred.

*D3.* This sample was taken from the boundary between the grey till and the upper saturated grey till layer, which also contained a pink clay band. The thin sections showed a heterogeneous mix of till and more sorted silts and sands. The silts and sands occurred as patches and wisps within the till. Silts were often found forming a thin casing around skeletal grains in all three oriented thin sections. A thin, highly sinuous, clay band cut across the VT section and appeared to have been injected into the till following deposition. Many grains had silts and clays in deep embayments, which had a graded contact with surrounding till. Section VL cut through the main clay band which was situated immediately below the saturated upper till. There was some evidence for ductile mixing and folding of the clays with silts and sands. Figure 6c shows one example where red clays had been mixed with surrounding till, which had then been brecciated into till pebbles. However, in most incidences graded silt–clay units remained undeformed with sharp contacts to more disrupted areas. Section VL also showed a limited number of grains with triangular tails of fines strung out on either side in a 'galaxy' form. This structure is indicative of grain rotation in ductile conditions and was not seen in VT or HL sections. There was also evidence for brecciation of the clay/silt sequence, which is evidence for brittle deformation.

*D4.* This sample was taken from the lower orange-coloured till layer in a mid-drumlin location, 40 cm above the bedrock. Sections VL and VT showed a weak banding of fine sands and coarser layers with both sharp and diffuse contacts. A distinct subhorizontal fissility was also present, which in thin section consisted of strings of unconnected elongated voids (Fig. 6d). Some voids were associated with a collection of fines around their extremities. A large grain had deep embayments that were partly filled by laminated

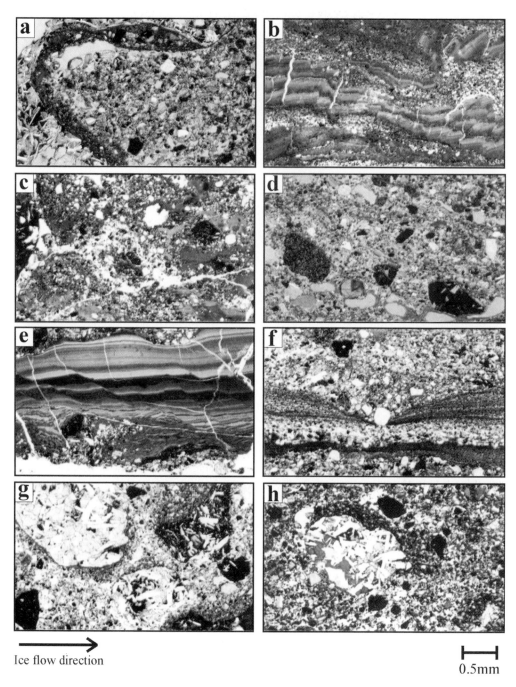

Ice flow direction

0.5mm

**Fig. 6.** Photographs of parts of thin-sections. Sample locations are shown in Figs 5 and 7. Scale bar is same for all sections. (**a**) D1 (VL). Large grain with deep embayement partly filled with silts. (**b**) D2 (VL). Sequence of silts and red clays at the top of the lower till layer, the clay band is broken into blocks by reverse and thrust faults. (**c**) D3 (VL). Ductile mixing of clays with till was followed by brecciation into till pebbles, which are delineated by voids (pink colour). (**d**) D4 (VL). Elongate voids occur in subhorizontal strings and are largely unconnected. (**e**) D5 (VL). Section through the clay band at the base of the saturated upper till layer. Loading occurs along the upper edge while a sand lens is intruded along the lower edge during water escape. (**f**) D6 (VL). A till pebble of high plasma content deforms graded silt laminations at the base of the sequence. (**g**) F2 (VL). Triangular areas of fines around grain extremities. (**h**) F2 (VL). A casing of silts occurs on the upper edge of a grain.

red clays and silts. Many larger grains were associated with thin casings of silt-sized material that did not show a birefringence. A small area of high porosity sands was also apparent. Section HL contained numerous patches of sorted silts and fine sands. As with the VL section there were casings of fines around grain extremities. Section VT contained weakly defined circular elements of till, which were largely delineated by voids, this may be indicative of brecciation of the till body by the movement of pressurized water through the sample.

*D5.* This sample was taken from the clay band that separates two sub-units of the upper till layer from stratigraphic section 2. Section HL sampled the material below the clay band and was composed of two distinct areas defined by a marked change in porosity and grain size. A fine-grained unit with discontinuous bands of silts and few sand grains contrasted with a unit of poorly sorted sand. In this part of the section the till was broken into subrounded elements which were delineated by voids. A small number of skeletal grains had fine casings of silt and some were associated with patches of sorted fine sands. In most cases there were sharp contacts with the grain casings/infills and the surrounding till. Sand grains often occurred in small clusters within this section, which showed isolated examples of grain fracture. Section VL was dominated by a laterally continuous laminated clay band (Fig. 6e), which was crossed with normal faults that had listric profiles. Towards the lower edge the clays were more finely laminated and disrupted by folding and mixing with underlying units. A lens of fine sands had been intruded upwards into the clays and some flowage of clays into sands had occurred. These structures probably resulted from water escape. Loading has occurred along the top edge of the clay band, which resulted in the pinching out of the topmost laminations. The base of the band followed a large void, which may have been a pathway for water, which was trying to escape upwards. The clays were capped by a poorly sorted sandy till with variable interstitial silts. No obvious structures were present within this till, which was saturated in the field. Below the clay band a few sand grains had deep embayments filled with clays and silts. Section VT shows that the till beneath the clays had a weak subhorizontal fissility, which is likely to be related to loading during ice advance-retreat.

*D6.* This sample was taken from the lower orange till layer in the distal part of the drumlin. Section VL showed a banded sequence of sorted fine sands and silts (Fig. 6f). In the lower part of the section, layers were weakly deformed into folds and also mixed in places. A laterally continuous silt band capped the lower folded layers and had a sharp contact with the underlying material. Above the prominent silt band there was a laminated sequence of silts and fine sands with largely planar contacts. There was a mix of graded and ungraded bands within this sequence. The uppermost fine layer was deformed by loading from a rounded till pebble containing many sand grains (Fig. 6f). The pebble had caused a pinching out of this layer. A more massive, poorly sorted, till unit overlay the banded sequence, which reflects a change from deposition from water to till deposition. The silt band was breached in one location where sands from the lower deformed sequence have mixed with silts and sands from the upper horizon probably during water escape.

*Interpretation*

Thin-sections from the drumlin showed a combination of clay, sand and diamicton units. Vertically oriented sections taken from close to the bedrock (D1, D4 and D6) showed a distinct subhorizontal fissility and textural banding. A fissile structure can be produced during loading and unloading cycles under a normal stress. The occurrence of graded units within the lower diamicton in stratigraphic section 3 sequences suggests periods of deposition from running and standing water prior to, and interspersed with, till deposition.

Two laminated clay bands were traced between sedimentary logs and occurred at the top of the lowest orange till layer and also below the upper saturated grey till layer. The clay bands show evidence for both compressional and extensional deformation in VL sections, which has resulted in high angle reverse faults, thrust faults and small folds in the former case and normal faults in the latter. The coherent nature of the clay bands suggests that deformation occurred when the units were relatively stiff, and that strain was low. Such laminated units represent an abrupt transition to sedimentation within standing water. One explanation may invoke a proglacial origin with subsequent incorporation into the drumlin during ice advance. However, the lateral continuity and small-scale internal deformation pattern is not consistent with an origin as a rafted sequence. An alternative explanation is provided by van der Meer (1987), who suggests that such clay layers are representative of deposition within subglacial cavities. If this is the case then these

layers would represent times where the ice was
separated from the sediment by thin water layer
over an area comparable to the size of the
drumlin. Transfer of basal shear stresses to
underlying sediments would be prevented by
presence of a water film (Piotrowski & Tulaczyk
1999) explaining why subsequent strain in the
clays was so low. Till deposition followed clay
deposition in all cases which also supports a
subglacial origin for the clay layers. Additionally
the upper surfaces of the clay bands are
deformed by clasts and till pebbles which also
suggests subglacial deposition. A high basal melt
rate, as would be expected in surge, may also
protect the clay units from basal shear stresses
by raising the deforming layer by the addition of
basal material. Loading of the sequence by the
ice resulted in water escape structures forming at
the base of the clay bands and load structures
at the top.

In the non-saturated grey till below the upper
clay band isolated examples of galaxy structures
(plasma accumulation in triangular shaped areas
adjacent to grains) were seen (e.g. Menzies &
Maltman 1992). This structure may be deformational or caused by differential pore water movement around patches of plasma (Menzies &
Maltman 1992). The upper saturated horizon
(10–16 cm thick) comprises a poorly sorted till
with no obvious deformation structures. It may
be that this layer has experienced little shear
deformation or that it has been deformed to high
bulk strain in a saturated state and as such no
structures were recorded. The notable lack of a
plasmic fabric may be indicative of a low strain
but this hypothesis is problematic because
formation of a plasmic fabric is also dependent
on clay content, which is <2% for these
sediments (Fig. 5c). However, the widespread
preservation of bedding, clay bands and low
strain structures below this layer suggest that
pervasive deformation of the whole sediment
thickness did not occur.

## Flute sedimentology

Flutes and till wedges were composed of a sandy
till, which was generally fine grained with few
large clasts. The long axes of initiating obstacles
of both flutes and till wedges were aligned either
in a flow parallel or perpendicular direction.
A majority of the obstacles were a spheroidal
shape (C:A axial ratios 0.4–0.9) and most (70%)
were subrounded or subangular. The remainder
were angular or very angular reflecting a passive
transport history. A limited number of obstacles
(<5%) were classic stoss-lee forms (Benn 1994)

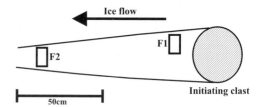

**Fig. 7.** Location of thin-section sample locations in
flute tail.

and few were striated. The micromorphology of
one flute is described in this section. Figure 7
shows the location of thin-sections taken from
this flute; F1 was cut close to the initiating clast
and sampled material approximately half the
width of the bedform. F2 was cut from the tail of
the flute where the sample block was of similar
size to the flute width. In the field, the lee-side till
was identical to the grey saturated upper till layer
and had no obvious deformation structures.

## Flute micromorphology

*F1.* Sections VL, HL and VT were characterized by distinct textural variability over
small areas. The till was poorly sorted and
interspersed with broken patches and wisps of a
more silt/clay-rich till. Contacts between different textural areas were diffuse which implies a
mixing and reworking of units during ductile
deformation. There were isolated wisps of red
clays distributed along a void system but there
were no obvious clay bands, in contrast to those
seen in the drumlin sections.

*F2.* Sections from the distal part of the flute tail
also showed a texturally heterogeneous till. Some
sand grains had deep embayments, which were
filled with plasma. Other grains were associated
with a 'casing' of finer grained material, which
may or may not surround the entire grain. In all
sections silt accumulations occasionally occurred
in triangular areas on the sides of grains in
variable orientations (Fig. 6g). Other sand grains
were associated with silts only on upper and
lee surfaces (Fig. 6h). Section VL contained a
number of isolated circular till pebbles of higher
silt content than the surrounding till. Other
elements were more angular shaped and were
delineated by voids – such elements could represent till brecciation by the passage of water at
pressure through the sample. In regions of high
silt content, short discrete high angle shears were
apparent. There was no evidence for the reorientation of plasma.

## Interpretation

Flute sediments were highly heterogeneous and comprised poorly sorted tills with an uneven distribution of fines. Inter-fingered with tills were patches of sorted unimodal sands and silts with diffuse contacts. These observations are consistent with pervasive mixing of till and meltout sediments. The occurrence of isolated rounded till pebbles is thought to be indicative of till brecciation with subsequent reworking and rounding of individual pebbles. Since the pebbles were usually isolated, the reworking was probably extensive. Casings of fines around the whole or part of grains are thought to be a result of grain rotation or the differential movement of pore water and selective flushing of fines. The occurrence of discrete shears in the distal part of the flute indicates a degree of brittle deformation during or subsequent to flute formation possibly when the till was frozen. The presence of isolated red clays along a void indicates water movement through the sediment, but the origin is uncertain. These structures suggest that the material comprising the flute sediments had been pervasively deformed.

## Discussion

Subglacial streamlined bedforms form at the ice-bed interface (e.g. Boulton 1976, 1987; Menzies 1987) and thus should provide information regarding the nature of this zone. Boulton (1976) showed that the sliding speed of the ice exerts control on the length of flutes. It follows that the morphometry of streamlined bedforms can be used to crudely assess the degree of ice–bed coupling, which has ramifications for the operation of a 'deforming bed' surge mechanism (e.g. Clarke et al. 1984). Table 1 shows the lengths and elongation ratios for flutes observed at other sites. In most cases the flutes are considerably longer and often have higher elongation ratios than flutes at Hagafellsjökull (where the longest flute measured was 7.6 m). Flutes more than 100 m in length have been described by Lawson (1976), Karlén (1981), Benn (1994) and Gordon et al. (1992). Rose (1989) observed flutes with comparable elongation ratios to the Hagafellsjökull flutes, but again the bedforms were of significantly greater length. Flutes of similar dimensions to Hagafellsjökull Vestari were observed at Tutmann Glacier, Switzerland but only a limited number of forms were measured (van der Meer 1997). Results from Hagafellsjökull Vestari suggest that ploughing was a dominant process at this site due to the widespread occurrence of till wedges. This process requires the ice to be moving faster than the till. The short length and generally low elongation ratios of all flutes suggests a low differential velocity between the ice and initiating obstacle. This discussion has assumed that the flutes were formed during the active phase of the surge cycle. It is also possible that the flutes might have formed right at surge termination, which would limit the time period for formation and hence their length, or during quiescence.

Flutes at Hagafellsjökull with the highest elongation ratios have been produced where the initiating clast was more retarded relative to the ice, probably by increasing clast burial depth within the substrate. In such cases obstacle size is seen to exert a control on bedform length, which is expected since a large boulder would produce a larger pressure shadow in its lee. The morphology of obstacles initiating flutes and till wedges also suggests that obstacles were relatively mobile. The majority of clasts lack features that indicate lodgement or in situ modification, i.e. there is a lack of stoss/lee forms or striated upper surfaces etc. (Benn 1994). A limited degree of edge wear during transport has occurred, which is shown by the large number of sub-rounded and sub-angular class

**Table 1.** *Lengths and elongation ratios of flutes from a selection of other sites*

| Author | Site | Length | Elongation ratio |
|---|---|---|---|
| Lawson (1976) | Spencer Glacier, Alaska | c. 300 m | c. 300 |
| Karlén (1981) | Svartisen ice cap, North Norway | 500 m | 500 |
| Rose (1989) | Austre Okstindbreen, South Norway | 39 m (mean) | 7.9 (mean) |
| | | 8–93 m (range) | 2.2–18 (range) |
| Benn (1994) | Slettmarkbreen, South Norway | Up to 150 m | ? |
| Gordon et al. (1992) | Southern Lyngen Peninsula, North Norway | 340 m | Minimum 68 |
| Eklund & Hart (1996) | Isfallsglaciaren, Sweden | 50 m+ | ? |
| Van der Meer (1997) | Turtmann Glacier, Switzerland | 3 m | 7.5 |
| This paper | Hagafellsjökull Vestari, Iceland | 1.49 m (mean) | 9.8 (mean) |
| | | 0.12–7.6 m (range) | 1–50 (range) |

boulders/clasts initiating bedforms. These are characteristics of a limited degree of transport or transport in a passive media. Overall the morphometry and features of initiating obstacles suggest strong coupling between the ice and its bed. Typically ice–bed coupling is thought to result in deformation of the underlying sediments (e.g. Alley *et al.* 1989).

Within the drumlins the macroscale sedimentology showed homogeneous tills with no structures so evidence of such deformation was sought using micromorphology. Thin-sections showed a mix of sorted and layered sediments together with homogeneous poorly sorted till. The poorly sorted till units showed isolated examples of both brittle (discrete shears, broken grains) and ductile (grain casings, unit mixing) deformation, but large areas showed no obvious structures. These units could be (1) undeformed sediments or (2) pervasively deformed material that has experienced homogenization at high shear strains. Although there are no accepted criteria to determine unequivocally that a sediment has undergone pervasive deformation, there is ample evidence at this site that suggests the drumlin tills have not experienced pervasive deformation following deposition. For example: (i) there was widespread preservation of bedded sediments and clay bands with low strain structures; (ii) there was no development of plasmic fabrics (although this could be a consequence of the low clay content of all samples rather than low strain). Pervasive shear deformation would be expected to result in intense mixing and remoulding which would highly distort original sediment components.

The lack of evidence for pervasive deformation at Hagafellsjökull Vestari shows that movement must have been concentrated at the glacier sole, and it appears that the till did not weaken sufficiently for pervasive deformation to occur. The upper till layer was dominated by short tapering flutes and till wedges associated with ploughed boulders. This evidence suggests that conditions for clast ploughing were widespread. It is seen that sediments in flutes do not contain sorted or layered sediments as seen in drumlin sections but instead are dominated by evidence for clast rotation and ductile mixing creating a highly heterogeneous till. We suggest that deformation was confined to an upper layer which was less than 16 cm thick, this result stands whenever the flutes were formed during the surge cycle. A band of clay occurs at this depth throughout the drumlin above which the till is water saturated. Deformation is unlikely to have penetrated to greater depths since the clay band is undeformed. This depth also corresponds to

the maximum depth to which flute/till wedge boulders are buried. If we assume that the bedforms relate to the most recent surge then these observations place a limit on the maximum depth of deformation. Hart (1997) asserts that the maximum thickness of the deforming layer is related to the height of subglacial bedforms. If this were true for Hagafellsjökull Vestari deformation would be required to the full depth of the drumlin (up to 2 m). Examination of drumlin internal structure suggests that this is not the case and this assertion is untrue for this site. Our estimate of the probable thickness of the deforming layer also has implications for the lodgement process at this site. We have shown that the maximum burial depth of clasts associated with flutes and till wedges is consistent with the depth of undeformed clay bands in the drumlin. A factor in clast lodgement during flute formation could be clasts bridging across the deforming layer to the more stable substratum (Benn 1994). Smaller material would be more mobile and move into the lee side low-pressure zone (Benn 1994).

Evidence presented above suggests that, although the glacier overlies a soft bed, the surge advance of Hagafellsjökull Vestari did not occur by pervasive sediment deformation as envisaged by Clarke *et al.* (1984). Instead at times of low effective pressure the strength of the ice-bed interface was lower than the strength of the underlying sediments. Clay bands between till units are interpreted as waterlain sediments deposited at the ice-bed interface during periods of decoupling. Such a water layer may also have prevented the transfer of basal shear stresses to the underlying sediments. The lack of evidence for pervasive deformation also supports the notion of ice-bed separation. Thus evidence suggests that cavitation and an associated distributed water system was important at least at certain times. We can thus tentatively suggest that the surge mechanism at this site may involve the transition of the quiescent phase basal water system, which consists of a few large subglacial conduits emerging at the ice margin to a linked cavity system as envisaged by Kamb (1987). Furthermore, the observation of flutes emerging frozen raises intriguing questions about the glacier thermal regime at least close to the glacier margin.

## Conclusions

The surge of Hagafellsjökull Vestari resulted in a frontal advance of around 700 m. Subsequent retreat has revealed a suite of streamlined bedforms and former subglacial sediments. The morphometry and sedimentology of these bedforms was used to assess the question of ice–bed coupling, which has implications for a deforming bed surge mechanism. The short and tapering nature of flutes at this site implies that the ice was well coupled to its bed and thus deforming bed conditions would be expected to be a strong possibility at times of low effective stresses. However, micromorphological evidence from a drumlin exposure revealed a layered till interspersed with laterally continuous and largely undisturbed clay bands. There are isolated examples of brittle and ductile deformation structures throughout the drumlin thickness but in general deformation structures are conspicuously absent. The preservation of bedding and clay bands with sharp boundaries to overlying till precludes deformation of the whole sediment pile. It is likely that at times of low effective pressure the strength of the ice–sediment interface was lower than the strength of the substrate and motion was concentrated in a narrow zone close to the sole. It may be the case that the presence of a soft bed is required for the surge behaviour of this glacier, but we suggest that fast flow in the active phase was not accomplished by pervasive deformation of the bed, at least in the marginal 400 m. Instead the ice advanced by deformation of a thin upper horizon accompanied by sliding. A thin water layer may have reduced the transfer of basal shear stresses to underlying sediments.

We would like to thank A. Evans for help in the field and J.K. Hart for advice regarding field site selection. Thanks also to T. Ridgeway at Aberystwyth for thin-section preparation. Personal support to S. Fuller was provided by a University of Leeds scholarship. Fieldwork in Iceland was carried out in association with NERC grant GR3/R114 and was partially funded by the School of Geography, University of Leeds.

# References

ALLEY, R. B., BLANKENSHIP, D. D., BENTLY, C. R. & ROONEY, S. T. 1987. Till beneath Ice Stream B. 3. Till deformation evidence and implications. *Journal of Geophysical Research*, **92(B9)**, 8921–8929.

——, ——, ROONEY, S. T. & BENTLEY, C. R. 1989. Water pressure coupling of sliding and bed deformation III. Application to Ice Stream B Antarctica. *Journal of Glaciology*, **35**, 130–139.

BEGET, J. E. 1986. Modelling the influence of till rheology on the flow and profile of the lake Michigan lobe, Southern Laurentide ice sheet, USA. *Journal of Glaciology*, **32**, 235–240.

BENN, D. I. 1994. Fluted moraine formation and till genesis beneath a temperate valley glacier: Slett-markbreen, Jotunheimen, Southern Norway. *Sedimentology*, **41**, 279–292

BLAKE, E., CLARKE, G. K. C. & GÉRIN, M. C. 1992. Tools for examining subglacial bed deformation, *Journal of Glaciology*, **38**, 388–396.

BOULTON, G. S. 1976. The origin of glacially fluted surfaces: Observation and theory. *Journal of Glaciology*, **17**, 287–309.

——1987. A theory of drumlin formation by subglacial sediment deformation. *In*: MENZIES, J. & ROSE, J. (eds) *Drumlin Symposium*. Balkema, Rotterdam, 25–80.

—— & HINDMARSH, R. C. A. 1987. Sediment deformation beneath glaciers, rheology and geological consequences. *Journal of Geophysical Research*, **92**, 9059–9082.

BROWN, N. E., HALLET, B. & BOOTH, D. B. 1987. Rapid soft bed sliding of the Puget glacial lobe. *Journal of Geophysical Research*, **92**, (B9), 8985–8997.

CLARKE, G. K. C., COLLINS, S. G. & THOMPSON, D. 1984. Flow, thermal structure and subglacial conditions of a surging glacier. *Canadian Journal of Earth Science*, **21**, 231–240.

CLARK, P. U. & HANSEL, A. K. 1989. Clast ploughing, lodgement and glacier sliding over a soft glacier bed. *Boreas*, **18**, 201–207.

EKLUND, A. & HART, J. K. 1996. Glaciotectonic deformation within a flute from the Isfallsglaciaren, Sweden. *Journal of Quaternary Science*, **11**, 299–310.

ENGELHARDT, H. & KAMB, B. 1998. Basal sliding of Ice Stream B West Antarctica. *Journal of Glaciology*, **44** (147), 207–230.

GORDON, J. E., WHALLEY, W. B., GELLATLY, A. F. & VERE, D. M. 1992. The formation of glacial flutes: Assessment of models with evidence from Lyngsdalen, North Norway, *Quaternary Science Reviews*, **11**, 709–731.7

HART, J. K. 1995. Drumlins, flutes and lineations at Vestari-Hagafellsjökull, Iceland. *Journal of Glaciology*, **41**, 596–606.

——1997. The relationship between drumlins and other forms of subglacial glaciotectonic deformation. *Quaternary Science Reviews*, **16**, 93–107.

—— & SMITH, B. 1997. Subglacial deformation associated with fast ice flow, from the Columbia glacier, Alaska. *Sedimentary Geology*, **111**, 177–197.

——, GANE, F. & WATTS, R. 1996. Evidence for deforming bed conditions on the Dänischer Wohld peninsula, northern Germany. *Boreas*, **25**, 101–113.

——, BAKER, R. W., HOOKE, R.LeB., HANSON, B. & JANSSON, P. 1999. Coupling between a glacier and soft bed:1. A relation between effective pressure and local shear stress determined from till elasticity. *Journal of Glaciology*, **45**, 31–40.

——, HANSON, B., HOOKE, R. LeB. & JANSSON, P. 1995. Flow mechanism of glaciers on soft beds. *Science*, **267**, 80–81.

JISKOOT, H., BOYLE, P. & MURRAY, T. 1998. The incidence of glacier surging in Svalbard: evidence from multivariate statistics. *Computers and Geosciences*, **24**, 387–399.

KAMB, B 1987. Glacier surge mechanisms based on linked cavity configuration of the basal water system conduit system. *Journal of Geophysical Research*, **92**, 9083–9100.

——1991. Rheological nonlinearity and flow instability in the deforming bed mechanism of ice stream motion. *Journal of Geophysical Research*, **96**, B10, 16 585–16 595.

KARLÉN, W. 1981. Flutes on bare bedrock, *Journal of Glaciology*, **27**, 190–192.

LAWSON, D. E. 1976. Observations on flutings at Spencer Glacier, Alaska, *Arctic and Alpine Research*, **8**, 289–296.

MENZIES, J. 1987. Towards a general hypothesis on the formation of drumlins. *In*: MENZIES, J. & ROSE, J. (eds) *Drumlin Symposium*. Balkema, Rotterdam, 9–24.

—— & MALTMAN, A. J. 1992. Microstructures in diamictons – evidence of subglacial bed conditions. *Geomorphology*, **6**, 27–40.

PIOTROWSKI, J. A. & KRAUS, A. M. 1997. Response of sediment to ice sheet loading in northwestern Germany: effective stresses and glacier bed stability. *Journal of Glaciology*, **43**, 494–502.

—— & TULACZYK, S. 1999. Subglacial conditions under the last ice sheet in northwest Germany: ice-bed separation and enhanced basal sliding? *Quaternary Science Reviews*, **18**, 737–751.

PORTER, P. R., MURRAY, T. & DOWDESWELL, J. A. 1997. Sediment deformation and basal dynamics

beneath a glacier surge front: Bakaninbreen, Svalbard. *Annals of Glaciology*, **24**, 21–26.

ROSE, J. 1987. Drumlins as part of a glacier bedform continuum. *In*: MENZIES, J. and ROSE, J. (eds) *Drumlin Symposium*. Balkema, Rotterdam, 103–116.

——1989. Glacier stress patterns and sediment transfer associated with the formation of superimposed flutes. *Sedimentary Geology*, **62**, 151–176.

SIGURSSON, O. 1998. Glacier variations in Iceland 1930–1995. *Jökull*, **45**, 3–26

THORARINSSON, S. 1969. Glacier surges in Iceland, with special reference to the surges of Bruárjökull. *Canadian Journal of Earth Sciences*, **6**, 875–881.

TULACZYK, S., KAMB, B. SCHERER, R. P. & ENGELHARDT, H. F. 1998. Sedimentary processes at the base of a West Antarctic Ice Stream: Constraints from textural and compositional properties of subglacial debris, *Journal of Sedimentary Research*, **68**, 487–496.

VAN DER MEER, J. J. M. 1987. Micromorphology of glacial sediments as a tool in distinguishing genetic varieties of till. *Geological Survey of Finland, Special Paper*, **3**, 77–89.

——1993. Microscopic evidence of subglacial deformation. *Quaternary Science Reviews*, **12**, 553–587.

——1997. Short-lived streamlined bedforms (annual small flutes) formed under clean ice, Turtmann Glacier, Switzerland. *Sedimentary Geology*, **111**, 107–118.

# Radar evidence of water-saturated sediments beneath the East Antarctic Ice Sheet

MARTIN J. SIEGERT

Bristol Glaciology Centre, School of Geographical Sciences, University of Bristol,
Bristol BS8 1SS, UK (e-mail: m.j.siegert@bristol.ac.uk)

**Abstract:** Ice-penetrating airborne radar data from East Antarctica were examined in order to identify regions of the ice-sheet base where water-saturated sediments are thought to occur. Distinctive radar returns are identified from three subglacial environments as follows. (1) A frozen ice–bedrock interface, which shows scattering of a weak radar signal. (2) An ice–water contact above a subglacial lake where the radar reflections are bright, and horizontally flat. (3) The surfaces of two regions of water-saturated basal sediment, one at the centre of the ice sheet and one near the ice margin, where the radar returns are almost as bright as those from the subglacial lake, but not as flat. Characterization of radar signals from these regions is valuable since the results can be compared with data from other areas of the ice-sheet base to establish the nature of the sub-ice contact at a continental scale. The subglacial geomorphology of sediments near the ice margin displays large-scale (c. 10 km long, c. 200 m high) features with slopes that are relatively steep up-glacier, and shallow down-glacier, plus small-scale (<1 km long, <50 m high) regularly spaced undulations. Such sub-ice physiography indicates that the radar data may display *in-situ* sedimentary structures.

The importance of deforming water-saturated subglacial sediment to glacier dynamics has been appreciated for several years (e.g. Boulton & Jones 1979; Boulton & Hindmarsh 1987; Blankenship et al. 1986). It is now widely accepted that the flow of glaciers is enhanced by the presence of a deforming base because this material causes the subglacial frictional drag that can inhibit ice flow to be reduced to very low values (e.g. Murray 1997). In North America, subglacial deforming sediments may have been instrumental in the glacial dynamics of the southern margin of the Laurentide Ice Sheet during the last ice age (e.g. Alley 1991; Clark 1994). There, proglacial sedimentary sequences provide evidence that the ice-sheet margin experienced a number of rapid advances aided (and possibly controlled) by a deforming glacier base. These geological data show that deforming subglacial sediment is vital to large-scale ice-sheet dynamism. If the flow of the Antarctic Ice Sheet is influenced by deforming subglacial material, it too may be susceptible to the relatively unstable ice flow that occurred in North America after 15 000 years ago. It is therefore important to determine where weak subglacial material occurs in Antarctica and assess how this material may influence ice-sheet flow.

Several field studies have shown that ice flow in West Antarctica is organized through a series of ice streams where the deformation of sediments is a major controlling factor (e.g. Blankenship et al. 1986; Alley et al. 1987; Anandakrishnan et al. 1998; Bell et al. 1998). However, to date there has been virtually no assessment of the role of subglacial sediments in East Antarctic ice-sheet flow because direct information from the ice-sheet base is difficult to obtain. For example, no ice core has yet penetrated to the base of the central East Antarctic Ice Sheet. However, about 40% of East Antarctica has been surveyed by ice-penetrating airborne radar at 60 MHz. These radar data, stored in archive at the Scott Polar Research Institute (SPRI) in Cambridge, represent the only resource for investigating the subglacial environment of much of East Antarctica.

This paper demonstrates how airborne radar data, in conjunction with satellite altimetry, can be used to identify regions of subglacial sediment in East Antarctica. The paper also provides an insight into the arrangement of the subglacial environment across several regions within the area of the SPRI radar survey. It should be noted that a similar investigation was undertaken in West Antarctica, where the nature of 60 MHz radar reflections from several

*From*: MALTMAN, A. J., HUBBARD, B. & HAMBREY, M. J. (eds) *Deformation of Glacial Materials*. Geological Society, London, Special Publications, **176**, 217–229. 0305-8719/00/$15.00 © The Geological Society of London 2000.

subglacial environments was determined (e.g. Reynolds & Whillans 1979). This work has led to recent attempts at mapping the radar characteristics of West Antarctica (e.g. Bentley *et al.* 1998). However, previous investigations of the radar characteristics of subglacial environments in East Antarctica were undertaken before the action of subglacial weak sediments on ice flow was fully appreciated and accepted (e.g. Drewry 1982; Steed & Drewry 1982; McIntyre 1983). Given this new understanding of sub-ice processes, and the advent of accurate satellite altimetry allowing detailed maps of the ice-sheet surface (and an assessment of the driving stress), it is now timely to reassess the SPRI 60 MHz radar database to ascertain the radio-wave reflection characteristics above subglacial sedimentary environments.

## Form of 60 MHz radar data and the identification of subglacial lakes

Radar data can be recorded in two different formats. The first comprises graphs of two-way travel time versus signal strength from single radio-wave pulses, known as A-scopes (Fig. 1). The second form represents a series of single radio-wave returns, collected in a time-dependent manner in which signal strength is identified by brightness rather than an axis variable. When plotted as two-way radio-wave travel-time against real time, these time-dependent data yield pseudo-cross sections of the ice sheet, or Z-scope data (Fig. 1).

Radio waves at 60 MHz penetrate through several kilometres of cold ice. Because of this, a high proportion of the SPRI radar database contains information on the ice-sheet base. The nature of the glacial radar reflector will be dependent on the dielectric constant of the ice ($\varepsilon = 3.2$) and of the subglacial substrate. Subglacial lakes can be identified easily in radar data because the dielectric constant of water ($\varepsilon = 81$) is very different to that of bedrock ($\varepsilon = 4$–9). These electrical properties lead to stronger radar reflections from an ice–water interface than from an ice–rock boundary. Moreover, radar energy is scattered at a rough boundary between ice and bedrock, but reflected well from the smooth flat ice–water interface above a lake. Thus, in radar data, subglacial lakes are distinguishable because their surfaces appear as bright echoes of constant strength along the flight track and appear horizontal and flat (Oswald & Robin 1973; Siegert *et al.* 1996). In actual fact, in order for hydrostatic equilibrium to be maintained, the surface slope of the

subglacial lake will be 11 times, and in the opposite direction to, the surface slope. If subglacial lakes are longer than about 4 km, the ice sheet above them will be noticeably smooth and flat in satellite altimetric data (e.g. Siegert & Ridley 1998). This is because the friction between the ice and water is negligible, which results in a change in ice dynamics from base-parallel shearing over rock to longitudinal extension over water.

It should be noted that if the ice–water interface above a lake is relatively rough, a subglacial lake might not be observable in radar data (Reynolds & Whillans 1979).

The best-known and largest subglacial lake is the 230 km long fresh-water body beneath Vostok Station, known commonly as Lake Vostok (Kapitsa *et al.* 1996). Here, radar data show the ice–water interface to be typically smooth and flat, ERS-1 altimetry indicates a very flat ice surface above the lake and seismic data reveal that the water depth beneath Vostok Station is around 500 m. Apart from analysis of seismic data, there are two other ways of estimating the water depths of subglacial lakes. For example, in some cases (including Lake Vostok), 60 MHz radar has been observed to penetrate through subglacial water, and reflect off the lake floor interface (Gorman & Siegert 1999). VHF radio-wave penetration through water is only possible if the water is extremely pure and fresh. Because water absorbs radio waves relatively well, only water depths less than 20 m are detectable by this method. The other means of estimating the water depth of subglacial lakes is from examination of the sub-ice topographic setting. Such analysis reveals that the side walls adjacent to subglacial lakes have gradients of up to 0.1 (Dowdeswell & Siegert 1999). Extrapolating these gradients beneath the ice water interface suggests that many subglacial lakes are at least several tens of metres deep, and of the order of hundreds of metres in a number of cases.

Conceivably, accumulations of water-saturated basal sediments may also yield relatively strong 60 MHz radar returns (because their overall dielectric-constant values will be influenced heavily by that of the water). Moreover, if the sediments are weak, there will be little friction between the sediment and water, causing the ice surface above them to be flat and smooth. Thus it is possible that some subglacial lakes identified in Siegert *et al.* (1996) may comprise pockets of water-saturated weak sediments at the ice base. It is also possible that where subglacial sediments are present, the ice–sediment interface may not be perfectly flat, but

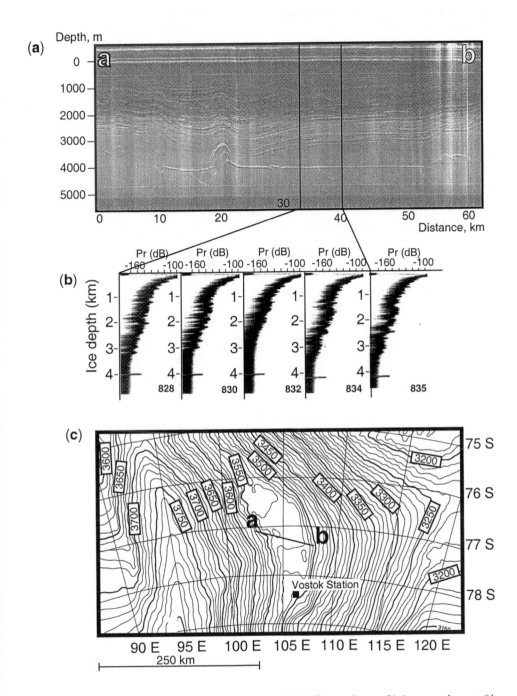

**Fig. 1.** 60 MHz radar data across the centre of Lake Vostok. (**a**) Z-scope image. (**b**) A-scopes along an 8 km section of the flight track. (**c**) ERS-1 altimetric surface elevation of the ice-sheet between Ridge B and Dome C. The smooth flat surface above Lake Vostok is clearly distinguishable from the general slope of the surrounding ice (adapted from Siegert & Ridley 1998).

undulating in nature, causing their radar character to be different to that of subglacial lakes.

## Evidence for sediments across the floor of subglacial lakes

Evidence of sediments within subglacial lakes comes from seismic data collected over Lake Vostok (Fig. 2). These data show a number of reflections directly beneath the floor of the lake, which indicate an accumulation of several hundred metres of sediment. Recent seismic investigations of Lake Vostok by Russian geophysicists have confirmed this finding (Popkov & Lukin pers. comm. 1998). Since Lake Vostok contains sediment, it is highly likely that other subglacial lakes will contain similar material. Given the large thickness of sediments at the floor of Lake Vostok, it is also likely that shallower subglacial lakes may, over thousands of years, 'fill up' with sediment (e.g. Zotikov 1987). This possibility means that pockets of weak, water saturated sediments at the base of the Antarctic Ice Sheet may be at least as common as subglacial lakes.

Around 70 subglacial lakes have been identified from the SPRI radar database (Siegert et al. 1996). These lakes are clustered within several locations in Antarctica. The majority of subglacial lakes exist in the vicinity of Dome C and other ice divides such as Ridge B and Titan Dome. In addition, several subglacial lakes exist near to the margin of the ice sheet, at the onset of enhanced flow of ice. Therefore, there is evidence of subglacial water beneath both the centre and the margins of the East Antarctic Ice Sheet (Dowdeswell & Siegert 1999). This evidence is essential to the identification of deforming sediments since the ice sheet is likely to be warm-based in the vicinity of subglacial lakes.

## Evidence of sediments beneath the centre of the East Antarctic Ice Sheet

Around 80% of known subglacial lakes are located beneath major ice divides (Dowdeswell & Siegert 1999). The presence of water means these areas of the ice sheet must be warm based. Because of this, the sub-ice environment near ice divides is also a potential area for water-saturated sediments. The location of several flat regions on the ice surface at Dome C coincides with subglacial lakes detected in 60 MHz radar data. In addition, radar data indicate that there are a number of other anomalous flat surfaces that do not have subglacial lakes beneath. These 'non-lake' surface features are caused by either (1) a topographic control on ice dynamics or (2) a change in the frictional qualities of the ice-sheet base. An example of the ice flow over unusual subglacial topography yielding a flat surface feature is where the ice-sheet flows across the Adventure Subglacial Trench (Fig. 3). However, a number of other flat surfaces around Dome C

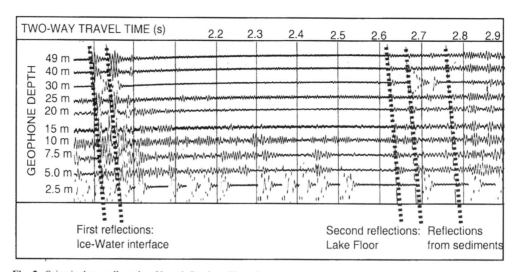

**Fig. 2.** Seismic data collected at Vostok Station. The seismogram was compiled from a series of geophones held within a 50 m deep borehole. The lake-floor reflector, after the ice surface arrival shows the water depth to be around 510 m. After the lake floor reflector, several other reflections are observed from sediments over the floor of the lake (data taken from Kapitsa et al. 1996).

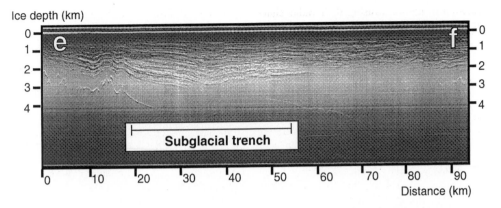

**Fig. 3.** 60 MHz radar Z-scope across transect 'ef' (Fig. 4). The data show a steep subglacial basin at around 20 km across the transect, which corresponds with the edge of a flat surface feature (Fig. 4b). It should be noted that the ice-sheet base reflector is lost as ice becomes deeper than about 4 km. However, the general shape of the trench can be inferred from the pattern of englacial internal layering.

are associated with relatively flat sub-ice topography. In such cases, only a change in the subglacial friction, like as happens over subglacial lakes, can cause the ice surface to become flat. An example of where this happens in Dome C is provided in Fig. 4. A 50 km wide flat ice-surface area has two radar lines across it. One of these lines indicates that two subglacial lakes are present, providing evidence that the ice sheet is warm based beneath the flat region. Between these lakes the radar returns are measurably stronger than those from nearby regions at higher elevations. The second radar line shows relatively strong reflections, but the radar signal from the ice base undulates slightly (i.e. there is no subglacial lake present). Since the flat ice-surface is continuous across and around both the sub-ice lakes, the sub-ice friction across both radar lines must be small. For the first line, negligible friction is expected between the ice–water contact above the lakes. However, across the second radar line, the low frictional contact can be best explained if weak water-saturated sediments are present. Thus, the entire flat-surface region (Fig. 4) may be underlain by both subglacial-lake water and weak water-saturated sediment. The effect of low friction across this region reduces the surface slope from 0.037° to 0.011°. Because basal friction is related to driving stresses at the ice base, and the basal stress is related to the ice surface slope, the average friction across this region can be estimated at one third of that upstream.

A second example of this type of subglacial sediment pocket is just south of Dome C (Fig. 5), where another flat surface region exists above the ice sheet base which is neither a deep subglacial trench (as in Fig. 3), or a subglacial lake. Siegert & Ridley (1998) concluded that this region could be interpreted as a zone of water-saturated subglacial sediments, because no alternative explanation for a low-friction subglacial environment was available. Further analysis of radar data from the Dome C region indicated that at least 13 flat ice surfaces, covering an area of c. 8000 km², occur above regions of water-saturated sediment (Siegert & Ridley 1998).

## Comparison between radar records from cold- and warm-based regions

Subglacial lakes have also been observed in radar data near to the margin of the ice sheet. One such location, where three subglacial lakes have been discovered, is within the Wilkes Subglacial Basin. This trough is 200 km wide and stretches for 1000 km across the foreland of the Transantarctic Mountain range (Fig. 6). The identification of subglacial lakes within the trough makes it very likely that the ice-sheet is warm based throughout this 3500 m thick region of the ice sheet. A recent model of Antarctic ice-sheet flux demonstrates that the Wilkes Subglacial Basin coincides with the onset of enhanced ice-sheet flow (Bamber et al. 2000). This suggests that the trough is associated with an alteration in ice-sheet dynamics, possibly where the ice base changes from cold to warm conditions.

Ice flows into the Wilkes Subglacial Basin from the Belgica Subglacial Highlands where ice thickness is only around 2.5 km (Fig. 7). It is likely that the ice sheet is cold based across these highlands. Therefore, Wilkes Subglacial Basin

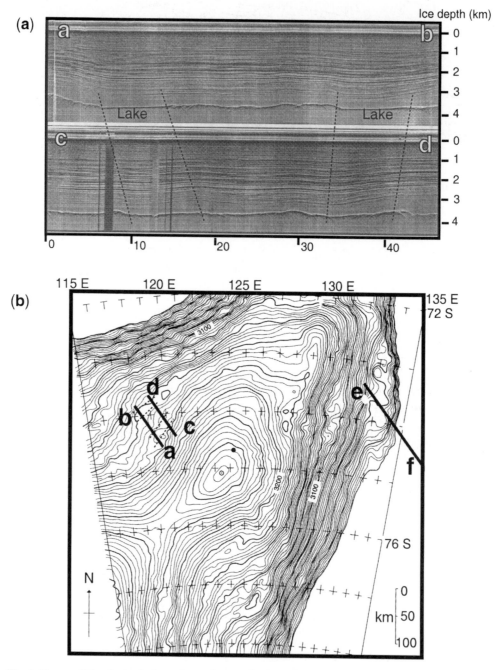

**Fig. 4.** Two parallel radar tracks located over the same flat ice surface feature. (**a**) Z-scope radar data at 60 MHz. (**b**) ERS-1 altimetric surface elevation of the ice sheet around Dome C central East Antarctica. Also shown is the location of a radar flightline 'ef' (Fig. 3).

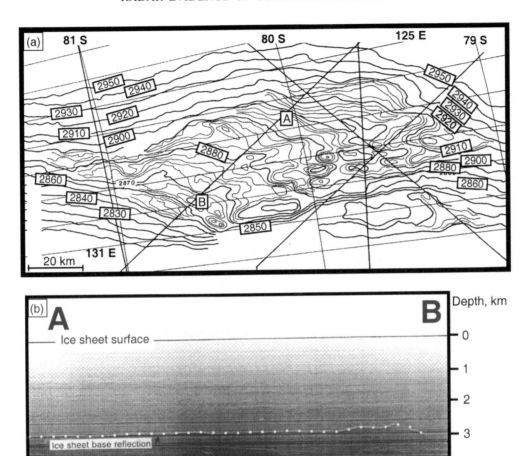

**Fig. 5. (a)** ERS-1 satellite altimeter map of the ice surface south of the Dome C region of East Antarctica. Note that the ice sheet is unusually flat, suggesting relatively low basal shear stress. Positions of radar flightlines are indicated. **(b)** 60 MHz radar across the flat surface feature. Note how the subglacial topography shows very little change in relief at the margins of the flat surface feature (adapted from Siegert & Ridley 1998).

and the Belgica Subglacial Highlands represent two distinctive neighbouring environments that can be examined using radar data.

The radar data collected over the Belgica Subglacial Highlands shows a characteristic response from ice frozen to bedrock (Fig. 7). The radar data show (1) evidence of significant scattering from a rough surface, because the signal fails to maintain a steady power and (2) that the maximum power of the radar reflection is actually less than reflections from thicker

regions over ice–water contacts. Radar data from within the Belgica Subglacial Highlands contrasts markedly with data collected downstream over the Wilkes Subglacial Basin (Figs 7 and 8). The radar signal downstream of the highlands shows (1) echoes of consistently high strength and (2) a slightly undulating surface (of the order of 10s of metres in amplitude over tens of kilometres in horizontal distance), indicative of a warm-based environment where water saturated sediments are present (Fig. 8).

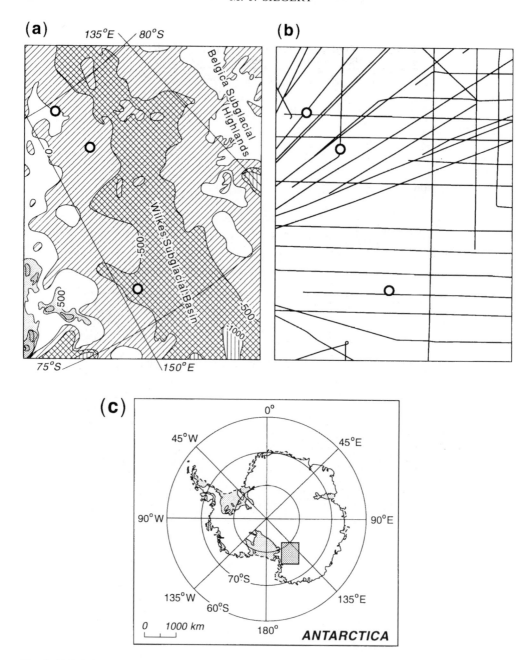

**Fig. 6.** (a) Subglacial topography and location of sub-ice lakes (denoted as circles) across the Wilkes Subglacial Basin, East Antarctica. (b) Location of radar flightlines. (c) Location of the map shown in (a) within Antarctica.

A warm-based ice–bedrock interface is likely to give a very different radar response from that over the Wilkes Subglacial Basin (Fig. 8). If the ice-sheet base is subject to sliding over bedrock, the ice-sheet base is likely to become 'fluted' as a result of abrasion over subglacial obstacles (e.g. Reynolds & Whillans 1979). This would lead to a noticeable variation in the signal strength from the ice base along the flight-track. The observations from the Wilkes Subglacial Basin show a

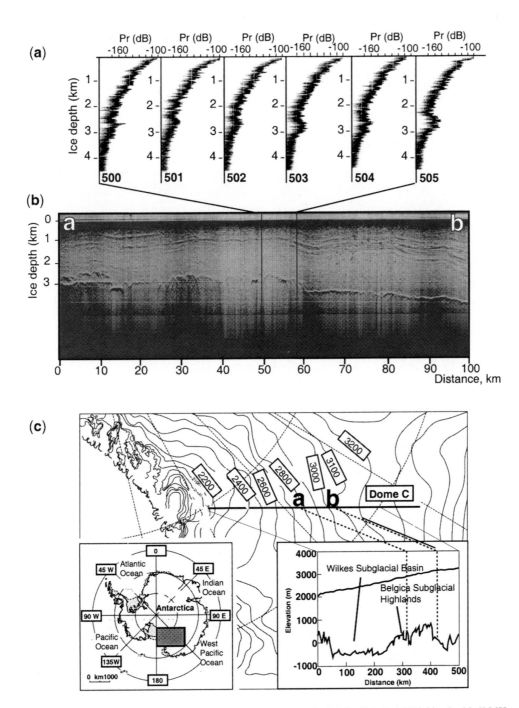

**Fig. 7.** Evidence for cold-based ice–rock subglacial contact over the Belgica Subglacial Highlands. (**a**) 60 MHz radar A-scopes along an 8 km section of flight track. (**b**) Z-scope radar data. These radar data show low power returns from the ice base and a severely rough subglacial environment that causes scattering of the radar signal. (**c**) Location of the radar transect and ice surface and subglacial elevation across the transect.

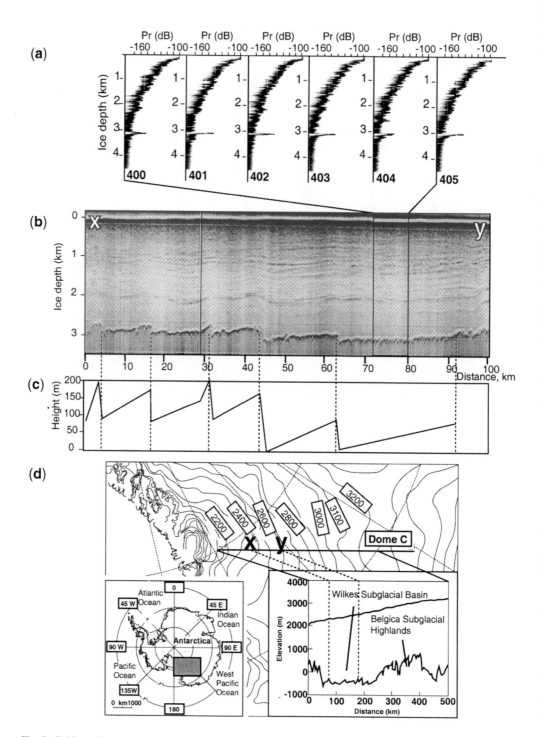

**Fig. 8.** Evidence for water saturated sediments within the Wilkes Subglacial Basin. (a) 60 MHz radar A-scopes along an 8 km section of flight track. The ice-base reflections are similar in appearance to those off an ice–water interface above a subglacial lake. (b) 60 MHz radar Z-scopes across a transect 'xy'. The data show that the ice base is not flat like a subglacial lake, but is similar to the subglacial environment next to the lake in Fig. 5. (c) Simplified subglacial morphology from the Z-scope image. The morphology of the ice base is characteristic of sedimentary features such as drumlinoid ridges, where a steep stoss face is adjacent to a shallow dipping less side. (d) Location, and surface/bedrock elevations across the radar flight track.

very constant power returned from the ice-sheet base (Fig. 8), implying that the subglacial interface is relatively smooth.

## Evidence for *in situ* subglacial sedimentary formations near the ice-sheet margin?

The subglacial morphology derived from the radar data suggests the formation of sedimentary structures along the line of ice flow (Fig. 8). In at least four cases along one single transect, the subglacial radar evidence shows topography that is steep sided on the stoss face and shallow on the lee side. These geomorphic features are about 15 km long (aligned with ice flow) and about 200 m high. The stoss face has an angle of *c.* 6° and the lee side dips at *c.* 1°. Given that the region is warm based, these topographic features may be caused by a number of possible processes. One is by the erosion of sedimentary units which dip in the general direction of ice flow. A second process is deformation of subglacial material to yield sedimentary features. It could be possible that these features are *in-situ* sedimentary structures such as drumlinoid, or crag-and-tail features (Fig. 8). If this interpretation is correct, the Wilkes Subglacial Basin may be a region where subglacial sediment deformation is important to ice flow and may provide a modern glacial analogue for regions of past glaciation where similar subglacial physiography occurs. Support for the interpretation of these features would be possible if radar information transverse to the transect in Fig. 8 were available. Unfortunately, no such lines were flown in the SPRI survey.

In addition to these large-scale sedimentary structures, smaller scale formations are also observed in the radar data within the Wilkes Subglacial Basin (Fig. 8). The ice-substrate interface is characterized in radar data as a series of relatively small hyperbolae. The crest to crest wavelength of these hyperbolae (between 65 and 90 km along the flightline 'xy'; Fig. 8) is between 500 m and 1 km. There are very few other reflections other than these well-organised hyperbolae. Welch *et al.* (1998) have demonstrated by processing radar data that hyperbola, such as those shown in Fig. 8 may originate from single, relatively small perturbations in the ice-base. The power returned from the ice-base is continuously high (Fig. 8a), and does not suggest there is significant scattering of the radar signal, despite the obvious undulation of the ice base indicated by the hyperbolae. The continuously high power of the radar reflections requires the ice-base interface to be smooth. It can therefore be contended that these sub-kilometre scale undula-

tions may be formed by sedimentary features such as drumlins. It should be noted that the radar beam used in the SPRI survey had a 120° width along the flight line, but only 20° transverse to the flight direction. It is, therefore, likely that the radar hyperbola are from subglacial morphology located directly below the flight line. It is also likely that similar subglacial features occur off-image either side of the radar line (Fig. 8).

## Discussion

This paper identifies two areas where water-saturated subglacial sediments are likely to occur. The first is located near the ice-sheet divide south of Dome C where subglacial lakes exist (Dowdeswell & Siegert 1999). Since it is known that Lake Vostok contains several hundred metres of sediments, it can be postulated that some former subglacial lakes may have now filled up with sediment. This is a possible explanation for the sediment accumulations identified around the Dome C region. The second region where water-saturated sediments may occur is located within the Wilkes Subglacial Basin, downstream of a zone of cold-based ice (Figs 7 and 8). Here, the radar data show strong subglacial reflections, and geomorphological characteristics indicative of sub-ice sedimentary structures, or eroded sedimentary units.

The identification of deformable subglacial sediments has implications for our understanding of ice-sheet dynamics. At present, numerical models of the Antarctic ice-sheet dynamics do not account well for subglacial deforming sediments. In order to replicate the true motion of the ice sheet, model results must be compared with glacial geological evidence, and adjusted to be compatible with this evidence. The identification of zones of subglacial water-saturated sediments is vital to this procedure. A thorough re-examination of the SPRI radar database is therefore required, under the guidelines presented in this paper, to establish the potential sub-ice extent of water-saturated deformable sediments beneath the Antarctic Ice Sheet. In addition, new radar data, which measure subglacial roughness to a centimetre-scale, being collected across the Wilkes Basin by US glaciologists will allow the morphological features described in this paper to be examined further.

## Summary

Airborne radar at 60 MHz has been used to classify three distinct subglacial environments within East Antarctica. The first is the ice-water interface above a subglacial lake. Second is the cold-based ice–rock contact beneath relatively thin ice. Third is the warm-based sub-ice regions, one near the ice divide and another near the ice margin, which have very little frictional force between the ice and substrate, allowing flat surfaces on the ice sheet to develop. The warm based regions near the ice margin show subglacial geomorphic features suggestive of either ice-moulded sedimentary structures, indicating deformation of basal sediments, or ice-scoured bedrock formations. These classifications can be used to identify regions of subglacial sediment at a continental scale within East Antarctica from airborne radar data.

No ice core has yet penetrated to the base of the central East Antarctic Ice Sheet, and so there is no direct evidence for subglacial sediments. However, seismic information from Lake Vostok indicates that hundreds of metres of sediment have accumulated on the lake floor (Kapitsa et al. 1996). This suggests that sub-glacial sediments may occur at the floor of the other 70 or so lakes that have been discovered. It also appears likely that shallow subglacial lakes may, over several thousands of years, fill up with sediment.

Identification of subglacial sediments is pos-sible when radar data show strong ice-base reflections (but not the mirror-type reflector from a subglacial lake) beneath a flat ice surface feature. The occurrence of these two features is possible if the basal shear stress is very low. Since there is no subglacial lake, the most likely explanation is that the ice base is covered by water-saturated deformable sediments.

Potential sources of sediment within and next to subglacial lakes have been identified at Dome C, Ridge B, Vostok, and Titan Dome. In addition, deformable sediments may also occur within the Wilkes Subglacial Basin. These sediment sources are important because sedi-mentary basins may represent an essential element in the initiation of enhanced ice flow. The sediment sources identified in this paper need to be acknowledged in models of Antarctic ice flow for past, present and future scenarios.

The characterization of 60 MHz radar data can be used in future to determine the locations of basal sediment across the 40% of East Antarctica covered by the SPRI database.

Thanks are given to the Director of the Scott Polar Research Institute in Cambridge for access to the airborne radar archive and to D. Blankenship, F. Ng and A. Smith for constructive and insightful reviews.

## References

ALLEY, R. B. 1991. Deforming bed origin for southern Laurentide till sheets? Journal of Glaciology, 37, 67–76.

——, BLANKENSHIP, D. D., ROONEY, S. T. & BENTLEY, C. R. 1987. Till beneath Ice Stream B. 4. A coupled ice-till flow model. Journal of Geophysical Research, 92, 8931–8940.

ANANDAKRISHNAN, S., BLANKENSHIP, D. D., ALLEY, R. B. & STOFFA, P. L. 1998. Influence of subglacial geology on the position of a West Antarctic ice stream from seismic observations. Nature, 394, 62–65.

BAMBER, J. L., VAUGHAN, D. G. & JOUGHIN, I. 2000. Widespread complex flow of ice in central regions of Antarctica. Science, 287, 1248–1250.

BELL, R. E., BLANKENSHIP, D. D., FINN, C. A., MORSE, D. L., SCAMBOS, T. A., BROZENA, J. M. & HODGE, S. M. 1998. Influence of subglacial geology on the onset of a West Antarctic ice stream from aerogeophysical observations. Nature, 394, 58–62.

BENTLEY, C. R., LORD, N. & LIU, C. 1998. Radar reflections reveal a wet bed beneath stagnant Ice Stream C and a frozen bed beneath ridge BC, West Antarctica. Journal of Glaciology, 44, 149–156.

BLANKENSHIP, D. D., BENTLEY, C. R., ROONEY, S. T. & ALLEY, R. B. 1986. Seismic measurements reveal a saturated porous layer beneath an active Antarctic ice stream. Nature, 332, 54–57.

BOULTON, G. S. & HINDMARSH, R. C. A. 1987. Sediment deformation beneath glaciers: rheology and geological consequences. Journal of Geophysical Research, 92, 9059–9082.

—— & JONES, A. S. 1979. Stability of Temperate Ice Caps and Ice Sheets resting on Beds with Deformable Sediment. Journal of Glaciology, 24, 29–42.

CLARK, P. U. 1994. Unstable behaviour of the Laurentide Ice Sheet over deforming sediment and its implications for climate change. Quaternary Research, 41 19–25.

DOWDESWELL, J. A. & SIEGERT, M. J. 1999. The dimensions and topographic setting of Antarctic subglacial lakes and implications for large-scale water storage beneath continental ice sheets. Geological Society of America Bulletin, 111, 254–263.

DREWRY, D. J. 1982. Ice flow, bedrock and geothermal studies from radio-echo sounding inland of McMurdo Sound, Antarctica. In: CRADDOCK, C. (ed.) Antarctic Geoscience. Symposium on Antarctic Geology and Geophysics. Madison, Wisconsin, August 1977, 977–983.

GORMAN, M. R. & SIEGERT, M. J. 1999. Penetration of Antarctic subglacial water masses by VHF electromagnetic pulses: estimates of minimum water depth and conductivity. Journal of Geophysical Research, 104(B12), 29 311–29 320.

KAPITSA, A., RIDLEY, J. K., ROBIN, G. DE Q., SIEGERT, M. J. & ZOTIKOV, I. 1996. Large deep freshwater lake beneath the ice of central East Antarctica. Nature, 381, 684–686.

McINTYRE, N. F. 1983. *The topography and flow of the Antarctic ice sheet*. PhD thesis, University of Cambridge.

MURRAY, T. 1997. Assessing the paradigm shift: deformable glacier beds. *Quaternary Science Reviews*, **16**, 995–1016.

OSWALD, G. K. A. & ROBIN, G. DE Q. 1973. Lakes beneath the Antarctic Ice Sheet. *Nature*, **245**, 251–254.

REYNOLDS, R. & WHILLANS, I. M. 1979, Glacial-bed types and their radar-reflection characteristics: *Journal of Glaciology*, **23**, 439–440.

SIEGERT, M. J. & RIDLEY, J. K. 1998. Determining basal ice sheet conditions at Dome C, central East Antarctica, using satellite radar altimetry and airborne radio-echo sounding information. *Journal of Glaciology*, **44**, 1–8.

——, DOWDESWELL, J. A., GORMAN, M. R. & McINTYRE, N. F. 1996. An inventory of Antarctic subglacial lakes. *Antarctic Science*, **8**, 281–286.

STEED, R. H. N. & DREWRY, D. J. 1982. Radio echo sounding investigations of Wilkes Land, Antarctica. *In*: CRADDOCK, C. (ed.) *Antarctic Geoscience. Symposium on Antarctic Geology and Geophysics*. Madison, Wisconsin, August 1977, 969–975.

WELCH, B. C., PFEFFER, W. T., HARPER, J. T. & HUMPHREY, N. F. 1998. Mapping subglacial surfaces of temperate valley glaciers by two-pass migration of a radio-echo sounding survey. *Journal of Glaciology*, **44**, 164–170.

ZOTIKOV, I. A. 1987. *The thermophysics of glaciers*. Reidel Publishing Co., Dordrecht.

# Laboratory investigations of the strength, static hydraulic conductivity and dynamic hydraulic conductivity of glacial sediments

BRYN HUBBARD & ALEX MALTMAN

*Institute of Geography and Earth Sciences, University of Wales, Aberystwyth, Ceredigion SY23 3DB, Wales, UK (e-mail: byh@aber.ac.uk)*

**Abstract:** We report the first data on how effective pressure and deformation influence the hydraulic conductivity of glacial sediments. Fifty eight static hydraulic conductivity tests, 28 triaxial deformation tests and 25 dynamic hydraulic conductivity tests (in which hydraulic conductivity is measured simultaneously with deformation) have been undertaken on seven samples of glacial sediments recovered from the margins of Haut Glacier d'Arolla, Switzerland, and from Traeth y Mwnt, mid Wales. Testing reveals that hydraulic conductivity is inversely related to effective pressure, particularly at effective pressures below $c.$ 100 kPa. Over the full range of effective pressures used (50–900 kPa), this relationship is best described by a negative power law above a base hydraulic conductivity value, termed $K_0$. The value of $K_0$ varies between samples by over three orders of magnitude, from $10^{-8}\,\mathrm{m\,s^{-1}}$ to $10^{-11}\,\mathrm{m\,s^{-1}}$. These values vary directly, but weakly, with the square of the effective grain-size of the samples tested.

Dynamic testing revealed a commonly repeated pattern of sample failure: axial stress approached a maximum value after which the sample deformed in a ductile manner from its initially cylindrical shape to a barrel shape. Most commonly, sample deformation and failure were accompanied by a decrease in hydraulic conductivity, although increases were also recorded. Dynamic testing also resulted in strongly linear relationships between effective pressure and the yield stress at failure. Such relationships are broadly consistent with a Mohr–Coulomb type model, revealing significant inter-sample variability in frictional resistance, and a cohesive term that is statistically indistinguishable from zero.

Subglacial sediment deformation and drainage are interdependent. This relationship is widely recognized, for example, in terms of the effect of sediment pore-water pressure on sediment strength (e.g. Boulton & Hindmarsh 1987), especially where overpressures arise. However, the impacts of factors such as deformation or changes in effective pressure on subglacial sediment hydraulic conductivity are largely unknown (e.g. Murray & Dowdeswell 1992). It has been realized for some time that the deformation of ocean-floor sediments exerts a fundamental control over their hydraulic conductivity, and experiments have indicated the complexity of this relationship. For example, Byrne *et al.* (1993) and Stephenson *et al.* (1994) investigated the syn-deformational (or *dynamic*) hydraulic conductivity of fine-grained, marine core mudstones from the Nankai accretionary prism, offshore Japan, and analogue materials. These studies revealed intricate variations, with different samples experiencing increases, decreases or insignificant changes in hydraulic conductivity during deformation and failure at a variety of sample orientations and confining pressures. More recent experiments, also in the accretionary prism context, have led to detailed scenarios involving cyclic variations of hydraulic conductivity with continuing deformation (e.g. Bolton *et al.* 1999). To date, no such work has been reported for glacial sediments. This deficiency is particularly significant since there appears to be some discrepancy between field-based estimates of the hydraulic conductivity of subglacial sediments, which are relatively high (generally $10^{-4}$ to $10^{-7}\,\mathrm{m\,s^{-1}}$), and static laboratory measurements of this parameter, which are relatively low (generally less than $10^{-7}\,\mathrm{m\,s^{-1}}$) (see review by Fountain & Walder 1998). This contrast may reflect one or more of a number of factors, including the following:

(1) *In-situ* determination of the hydraulic conductivity of subglacial sediments from proxy data may systematically overestimate true, Darcian hydraulic conductivity. Such an effect may be caused by the presence of numerous non-Darcian flow pathways beneath actual ice masses that cannot be discriminated from

*From*: MALTMAN, A. J., HUBBARD, B. & HAMBREY, M. J. (eds) *Deformation of Glacial Materials.* Geological Society, London, Special Publications, **176**, 231–242. 0305-8719/00/$15.00 © The Geological Society of London 2000.

Darcian flow *sensu stricto* (which occurs through the matrix of the host material) (e.g. Fountain 1992; Hubbard *et al.* 1995). It may therefore be more accurate to refer to hydraulic conductivities reconstructed from proxy data as 'effective' values, which may represent an upper limit to true permeating matrix flow.

(2) Systematic differences may exist between sediments tested in the laboratory and those actually present beneath contemporary ice masses. For example, restrictions on the size of laboratory apparatus commonly requires that pebble-sized clasts are excluded from analysis, despite their frequent presence within actual glacial sediments. Subglacial sediment sampling from glaciers may also be size-specific. For example, borehole-based subglacial sediment samplers are restricted in size, typically excluding clasts greater than some millimetres in diameter. Borehole corers and samplers also operate subaqueously, potentially exposing the sample to some degree of sorting by flowing water. Further, sediment thickness and character may be spatially variable, particularly at the beds of temperate glaciers (e.g. Harper & Humphrey 1995; Hooke *et al.* 1997; Harbor *et al.* 1997; Fischer *et al.* 1999). The net flow pathways followed by subglacial waters in such environments will be more complex than those followed in laboratory tests of permeating flow through discrete sediment samples.

(3) Laboratory-based determinations of hydraulic conductivity may have failed to account fully for variability in the properties of sediments located at the glacier bed. These include variations in rates of sediment deformation (e.g. Iverson *et al.* 1995; Fischer *et al.* 1999) and in pore-water pressure (e.g. Hubbard *et al.* 1995; Fischer *et al.* 1998).

In this paper we record the static and dynamic hydraulic conductivity of subglacial sediment samples prior to, and during, triaxial deformation. Samples have been deformed under a variety of confining pressures in order to investigate relationships between effective pressure and sediment strength and hydraulic conductivity.

## Sediment provenance and texture

Five subglacial sediment samples were recovered from ice-marginal and subglacial locations at Haut Glacier d'Arolla, Switzerland, and two samples of diamicton of Late Devensian age from a coastal cliff exposure at Traeth y Mwnt, Wales. The character and setting of each of these samples is presented in Table 1, the Haut Glacier d'Arolla samples having been analysed for their grain-size characteristics in some detail in an earlier study (Fischer & Hubbard 1999). Sediment texture was determined by dry

**Table 1.** *Summary of sample grain-size characteristics and setting*

| Sample | Gravel (%) | Sand (%) | Silt & clay (%) | $d_{10}$ (mm) | Facies* | Setting† |
|---|---|---|---|---|---|---|
| Arolla 2 | 42.2 | 43.0 | 14.8 | 0.014 | Clast-rich intermediate diamicton | Deformed subglacial sediment |
| Arolla 3 | 59.5 | 31.5 | 9.0 | 0.051 | Sandy gravel | Undeformed subglacial sediment underlying unit containing Arolla 2 |
| Arolla 7 | 71.0 | 23.0 | 6.0 | 0.099 | Sandy gravel | Undeformed subglacial diamict from glacier headwall |
| Arolla 12 | 51.7 | 37.9 | 10.4 | 0.035 | Clast-rich intermediate diamicton | Extensively deformed subglacial diamict from glacier terminus |
| Arolla 13 | 31.7 | 56.7 | 11.6 | 0.031 | Clast-rich intermediate diamicton | Basal diamict from beneath dead ice at glacier terminus |
| Mwnt 1 | 15.5 | 56.1 | 28.4 | 0.011 | Clast-rich intermediate diamicton | Glaciomarine/ glaciolacustrine diamicton |
| Mwnt 3 | 28.2 | 66.1 | 5.7 | 0.003 | Clast-rich sandy diamicton | Glaciomarine/ glaciolacustrine diamicton |

*Facies classification is based on that of Hambrey (1994, p. 9).
†Haut Glacier D'Arolla sample interpretations are based on Fischer & Hubbard (1999).

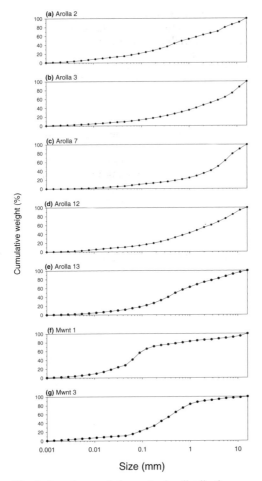

**Fig. 1.** Sample cumulative grain-size distributions. Sample characteristics are summarized in Table 1.

sieving between the maximum cut-off clast size, $-4\phi$ (16 mm), and $4\phi$ (0.063 mm), and by settling analysis (Sedigraph 5100, Micrometrics) between $4\phi$ and $<10\phi$. Grain-size distributions for the samples are given in Fig. 1 and summarized in Table 1.

## Laboratory methods

### Laboratory equipment

The testing apparatus consists of a triaxial deformation load frame that contains a pressurised cell housing a latex-sheathed cylindrical sample (Fig. 2). Confining pressure (i.e. the water pressure within the cell) is regulated by an operator-programmed hydraulic actuator. A syringe infusion pump is used to drive permeant (water) into the top of the sample at a

precisely determined constant rate, and the associated pressure head across the sample is continually monitored by a differential pressure transducer accurate to $c.0.5$ kPa. Background pore-fluid pressure, which defines the back-pressure at the base of the sample and ensures saturation, is driven by pressurized nitrogen gas. This operates in conjunction with a volume change device, which measures the volume of water exiting from the base of the sample. Sediments may be subjected to a large range of pre-determined isotropic effective stresses by adjusting the pore-fluid pressure and/or the cell confining pressure.

Applied axial stress is measured by a displacement transducer, which continually monitors the motion of a calibrated proving ring. For each incremental change in consolidation stress, a precise measurement of cell (and therefore sample) volume change is made with the hydraulic actuator, enabling simultaneous measurement of sample volumetric strain and bulk variations. Samples are normally deformed under fully-drained conditions, allowing equilibrium dissipation of pore-fluid pressure and hence the continual measurement of dynamic hydraulic conductivity during strain.

### Sample preparation and testing

Samples were prepared and tested according to the following procedure.

Samples were trimmed to cylinders of diameter 38 mm and length 76 mm. Clasts with $b$-axes longer than 4 mm that protruded from the sides of the sample cylinder were removed and replaced with matrix sediment. The samples were pre-consolidated to 1000 kPa over about 48 hours.

Sample cylinders were sleeved with a thin rubber membrane, and positioned within the triaxial cell.

Samples were saturated at a pore-water pressure of 300 kPa and subjected to a confining pressure of 450 kPa, typically taking between 12 and 24 hours.

Static hydraulic conductivity tests were undertaken at a variety of confining pressures. These confining pressures were typically (in order): 450, 500, 550, 600, 650, 700, 800, 900, 1000, 350 and 400 kPa, although other confining pressures were used (e.g. samples Arolla 2 and Arolla 13 were additionally subjected to confining pressures of 1100 and 1200 kPa) (Table 2). The success of these tests depended on a variety of factors, including sample hydraulic conductivity and integrity, such that not all samples could

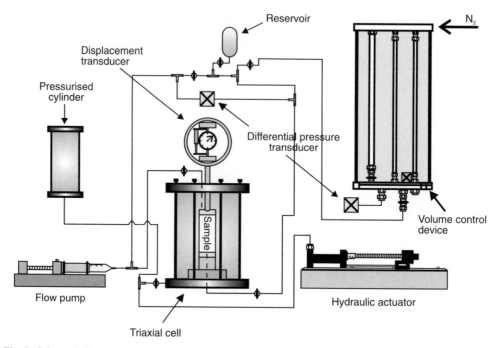

**Fig. 2.** Schematic illustration of the experimental apparatus, comprised of a flow pump permeameter and a triaxial deformation frame. Individual components are described in the text. Adapted from Bolton & Maltman (1998).

be tested at all pressures. At each pressure the permeant flow-rate was adjusted until the differential pressure across the sample was <15 kPa, and the test continued until the differential pressure equilibrated at a steady value, normally requiring some minutes to some hours.

The motor on the triaxial load frame was started, and samples were deformed axially at

**Table 2.** *Sample testing schedule*

| $P_{eff}$ | Arolla 2 | Arolla 3 | Arolla 7 | Arolla 12 | Arolla 13 | Mwnt 1 | Mwnt 3 |
|---|---|---|---|---|---|---|---|
| 50 | † | † | † | † | | | |
| 100 | † | †* | * | * | | | * |
| 150 | † | † | † | † | | † | |
| 200 | †* | †* | †* | †* | | †* | * |
| 250 | † | † | † | † | | † | |
| 300 | * | †* | †* | †* | † | * | †* |
| 350 | † | | † | † | | † | † |
| 400 | †* | †* | | †* | | * | † |
| 450 | | | | | | † | |
| 500 | †* | †* | † | †* | † | | † |
| 550 | | | | | | † | |
| 600 | | †* | † | †* | | * | † |
| 700 | † | † | † | †* | † | †* | † |
| 800 | † | | | | † | | |
| 900 | † | | | | † | | |

Sample textures are presented in Fig. 1 and summarized, along with their inferred provenance, in Table 1. Static hydraulic conductivity tests are marked by †, and strength tests by *. Dynamic hydraulic conductivity data were acquired for all but three of the strength tests.

a constant strain rate $(3 \times 10^{-6}\,\text{s}^{-1})$ until and beyond the point at which recorded axial stress peaked or leveled off.

The motor and pump were stopped, the confining pressure changed and the deformation procedure repeated.

The testing schedule of each sample is given in Table 2.

## Results

### Static hydraulic conductivity

Results from static hydraulic conductivity tests (Fig. 3) reveal significant variability in hydraulic conductivity, both between different samples and at different effective pressures for a given sample. The latter pattern is relatively consistent across all samples: hydraulic conductivity is relatively constant at high effective pressures, but increases markedly as effective pressure falls, particularly below c. 100 kPa. These relationships (Fig. 3) suggest that hydraulic conduct-

ivity $(K)$ varies as an inverse power function of effective pressure $(P_e)$ above a base hydraulic conductivity value $(K_0)$:

$$K = K_0 + A P_e^{-n} \qquad (1)$$

where $A$ and $n$ are empirically-defined, material constants. Curve fitting reveals that Eqn 1 matches the sample data sets presented in Fig. 3 closely (Table 3), with values of the exponent, $n$ (which describes the rate of increase in $K$ at low $P_e$), varying between 1.6 for sample Arolla 12 and 9.6 for sample Arolla 7 (Table 3). Values of $K_0$, which describes the relatively constant base hydraulic conductivity at high effective pressures for each sample, varies between $8.8 \times 10^{-9}\,\text{m s}^{-1}$ for Arolla 13 and $8.3 \times 10^{-12}$ $\text{m s}^{-1}$ for Arolla 2. Freeze & Cherry (1979) reported that the hydraulic conductivity, $K$, of a sediment typically varies with the square of its effective grain size (the upper diameter of the finest 10% of the sample by weight), $d_{10}$, according to:

$$K = B(d_{10})^2 \qquad (2)$$

where $B$ is a constant, $K$ is in $\text{cm s}^{-1}$ and $d_{10}$ is in mm (Freeze & Cherry 1979, p. 350). Plotting $d_{10}$ against $K_0$ for each sample in the present study (Fig. 4) reveals a general correspondence with Eqn 2 although the data are considerably scattered around the best-fit curve given by this relation.

### Dynamic hydraulic conductivity

During triaxial testing, initial sample shortening was generally accompanied by a relatively sharp increase in axial stress. After some time, the stress either leveled off or decreased, at which point the sample was considered to have reached steady-state deformation or to have failed. Testing continued for some time following failure, and sediment dynamic hydraulic conductivity was, where possible, measured throughout the deformation and failure events. Twenty eight such deformation tests were conducted in total (Table 2), of which 25 yielded stable dynamic hydraulic conductivity data. Changes in hydraulic conductivity during individual deformation tests were marked, but generally less than one order of magnitude. The nature of these responses is summarized in Table 4 and Fig. 5 in which net hydraulic conductivity change (%) during deformation is plotted against the effective pressure at which the test was conducted. These data reveal that hydraulic conductivity varies during deformation by <100% (with the

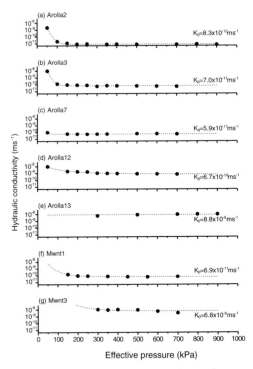

**Fig. 3.** Bivariate plot of pre-deformation static hydraulic conductivity (closed circles) against effective pressure for all samples. The best fit power-law relation (Equation 1 described in the text and summarized in Table 3) is presented as a dotted line.

**Table 3.** *Summary of best-fit constants in Equation 1, describing the relation between static hydraulic conductivity and effective pressure for each sample, as illustrated in Fig. 3*

| Sample | $K_0$ (m s$^{-1}$) | $A$ | $n$ | $\chi^2$ |
|--------|--------------------|-----|-----|----------|
| Arolla 2 | $8.3 \times 10^{-12}$ | $8.4 \times 10^{3}$ | 7.4 | $1.6 \times 10^{-24}$ |
| Arolla 3 | $7.0 \times 10^{-11}$ | $2.6 \times 10^{4}$ | 7.3 | $1.8 \times 10^{-22}$ |
| Arolla 7 | $5.9 \times 10^{-11}$ | $9.4 \times 10^{5}$ | 9.6 | $1.6 \times 10^{-23}$ |
| Arolla 12 | $6.7 \times 10^{-10}$ | $5.3 \times 10^{-6}$ | 1.6 | $3.1 \times 10^{-20}$ |
| Arolla 13 | $8.8 \times 10^{-9}$ | $9.1 \times 10^{-10}$ | 6.1 | $5.9 \times 10^{-18}$ |
| Mwnt 1 | $6.9 \times 10^{-11}$ | $2.4 \times 10^{-2}$ | 3.9 | $2.7 \times 10^{-23}$ |
| Mwnt 3 | $6.8 \times 10^{-9}$ | $2.3 \times 10^{3}$ | 4.7 | $9.0 \times 10^{-18}$ |

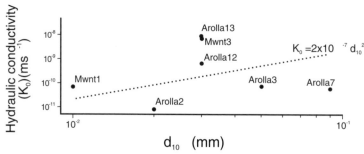

**Fig. 4.** Bivariate plot of static hydraulic conductivity against effective grain size ($d_{10}$). The best square relation (Equation 2 described in the text) is presented as a dotted line.

notable exception of sample Arolla 3 at an effective pressure of 100 kPa, which registered a hydraulic conductivity increase of 277%). Figure 5 also reveals that most samples registered a net hydraulic conductivity decrease during deformation. There is also no clear trend in either the direction or the magnitude of the hydraulic conductivity response to deformation with effective pressure.

The detailed nature of the dynamic hydraulic conductivity responses is illustrated through the two contrasting examples presented in Fig. 6. The most common response pattern is illustrated by Fig. 6a (sample Mwnt 1 deformed at an effective pressure of 600 kPa), in which hydraulic conductivity decreases steadily as the applied axial stress increases and the sample strains. Responses such as this accounted for 20 of the 25 tests conducted (Table 4). The response of the remaining five tests is illustrated by Fig. 6b (sample *Arolla 2* deformed at an effective pressure of 200 kPa), in which hydraulic conductivity increases steadily during deformation. Observations made during testing suggested that such increases in dynamic hydraulic conductivity were preferentially associated with more brittle modes of sample failure, typically along a set of inclined shear planes.

*Sediment strength*

As expected from theory, a positive linear relationship exists between effective pressure and axial stress at failure ('peak strength') for each sample (Fig. 7). Least squares linear regression indicates that the slopes of these relationships vary between 3.40 and 4.38 for the Arolla samples and between 1.60 and 1.78 for the Traeth y Mwnt samples, suggesting that the former are more than twice as strong as the latter at a given effective pressure. Although most samples appeared to deform macroscopically in a ductile manner (changing from a straight-sided cylinder to a barrel shape), it is likely that failure took place even in these samples on microscopic shear surfaces (e.g. Maltman 1987). Such essentially brittle failure allows the plots to be reformulated in terms of Mohr's circles, which provide a graphical representation of the stress field sustained by each sample during testing. Mohr's circles are constructed on plots of shear stress ($\tau$) (abscissa) against principal normal stress ($\sigma$) (ordinate) by plotting the major and minor principal normal stresses (conventionally denoted $\sigma_1$ and $\sigma_3$) as the points at which the circle intersects the abscissa. In our plots, $\sigma_1$ is the axial stress at failure and $\sigma_3$ the least

**Table 4.** Absolute ($m\,s^{-1}$) and relative (%) changes in sample hydraulic conductivity with deformation at a variety of effective pressures

| Sample | 100 kPa | | 200 kPa | | 300 kPa | | 400 kPa | | 500 kPa | | 600 kPa | | 700 kPa | |
|---|---|---|---|---|---|---|---|---|---|---|---|---|---|---|
| | $m\,s^{-1}$ | % | $m\,s^{-1}$ | % | $m\,s^{-1}$ | % | $m\,s^{-1}$ | % | $m\,s^{-1}$ | % | $m\,s^{-1}$ | % | $m\,s^{-1}$ | % |
| Arolla 2 | | | $6.1 \times 10^{-12}$ | +46.2 | $2.3 \times 10^{-12}$ | +19.7 | $-3 \times 10^{-13}$ | −2.6 | $1.4 \times 10^{-12}$ | +12.7 | | | | |
| Arolla 3 | $1.74 \times 10^{-10}$ | +277 | $3.4 \times 10^{-11}$ | +31.8 | $-3.4 \times 10^{-11}$ | −35.2 | $-2.0 \times 10^{-11}$ | −25.2 | $9.7 \times 10^{-12}$ | −14.1 | $-3.5 \times 10^{-12}$ | −5.4 | | |
| Arolla 7 | $-2.8 \times 10^{-11}$ | −30.4 | $-4.7 \times 10^{-12}$ | −15.8 | $-2.0 \times 10^{-11}$ | −48.2 | | | | | | | | |
| Arolla 12 | $-5.3 \times 10^{-10}$ | −57.8 | $-3.5 \times 10^{-10}$ | −39.2 | $-4.4 \times 10^{-10}$ | −83.5 | $-2.9 \times 10^{-11}$ | −31.1 | $-8.3 \times 10^{-12}$ | −13.5 | $-1.0 \times 10^{-13}$ | −0.2 | $-4.0 \times 10^{-13}$ | −1.2 |
| Mwnt 1 | | | | | $-9.3 \times 10^{-12}$ | −10.7 | $-1.4 \times 10^{-11}$ | −17.3 | | | $-2.0 \times 10^{-11}$ | −29.8 | $-2.3 \times 10^{-11}$ | −38.0 |
| Mwnt 3 | $-1.5 \times 10^{-8}$ | −60.5 | | | | | | | | | | | | |

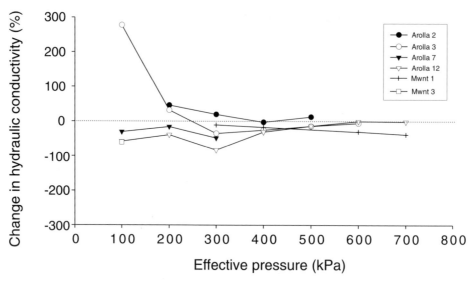

**Fig. 5.** Bivariate plot of the net change in hydraulic conductivity recorded over each deformation test against the effective pressure at which the tests were conducted.

principal stress, equivalent in value to the confining pressure in the cell. Thus, the centre of each Mohr's circle is located on the abscissa, and its value is given by $(\sigma_1 + \sigma_3)/2$. The diameter of each circle is given by $(\sigma_1 - \sigma_3)$. Because we are plotting the stress values at failure, shear strength envelopes are defined for each sample by the tangent to its set of Mohr's circles, each of which represents an individual strength test at a certain confining pressure, $\sigma_3$ (Fig. 8). As with the plots of peak strength against effective pressure, the failure envelopes define slopes that are markedly steeper for the Arolla samples than for the Traeth y Mwnt samples, reflecting their greater frictional resistance to shear (varying between 0.94 and 1.16 for the former and 0.43 and 0.47 for the latter) (Fig. 8). With the exception of sample Mwnt 3, all failure envelopes also pass approximately through the origin, indicating very low sample cohesion.

## Discussion

Our data indicate that the hydraulic conductivity of glacial sediments cannot be considered independently of effective pressure, particularly at effective pressures of less than $c$. 100 kPa, where hydraulic conductivity increases markedly. The glacial sediments tested here may well be overconsolidated. This situation is similar to that reported by Bolton & Maltman (1998) on the basis of experiments on ocean-floor sediments, where overconsolidation (where antece-

dent natural loading is significantly greater than that employed in a test) was found to be highly influential. The enhanced packing of the sediment grains induced by a previous load improves mechanical stability, but, with increasing fluid pressures, there comes a point where inter-particle attraction is exceeded and the pores abruptly dilate. Not only are the pores themselves now able to store more fluid but the pore-throats dilate and their interconnectivity grows markedly, greatly facilitating fluid flow.

Hydraulic conductivity varies less sensitively with effective pressures above $c$. 100 kPa, allowing samples to be characterized by a single hydraulic conductivity value ($K_0$ in Eqn 1). The values of $K_0$ for all 7 samples analysed in this study fall within the published range of subglacial sediment hydraulic conductivities determined in the laboratory. However, variations in sample material properties induce three orders of magnitude variability in inter-sample hydraulic conductivity. While effective grain size ($d_{10}$) exerts some control over hydraulic conductivity, the considerable scatter that characterises this relationship (Fig. 4) suggests the operation of additional controlling variables. Preliminary investigations reveal that alternative grain-size parameters such as $d_{50}$ and $d_{90}$ do not improve on the explanation provided by $d_{10}$. The most likely explanation for this variability in $K_0$ therefore lies in variations in sample microstructure and fabric. These material characteristics, however, were not recorded in the present study.

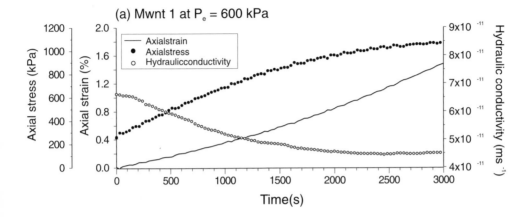

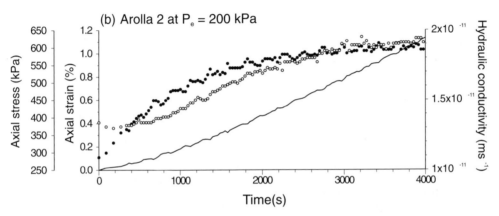

**Fig. 6.** Axial strain, axial stress and dynamic hydraulic conductivity time series during two example deformation tests: (**a**) sample Mwnt 1 deformed at an effective pressure of 600 kPa, and (**b**) sample Arolla 2 deformed at an effective pressure of 200 kPa. In the case of (**a**), recorded peak axial stress is $c$. 1000 kPa, and the change in hydraulic conductivity over the deformation test is $c. -2 \times 10^{-11}$ m s$^{-1}$. This represents a net reduction in sample hydraulic conductivity for this test of $c$. 30% (Table 4; Fig. 5). In the case of (**b**), recorded peak axial stress is $c$. 600 kPa, and the change in hydraulic conductivity over the deformation test is $c. 6 \times 10^{-12}$ m s$^{-1}$. This represents a net increase in sample hydraulic conductivity for this test of $c$. 46% (Table 4; Fig. 5).

In contrast to (static) hydraulic conductivity, dynamic hydraulic conductivity did not respond in a predictable or uniform manner during testing. Dynamic hydraulic conductivity decreased by up to 100% in the majority of tests, although increases, of between 10 and (exceptionally) 300% were also measured. These results do not therefore provide support for the hypothesis that the hydraulic conductivity of actively deforming subglacial sediment is significantly greater than that of non-deforming sediment. However, our experimental apparatus deforms samples only to relatively low total strains ($c$. 10%), resulting in minimal grain realignment (e.g. Murray & Dowdeswell 1992).

It is therefore possible that significant grain alignment at higher total strains (e.g. Benn 1995; Dewhurst *et al.* 1996; Iverson *et al.* 1998, Hooyer & Iverson in press) could increase hydraulic conductivity through reducing the tortuosity of individual flow pathways. It may therefore be relevant that dynamic permeability increases appeared to be preferentially coincident with those tests where deformation was achieved by failure along multiple, inclined shear surfaces rather than by ductile barrelling.

The role of effective pressure may be even more crucial in deformed sediments, especially where discrete, through-going shear zones have been produced. Such shear zones may undergo

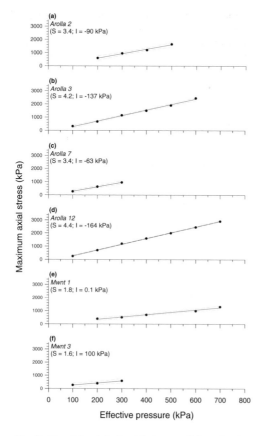

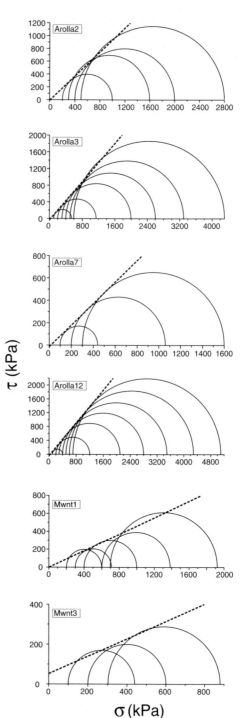

**Fig. 7.** Bivariate plot of axial stress at failure (peak strength) against effective pressure for each sample. LSL regression lines (with 95% confidence bounds) and statistics are presented ($S$ = slope, $I$ = intercept).

the dilating effect noted above for pores in static tests, providing long, low tortuosity pathways that will allow rapid dewatering. The phenomenon has been termed *fracture permeability* in ocean-floor sediment studies (e.g. Brown *et al.* 1994), and laboratory tests have confirmed its effectiveness below a certain threshold of effective pressure (Bolton *et al.* 1998). It is possible that those static tests summarized above that showed the most non-linear response of hydraulic conductivity to low effective pressures may have involved samples already containing natural shear zones and were exhibiting just this behaviour. Moreover, Bolton *et al.* (1999) argued that the interaction of shear zones and fluid flow is likely to be cyclic. Preferred flow along dilated shear zones will efficiently drain the nearby material prompting pore collapse; as fluid is now trapped, pore pressures again

**Fig. 8.** Sample strength represented in terms of Mohr's circles (see text for explanation).

increase until the threshold for fracture permeability is again reached.

Our results further indicate that the peak strength, $s$, of the samples increases linearly with effective pressure ($P_e$), as expected on the basis of the classic Mohr–Coulomb failure relation:

$$s = c' + P_e \tan \varphi' \qquad (3)$$

where $c'$ is the sample cohesion and $\varphi'$ the angle of internal friction at that effective pressure (Fig. 7). Our Mohr's circles diagrams of shear strength (maximum sustained shear stress, $\tau$) against $P_e$ (Fig. 8) reveal negligible cohesion ($c'$) (with the exception of sample Mwnt 3) and values of $\varphi'$ that vary from 24° (Mwnt 3) to 49° (Arolla 12). These values are consistent with those measured for subglacial sediments elsewhere (Paterson 1994, p. 169). However, intersample variability in strength is not easily explained simply by the grain-size data reported here. Clearly, additional factors are operating and, judging by work on ocean-floor sediments, these may include diagenetic changes, microfabric characteristics (e.g. Dewhurst et al. 1996), fabric anisotropy (e.g. Arch & Maltman 1990; Brown & Moore 1993) and, perhaps most importantly, antecedent stress history (e.g. Bolton et al. 1998). Since consolidation generally increases sediment strength by bringing particles closer together (e.g. Maltman 1994), varying degrees of previous consolidation will therefore at least partially account for differences in present sample strength, especially where overconsolidation has occurred. Samples that are overconsolidated not only tend to show anomalously high strengths, but can also display markedly non-linear fluid-flow behaviours (Bolton & Maltman 1998).

## Conclusions

Fifty eight static hydraulic conductivity tests, 28 deformation tests and 25 dynamic hydraulic conductivity tests have been undertaken on seven samples of glacial sediments recovered from the margins of Haut Glacier d'Arolla, Switzerland, and from Traeth y Mwnt, Wales. A number of conclusions may be drawn on the basis of these tests.

Static hydraulic conductivity scales inversely with effective pressure, particularly at effective pressures lower than $c.\,100\,\text{kPa}$. This relationship is best described over the full range of effective pressures imposed by a negative power law (with exponents of between $-2$ and $-10$) above a baseline hydraulic conductivity that may be considered as a characteristic material property at high effective pressures.

Inter-sample static hydraulic conductivity varies between $10^{-8}$ and $10^{-11}\,\text{m}\,\text{s}^{-1}$ at high effective pressures. This hydraulic conductivity scales directly, but weakly, with the square of the effective grain size of the samples tested.

In most cases, dynamic hydraulic conductivity decreased during deformation by macroscopic ductile barrelling. In the small number of cases where dynamic hydraulic conductivity increased during deformation, sample failure appeared to be achieved by more brittle shearing.

Dynamic testing results in strongly linear relationships between effective pressure and stress at failure. These relationships are consistent with a Mohr–Coulomb type failure model, revealing significant inter-sample variability in shear strength, largely due to frictional resistance.

We gratefully acknowledge the assistance of B. Anderson, D. McCarroll, M. Peters, and K. Rijsdijk; U.H. Fischer and D. Van der Wateren are thanked for helpful reviews. This research was funded by the University of Wales Academic Support Fund and the UK Natural Environment Research Council (GR9/02530).

## References

Arch, J. & Maltman, A. J. 1990. Anisotropic permeability and tortuosity in deformed wet sediments. *Journal of Geophysical Research*, **95**, 9035–9046.

Benn, D. 1995. Fabric signature of subglacial till deformation, Breidamerkurjökull, Iceland. *Sedimentology*, **42**, 735–747.

Bolton, A. & Maltman, A. J. 1998. Fluid-flow pathways in actively deforming sediments: the role of pore fluid pressures and volume change. *Marine and Petroleum Geology*, **15**, 281–297.

——, Clennell, M. B. & Maltman, A. J. 1999. Nonlinear stress dependence of permeability: a mechanism for episodic fluid flow in accretionary wedges. *Geology*, **27**, 239–242.

——, Maltman, A. J. & Clennell, M. B. 1998. The importance of timing in the consolidation – permeability evolution of deforming clayey sediments. *Journal of Structural Geology*, **20**, 1013–1022.

Boulton, G. S. & Hindmarsh, R. C. A. 1987. Sediment deformation beneath glaciers: rheology and geological consequences. *Journal of Geophysical Research*, **92** (B9), 9059–9082.

Brown, K. M. & Moore, J. C. 1993. Comment on "Anisotropic permeability and tortuosity in deformed wet sediments" by J. Arch & A. J. Maltman. Deformation structures and fluid flow in the toe region of the Nankai accretionary prism. *Journal of Geophysical Research*, **98**, 17 859–17 864.

——, Bekins, B., Clennell, M. B., Dewhurst, D. & Westbrook, G. 1994. Heterogeneous hydrofracture development and accretionary prism dynamics. *Geology*, **22**, 259–262.

BYRNE, T., MALTMAN, A. J., STEPHENSON, E., SOH, W. & KNIPE, R. 1993. Deformation structures and fluid flow in the toe region of the Nankai accretionary prism. *Proceedings of the Ocean Drilling Program, Scientific Results*, **131**, 83–101.

DEWHURST, D., CLENNELL, M. B., BROWN, K. M. & WESTBROOK, G. 1996. Fabric and hydraulic conductivity of sheared clays. *Géotechnique*, **46**, 761–768.

FISCHER, U. H. & HUBBARD, B. 1999. Subglacial sediment textures: character and evolution at Haut Glacier d'Arolla, Switzerland. *Annals of Glaciology*, **28**, 241–246.

——, CLARKE, G. K. C. & BLATTER, H. 1999. Evidence for temporally varying 'sticky spots' at the base of Trapridge Glacier, Yukon Territory, Canada. *Journal of Glaciology*, **45**, 352–360.

——, IVERSON, N. R., HANSON, B., HOOKE, R. LeB. & JANSSON, P. 1998. Estimation of hydraulic properties of subglacial till from ploughmeter measurements. *Journal of Glaciology*, **44**, 517–522.

FOUNTAIN, A. G. 1992. Subglacial water flow inferred from stream measurements at South Cascade Glacier, Washington, USA *Journal of Glaciology*, **38**, 51–64.

—— & WALDER, J. S. 1998. Water flow through temperate glaciers. *Reviews of Geophysics*, **36**, 299–328.

FREEZE, R. A. & CHERRY, J. A. 1979. *Groundwater*. Prentice-Hall, Englewood Cliffs, N.J.

HAMBREY, M. 1994. *Glacial Environments*. UCL Press, London.

HARBOR, J., SHARP, M., COPLAND, L., HUBBARD, B., NIENOW, P. & MAIR, D. 1997. Influence of subglacial drainage conditions on the velocity distribution within a glacier cross section. *Geology*, **25**, 739–742.

HARPER, J. T. & HUMPHREY, N. F. 1995. Borehole video analysis of a temperate glacier's englacial and subglacial structure: implications for glacier flow models. *Geology*, **23**, 901–904.

HOOKE, R. LeB., HANSON, B., IVERSON, N. R., JANSSON, P. & FISCHER, U. H. 1997. Rheology of till beneath Storglaciären, Sweden. *Journal of Glaciology*, **43**, 172–179.

HOOYER, T. S. & IVERSON, N. R. Clast-fabric development in a shearing granular material: implications for subglacial till and fault gouge. *Geological Society of America Bulletin*, in press.

HUBBARD, B., SHARP, M., WILLIS, I. C., NIELSEN, M. & SMART, C. C. 1995. Borehole water-level variations and the structure of the subglacial hydrological system of Haut Glacier d'Arolla, Switzerland, *Journal of Glaciology*, **41**, 572–583.

IVERSON, N. R., HANSON, B., HOOKE, R. LeB. & JANSSON, P. 1995. Flow mechanism of glaciers on soft beds. *Science*, **267**, 80–81.

——, HOOYER, T. S. & BAKER, R. W. 1998. Ring-shear studies of till deformation: Coulomb-plastic behavior and distributed strain in glacier beds. *Journal of Glaciology*, **44**, 634–642.

MALTMAN, A. J. 1987. Shear zones in argillaceous sediments – an experimental study. *In*: JONES, M. E. & PRESTON, R. M. F. (eds) *Deformation of Sediments and Sedimentary Rocks*. Geological Society, London, Special Publications, **29**, 77–87.

—— (ed.) 1994. T*he Geological Deformation of Sediments*. Chapman & Hall, London.

MURRAY, T & DOWDESWELL, J. A. 1992. Water throughflow and the physical effects of deformation on sedimentary glacier beds. *Journal of Geophysical Research*, **97** (B6), 8993–9002.

PATERSON, W. S. B. 1994. *The Physics of Glaciers* (3rd Edition). Pergamon Press.

STEPHENSON, E. L., MALTMAN, A. J. & KNIPE, R. J. 1994. Fluid flow in actively deforming sediments: 'dynamic permeability' in accretionary prisms. *In*: PARNELL, J. (ed.) *Geofluids: Origin, Migration and Evolution of Fluids in Sedimentary Basins*. Geological Society, London, Special Publications, **78**, 113–125.

# Subglacial Deformation

(Overleaf) A stacked sequence of reverse thrusts formed within proglacial sediments exposed by stream-cut section, Axel Heiberg Island, Canadian Arctic. (Photo: B. Hubbard).

# Micromorphological analyses of microfabrics and microstructures indicative of deformation processes in glacial sediments

JOHN MENZIES

*Departments of Earth Sciences & Geography, Brock University, St Catharines, Ontario, L2S 3A1, Canada (e-mail: jmenzies@brocku.ca)*

**Abstract**: Various forms of sediment deformation can be detected within glacial sediments at the microscopic scale. Analyses of these forms leads to a preliminary classification of microfabrics and microstructures of brittle, ductile and polyphase modes of deformation in glacial sediments. With the development of a taxonomy of different microfabrics and microstructures these processes, once differentiated, permit insights into glacial sediments to a scale and level of detail hitherto unknown. Examples are presented that illustrate some of these different forms of deformation often within a single glacial sediment sample. This research suggests many of the past ideas with regard to details of glacial depositional processes, especially in terms of diamicton deposition and subsequent classifications, need to be re-evaluated.

In both macroscopic and microscopic investigations of glacial sediments, it is readily apparent that these sediments typically contain a very large number of different structures, artefacts and sedimentary forms that are directly related to processes which are pre-depositional, emplacement, depositional or post-depositional in origin. Sedimentological characteristics can be often differentiated in terms of inherited, actual *in situ*, or post-depositional diagenetic forms. However, as yet, the significance and relevance of individual structures and artefacts often remain vague or indeterminate (see Rappol 1987; Piotrowski 1991; Dreimanis 1993). For example, in studying glacial sediments at the microscale, investigations remain relatively limited and our understanding of the many microstructures observed in these sediments continues to be comparatively rudimentary; only a few studies have examined glacial sediments at the micro-level. However, increasing awareness of the value of micromorphological analyses in understanding the mechanics of glacial deposition has been demonstrated in several recent studies (e.g. van der Meer *et al.* 1983; van der Meer 1987, 1993; Menzies & Maltman 1992; Bordonau 1993; Menzies & Woodward 1993; Carr 1999; Menzies, 2000).

In the examination of individual glacial diamicton lithofacies, for example, the search continues for diagnostic criteria that would allow more precise differentiation of individual facies types, namely the ability to distinguish subglacial from subaqueous till, or subglacial from proglacial till; as well as lodgement from melt-out tills, or waterlain from flow tills (Dreimanis 1988; van der Meer 1993; Carr 1999). It is in this scientific pursuit that micromorphological analysis has a key role to play. A perpetual problem is that of equifinality since many different sets of processes may lead to sediments that are remarkably similar in many aspects making differentiation arduous. However, micromorphological analyses provides a means for the reconstruction of transport and depositional processes as indicated by the presence, disposition, size, and interrelationships between individual microstructures in these sediments.

It is the objective of this paper to develop a preliminary taxonomy of many of the microfabrics and microstructures observed in examination of glacial sediments, within the context of microstructures in glacial sediment transport, deposition and post-depositional influences (see van der Meer 1993).

## Microfabric and microstructure development within glacial sediments

Any structure or fabric, whether macro or micro in scale, is diagnostic of the depositional and/or post-depositional process(es) and ultimately of a particular depositional environment (Allen 1982). As a diagnostic tool, microstructures and plasmic microfabrics (see glossary) are especially valuable within glacial sedimentology. In general, sediments being deposited at the base

*From*: MALTMAN, A. J., HUBBARD, B. & HAMBREY, M. J. (eds) *Deformation of Glacial Materials*. Geological Society, London, Special Publications, **176**, 245–257. 0305-8719/00/$15.00 © The Geological Society of London 2000.

of an ice sheet, or below a mobile sediment layer sandwiched between the immobile glacier bed and the moving ice mass are subject to large shear stresses. Associated with such stresses are significant fluctuations in porewater content as a function of meltwater production, varying thermal conditions and wide variations in the rheology of the sediments (Alley *et al.* 1986, 1987; Boulton 1987, 1996; Boulton & Hindmarsh 1987; Menzies 1987; Hart 1994). The conditions at the major discontinuity at the boundary between ice and sediment, or mobile and immobile sediment are, therefore, conducive toward the development of a wide range of microstructures and microfabrics spawned from the generating surfaces in this zone of discontinuity. Where clay content, consolidation levels, and porewater conditions permit, ductile flow conditions can be inferred, but where lower porewater, or clay content, or diminished levels of stresses occur then brittle fracture events may take place within comparatively very limited areas within the larger zone of discontinuity or décollement (see Maltman 1994; Murray 1994). Where in the past an all pervasive form of deformation has been favoured it is apparent from thin section examination that localized non-pervasive deformation occurs within this particular subglacial environment as a function of the varying sediment rheology that is typical of glacial diamictons (see Menzies & Maltman 1992; van der Meer 1993, 1996). The rheology of the sediment is subject to continual fluctuation such that rheological conditions may transmute from ductile to brittle and back again thus permitting polyphase deformation to take place.

## Deformation structures

Examination of glacial sediments demonstrates that deformation structures found in most of these sediments can be subdivided into structures indicative of brittle, ductile and polyphase modes of deformation (van der Meer 1993, 1996; Maltman 1994; Murray 1994; Passchier & Trouw 1996; Menzies 1998, 2000).

Deformation within a sediment, following Maltman (1994, p. 1), can be defined as a change in bulk shape principally in a particulate or granular manner, by frictional grain-boundary sliding. Since particles are likely the strongest part of any aggregate, deformation is dominantly the result of 'independent particulate flow' (Borradaile 1981). Deformation occurs within glacial sediments when a set of factors that include grain size distribution, porewater content, local stratigraphy, thermal conditions

of the sediment under stress, the level of stress, and the rate and length of stress application interact such that the critical stress level within the sediment is surpassed permitting a reorganization of grains one to another (Maltman 1994). Deformation within glacial environments is dynamic; the changing frequency and magnitude of factors noted above result in stress conditions that are ever changing, at times recurrent, and never constant. Therefore, as sediment undergoes deformation within a glacial environment, virtually no single factor can be regarded as 'fixed' and, as a consequence, the variability, size, number and very type of microstructures developed during, and after phases of deformation are enormous. In simple terms, as shown in Fig. 1, failure within a sediment will pass from a brittle to a ductile form with a range of variations between (Passchier & Trouw 1996).

Brittle deformation occurs when either stress applied to a sediment is so rapid and/or sufficiently large that the sediment particulates and matrix cannot adjust in sufficient time to the applied stress and therefore fractures in the sediment occur (Twiss & Moore 1992).

Ductile deformation takes place typically either under low stress application permitting particle and matrix adjustment in the form of flow and sliding, or where stress levels are applied within a high confining stress environment permitting flow to occur. The usage of the term ductile is, as pointed out by Rutter (1986) and (Twiss & Moore 1992), fraught with considerable controversy. In this paper the term is used to describe a spectrum of flow conditions that range from cataclastic to granular flow (see below).

Polyphase deformation is typical of most glacial sediments since both brittle and ductile phases of deformation may be interposed upon each other during the development and emplacement of these sediments. In many instances glacial sediments appear to transmute between brittle and ductile deformation behaviour such that both deformation forms can occur very close to one another both in time and space within any small part of the actively deforming sediment.

A final category that is often difficult to temporally identify are inherited or remnant microstructures and microfabrics that persist either since post-formational stress levels never reach the previous formative conditions or destruction of these features was never totally accomplished.

In the analyses of these sediments it must be clearly pointed out that while ductile deformation behaviour may be manifest in the macroscale, at the microscale brittle deformation may occur and *vice versa* (Williams *et al.* 1994).

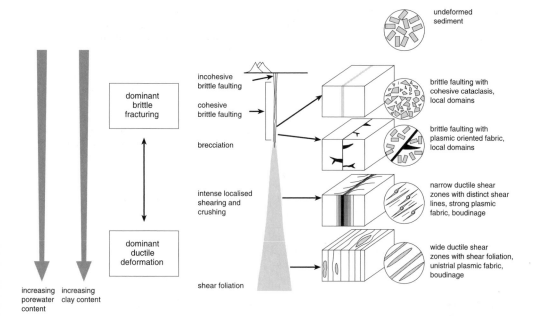

**Fig. 1.** Ductile to brittle deformation transition within sediment under stress as a function of effective stress levels, clay content and porewater content (modified after Passchier & Trouw 1996).

In discussing microstructure and microfabric types the reader is directed to Brewer (1976) and van der Meer (1993, 1996).

## A preliminary taxonomy of microstructures in glacial sediments

As stated above, the processes of glacial sediment emplacement, deposition and post-depositional diagenesis are exceedingly complex. Here an attempt is made to place the differing microstructures within the context of the style of deformation they have undergone. Likewise plasmic microfabrics indicate certain sedimentological conditions and formative environments. On this basis the presence of specific microstructures and plasmic microfabrics, in association, may be useful in a preliminary assessment of glacial lithofacies types apart from other macro-scale attributes. Depending on the nature of the deformation conditions during and immediately following deposition, specific microstructures and microfabrics are probably more likely to be found in certain glacial lithofacies types more than in others (Menzies 1998). It must be clearly stated at the outset that Fig. 2 is a preliminary taxonomic diagram that can be expected to change and enlarge as further investigations

of this nature and more glacial sediments are investigated using micromorphology.

In describing soils and sediments in thin-section, Brewer (1976), following Kubiena (1938) and Shrock (1948), distinguished between the general ground mass or matrix, and particles or skeleton grains. Skeleton grains are essentially any lithic fragment greater in thickness than the thin-section itself (approx >25–30 $\mu$m). Plasma, in contrast, are typically clay-sized particles >25–30 $\mu$m in thickness. Typically, under cross-polarized light, plasma will often exhibit distinct birefringence, indicative of specific preferred orientations of clay-sized particles, and therefore is said to possess a microfabric.

Microstructures, in contrast, can be seen as a combination of plasma, skeleton grains and voids in some constituent arrangement. This combination of skeleton grains, voids and plasma is termed the S-matrix (Brewer 1976). Microstructures are formed, therefore, within the S-matrix. In introducing the above terminology considerable thought has been given to finding equivalencies in geology but since this work deals with soft sediments and unlithified rocks no other satisfactory terms appear to exist.

*Plasmic microfabrics.* Within thin-sections distinct areas of clay-sized material can, sometimes if present, be recognized based upon certain unique characteristics associated with clay particle orientations. These areas or 'domains' can be seen with rotation of the petrological microscope. Where clay particles have 'settled' as a result of depositional and post-depositional mechanisms the clay particles can be arranged in a variety of arrangements (fabric) that often indicate a spectrum of shear and normal stress applications and/or zero stress , as well as many other geological, and/or pedological agencies. However, within glacial sediments, below the level of pedogenic processes, the range of plasma fabrics appears to be, at this time, generally restricted to those illustrated in Fig. 2. Definitions of each of the plasma fabrics are given at the end of this paper; however the significance and interpretative value of each microfabric needs elucidation. In general, where single or multiple preferred orientated fabrics exist they can be interpreted as indicative of an externally applied stress. The dominance of one or several directions implies possibly one or several stress applications the result of the plasma domain having been affected by stresses of sufficient magnitude to result in clay particle orientation. In order to effect such orientation the plasma may have been saturated and under some form of hydrostatic pressure system or systems. The stresses involved can be by direct or indirect application the former due to overlying ice mass or mobile soft sediment drag, while the latter most likely through porewater diffusion under a stress field. Of course other sources of stress within complex glacial environments are myriad in number but the above stress applications are the likely dominant ones in plasma microfabric development (Carr & Rose 1997; Carr 1999). The strength of plasma microfabrics vary enormously from very strong preferred orientations as in unistrial microfabrics to much weaker almost imperceptible fabrics in other microfabric sets. Insepic plasma microfabrics being recognized as clusters of plasma, and therefore separate plasma domains, are indicative of sediment rafting or fragmentation of plasma units occurring during transport and prior to deposition. Likewise omnisepic fabrics indicate a generally random set of orientations suggestive that no significant external stress application has occurred or that deformation stresses have been so high as to completely destroy any existing microfabric. Where banded plasma fabrics occur evidence points to strong intercalation processes being involved. A similar scenario for

development can be suggested for kinking in plasmic fabrics. Finally, only in the case of skelsepic fabrics does a combination of both plasma and S-matrix lead to a special form of microfabric where distinct encapsulation of individual skeleton grains by plasma can be observed. The cause of such a fabric remains under some debate but appears indicative of stress application leading to either dilatancy associated with porewater induced translocation of clays or possibly rotation of grains within the plasma while the sediment package is under stress.

## S-matrix microstructures

In examining thin-sections for microstructures it can be shown that all of these artefacts are found within the S-matrix. In classifying microstructures within glacial sediments a division based upon the style of deformation and the overarching effects of porewater on sediment structures can be made. Microstructures can be differentiated into those structures that are indicative of dominantly brittle, ductile, and polyphase (brittle/ductile) styles of sediment deformation. A further category of porewater-influenced or -induced structures can also be readily identified. The categorization is not definitive since several structures combine elements of both ductile and brittle deformation but with a dominance of one over the other. For example, shear zones are recognized as inherently brittle yet contain within the discrete shear zone ductile deformed sediments.

The style of deformation within glacial sediments as discussed by van der Meer (1987, 1993), Menzies & Maltman (1992), Murray (1994) and Carr (1999), for example, suggests that under complex and varying rheological conditions glacial sediments may react in either a brittle or ductile fashion. Under brittle conditions induced by localised sediment dewatering, or freezing, or clay winnowing, related possibly to rapid fluctuations in confining pressures and relatively high effective pressures, sediments fracture and disaggerate along fracture planes or zones of weakness.

In contrast, ductile deformation takes place usually under high confining pressures often with associated high porewater pressures with effective pressures approaching zero. The term ductile is one in geology wherein lies considerable variation, confusion and misunderstanding (see

**Fig. 2.** A preliminary taxonomy of microfabrics and microstructures within glacial sediments (modified after van der Meer 1993).

Rutter 1986; Twiss & Moore 1992; Maltman 1994). One of the problems inherent in defining ductile deformation within soft, unlithified sediments is that comparsions with crystalline rocks, and the generally accepted terminology in structural geology invite conflicting opinions as to meaning. In structural geological terms, ductile deformation in this paper can be compared with a spectrum of flow conditions that extends from cataclastic to granular flow; both states of deformation tend to be close to the ductile–brittle transition. Undoubtably, in soft sediment of this nature within glacial environments, the nature of failure and deformation, as indicated by the presence in Fig. 2 of many polyphase structures, substantiates this more particular definition of ductile deformation. Cataclastic flow is a process of deformation and allied comintion of particles that involves progressive brittle fracturing of grains while in motion under relatively high confining pressures. At the other end of the spectrum, granular or particulate flow involves only sliding and rotation of particles past one another under relatively much lower confining pressures. It seems most likely that, except under conditions of porewater drainage or rapid localized sediment freezing where a reduction in liquid water would occur, most instances of glacial sediment deformation are of the type generally classified as granular flow.

*Brittle deformation microstructures.* Evidence of brittle failure within glacial sediments consists of angular fragmentation, typically angular clasts, angular plasma domains, brecciation, short-distance fracturs, random plasmic fabric and faulting. Brittle microstructures (Fig. 2) consist, for example, of edge to edge grains, with crushed contacts, discrete shear lines, various faults within the S-matrix, faulted plasma domains, kink band arrays, and discrete areas of highly fractured S-matrix.

*Ductile deformation microstructures.* In contrast, ductile microstructures (Fig. 2) can be distinguished as microstructures that exhibit, for example, fold structures, foliation structures, rotational elements, strain caps and shadows, banded or intercalated units of differing plasma, necking structures, and various types of layering, plasma flowage around clasts, various forms of foliation, and boudinage.

*Polyphase deformation microstructure sets.* More commonly glacial sediments contain combinations of juxtaposed brittle and ductile deformation microstructures (Fig. 2). It is not uncommon to find intense intercalation, interbedding and various geometric layering indicative of multiple deformation events – the consequence of polyphase deformation episodes.

*Porewater-influenced and -induced microstructure sets.* As glacial sediment deform under various confining pressures usually associated with fluctuating porewater contents and, therefore, porewater pressures, a range of microstructures peculiar to porewater movement can also be detected. Porewater in many instances moves in front of an advancing zone of high stress or in the case of an advancing freezing front may move towards or away from that front (Menzies 1981). In all cases the effect of porewater motion is to transport clays, silts and solutes through the S-matrix. Closely linked to porewater movement are a number of structures related to water escape, bedding unit fragmentation and disruption.

### Glacial sediment examples

In all the examples discussed below it should be borne in mind that an array of microstructures are typically found thus no one structure can ever be used as directly indicative of a particular stress environment, or set of glacial or non-glacial processes, or particular lithofacies environment. Instead it is rather the dominant associations of microstructures and microfabrics that lend support for a specific glacial environment.

The examples used here are purely illustrative rather than exhaustive of the many microfabrics, microstructures and porewater-related structures that can be observed in glacial sediment thin-sections. Examples for this paper have been drawn from an extensive collection of diamicton and other glacigenic and non-glacigenic sediment thin-sections.

In Fig. 3 two examples of subaqueous sediment are presented that reveal typical characteristics for diamicton lithofacies types. These sediments are taken from cores off the SE coast of Nova Scotia, Canada within Wisconsinan glacial sediments deposited close to the maximum extension of the Laurentide Ice Sheet in eastern Canada (Piper *et al.* 1990). The sediment in Fig. 4a shows a typical diamicton with angular and subangular clasts within a fine-grained matrix. Those clasts, in general, are of two dominant size classes ($\leq 0.5$ mm and $\geq 1$ mm). Clasts at the $\geq 1$ mm size exhibit some preferred orientation with distinct lineation zones in a few places. Smaller clasts can be seen 'flowing' around larger clasts such as around the large clast centre left in the photomicrograph.

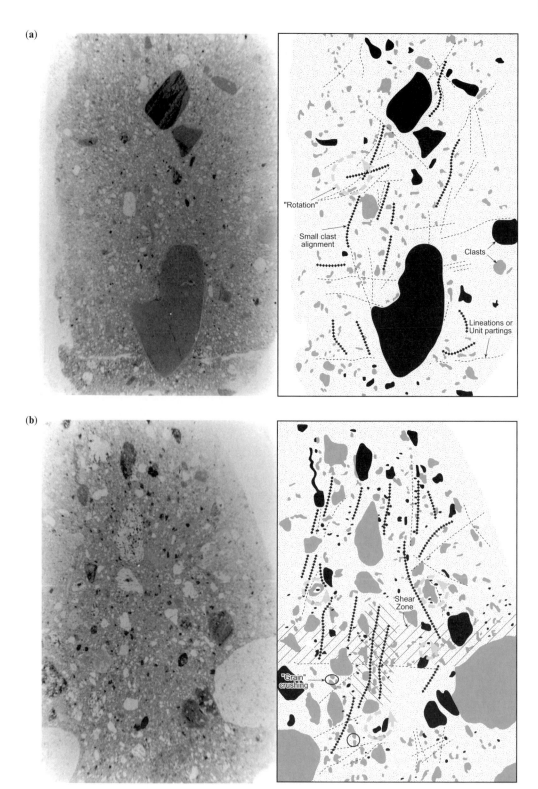

**Fig. 3.** Micrographs and associated sketches of glaciomarine diamictons from the Scotian Shelf, eastern Canada (thin-sections are approximately $40 \times 70$ mm). In (**a**) lineations or unit partings refer to fractures, joints or shear lineations, the latter normal to the plane of the thin-section.

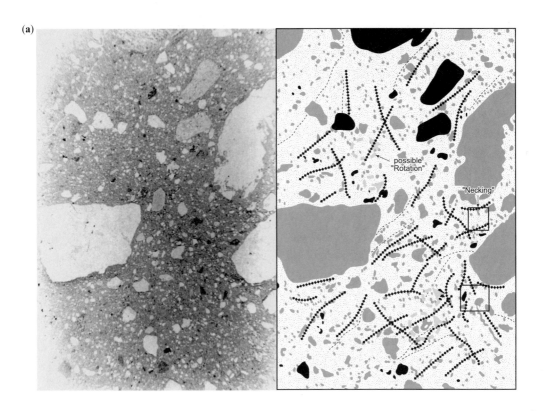

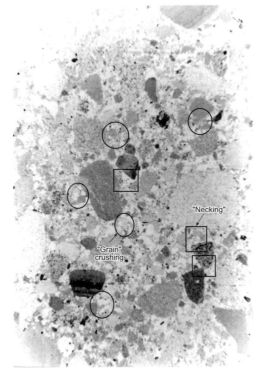

**Fig. 4.** (a) Micrograph and associated sketches of subglacial diamicton from Mohawk Bay, southern Ontario, Canada and (b) a micrograph from Chimney Bluffs, New York State, USA both sampled from streamlined subglacial bedforms (thin-sections are approximately 40 × 70 mm).

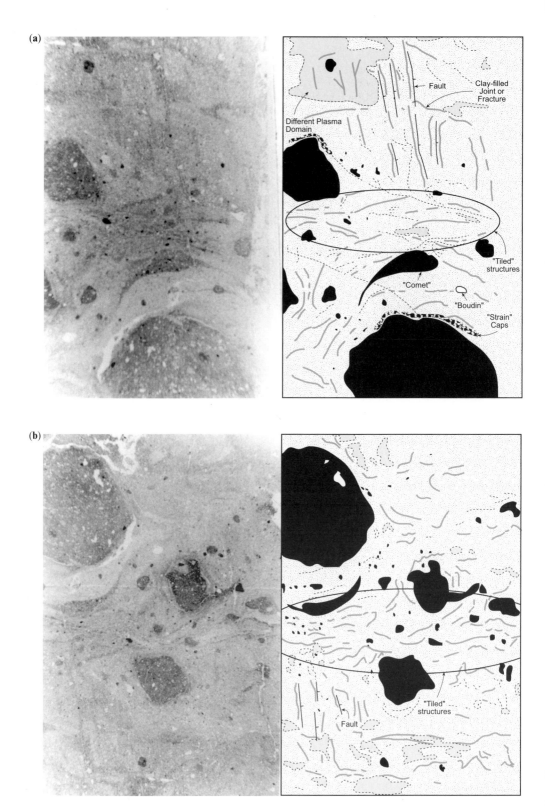

**Fig. 5.** Debris flow micrographs and associated sketches from Mohawk Bay, southern ontario, Canada (thin sections are approximately 40 × 70 mm).

The photomicrograph illustrated in Fig. 3b exhibits a greater degree of small clast fabric lineations. Across the centre of the figure from left to right is probably a diffuse shear zone. A secondary lineation occurs at approximately 80° to the main shear zone. Further shallow shear lines occur at around 22–25° possibly remnants of Riedel shearing. Unlike Fig. 3a, this diamicton exhibits clearer evidence of internal deformation.

Figure 4a is an example of sediment sampled from a large mega-flute at Mohawk Bay, southern Ontario (Menzies 1990), while Fig. 4b is from an exposed coastal bluffs of a prominent drumlin at Chimney Bluffs, east of Rochester, New York State (Menzies *et al.* 1997). Both sediments are of Late Wisconsinan age, both from within streamlined subglacial bedforms. The subglacial diamicton illustrated in Fig. 4a shows multiple lineations around large clasts, with small clast alignments and ground mass (plasma) shear lineation. Small clast flow between large clasts, as at the left centre, indicates 'necking'. Possible 'rotation' structures can be seen above the large clast on the left side of the photomicrograph. The diamicton appears to exhibit a 'rhombosepic' fabric typical of one dominant and a later secondary shear deformations. In Fig. 4b a coarser-grained diamicton can be seen that exhibits edge to edge grain crushing, 'necking' structures and clast orientation (see Fig. 7b).

In contrast, but originating from the same sediment as shown in Fig. 4a, are two examples of modern debris flow deposits. The sediment illustrated in Fig. 5a exhibits characteristics distinctly different from the microstructures within the original sediment (Fig. 4a). It is assumed that these microstructures are the result of the original diamicton having 'flowed' as a debris flow. The major distinguishing microstructures are 'tiled' units that appear to represent plasma flow and deceleration as the flow subsided and slid to a halt (see Martinsen 1994). These structures can be seen above the large clast at the bottom of the photomicrograph where they present bands transverse to the direction of the debris flow. These structures have only been recognized within debris flows (see Bertran *et al.* 1995; Bertran & Texier 1999) and not, at present, in diamictons. Strain caps can be seen above large clasts, and a small 'comet structure' occurs in the centre right of the photomicrograph. Within multiple domains, in the upper half of the photomicrograph, both normal and wrench faulting can be noted. In Fig. 5b, similar in many respects to Fig. 5a, flow structures can be observed in the centre of the photomicrograph. Multilple domains in themselves are indicative of intense intercalation due to discrete non-pervasive deformation (van der Meer 1996; Menzies 1998).

South of the Highland Boundary Fault in eastern Scotland extensive Late Devensian (Wisconsinan) outwash fans intrude into the valley of Strathmore, in Perthshire and Angus. At the site (Meikleour) the samples illustrated in Fig. 6 were taken from what has been suggested as a location where neo-seismic activity has deformed the sediment pile (Davenport & Ringrose 1987). Both Figs 6a and 6b illustrate interesting and extensive folding at a scale and degree hitherto unobserved. The site from which the samples were taken lies close to the River Tay in a high terrace above the river. No evidence exists of post-depositional overburden and subsequent removal nor of later ice movements. The site is laterally stable and therefore it must be concluded that the microstructures observed are either penecontemporaneous with primary deposition of the outwash sediments, or are post-depositional.

Figures 7a and 7b are illustrative of the longevity and stability of many of these microstructures found within glacigenic sediments. The examples shown here are derived from the Gowganda diamictites of Precambrian, Huronian age sampled some 10 km north of Elliot Lake, Ontario (Young & Nesbitt 1985; Menzies 2000). These diamictites exhibit essentially the same micromorphic characteristics as do more modern Quaternary terrestrial and marine diamictons as illustrated in Figs 3a, b and 4a, b.

## Summary

In conclusion, it is apparent that in developing a taxonomy of microfabrics and microstructures from which differentiation of glacial environments can be attempted that many persistent problems remain in terms of microfabric recognition and interpretation, and microstructure identification, their significance and mechanics of formation, the systemic concern of equifinality and the, as yet, relatively unexplained aspects of diagenesis within many glacial sediments.

The manifestation of plasmic microfabrics inherently indicates the presence of clays and the impact of stress on both the plasma domain and the S-matrix. Since individual microfabrics give some indication of the nature and magnitude of the stresses involved, then such microfabrics are intuitively diagnostic. With recent advances in image analysis the strength of these microfabrics can be evaluated and quantified, thus

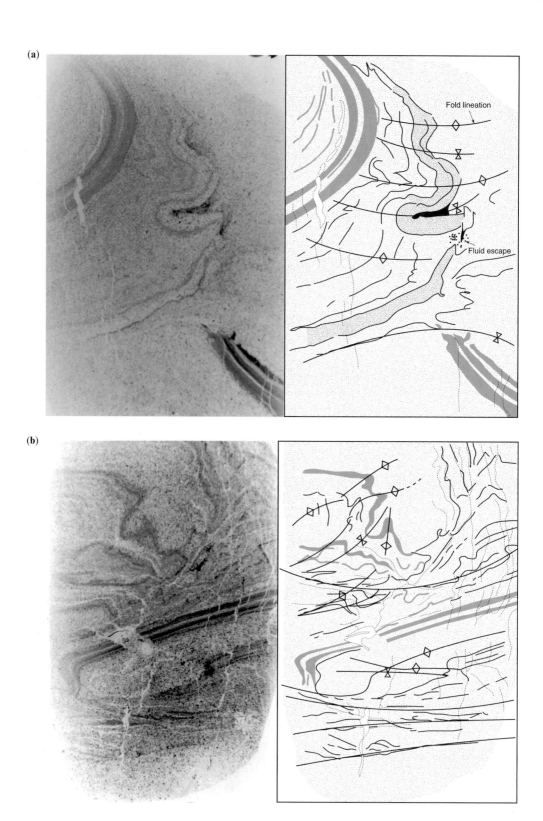

(a)

Fold lineation

Fluid escape

(b)

**Fig 6.** Micrographs and associated sketches of glacial debris, possibly neo-seismically deformed, from Meikleour, Scotland (thin-sections are approximately 40 × 70 mm).

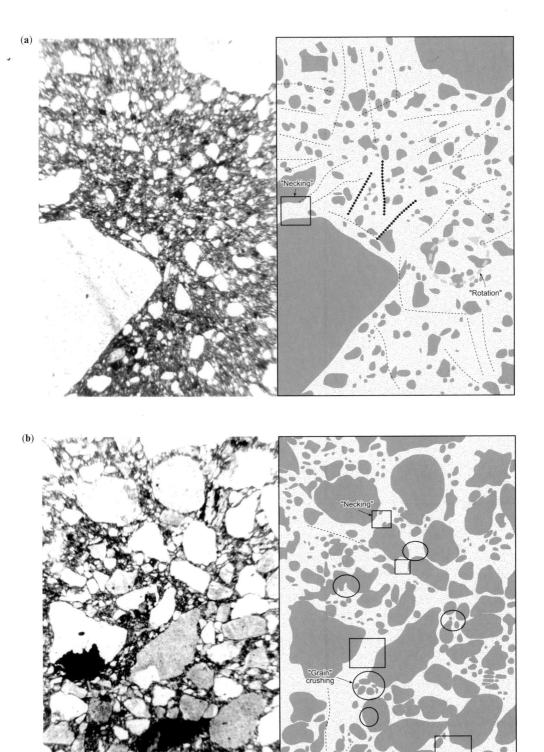

**Fig. 7.** Micrographs and associated sketches of Precambrian diamictites from near Elliot Lake, northern Ontario, Canada (thin-sections are approximately $40 \times 70$ mm).

their value as a diagnostic tool is greatly heightened (Zaniewski pers. comm.)

The microstructures discussed within this paper can be viewed as common or persistent in the sense that they are typical almost endemic to most glacial environments and thus glacial sediments. Other microstructures do occur but are seldom or rarely encountered and, therefore, are viewed as distinctive of a particular sediment from a specific location or are specific to a rare event either during emplacement, deposition, post-depositional, or inherited. These rare microstructures are probably not inherently useful in diagnostic terms, but that, in itself, is subjective and judgmental based only on past knowledge. Many glacial sediments remain to be examined microscopically and as such many microstructures may still remain to be discovered, described and understood.

The pervading presence, in most thin-sections of glacial sediment, of porewater-related structures and artefacts is indicative of the key importance of porewater in these sediments and the close association of porewater and sediment rheology in processes of deposition, deformation and diagenesis.

It is essential that microfabrics and microstructures are perceived as not being unique and therefore singularly diagnostic of specific conditions or environments but only in combination or association can glacial lithofacies discrimination and classification be achieved. In terms of glacial stratigraphy, much interpretation relies on process/sediment type designation into specific lithofacies types and subtypes; it therefore critical that the correct designation of particular lithofacies be achieved. What micromorphology highlights is another tool in lithofacies selection. It also indicates that lithofacies assessments in the past may have been erroneous thus requiring the reinterpretation or re-evaluation of past glacial stratigraphies.

Over the years many students are to be thanked for their collective help in sampling diamictons and diamictites especially K. Zaniewski, S. Habib, T. van Dyck, J. Taylor, and P. Tsementzis. The author also wishes to thank S. Carr, D. Cowan, A. Maltman, and E. Phillips for their extremely useful and insightful comments. This work was supported by Canadian NSERC Research Fund Grant A6900.

## Glossary

(also see Figure 2)

| | |
|---|---|
| *Argillans* | accumulation of clay particles within voids or other separations within the plasma. |
| *Domain* | small zone of sediment matrix different from surrounding matrix. |
| *Plasma* | colloidal size particles ($<2\,\mu$m) consisting of clay particles, individual particles not visible under microscopic examination, may exhibit birefringence; can be termed *matrix* or *groundmass*. |

### Plasmic fabrics

| | |
|---|---|
| *Lattisepic* | plasma separations occurring in two sets of orientations often normal to each other. |
| *Insepic* | plasma separations in clusters each separated unit may have a striated orientation but no unifying orientation can be observed between clusters. |
| *Masepic* | plasma particles oriented in one direction. |
| *Omnisepic* | plasma particles oriented randomly revealing many orientation directions. |
| *Rhombosepic* | plasma particles oriented in a rhombic pattern. |
| *Skelsepic* | plasma particles oriented around skeleton grains. |
| *Unistrial* | anisotropic clay (plasma) with sharp birefringence bands in one direction. |

## References

ALLEN, J. R. L. 1982. *Sedimentary Structures: Their Character and Physical Basis.* 2 Volumes. Elsevier Science Publishing, Amsterdam.

ALLEY, R. B., BLANKENSHIP, D. D., BENTLEY, C. R. & ROONEY, S. T. 1986. Deformation of till beneath ice stream B, West Antarctica. *Nature*, **322**, 57–59.

——, ——, ROONEY, S. T. & BENTLEY, C. R. 1987. Continuous till deformation beneath ice sheets. *International Association of Hydrological Sciences Bulletin*, **170**, 81–91.

BERTRAN, P. & TEXIER, J.-P. 1999 Facies and microfacies of slope deposits. *Catena*, **35**, 99–122.

——, FRANCOU, B. & TEXIER, J.-P. 1995. Stratified slope deposits: the stone-banked sheets and lobes model. *In*: SLAYMAKER, O. (ed.) *Steepland Geomorphology.* John Wiley, Chichester, 147–169.

BORDONAU, J. 1993. The Upper Pleistocene ice-lateral till complex of Cerler (Esera Valley, central southern Pyrenees: Spain). *Quaternary International*, **18**, 5–14.

BORRADAILE, G. J. 1981 Particulate flow of rock and the formation of cleavage. *Tectonophysics*, **71**, 305–321.

BOULTON, G. S. 1987. A theory of drumlin formation by subglacial sediment deformation. *In*: MENZIES, J. & ROSE, J. (eds) *Drumlin Symposium.*, A. A. Balkema, Rotterdam, 25–80.

——1996. Theory of glacial erosion, transport and deposition as a consequence of subglacial deformation. *Journal of Glaciology*, **42**, 43–62.

—— & HINDMARSH, R. C. A. 1987. Sediment deformation beneath glaciers: rheology and geological consequences. *Journal of Geophysical Research*, **92B**, 9059–9082.

BREWER, R. 1976. *Fabric and Mineral Analysis of Soils.* (2nd Edition). Krieger, Huntington, New York.

CARR, S. J. 1999. The micromorphology of Last Glacial maximum sediments in the Southern North Sea. *Catena*, **35**, 123–145.

—— & ROSE, J. 1997. An evaluation of variations in the orientation of different size particles in tills, and the significance of these patterns fore the movement of grains during till formation and processes of till deposition and deformation. *In*: *Developments in micromorphology.* Fysisch Geografisch en Bodemkundig Laboratorium. Report **65**, University of Amsterdam.

DAVENPORT, C. A. & RINGROSE, P. S. 1987. Deformation of Scottish Quaternary sediment sequences by strong earthquake motions. *In*: JONES, M. E. & PRESTON, R. M. F. (eds) *Deformation of Sediments and Sedimentary Rocks.* Geological Society, London, Special Publications, **29**, 299–314.

DREIMANIS, A. 1988. Tills: their genetic terminology and classification. *In*: GOLDTHWAIT, R. P. & MATSCH, C. L. (eds) *Genetic classification of glacigenic deposits.* A. A. Balkema, Rotterdam, 17–83.

——1993. Small to medium-sized glacitectonic structures in till and in its substratum and their comparison with mass movement structures. *Quaternary International*, **18**, 69–80.

HART, J. K. 1994. Till fabric associated with deformable beds. *Earth Surface Processes and Landforms*, **19**, 15–32.

KUBIENA, W. L. 1938. *Micropedology.* Collegiate Press, Ames, Iowa.

MALTMAN, A. J. (ed.) 1994. *The Geological Deformation of Sediments.* Chapman and Hall, London.

MARTINSEN, O. 1994. Mass Movements. *In*: MALTMAN, A. J. (ed.) *The Geological Deformation of Sediments.* Chapman and Hall, London, 127–165.

MENZIES, J. 1981. Freezing fronts and their possible influence upon processes of subglacial erosion and deposition. *Annals of Glaciology*, **2**, 52–56.

——1987. Towards a general hypothesis on the formation of drumlins. *In*: MENZIES, J. & ROSE, J. (eds) *Drumlin Symposium.* A. A. Balkema, Rotterdam, pp. 9–24.

——1990. Brecciated diamictons from Mohawk Bay, S. Ontario, Canada. *Sedimentology*, **37**, 481–493.

——1998. Microstructures within subglacial diamictons. *In*: KOSTRZEWSKI, A. (ed.) *Relief and Deposits of Present-day and Pleistocene Glaciation of the Northern Hemishpere – selected problems.* Adam Michiewicz University Press, Poznan, Geography Series, **58**, 153–166.

——2000 Microstructures within diamictites of the Lower Gowganda Formation (Huronian), near Elliot Lake, Ontario. *Journal of Sedimentary Research*, **35**, 210–216.

—— & MALTMAN, A. J. 1992. Microstructures in diamictons – evidence of subglacial bed conditions. *Geomorphology*, **6**, 27–40.

—— & WOODWARD, J. 1993. Preliminary study of subglacial diamicton microstructures as reflected

in drumlin sediments at Chimney Bluffs, New York. *In*: ABER, J. S. (ed.) *Glaciotectonics and mapping glacial deposits.* Proceedings of the INQUA Commission on Formation and Properties of Glacial Deposits, **1**. Canadian Plains Research Centre, University of Regina, Regina, 36–45.

——, ZANIEWSKI, K. & DREGER, D. 1997 Evidence, from microstructures, of deformable bed conditions within drumlins, Chimney Bluffs, New York State. *Sedimentary Geology*, **111**, 161–176.

MURRAY, T. 1994. Glacial Deformation. *In*: MALTMAN, A. J. (ed.) *The Deformation of Sediments.* Chapman and Hall, London, 74–93.

PASSCHIER, C. W. & TROUW, R. A. J. 1996. *Microtectonics.* Springer-Verlag, Heidelburg.

PIOTROWSKI, J. A. 1991. *Quartär- und hydrogeologische Untersuchungen im Bereich der Born höveder Seenkette, Schleswig-Holstein.* Reports Geologisch-Paläontologisches Institut und Museum, **43**.

PIPER, D. J. W., MUDIE, P. J., FADER, G. B., JOSENHANS, H. W., MACLEAN, B. & VILKS, G. 1990. Quaternary geology. *In*: KEEN, M. J. & WILLIAMS, G. L. (eds) *Geology of the continental margin of eastern Canada.* Geological Survey of Canada, Geology of Canada, **2**, 473–607.

RAPPOL, M. 1987. Saalian till in The Netherlands: a review. *In*: VAN DER MEER, J. J. M. (ed.) *Tills and Glaciotectonics.* A. A. Balkema, Rotterdam, 3–21.

RUTTER, E. H. 1986. On the nomenclature of mode of failure transitions in rock. *Tectonophysics*, **122**, 381–387.

SHROCK, R. F. 1948. *Sequence in Layered Rocks.* McGraw-Hill, London.

TWISS, R. J. & MOORE, E. M. 1992. *Structural Geology*, W. H. Freeman & Company, New York.

VAN DER MEER, J. J. M. 1987. Micromorphology of glacial sediments as a tool in distinguishing genetic varieties of till. *In*: KUJANSUU, R. & SAARNISTO, M. Geological Survey of Finland. Special Paper, **3**, 77–89.

——1993. Microscopic evidence of subglacial deformation. *Quaternary Science Reviews*, **12**, 553–587.

——1996. Micromorphology. *In*: MENZIES, J. (ed.) *Past Glacial Environments – sediments, forms and techniques*, **2**. Butterworth-Heineman, Oxford, 335–356.

——, RAPPOL, M. & SEMEIJN, J. N. 1983. Micromorphological and preliminary X-ray observations on a basal till from Lunteren, The Netherlands. *Acta Geologica Hispanica*, **18**, 199–205.

WILLIAMS, P. F., GOODWIN, L. B. & RALSER, S. 1994. Ductile deformation processes. *In*: HANCOCK, P. L. (ed.) *Continental Deformation.* Pergamon Press, Oxford, 1–27.

YOUNG, G. M. & NESBITT, H. W. 1985. The Gowganda Formation in the southern part of the Huronian Outcrop Belt, Ontario, Canada: Stratigraphy, depositional environments and regional tectonic significance. *Precambrian Research*, **29**, 265–301.

# Kinematic indicators of subglacial shearing

FREDERIK M. VAN DER WATEREN[1], SJOERD J. KLUIVING[2]
& LOUIS R. BARTEK[3]

[1] *Faculty of Earth Sciences, Vrije Universiteit, de Boelelaan 1085, 1081 HV Amsterdam, Netherlands (e-mail: wateren@xs4all.nl)*
[2] *Department of Geology, Box 870338, University of Alabama, Tuscaloosa, AL 35487–0338, USA*
[3] *Department of Geological Sciences, University of North Carolina, Chapel Hill, NC 27599-3315, USA*

**Abstract:** Criteria to distinguish between sediments that have been subglacially deformed and those that are undeformed, or deformed by other mechanisms, are sparse. In this paper we develop structural criteria to reconstruct the deformation history of glacial sediments that can be readily applied in the field as well as to analyses of thin sections of tills and related materials.

Progressive simple shear is the simplest model to describe the deformation history of subglacially deformed sediments. It includes most of their characteristic structural aspects and provides tools for the kinematic analysis of subglacially deformed sediments. Progressive simple shear generates asymmetric structures, in which the principal direction of finite extension is subparallel to the direction of shearing. This is the simple shear fabric's most distinctive characteristic, and that which most reliably defines the palaeo-ice flow direction. At a moderately strong intensity of deformation a typical shear zone in unlithified sediments may contain folded and strongly attenuated sediment layers, producing a transposed foliation which must not be mistaken for a sedimentary layering. Original sedimentary and deformation structures may completely disintegrate in the most intensely deformed sediments leading to its homogenization, although the typical shear zone fabric may still be identified in thin section.

There has been considerable debate about the nature of the geological substrate over which the Fennoscandian and Laurentide Ice Sheets advanced and its consequences for ice sheet volumes, advance and retreat rates, sea level change and atmospheric circulation (e.g. Boulton 1996b; Clark *et al.* 1999; Kleman & Hättestrand 1999, and references therein). Arguments for widespread deformable bed conditions beneath temperate ice sheets include features which are regarded as typical of subglacial deformation of saturated sediments. These range in scale from megaflutes and drumlins (Boulton 1987; Clark 1997) to smaller-scale boudins, augen, 'pods', and tectonic laminations developed within tills (Hart 1994; Hart & Roberts 1994). Those in favour of frozen bed conditions present evidence for the preservation of fragile landforms beneath large parts of the Pleistocene ice sheets (Kleman & Hättestrand 1999). Other arguments against a deformable bed model originate from the inter-pretation of laminated diamicts as glaciomarine deposits and subaquatic (meltout) tills (Eyles & McCabe 1991; Piotrowski & Kraus 1997; Piotrowski & Tulaczyk 1999; Piotrowski *et al.* in press).

Boulton (1996b) argued that it would be highly unlikely for a temperate ice sheet not to deform the underlying sediments. Large volumes of meltwater are produced at the base of temperate ice sheets which are drained through the bed. Over large areas subglacial drainage is likely to produce high pore-water pressures in the sub-glacial sediments, reducing the effective stresses to levels low enough for these sediments to deform, or even fluidizing them. Identification of features resulting from subglacial deformation may therefore help to reconstruct the dynamics of former ice sheets. Consequently, it is of critical importance to establish an unambiguous set of criteria that can be used to identify subglacial deformation.

*From:* MALTMAN, A. J., HUBBARD, B. & HAMBREY, M. J. (eds) *Deformation of Glacial Materials.* Geological Society, London, Special Publications, **176**, 259–278. 0305-8719/00/$15.00 © The Geological Society of London 2000.

The necessary first step is to understand the kinematics (movement history) of subglacial deformation. We will then be able to recognize and analyse the structures which are left behind by former ice sheets. The next step is to evaluate the conditions which prevailed within the bed during deformation.

Slater (1926), Gry (1942) and Banham (1975, 1977) were among the first to recognize the similarities between the deformation structures present in glacial tills and those in regionally deformed metamorphic rocks (e.g. slates, schists) and shear zones (e.g. mylonites). Shear tests with clay by Maltman (1987) produced structures very similar to cleavages in low-grade metamorphic rocks (e.g. slates), while later studies of actively deforming accretionary wedges (Byrne 1994; Maltman 1994) demonstrated that these structures indeed form in unlithified sediments under natural conditions. Alley (1991), Boulton (1996*a, b*), Boulton & Hindmarsh (1987) and Boulton & Jones (1979) have shown that the dynamics of mid-latitude ice sheets are, to a high degree, controlled by deformation of the bed. Outside their core regions, these ice sheets advance over a bed of unlithified and weakly lithified sediments.

Although the work by Danish Quaternary geologists (Berthelsen 1978, 1979; Houmark-Nielsen 1987, 1994; Houmark-Nielsen & Berthelsen 1981) has shown the usefulness of structural analysis as an additional tool in reconstructing past glacial environments, its use in a way that is common in orogenic geological studies has hardly ever been applied to glacial sediments. This has led to a gross underestimate of the distribution of subglacially deformed sediments.

A recent debate on the distribution and relative importance of subglacial deformation (Hart *et al.* 1996, 1997; Piotrowski *et al.* 1997, in press; Piotrowski & Kraus 1997; Piotrowski & Tulac-zyk 1999) has made it clear that widely accepted criteria to distinguish sediments deformed by subglacial simple shear from sediments which are undeformed, or deformed by other mechanisms, are largely lacking. Part of the disagreement stems from the lack of understanding of the various deformation histories and the resulting till fabrics. It is also the consequence of not consistently applying a structural geological approach to the analysis of subglacial tills, as is clear from the statement by Piotrowski *et al.* (in press) that 'it is easier to show that some tills were not pervasively deformed, than to demonstrate that others were'.

In this paper we apply a structural approach to analyse subglacially deformed unlithified to weakly lithified sediments. The primary objective

is to establish a set of criteria to distinguish undeformed sediments from those that have been subglacially sheared or otherwise deformed, and produce reliable tools for the reconstruction of past ice movement directions. The former is of critical importance for studies of piston or drill cores where the structural and sedimentological context is much less clear than it is in a good exposure. Structural analysis is not a stand-alone tool; we strongly favour a multi-disciplinary approach including sedimentological, morphological and geophysical data (Kluiving *et al.* 1999*a*).

The simplest model to describe subglacial deformation is progressive simple shear. This model has the advantage that, in general terms, it describes the deformation history of most subglacial tills with sufficient accuracy and that the geometry of the resultant fabric is relatively easy to understand. In reality both simple and pure shear usually act together. Compaction by dewatering and ice loading are examples of shear-plane normal compression resulting in volume loss. Flow within the deforming bed is generally not uniform: e.g. shear plane parallel/subhorizontal compression occurs at the ice sheet margin.

In subglacial shear zones these departures from the ideal simple shear model are relatively minor. Below we will present further arguments for favouring simple shear over more complicated shear zone models. For discussions of flow models that are applicable to a wider range of shear zones, including monoclinic and triclinic shear zone models, the reader is referred to Jiang & Williams (1998), Lin *et al.* (1998), Passchier (1998) and Passchier *et al.* (1997) and a discussion on transpression and transtension zones in Jiang *et al.* (1999).

We will use examples from past glacial environments in Europe. Using the same criteria, case studies from the Ross Sea, Antarctica, are discussed in Kluiving *et al.* (1999*a, b*). However, the same method can be applied to deformed non-glacial sediments such as slumps and landslides.

## Theoretical model of subglacial deformation

As mentioned previously, the same dynamics and kinematics underlie deformation of metamorphic rocks and unlithified sediments. This has been the rationale behind the use of unconsolidated materials in some early analogue models of mountain building, and even today, sandbox experiments are a useful tool to understand the development of geological structures. (For a structural analysis of push moraines and other ice-marginal compression features as analogues for orogenic structures, see Van der Wateren 1985, 1995.) Thus, rather than treating them as a separate entity, we will treat structures in the subglacial deforming layer as part of a continuum of deformation structures ranging from unlithified sediments to high-grade metamorphic rocks. The difference obviously is the time scale and the absence of crystal lattice deformations and recrystallization features in sediments. A number of structures may be recognized which make it possible to determine finite shear strain, if not quantitatively then at least to distinguish between fabrics produced at different intensities of deformation and determine the orientation of the finite strain axes and the tectonic transport direction (Chester & Logan 1987; Malavieille 1987; Petit 1987; Platt 1986).

**Table 1.** *Definitions used in this paper (mainly after Passchier & Trouw 1996)*

| | |
|---|---|
| Kinematics | Movement history. |
| Dynamics | Force distribution. |
| Fabric | The complete spatial and geometrical configuration of all those components that are contained in a rock, and are penetratively and repeatedly developed throughout the volume of rock at the scale of observation. Includes features such as foliation, lineation, preferred orientation and grain size. |
| Foliation | Planar fabric consisting of a compositional layering or a preferred orientation of planar discontinuities (fractures, platy minerals). |
| Bedding | Primary foliation, alternating bands with distinct lithologies. |
| Cleavage | Secondary foliation defined by a preferred orientation of inequant fabric elements, a penetrative set of discrete fracture surfaces (fracture cleavage; term to be avoided), shear bands (shear band cleavage), or surfaces along which platy minerals are bent (crenulation cleavage). |
| Stress | Tensorial quantity describing the orientation and magnitude of force vectors acting on planes of any orientation at a specific point in a volume of rock. |
| | Force per unit area acting in a given direction on a body at an instant in time. |
| Normal stress | Stress acting perpendicular to a material plane. |
| Shear stress | Stress acting parallel to a material plane. |
| Strain | The change in shape or internal configuration of a body resulting from certain types of displacement; tensorial quantity including features of distortion and rotation. |
| | A strained situation is commonly represented by an ellipsoid, compared with the unstrained sphere. |
| Incremental (infinitesimal) strain | Imaginary, infinitely small strain. |
| Finite strain | Strain accumulated over a finite period of time. |
| Coaxial deformation | Material lines that are instantaneously parallel to the incremental stretching axes remain fixed with respect to these axes throughout the deformation history. |
| Non-coaxial deformation | Material lines that are instantaneously parallel to the incremental stretching axes rotate with respect to these axes. |
| Style | Embodies all morphological features of a structure or group of structures (e.g. fold shape, presence/absence of foliations). |
| Brittle deformation | Failure of a stressed body along discrete discontinuous dislocations (cracks and faults), when the elastic limit is exceeded. |
| Ductile deformation (ductile flow) | In structural geology, usually reserved for deformations that are continuous throughout the rock body, such as crystal lattice deformations, crystal flow, grain boundary diffusion etc. |
| | Since these are obviously lacking in deformed unlithified sediments, we use the term ductile, lacking a better term, to denote a permanent deformation without fracturing on the observation scale. |
| Rheology | The relationship of stress and strain within fluids in motion (in the strict sense). Used here in a wider sense, referring to the mechanical behaviour of materials. |

Basically, two modes of flow occur in ice sheets and glaciers. A compressive flow regime occurs within a relatively narrow marginal zone between the equilibrium line and the ice sheet margin, while extensional flow occurs further upstream (Boulton 1996*b*). Similar conditions occur in deforming sediments beneath an ice sheet. Sediments are eroded by subglacial shearing in the region of extending flow and transported towards the margin where they accumulate. Deformation structures in these sediments record the conditions prevailing during deposition, although it must be realized that the majority of the sediments accumulated near the margin, including most of the compression structures, will be removed by meltwater streams. The style of deformation recorded by the sediments can be modelled using progressive simple shear, that is, in broad terms, subglacial deformation can be described as a shear zone. In sediments within a glaciated region we can thus distinguish two glaciotectonic regimes, the subglacial shear zone and the marginal compressive belt (Van der Wateren 1994, 1995).

Strain rates in saturated sediment beneath a temperate glacier are of the order of $10\,a^{-1}$ (Boulton 1987; Boulton & Hindmarsh 1987; Boulton *et al.* 1996), meaning that a vertical marker line would after one year be stretched to about ten times its original length and rotated to only a few degrees from horizontal. This strain rate equals $10^{-7}\,s^{-1}$ in a more common notation, to compare with strain rates in rocks which are of the order of $10^{-12}$ to $10^{-14}\,s^{-1}$. In other words, the intensity of deformation attained beneath an ice sheet in one year is reached in rocks only after 100 ka to 10 Ma. Subglacially deformed sediments will therefore generally show extremely high shear strains. This notion is the basis of our discussion of structures in subglacially deformed sediments.

Below we discuss the kinematics of progressive simple shear and structural style of the resulting fabrics, while geometric principles and terms are defined in Table 1. The validity of our assumption of the progressive simple shear model to describe the subglacial shear zone is discussed in the section 'Simultaneous pure shear and simple shear'.

The striking kinematic and geometric similarity with structures in cataclastic fault rocks and mylonites justifies the use of a descriptive terminology which is common in structural geology (Passchier & Trouw 1996; Ramsay & Huber 1983, and references therein). We therefore propose using a structural terminology, stressing the continuum of deformation processes in a wide range of materials, and for a large variation of time and spatial scales, as an alternative to Van der Meer's (1993, 1996, 1997) micromorphological terminology. Since the latter has originally been developed to describe features produced by pedological processes it is, in our opinion, not ideally suited for a kinematic analysis of deformation structures.

### Progressive simple shear

Figure 1 illustrates pure shear and simple shear, the two main deformation histories we will consider here. Simple shear has a monoclinic symmetry, in contrast with pure shear, which has an orthorhombic symmetry (Passchier & Trouw 1996). For the present we assume an idealized shear zone deforming by progressive simple shear, where all displacements occur within the plane of shearing (plane strain) (Fig. 2). We also

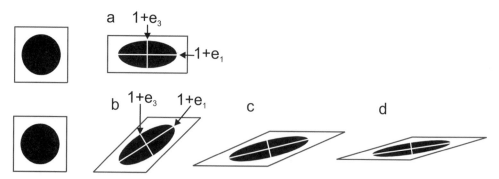

**Fig. 1.** Pure shear and simple shear. Principal axes of the finite strain ellipse are labeled $1 + e_1$, $1 + e_3$. (**a**) Pure shear (flattening = 0.67, extension = 1.50). (**b**) Simple shear ($\gamma = 1$); no volume loss. (**c**) Simple shear + 67% flattening; no volume loss. (**d**) Simple shear + flattening + consolidation, 33% volume loss.

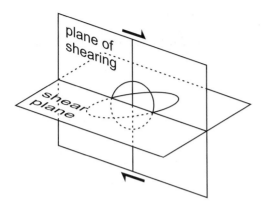

**Fig. 2.** Geometry conventions of the shear plane and the plane of shearing used in this paper. In plane strain, assumed for the idealized shear zone model, all deformation occurs within the plane of shearing.

assume that strain is homogeneous, both in the vertical and in the horizontal direction. The following discussion of fabric development in progressive simple shear is derived from Ramsay & Huber (1983) and Passchier & Trouw (1996).

An originally circular passive marker (unit circle) in the plane of shearing (Fig. 3a) deforms into an ellipse after one strain increment, with the instantaneous stretching axes ($1 + e_{1i}$ and $1 + e_{3i}$) of the incremental strain ellipse dipping at 45° to the shear direction. With each subsequent strain increment the ellipticity of the finite strain ellipse increases and its long axis ($1 + e_{1f}$) rotates in the direction of shearing (Fig. 3b), gradually stretching and rotating the strain ellipse. The fabric as we observe it, reflects finite strain, accumulated through a large number of incremental strain steps. One of the lines of no incremental extension is perpendicular, the other parallel to the shear plane (Fig. 3a). In progressive simple shear, the shear plane contains both the line of no incremental extension and the line of no finite extension Fig. 3b).

Progressive simple shear is thus a non-coaxial deformation in which material lines that are instantaneously parallel to the incremental stretching axes rotate with respect to these axes (Figs 1, 3b). With increasing finite strain the principal direction of finite extension $\lambda_1$ ($\lambda =$ quadratic extension $= 1 + e^2$) becomes increasingly parallel to the shear zone boundary (Choukroune *et al.* 1987; Ramsay 1967).

Both the incremental and the finite strain ellipses can be divided by lines of no longitudinal strain into four sectors, two in which material lines having orientations within this sector are shortened (incremental and finite

extension, $e_i$ and $e_f$, respectively, are negative), and two in which material lines have become longer ($e_i$ and $e_f$ positive) (Figs 3a, b). The plane of shearing contains the longest and the shortest principal axes of the strain ellipsoids ($1 + e_1$ and $1 + e_3$).

Superposition of the incremental strain ellipse on the finite strain ellipse yields three sectors in progressive simple shear (Fig. 3b) and we will consider the effects on a competent layer in an incompetent matrix (Fig. 3c).

(1) Finite longitudinal strain ($e_f$) and incremental longitudinal strain ($e_i$) are negative. Fabric elements, e.g. sediment layers, having this orientation are shortened and continue to shorten during the next increment. Shortening is accomplished by thickening and/or folding of competent layers (sector 1 in Fig. 3c).

(2) $e_f$ is negative and $e_i$ positive. Fabric elements in this sector have been initially shortened and will be stretched during the next increment. This may lead to unfolding of previously folded competent layers, but thinning and boudinage of fold limbs will be more common (sector 2 in Fig. 3c).

(3) Both $e_f$ and $e_i$ are positive. For orientations in this sector the competent layer has undergone extension throughout its deformation history. Competent layers suffer thinning and boudinage (sector 3 in Fig. 3c).

Folded boudins (Fig. 3d) are not normally produced in the shear zone under uniform flow (horizontal strain rate $\partial u/\partial x = 0$). They may be an indication of interference of different glacial advances, but are generally produced where longitudinal compression and extension alternate (non-uniform flow, $\partial u/\partial x \neq 0$), due to inhomogeneities (e.g. lithological contrast, pore water pressure variations), or near the ice margin, where compressive flow ($\partial u/\partial x < 0$) dominates.

It can be concluded that a conspicuous fabric asymmetry in the plane of shearing with evidence of stretching in a subhorizontal direction (subparallel to the boundaries of the till) will be evident in the field. Boudins and detached folds are common features of shear zones.

## Macro- and microstructures developed in response to subglacial deformation

Since strain rates are highest at the ice-bed interface (Boulton 1987; Boulton & Hindmarsh 1987) the intensity of subglacial deformation can be expected to increase from the *in situ* footwall sediments to the top of the deforming layer which carries the most allochthonous elements.

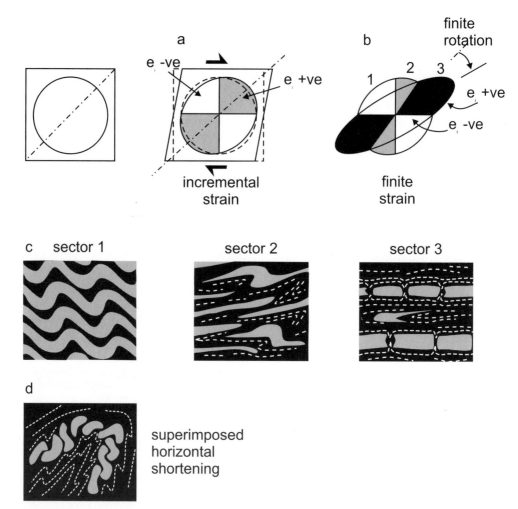

**Fig. 3.** Progressive simple shear (after Ramsay & Huber 1983 and Passchier & Trouw 1996). (**a**) The incremental strain ellipse divided by two lines of no instantaneous stretching into two sectors. Material lines with orientations falling within the sector in which the incremental extension ($e_i$) is positive (shading) will be stretched. Those within the sector of negative $e_i$ will be shortened. (**b**) Superposition of incremental and finite strain ellipses defines three sectors, one of continuous shortening of material lines oriented within this sector (1), one of shortening followed by extension (2) and one of continuous extension (3). (**c**) Illustrations of the deformation histories of competent layers within the three sectors of the finite strain ellipse: buckling and folding in sector 1, unfolding, stretching and boudinage of fold hinges and fold limbs in sector 2 and stretching and boudinage subparallel to the shear plane in sector 3. (**d**) Folding of shear zone fabrics (resulting in e.g. folded boudins) is an indication that simple shear is overprinted by horizontal shortening.

Therefore, our idealized shear zone needs to be modified to accommodate heterogeneous deformation (Fig. 4). Observations of deformation tills in Germany and the Netherlands indicate that in a vertical section it is quite common for the intensity of deformation to increase upwards from undeformed sediments to strongly sheared and homogenized diamict at the top (Kluiving *et al.* 1991; Van der Wateren 1987, 1995). In layered subglacial sediments heterogeneous flow may lead to deformation partitioning whereby the deformation concentrates in the weakest layers, and domains of low strain (e.g. containing boudins and augen) are separated by high strain domains (shear bands and narrow shear zones; Fig. 4d). However, depending on the scale of

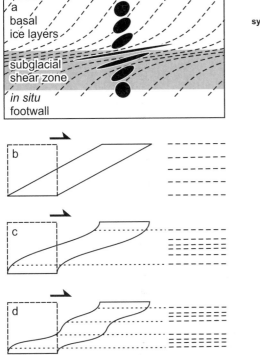

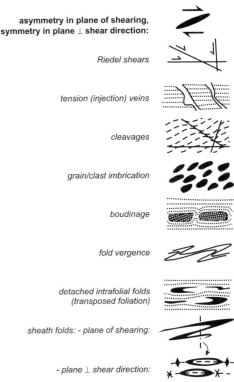

**Fig. 4.** Homogeneous and heterogeneous simple shear.
(**a**) Subglacial shear zone with finite strain ellipses and sigmoidally shaped strain trajectories ($\lambda_1$) in the basal ice layers and deforming subglacial sediments. $\lambda_1$ trajectories may define an S cleavage in the deformed sediments. (**b**) Homogeneous simple shear.
(**c**) Heterogeneous simple shear. (**d**) Heterogeneous simple shear where deformation is partitioned in domains of high strain. Low strain domains contain the most recognizable structures, thus giving a false impression of the true finite strain of the shear zone.

**Fig. 5.** Summary of structures produced by progressive simple shear, indicating how these can be used as kinematic indicators.

observation, within each domain strain can be considered homogeneous, carrying the distinctive features of simple shear.

Kinematic analysis of deformed sediments includes observations of the structural style (brittle or ductile, fold style etc.), and frequency of, and angular relationships between fabric elements. From such observations estimates of the shape and orientation of the strain ellipsoid can be made. Generally, however, this will be a very rough estimate because strain markers are lacking in typical glacial sediments.

The descriptive terminology used in the present approach has the advantages that (a) it is in agreement with terms used commonly in structural geology, stressing the continuity of deformation structures in different geological materials and (b) it describes morphological features which relate to the kinematic history of the structures.

The continuous rotation of material lines which are not parallel to the shear plane produces a clear structural asymmetry in the plane of shearing, while the section perpendicular to the shear vector contains more symmetric structures. This provides the most obvious and reliable indicator of tectonic transport and therefore palaeo-ice flow direction. Figure 5 summarizes in a schematic way the structures discussed below which are typical of subglacial shear zones, indicating how they can be used as kinematic indicators.

### Structural style

Two of the more peculiar aspects of shear zones in unlithified sediments are the coexistence of brittle and ductile style structures (as defined in Table 1), and the reversal of competence relationships on the micro- and the macroscale.

Brittle style structures usually form in coarse-grained sediments, while clays show more ductile behaviour. Generally this is a matter of scale, since macroscopic ductile deformation may be the cumulative effect of numerous microscale discrete displacements. On the other hand what appears to be a fault may in fact be a very narrow shear zone in which material is extremely attenuated but has not failed (Davis *et al.* 2000).

In thin section, clay laminae sometimes appear to be broken up into boudins, whereas sandy laminae show fluidal structures in the spaces between the clay boudins, in contrast with the usual competence relationships. The brittle response of strongly sheared clay probably is the effect of strain hardening due to compaction/dewatering during progressive deformation, maybe even inherited from a pre-existing deformation event.

To avoid confusion, we treat brittle and ductile microstructures separately, even though the two may develop side by side, and probably simultaneously.

*Brittle shear-zone structures.* Figure 6 (after Passchier & Trouw 1996) shows typical brittle shear zone structures comprising shear planes (Y) parallel to the shear zone boundary, synthetic Riedel shear planes (R), antithetic Riedels (R'), synthetic low angle P shears, and tension veins (T) usually filled with fluidized structureless sediment. Till wedges and clastic dikes (Åmark 1986; Dreimanis 1992; Dreimanis & Rappol 1997; Larsen & Mangerud 1992; Rijsdijk *et al.* 1999; Van der Wateren 1999), which are commonly found in and beneath tills, probably originate as tension veins.

Due to their orientation and opposite shear sense, R' shears are less frequently developed than the R planes. R may develop into R' as a result of rotation during progressive deformation, while older generations of shear planes are cut by later structures.

This association of brittle structures is typical of sheared sands, e.g. beneath tills or thrust sheets in push moraines (Van der Wateren 1987, 1995).

*Ductile shear zone structures.* Among the typical mesoscopic and microscopic ductile shear zone structures are initially buckle folds and compressive crenulations (Fig. 3c) which are produced by finite shortening of material lines in sector 1 of the strain ellipse (Fig. 3b). Boudins and extensional crenulations are produced in sector 3 by finite extension in the direction of shearing.

In progressive simple shear the fold axial surfaces and the folds become overturned and fold limbs are attenuated in the shear direction. At higher strains they are disrupted and develop into rootless intafolial folds (Fig. 7a). They may develop into sheath folds with their axes parallel to the bulk shear direction (Fig. 7b). In shear-parallel sections they appear as low-angle attenuated (and often detached) isoclinal folds, while in sections perpendicular to the shear direction they show as eye or concentric ring shapes (see also Kluiving *et al.* 1991). Sheath folds typically form at high shear strains ($\gamma > 10$) (Cobbold & Quinquis 1980; Grujic & Mancktelow 1995).

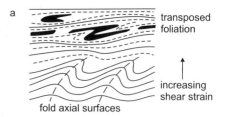

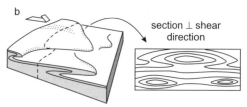

**Fig. 7.** Folding in a ductile shear zone. (**a**) With increasing shear strain (e.g. in a heterogeneous shear zone such as in Fig. 4) fold axial surfaces are overturned and folds are detached (upper part of diagram). At very high finite shear strains detached folds are widely separated in a layered matrix (transposed foliation) and may be difficult to detect. (**b**) Sheath folds forming at high shear strain (above 10). After Cobbold & Quinquis (1980).

**Fig. 6.** Geometry and shear direction of Riedel shears (Y, P R and R') in a dextral brittle shear zone (after Passchier & Trouw 1996). The diagram also shows a strain ellipse produced by brittle fracture (compare with ductile strain ellipse in Fig. 8).

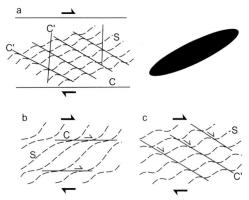

Fig. 8. (a) Geometry of shear band surfaces and symmetry of a dextral ductile shear zone. C and C' planes also known as extensional crenulation cleavages (ECC). (b) S–C fabric. (c) S–C' fabric.

They appear to be more common in subglacial shear zones than in metamorphic rocks.

Another set of ductile microstructures is similar to the shear-band cleavages found in mylonites, particularly in clay-rich materials (Fig. 8a). Rotation of platy and prismatic minerals towards the direction of finite extension

($\lambda_1$) produces a penetrative cleavage (S planes), which in thin section shows as a dominant bire-fringence direction. Rotation of skeletal grains (sand to pebble size) leads to preferred orientations toward the shear direction. The common method of measuring clast orientations to determine the till fabric is based on this principle.

Narrow shear bands parallel to the shear plane are termed C planes. C' planes may form conjugate sets of extensional crenulation clea-vages (ECC fabric in mylonitic rocks), although the synthetic C' planes are developed more frequently. Typical combinations are S–C and C–C' fabrics (Figs 8b, c). Figure 5 indicates how the structures discussed above can be used as kinematic indicators. Figure 9 shows an example of the association of brittle and ductile struc-tures in a shear zone beneath a nappe in the Dammer Berge push moraine, Germany (Van der Wateren 1987, 1995). Similar mixed brittle/ductile shear zones are common in tills.

## Simultaneous pure shear and simple shear

Simultaneous pure and simple shear may produce structural geometries that are signifi-cantly different from those produced by simple

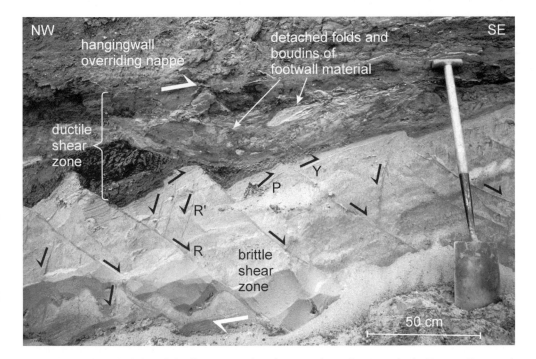

Fig. 9. Association of brittle and ductile structures in a shear zone beneath a nappe in the Dammer Berge push moraine (Van der Wateren 1987, 1995).

shear alone (Ghosh & Ramberg 1976; Hanmer & Passchier 1991; Jiang & Williams 1998; Passchier 1998; Passchier *et al.* 1997). Since, as we argue below, this effect is generally small, the characteristic asymmetry in the plane of shearing and strong flow-parallel extension will be consistent with the idealized shear zone model we proposed above.

To evaluate our assumption of a progressive simple shear model we need to consider the effect of pure shear, mainly shortening normal to the shear plane due to consolidation/dewatering of till and/or loading by the overlying ice sheet or glacier. The effect of shear-plane normal pure shear becomes significant where the rate of natural strain parallel to the shear plane is equal to or larger than the simple shear-strain rate (Ghosh & Ramberg 1976).

Other departures from the ideal simple shear model occur in zones of accelerating or decelerating flow (non-uniform flow). Here, an amount of horizontal stretching or shortening, respectively, is added to the extension resulting from progressive simple shear. On the scale of a glacier or ice sheet, accelerating flow occurs upstream from the equilibrium line where the bed is eroded and material is transported downstream (Boulton 1996*a, b*; Van der Wateren 1994, 1995). Structures in the relatively thin till layers left behind will show strong horizontal extension and overlie a deeply eroded substrate. We will usually be dealing with structures from the marginal zone where tills accumulate. Here, the shear zone fabric of asymmetric and attenuated structures is overprinted by horizontal shortening.

The most common effect is folding and stacking of till layers. Examples of these are Møns Cliff (Slater 1927; Aber 1979; Berthelsen 1979) and sites elsewhere in Denmark (Jessen 1931; Sjørring 1977, 1983), the chalky tills of East Anglia (Slater 1926; Banham 1975; Hart 1987;

Hart *et al.* 1990), a till section near the Dutch–German border (Kluiving *et al.* 1991), and the Heiligenhafen cliff section (Van der Wateren 1999). All characteristics of the original simple shear fabric are preserved within the deformed till units and may still give clues about the transport direction prior to deposition (Fig. 10).

Flow may be non-uniform on exposure or smaller scales due to rheological contrasts and the presence of large clasts. Non-uniform flow has the same effect here as on the larger scale: folding and thrusting of simple shear fabrics. For the following reasons, therefore, except on a local scale, we do not expect simultaneous pure and simple shear to significantly alter the structural geometries produced by our idealized shear zone model.

(1) Observations of actively deforming till layers as well as theoretical considerations (Alley *et al.* 1986, 1987, 1989*a*; Blankenship *et al.* 1987; Boulton 1987, 1996*a, b*; Boulton & Hindmarsh 1987; Hart & Smith 1997; Murray 1994) indicate that shear strain rates in deforming tills can be very high – much higher indeed than those reported from any other naturally occurring shear zone.

(2) On the scale of the subglacial shear zone it is unlikely that pure shear strain rates will be of the same order of magnitude as simple shear strain rates. Since the drained shear strength of most sediments exceeds the basal shear stress of glaciers and ice sheets, which is in the range of 10–100 kPa, subglacial shearing occurs mainly at low effective stresses, i.e. when the sediments are saturated. Dewatering of the till locks the deformation and stabilizes the deforming layer (Boulton 1996*b*). Therefore, pure shear as a result of dewatering and consolidation will only play a role during the very last stages of subglacial deformation.

(3) Loading by the overlying ice sheet may be a source of shear plane normal shortening. However, since there is a fundamental difference between subglacial shear zones and regional shear zones such as beneath nappes, this is not likely to play a major role. Ice sheets and glaciers maintain their surface profiles by continuous simple shear deformation, even when the ice margin is not advancing (steady state). In the case of steady state or slowly advancing ice sheets the pure strain rate in the bed due to ice loading will be very low compared to the simple shear strain rate, for the rate of sediment collapse will be several orders of magnitude slower than the deformation rate of a till. In other geological contexts it is far more likely for the pure shear strain rate to equal or exceed the simple shear strain rate. Consequently, simple shear strain rate of

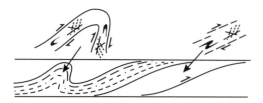

**Fig. 10.** Overprinting of shear zone fabric by horizontal shortening. This may occur e.g. near the ice margin and results in folded and stacked till sheets. On the scale of an outcrop or thin section the simple shear fabric with its characteristic geometry and symmetry is still clearly identifiable.

deforming saturated subglacial sediments generally outruns the rate of pure shear strain due to ice loading by several orders of magnitude.

(4) Finite simple shear strains in tills are very high, as can be deduced form the very large transport distances (tens to hundred of kilometres) of erratic material. Bulk simple shear strains are probably of the order of $10^4$ to $10^5$ in tills. Such values can never be obtained, except locally and on a very small scale, by horizontal extension under pure shear.

We conclude that the bulk strain of deformed subglacial sediments is the result of simple shearing while simultaneous pure shear strain probably contributes a few per cent at most. The subglacial shear zone can therefore be realistically modelled as a progressive simple shear zone.

## Estimation of the relative intensity of deformation

The very nature of subglacial deformation histories implies that the sediments evolve from nearly intact sedimentary layering to (macroscopically) complete homogenization. Far-traveled material will generally be more strongly homogenized than more locally derived material and will usually move at higher levels in the shear zone (Rappol *et al.* 1989; Rappol & Stoltenberg 1985; Van der Wateren 1999). At medium finite strains the shear zone commonly contains folded and strongly attenuated sediment lenses. These gradually disintegrate and wear down at higher shear strains and are macroscopically not discernable in the homogenized matrix. At higher magnifications in thin section they may still be recognized as sediment clasts, microboudins and rootless folds.

Many shear zones show a conspicuous layering, which should not to be confused with sedimentary layering. It is a tectonic foliation produced by transposition, converting a primary (sedimentary) lamination into a tectonic (transposed) foliation (Fig. 7a and case studies discussed below). Positive proof that this is indeed the result of very high shear strain and not some minor syndepositional deformation is given by the typical association of shear zone fabric elements like those discussed above (Figs 6, 8). Detached recumbent and sheath folds, strongly attenuated and boudinaged fold limbs indicate shear strains far in excess of 10, while the layering itself (transposed foliation) can only be produced by numerous repetitions of folding, extension, folding, extension, etc. Good examples of transposed foliations are in the chalk

bands in tills from East Anglia (e.g. Banham 1975; Hart 1987), or the Danish chalky and sandy tills (Ehlers 1983).

Transposed foliation is the intermediate stage between sedimentary layering and the end product of subglacial shearing: a completely homogenized diamict of deceptively low strain in which primary sedimentary structures have completely disintegrated (Boulton 1996*a*,*b*). Only the rare occurrence of isolated boudins or detached folds and the typical brittle/ductile microfabric are witness of the extremely high finite strains which are typical of most subglacial tills. These features, however rare, can still be used as kinematic indicators in addition to clast fabric analyses (Kluiving *et al.* 1991).

While the consistent fabric asymmetry within the plane of shearing compared to differently oriented cross-sections is the main criterion to distinguish sediments deformed by simple shear from non-deformed sediments or those deformed by pure shear, the following criteria have proven to be useful guidelines to determine the relative magnitude of finite shear strain in thin sections and outcrops:

- intensity of birefringence of the dominant cleavage, although this depends also on clay content;
- spacing between or frequency of cleavage planes: high frequency = high finite shear strain;
- angle between S and C planes: low angle = high finite shear strain;
- intensity of mixing of lithologies, ranging from minimum to very high strain: recognizable sedimentary structures → transposed foliation → homogeneous diamict;
- homogeneous diamicts of very high finite shear strains may show diffuse, undulating lamination with strongly sheared clay completely filling pore spaces between sand grains (rotational fabric: Van der Meer 1993).

Since the principal direction of finite extension ($\lambda_1$) tends to rotate towards the shear plane, the angle between an extensional fabric (e.g. S planes) and fabric elements parallel to the shear plane (e.g. C planes) could potentially give an indication of the magnitude of finite strain. So far, we have not been very successful using this criterion. S surfaces in the tills we investigated are generally oriented between 25° and 45° to the main shear plane, even within one sample. Although this may indicate strain partitioning whereby some domains have been more intensely deformed than others, this clearly needs more attention. Part of the problem is the arbitrary orientation of thin section samples from piston cores (Kluiving *et al.* 1999*a, b*). Additional thin sections at high angles to each other may help to identify the real orientation of the S fabric.

To avoid missing evidence of subglacial deformation or misinterpreting the structural style we propose that the following procedure be routinely taken for every till outcrop, even when only minor signs of deformation are evident. Since these can be an indication of very intense subglacial shearing (as discussed above) it would be a mistake to simply dismiss them as mere evidence of small-scale or even lack of subglacial deformation (Piotrowski & Tulaczyk 1999; Piotrowski *et al.* in press). Therefore, after noting the presence of deformation structures, the next step must be to determine what caused them. This requires recording fabric geometries, style and relative intensity of deformation.

This relatively simple field procedure also includes examining cross sections of different orientations to compare their symmetries. It is often enough to cut a few small holes in a cliff to observe the three-dimensional shape of a structure and measure the relevant fabric elements. Samples need to be taken (and their orientations with respect to the observed macrofabric recorded) from which thin sections are prepared. For sediment cores, which are usually not oriented, it may be helpful to prepare two or three thin sections which are oriented at 60 or 90 degrees to each other. Possible changes in symmetry between sections indicate horizontal simple shear, while a consistent symmetry about the vertical axis in all sections indicates pure shear due to loading, or even deformation during coring.

Dropstones in a laminated diamict are among the most commonly cited evidence of glaciolacustrine or marine deposition without subglacial shearing. Yet, at first sight, similar-looking features occur in strongly deformed sediments (see examples below) that may be, and have regularly been, mistaken for subaqueous deposits. Both true dropstones and clasts in a subglacially deformed matrix may show flexing of the underlying laminae. However, in a sediment deformed by simple shear the matrix surrounding the clasts should show evidence of flow consistently in one (sub)horizontal direction (Hart & Roberts 1994), whereas the deformation around the dropstones should have a symmetry indicating vertical loading (Fig. 11a, b). These, and other, criteria can be readily applied in the field as well as in thin-section analysis.

The strength of the fabric, e.g. intensity of the birefringence or frequency of cleavage planes, is not a measure of the ambient stresses during deformation, as has been suggested (Van der

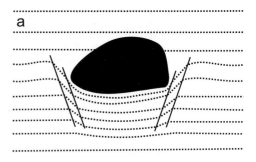

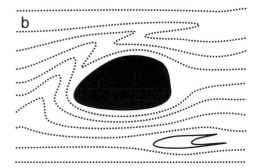

**Fig. 11.** Difference between dropstones and clasts in a subglacial shear zone. (**a**) Symmetric deformation due to vertical pure shear in glaciomarine and glaciolacustrine sediments beneath and adjacent to a dropstone. The clast is overlain by undisturbed lamination. (**b**) Asymmetric (monoclinic) structures surrounding a clast in a dextral simple shear zone. The diagram shows the plane of shearing. In a vertical section perpendicular to this structures will show a higher symmetry.

Meer 1993, 1997). There is no direct relationship between the orientation and magnitude of the stress field and the orientation of the finite strain axes as well as the magnitude of the finite strain. Finite strain is mainly determined by the rheological properties of the deformed material and the duration of the deformation under a given stress field. Thus, a relatively weak material may be deformed to very high finite strains if the deformation lasted sufficiently long even under very low effective stress conditions. Since subglacial sediments beneath a temperate ice sheet conduct large volumes of meltwater, high pore water pressures and therefore low effective stresses are very common, particularly in confined aquifers. Any attempt to relate fabric strength to effective stress, or even to ice thickness, is therefore bound to fail.

## Why are observed finite strains often less than expected finite strains?

The intensity of deformation of sheared sediments may appear to be deceptively low, although it can be deduced (e.g. because materials incorporated in the shear zone can be traced back to their origin) that the accumulated strains are much higher. Because dynamic recrystallization (as in mylonites) plays no role in the deformation of unlithified sediments, we would expect the total strain history to be preserved in the fabric. Yet, this is only rarely the case. Several mechanisms may account for this.

Deformation of the subglacial bed in most cases is the result of two processes simultaneously operating in the same direction, subglacial shearing and subglacial meltwater drainage. Most of the subglacial deformation therefore occurs while the sediment is saturated with water (Boulton 1996b; Boulton & Hindmarsh 1987). This may lead to liquefaction of the sediments, particularly where drainage is impeded by layers of low permeability. Liquefied granular material does not take up deformation and a fabric is only formed when a deformable framework begins to form following dewatering of the sediments. The observed structures therefore reflect a relatively late deformation stage when most of the pore water was expelled (strain hardening) and the sediment became less mobile.

In layered sequences, depending on the distribution of fluid pressures, deformation will be partitioned in the weakest layers, leading to detachments within the sequence.

The subglacial shear zone is an area of continuous erosion, reworking and redeposition of previously deposited sediment. Tills commonly contain domains of low as well as of high strain, lenses with original sedimentary structures and till clasts, respectively. The inclusion of till clasts (sometimes even clasts within clasts) is a particularly good indicator of very high strains. Clay-rich clasts regularly show a much more strongly developed and continuous cleavage than the surrounding matrix, identifying them as high strain domains (provided they are not reworked from older tills).

Progressive simple shearing tends to produce new fabrics overprinting and gradually erasing older fabrics. Estimates of the orientation and shape of the finite strain ellipsoid based on these younger structures may not be representative of the bulk finite strain. Thus, lenses and domains of higher strain may give better clues about the real strain.

These considerations may be helpful to distinguish high-strain sediments like subglacial tills from low-strain sediments like debris flows which in principle have experienced similar conditions of simple shear, albeit to a much lower degree. Also, kinematic indicators in subglacial tills will be more consistently unidirectional reflecting palaeo-ice flow.

## Case studies

### Sheared clay in till

The Heiligenhafen till section on the Baltic coast, Northern Germany (Van der Wateren 1999), comprises three tills; the lower two are probably Saalian in age. The lowest till contains numerous sediment boudins which can be interpreted as strongly attenuated fragments of one or several glaciofluvial deltas. These subaqueous sediments show a trend of decreasing grain size and stream flow energy in a westward direction, parallel to the palaeo-ice flow direction. At 510 m (from the eastern limit of the section) a clay lens, because of its textural and sediment-structural characteristics probably belonging to the distal delta facies, is tilted as a consequence of folding of both the lower and the middle tills following deposition (Fig. 12).

The glaciolacustrine clay is underlain and overlain by the massive chalk-rich lower till and shows conspicuous sets of fracture planes. One set, interpreted as the main shear plane, C is parallel to the boundaries of the clay layer. Another set, interpreted as the S planes, shows sigmoidal deflection towards the C planes. A thin section from this clay (indicated in Fig. 12) was taken parallel to the main shear direction as determined from outcrop scale structures.

These structures stand out clearly in thin section (Fig. 13a, b): continuous sigmoidal S planes, and sets of spaced C and C' shear bands, enhanced by desiccation cracks. The photograph is oriented with its long side parallel to the main shear direction, which is tilted due to folding of the till. The S planes show as a bright illumination in cross-polarized light (20–30° 'NE–SW'). Silty laminae are boudinaged due to dissection along C' shears leading to apparent back rotation of microboudins.

### Tension veins

The massive middle till in the Heiligenhafen cliff section (Van der Wateren 1999) contains numerous tension veins or clastic dykes filled with structureless sand. They are interpreted as cracks which were formed perpendicular to the direction of instantaneous stretching, opened under

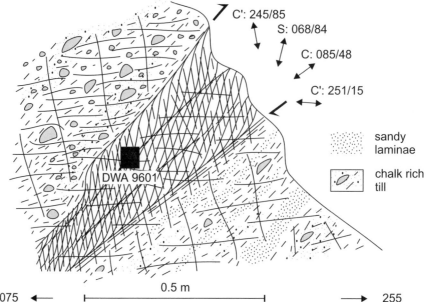

**Fig. 12.** Heiligenhafen, Germany, Hohe Ufer cliff section at 510 m (Van der Wateren 1999). Saalian till. A layer of glaciolacustrine clay surrounded by till is tilted due to folding following deposition of the till. Dessication cracks in the clay and the till define a S–C–C′ shear zone fabric. Sandy laminae in the chalk-rich till are mainly boudins of glaciofluvial sand, interpreted as part of an ice-marginal delta which was overridden by a Saalian ice advance. Black rectangle indicates photomicrograph (Fig. 13a).

tension and filled with fluidized sand migrating from overpressured sand lenses within the till. Figure 13c shows sets of tension veins which are offset by C surfaces, while younger veins cut older ones which have rotated clockwise. Together, these structures are clearcut evidence of dextral shearing related to E–W palaeo-ice movement.

## From transposed foliation to massive till

The Gliedenberg sandpit is on the southeastern flank of the Saalian age Dammer Berge push moraine, Germany at *c.* 80 m a.s.l. (Van der Wateren 1987, 1995). Here, an outcrop of strongly sheared deformation till could be linked with its source sediments (Fig. 14). These comprise: (1) Tertiary glauconitic clay, (2) Early Pleistocene white fluvial sands, (3) Middle Pleistocene pink fluvial sands and (4) green glauconitic glaciofluvial sands coeval with the formation of the push moraine. These were exposed in a diapiric fold at the front of a thrust sheet, approximately 50 m north of the till outcrop. Since the source sediments have such strongly contrasting colours and textures, they could easily be identified, even where the sediments

were strongly interfolded producing a transposed foliation (Fig. 13d). The till evidently formed by reworking of sediments which build the push moraine while it was overridden by the advancing Saalian ice sheet (Van der Wateren 1987, 1995).

Intensity of folding and sediment mixing increases rapidly towards the south, producing, within less than 100 m from the source materials, a macroscopically homogeneous diamict. However, on a microscale, this brown diamict still shows a diffuse compositional layering of clay-rich, sandy and glauconite-rich laminae (Fig. 13e): a transposed sedimentary foliation. At higher magnification it is clear that the diamict is the product of intense shearing: the matrix comprises well-developed continuous S planes and spaced C′ planes (Fig. 13f).

## Thickness of the deforming layer

There has been considerable debate on the role and thickness of the deforming layer underneath ice streams and ice sheets. Results from seismic studies at Ice Stream B yielded evidence for the existence of a 6 m thick layer of poorly consolidated till (Blankenship *et al.* 1986, 1987;

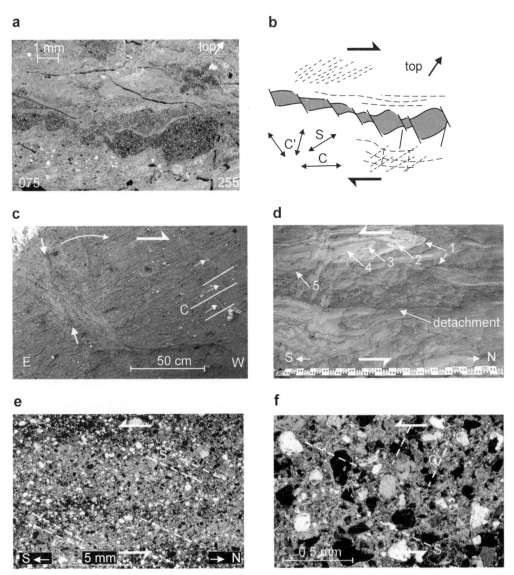

**Fig. 13.** Examples of kinematic indicators. (**a–c**). Glaciolacustrine clay from Heiligenhafen section, North Germany (Van der Wateren 1999). (**a**) Photomicrograph of boudinaged silt lamina within a shear zone showing S–C–C′ geometry. S planes are responsible for the bright illumination of most of the field of view. Crossed polarized light. (**b**) Diagram explaining the shear zone surfaces in (a). (**c**) Tension veins (clastic dykes) indicated by short arrows in chalk-rich till. Rotation and displacement across C shear bands indicate dextral (east–west) shearing. (**d–f**) Gliedenberg sandpit, Dammer Berge push moraine, Germany. (**d**) Transposed foliation composed of (1) Tertiary glauconitic clay, (2) white Lower Pleistocene fluvial sand, (3) pink Middle Pleistocene fluvial sand, (4) green glauconitic glaciofluvial sand, (5) brown diamict. Source materials could be identified from outcrop, 50 m to the north (upstream) (Van der Wateren 1995). Folds are non-cylindrical isoclinal folds and detached sheath folds with their axes obliquely cut by the outcrop surface. Flow is towards the SW. (**e**) Brown diamict (unit 5 in d) showing transposed foliation of clay-rich and sandy laminae (illuminated and dark bands, respectively). View in crossed polarized light. Brightly illuminated bands are clay-rich laminae with continuous S surfaces, separating clay-poor laminae with more widely spaced S planes. (**f**) Detail of (e) showing asymmetric S fabric and spaced C′ surfaces.

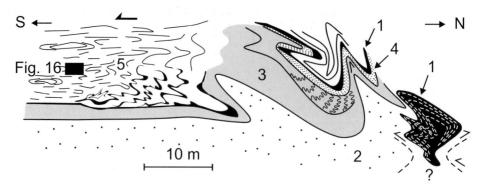

**Fig. 14.** Schematic diagram showing the deformation till and its source sediments at Gliedenberg, Dammer Berge push moraine, Germany (Van der Wateren 1995). The till (5) is a mixture of four source sediments which were exposed 50 m upstream (north) and which were folded and extruded in a diapiric structure cored with Tertiary clay (black). Annotation of the lithology is explained in Fig. 13d.

**Fig. 15.** Sheath folds in Hvideklint section, Møn, Denmark. Cliff face is roughly perpendicular to the fold axes which are parallel to the shear (ice flow) direction (towards north). The structures are overlain by the Late Weichselian Young Baltic till. Thickness of the subglacial deforming layer must have been at least the size of the sheath folds, probably 4 to 5 m.

Rooney *et al.* 1987). This observation led Alley *et al.* (1986, 1987, 1989*a*, *b*) to hypothesize that the fast motion of Ice Stream B is accommodated in this 6 m thick till layer, which behaves as a viscous fluid that deforms throughout its thickness and erodes the underlying sediments.

Hart *et al.* (1990) and Hart & Boulton (1991) suggested that the thickness of the deforming layer varies with basal shear stress, ranging from thin (10 cm) to thick (10 m), and that there

are two styles of subglacial deformation. Constructional deformation occurs where a thin deforming layer moves up the sequence, and excavational deformation where a thick deforming layer cuts down into the underlying sediment.

This view has been challenged by observations from bore holes and sediments collected from beneath Ice Stream B indicating that sliding, accommodated over a few centimetres thickness, on the top of the weak sediment layer, may be the predominant mode of ice stream motion (Engel-hardt & Kamb 1997; Scherer *et al.* 1998; Tulaczyk *et al.* 1998), rather than pervasive sediment defor-mation throughout the sediment layer as pro-posed by Alley *et al.* (1986, 1987, 1989*a*).

Our observations of Pleistocene till sections in NW Europe suggest that the deforming layer may reach a thickness of several metres with a steady (albeit not continuous) increase of finite strain from undeformed footwall sediments to a strongly homogenized diamict at the top.

In a section near De Lutte, the Netherlands (Kluiving *et al.* 1991), three overlying tills were exposed. The upper till showed a NW–SE transport direction. The lower till, up to a level of 4 m below the upper till, occasionally showed reorientation of structures along the NW–SE trend of the upper till. Visible reorientation of structures is expressed in detached folds and sheath folds.

More than 1 m thick inclusions of bodies of undeformed sediment such as boudins and sheath folds have been observed in the Heili-genhafen cliff (Van der Wateren 1999) discussed above. These structures are in till layers varying in thickness from a few metres to tens of metres.

More than two metres thick sheath folds are exposed in the Hvide Klint cliff on the island of Møn, Denmark (Fig. 15). The total thick-ness of the layer containing these structures is 4–5 m and it is overlain by more than 2 m of massive till in which clast fabrics (Aber 1979) show the same palaeo-ice flow direction as is indicated by the kinematics of the sheath folds. In an upward migrating A horizon of only a few decimetres in thickness, as has been suggested by Boulton (1996*a*), Boulton & Hindmarsh (1987) and Hart & Boulton (1991), such large struc-tures would not have survived intact, which suggests that at least in these cases the deform-ing layer was several metres thick. These observations are even harder to reconcile with the very thin deforming layer suggested by Engelhardt & Kamb (1997), Scherer *et al.* (1998) and Tulaczyk *et al.* (1998).

## Conclusions

Kinematic analysis of subglacially deformed sediments is a powerful tool to determine the structural style of shear zones, and to make qualitative estimates of the intensity of deforma-tion. Arguments have been presented to show that progressive simple shear is a realistic model to describe subglacial deformation. The most outstanding features of the resultant fabric are a prominent asymmetry and high degree of stretch-ing in the plane of shearing. Although on all

scales subglacial shear zone fabrics may depart from the ideal simple shear model, more often than not the characteristic asymmetric and atten-uated structures can be identified and used to re-construct the kinematic history of the sediments.

Features of subglacial shear zones have been demonstrated in exposures, as well as thin sec-tions. Using structural criteria, in combination with sedimentological and seismic observations (Kluiving *et al.* 1999*a, b*), we can ultimately decide whether sediments have been subglacially deformed or not, since in the latter case shear zone fabrics are either lacking, only weakly developed or overprinted by a dominantly non-simple shear fabric. Subglacial deformation is the physically most plausible mechanism to explain the widespread thick accumulations of sediments with very high finite shear strains. To us this is the most convincing evidence of deforming bed conditions beneath large parts of the Laurentide and Fennoscandian Ice Sheets. The deforming layer may have been as much as several metres in thickness.

Advantages of a structural analysis for glacial geological studies are:

(a) identification of deformation structures – distinction between undeformed (e.g. gla-ciomarine) sediments and tills;
(b) distinguishing core damage and subglacial deformation;
(c) distinction between deformation styles, e.g. subglacial/marginal, tills/debris flows;
(d) identification of zones of non-uniform flow, e.g. folded and stacked till sheets near former ice margins;
(e) reconstruction of former ice flow direc-tions, using symmetry/asymmetry of differ-ent cross sections.

We thank E. Phillips and S. Hanmer for their critical and detailed reviews which greatly improved our paper. We thank C. Biermann for his advice on strain theory.

F. M. Van der Wateren was supported by the Netherlands GeoSciences Foundation (GOA) with financial aid from the Netherlands Organization for Scientific Research (NWO). S. Kluiving and L. Bartek were supported by NSF grants OPP-9220848 and OPP-9316710.

## References

ABER, J. S. 1979. Kineto-stratigraphy at Hvideklint, Mön, Denmark and its regional significance. *Bulletin of the geological Society of Denmark*, **28**, 81–93.
ALLEY, R. B. 1991. Deforming-bed origin for southern Laurentide till sheets? *Journal of Glaciology*, **37**, 67–76.

——, BLANKENSHIP, D. D., BENTLEY, C. R. & ROONEY, S. T. 1986. Deformation of till beneath ice stream B, West Antarctica. *Nature*, **322**, 57–59.

——, ——, —— & ——1987. Till beneath ice stream B; 3 Till deformation; evidence and implications. *Journal of Geophysical Research*, **B92**, 8921–8929.

——, ——, ROONEY, S. T. & BENTLEY, C. R. 1989*a*. The sedimentary signature of deforming glacier beds in the Ross Embayment, Antarctica. *Annals of Glaciology*, **12**, 210–211.

——, ——, —— & ——1989*b*, Sedimentation beneath ice shelves – the view from Ice Stream B. *Marine Geology*, **85**, 101–120.

ÅMARK, M. 1986. Clastic dikes formed beneath an active glacier. *Geologiska Foereningen i Stockholm Foerhandlingar*, **108**, 13–20.

BANHAM, P. H. 1975. Glacitectonic structures: a general discussion with particular reference to the Contorted Drift of Norfolk. *In*: WRIGHT, A. E. & MOSELEY, F. (eds) *Ice Ages: Ancient and Modern. Geological Journal Special Issue*, **6**, 69–94.

——1977. Glacitectonites in till stratigraphy. *Boreas*, **6**, 101–106.

BERTHELSEN, A. 1978. The methodology of kinetostratigraphy as applied to glacial geology. *Bulletin of the geological Society of Denmark*, **27**, 25–38.

——1979. Recumbent folds and boudinage structures formed by subglacial shear: an example of gravity tectonics. *Geologie en Mijnbouw*, **58**, 253–260.

BLANKENSHIP, D. D., BENTLEY, C. R., ROONEY, S. T. & ALLEY, R. B. 1986, Seismic measurements reveal a saturated porous layer beneath an active Antarctic ice stream. *Nature*, **322**, 54–57.

——, ——, —— & ——1987. Till beneath ice stream B, 1, properties derived from seismic travel times. *Journal of Geophysical Research*, **92**, 8903–8912.

BOULTON, G. S. 1987. A theory of drumlin formation by subglacial sediment deformation. *In*: MENZIES, J. & ROSE, J. (eds) *Drumlin Symposium*. Balkema, Rotterdam, 25–80.

——1996*a*. The origin of till sequences by subglacial sediment deformation beneath mid-latitude ice sheets. *Annals of Glaciology*, **22**, 75–84.

——1996*b*. Theory of glacial erosion, transport and deposition as a consequence of subglacial sediment deformation. *Journal of Glaciology*, **42**, 43–62.

—— & HINDMARSH, R. C. A. 1987. Sediment deformation beneath glaciers; rheology and geological consequences. *Journal of Geophysical Research*, **B92**, 9059–9082.

—— & JONES, A. S. 1979. Stability of temperate ice sheets resting on beds of deformable sediment. *Journal of Glaciology*, **24**, 29–43.

——, VAN DER MEER, J. J. M., HART, J., BEETS, D., RUEGG, G. H. J., VAN DER WATEREN, F. M. & JARVIS, J. 1996. Till and moraine emplacement in a deforming bed surge; an example from a marine environment. *Quaternary Science Reviews*, **15**, 961–987.

BYRNE, T. 1994. Sediment Deformation, dewatering and diagenesis: illustrations from selected mélange zones. *In*: MALTMAN, A. (ed.) *The Geological Deformation of Sediments*. Chapman & Hall, London, 240–260.

CHESTER, F. M. & LOGAN, J. M. 1987. Composite planar fabric of gouge from the Punchbowl Fault, California. *Journal of Structural Geology*, **9**, 621–634.

CHOUKROUNE, P., GAPAIS, D. & MERLE, O. 1987. Shear criteria and structural symmetry. *Journal of Structural Geology*, **9**, 525–530.

CLARK, C. D. 1997. Reconstructing the evolutionary dynamics of former ice sheets using multi-temporal evidence, remote sensing and GIS. *Quaternary Science Reviews*, **16**, 1067–1092.

CLARK, P. U., ALLEY, R. B. & POLLARD, D. 1999. Northern Hemisphere Ice-Sheet Influences on Global Climate Change. *Science*, **286**, 1104–1111.

COBBOLD, P. R. & QUINQUIS, H. 1980. Development of sheath folds in shear regimes. *Journal of Structural Geology*, **2**, 119–126.

DAVIS, G. H., BUMP, A. P., GARCÍA, P. E. & AHLGREN, S. G. 2000. Conjugate Riedel deformation band shear zones. *Journal of Structural Geology*, **22**, 169–190.

DREIMANIS, A. 1992. Downward injected till wedges and upward injected till dikes. *Sveriges Geologiska Undersoekning*, Serie Ca. Avhandlingar och Uppsatser i **4**, 91–96.

—— & RAPPOL, M. 1997. Late Wisconsinan subglacial clastic intrusive sheets along Lake Erie bluffs, at Bradtville, Ontario, Canada. *Sedimentary Geology*, **111**, 1–4.

EHLERS, J. (ed.) 1983. *Glacial Deposits in Northwest Europe*. Rotterdam, A. A. Balkema.

ENGELHARDT, H. F. & KAMB, B. 1997. Basal hydraulic system of a West Antarctic ice stream: constraints from borehole observations. *Journal of Glaciology*, **43**, 207–231.

EYLES, N. & MCCABE, A. M. 1991. Glaciomarine deposits of the Irish Sea Basin; the role of glacio-isostatic disequilibrium. *In*: EHLERS, J., GIBBARD, P. L. & ROSE, J. (eds) *Glacial deposits in Great Britain and Ireland*. Balkema, Rotterdam, 311–331.

GHOSH, S. K. & RAMBERG, H. 1976. Reorientation of inclusions by combination of pure shear and simple shear. *Tectonophysics*, **34**, 1–70.

GRUJIC, D. & MANCKTELOW, N. S. 1995. Folds with axes parallel to the extension direction: an experimental study. *Journal of Structural Geology*, **17**, 279–291.

GRY, H. 1942. Diskussion om vore dislocerede Klinters Dannelse. *Meddelelser fra dansk geologisk Forening*, **10**, 39–51.

HANMER, S. & PASSCHIER, C. 1991. Shear-sense indicators: a review. *Geological Survey of Canada Paper*, **90**, 1–72.

HART, J. K. 1987. *The genesis of the North East Norfolk drift*. PhD thesis, University of East Anglia, Norwich.

——1994. Till fabric associated with deformable beds. *Earth Surface Processes and Landforms*, **19**, 15–32.

—— & BOULTON, G. S. 1991. The interrelation of glaciotectonic and glaciodepositional processes within the glacial environment. *Quaternary Science Reviews*, **10**, 335–350.

—— & ROBERTS, D. H. 1994. Criteria to distinguish between subglacial glaciotectonic and glaciomarine sedimentation. I. Deformation styles and sedimentology. *Sedimentary Geology*, **91**, 191–213.

—— & SMITH, B. 1997. Subglacial deformation associated with fast ice flow, from the Columbia Glacier, Alaska. *Sedimentary Geology*, **111**, 1–4.

——, GANE, F. & WATTS, R. J. 1996. Deforming bed conditions on the Dänischer Wohld Peninsula, northern Germany. *Boreas*, **25**, 101–113.

——, —— & ——1997. Deforming bed conditions on the Dänischer Wohld Peninsula, northern Germany: Reply to comments. *Boreas*, **26**, 79–80.

——, HINDMARSH, R. C. A. & BOULTON, G. S. 1990. Different styles of subglacial glaciotectonic deformation in the context of the Anglian ice sheet. *Earth Surface Processes and Landforms*, **15**, 227–242.

HOUMARK-NIELSEN, M. 1987. Pleistocene stratigraphy and glacial history of the central part of Denmark. *Geological Society of Denmark Bulletin*, **36**, 1–189.

——1994. Late Pleistocene stratigraphy, glaciation chronology and middle Weichselian environmental history from Klintholm, Mon, Denmark. *Bulletin of the Geological Society of Denmark*, **41**, 181–202.

—— & BERTHELSEN, A. 1981. Kineto-stratigraphic evaluation and presentation of glacial-stratigraphic data, with examples from northern Samsö, Denmark. *Boreas*, **10**, 411–422.

JESSEN, A. 1931. *Lønstrup Klint*. Danmarks Geologiske Undersøgelse, II Række, **49**.

JIANG, D. & WILLIAMS, P. F. 1998. High-strain zones: a unified model. *Journal of Structural Geology*, **20**, 1105–1120.

——, LIN, S., WILLIAMS, P. F., DEWEY, J. F., HOLDSWORTH, R. E. & STRACHAN, R. A. 1999. Discussion on transpression and transtension zones. *Journal of the Geological Society, London*, **156**, 1051–1055.

KLEMAN, J. & HÄTTESTRAND, C. 1999. Frozen-bed Fennoscandian and Laurentide ice sheets during the Last Glacial Maximum. *Nature*, **402**, 63–66.

KLUIVING, S. J., BARTEK, L. R. & VAN DER WATEREN, F. M. 1999a. Multi-scale analyses of subglacial and glaciomarine deposits from the Ross Sea continental shelf, Antarctica. *Annals of Glaciology*, **28**, 90–96.

——, —— & ——1999b. Sedimentology and glaciotectonics of Cenozoic glacial and glaciomarine sediments from the Ross Sea continental margin, Antarctica: the use of microstructures. *Terra Antarctica*, **3**, 167–171.

——, RAPPOL, M. & VAN DER WATEREN, F. M. 1991. Till stratigraphy and ice movements in eastern Overijssel, The Netherlands. *Boreas*, **20**, 193–205.

LARSEN, E. & MANGERUD, J. 1992. Subglacially formed clastic dikes. *Sveriges Geologiska Undersoekning*, Serie Ca. Avhandlingar och Uppsatser i, **4**, 163–170.

LIN, S., JIANG, D. & WILLIAMS, P. F. 1998. Transpression (or -transtension) zones of triclinic symmetry: natural example and theoretical modelling. *In*: HOLDSWORTH, R. E., STRACHAN, R. & DEWEY, J.

(eds) *Continental Transpressional and Transtensional Tectonics*. Geological Society, London, Special Publications, **135**, 41–57.

MALAVIEILLE, J. 1987. Kinematics of compressional and extensional ductile shearing deformation in a metamorphic core complex of the northeastern Basin and Range. *Journal of Structural Geology*, **9**, 541–554.

MALTMAN, A. 1987. Shear zones in argillaceous sediments – an experimental study. *In*: JONES, M. E. & PRESTON, R. M. (eds) *Deformation of Sediments and Sedimentary Rocks*. Geological Society, London, Special Publications, **29**, 77–87.

——1994. Deformation structures preserved in rocks. *In*: MALTMAN, A. (ed.) *The Geological Deformation of Sediments*. Chapman & Hall, London, 261–307.

MURRAY, T. 1994. Glacial deformation of sediments. *In*: MALTMAN, A. J. (ed.) *Geological Deformation of Sediments*. Chapman and Hall, London, 73–93.

PASSCHIER, C. W. 1998. Monoclinic model shear zones. *Journal of Structural Geology*, **20**, 1121–1137.

—— & TROUW, R. A. J. 1996. *Microtectonics*. Springer, Berlin.

——, DEN BROK, S. W. J., VAN GOOL, J. A. M., MARKER, M. & MANATSCHAL, G. 1997. A laterally constricted shear zone system – the Nordre Strømfjord steep belt, Nagssugtoqidian Orogen, W. Greenland. *Terra Nova*, **9** 199–202.

PETIT, J.1987. Criteria for the sense of movement on fault surfaces in brittle rocks. *Journal of Structural Geology*, **9**, 597–608.

PIOTROWSKI, J. A. & KRAUS, A. M. 1997. Response of sediment to ice-sheet loading in northwestern Germany; effective stresses and glacier-bed stability. *Journal of Glaciology*, **43**, 495–502.

—— & TULACZYK, S. 1999. Subglacial conditions under the last ice sheet in northwest Germany: ice-bed separation and enhanced basal sliding? *Quaternary Science Reviews*, **18**, 737–751.

——, DÖRING, U., HARDER, A., QADIRIE, R. & WENGHÖFER, S. 1997. Deforming bed conditions on the Dänischer Wohld Peninsula, northern Germany: Comments. *Boreas*, **26**, 73–77.

——, MICKELSON, D. M., TULACZYK, S. & KRZYSZKOWSKI, D. in press. Were deforming subglacial beds beneath past ice sheets really widespread? *Quaternary International*.

PLATT, J. 1986. Dynamics of orogenic wedges and the uplift of high-pressure metamorphic rocks. *Geological Society of America Bulletin*, **97**, 1037–1053.

RAMSAY, J. G. 1967. *Folding and Fracturing of Rocks*. McGraw Hill, New York.

—— & HUBER, M. I. 1983. *The techniques of modern structural geology. Volume 1: Strain Analysis*. Academic Press, London.

RAPPOL, M. & STOLTENBERG, H. M. 1985. Compositional variability of Saalian till in The Netherlands and its origin. *Boreas*, **14**, 33–50.

——, HALDORSEN, S., JÖRGENSEN, P., VAN DER MEER, J. J. M. & STOLTENBERG, H. M.1989. Composition and origin of petrographically-stratified thick till in the northern Netherlands and a Saalian glaciation model for the North Sea basin. *Mededelingen van de Werkgroep voor Tertiaire en Kwartaire Geologie*, **26**, 31–64.

RIJSDIJK, K. F., OWEN, G., WARREN, W. P., McCARROLL, D. & VAN DER MEER, J. J. M. 1999. Clastic dykes in over-consolidated tills: evidence for subglacial hydrofracturing in Killeney Bay, eastern Ireland. *Sedimentary Geology*, **129**, 111–126.

ROONEY, S. T., BLANKENSHIP, D. D., ALLEY, R. B. & BENTLEY, C. R. 1987. Till beneath ice stream B; 2, Structure and continuity. *Journal of Geophysical Research*, **B92**, 8913–8920.

SCHERER, R. P., ALDAHAN, A., TULACZYK, S., POSSNERT, G., ENGELHARDT, H. & KAMB, B. 1998, Pleistocene collapse of the West Antarctic Ice Sheet. *Science*, **281**, 82–85.

SJØRRING, S. 1977. The glacial stratigraphy of the island of Åls, southern Denmark. *Zeitschrift für Geomorphologie, N.F.*, Suppl.-Bd. **27**, 1–11.

——1983. Ristinge Klint. *In*: EHLERS, J. (ed.) *Glacial deposits in North-west Europe*. Balkema, Rotterdam, 219–226.

SLATER, G. 1926. Glacial tectonics as reflected in disturbed drift deposits. *Proceedings of the Geologists' Association*, **37**, 392–400.

——1927. The structure of the disturbed deposits of Møens Klint, Denmark. *Transactions of the Royal Society of Edinburgh*, **55**, 289–302.

TULACZYK, S., KAMB, B., SCHERER, R. P., ENGELHARDT, H. F. 1998. Sedimentary processes at the base of a West Antarctic Ice Stream: Constraints from textural and compositional properties of sub-glacial debris. *Journal of Sedimentary Research*, **68**, 487–496.

VAN DER MEER, J. J. M. 1993. Microscopic evidence of subglacial deformation. *Quaternary Science Reviews*, **12**, 553–587.

——1996. Microscomorphology. *In*: MENZIES, J. (ed.) *Glacial Environments: Volume 2; Past Glacial Environments – processes, sediments and landforms.* Butterworth–Heinemann, Oxford, 335–355.

——1997. Subglacial processes revealed by the microscope: particle and aggregate mobility in till. *Quaternary Science Reviews*, **16**, 827–831.

VAN DER WATEREN, F. M. 1985. A model of glacial tectonics, applied to the ice-pushed ridges in the central Netherlands. *Bulletin of the Geological Society of Denmark*, **34**, 55–74.

——1987. Structural geology and sedimentation of the Dammer Berge push moraine, FRG. *In*: VAN DER MEER, J. J. M. (ed.) *Tills and Glaciotectonics*. Balkema, Rotterdam, 157–182.

——1994. Processes of Glaciotectonism. *In*: MENZIES, J. (ed.) *Glacial Environments. Processes, Sediments and Landforms*. Pergamon Press, Oxford, 309–335.

——1995. Structural Geology and Sedimentology of Push Moraines. – Processes of soft sediment deformation in a glacial environment and the distribution of glaciotectonic styles. *Mededelingen Rijks Geologische Dienst*, **54**, 1–168.

——1999. Structural geology and sedimentology of Saalian tills near Heiligenhafen, Germany. *Quaternary Science Reviews*, **18**, 1625–1639.

# Micromorphological evidence for polyphase deformation of glaciolacustrine sediments from Strathspey, Scotland

## E. R. PHILLIPS & C. A. AUTON

*British Geological Survey, Murchison House, West Mains Road, Edinburgh EH9 3LA*
*(e-mail: erp@bgs.ac.uk)*

**Abstract:** A combination of field investigation and micromorphological analysis has been applied to polydeformed Late Devensian rhythmites and glacigenic diamicton, exposed in Strathspey, Scotland. This provided information on the geometry, kinematics and relative ages of ductile and brittle structures, and records a complex subglacial deformation history. The deformation is interpreted as resulting from a single progressive event, associated with over-riding of proglacial lake sediments by wet-based ice. The earliest deformation ('$D_1$') resulted from compaction/loading (pure shear) and imposed a bedding-parallel ($S_1$) fabric throughout the rhythmites. $S_1$ was subsequently deformed by kink bands and minor ductile shearing during '$D_2$'. A later '$D_3$' event, characterized by soft-sediment deformation and fluidization of matrix-poor sands, was accompanied by an increase in pore water pressure. This lead to hydrofracturing of the rhythmites. The most intense deformation ('$D_4$'), which resulted from simple shear, was partitioned into the upper part of the sequence. It produced folding, thrusting and brittle microfaulting in response to NNW-directed ice-push. These findings indicate that, in general, subglacial deformation is not homogeneous and can extend to depths of >3 m below the presumed ice-sediment interface.

The analysis of thin sections in geological structural and metamorphic studies is a long established technique (e.g. Spry 1969; Vernon 1989; Passchier & Trouw 1996 and references therein), providing a wealth of data on the geometry and kinematics of deformation structures. These data, combined with field measurements, are widely used in modelling the tectonic evolution of polydeformed and metamorphosed terranes. In contrast, the application of micromorphology in recognition and interpretation of microstructures associated with subglacial deformation is a relatively new technique (van der Meer & Laban 1990; van der Meer *et al.* 1992, 1994; van der Meer 1993; van der Meer & Warren 1997; Menzies this volume).

In this paper we apply a multidisciplinary approach, involving geological mapping, lithological logging and micromorphological analysis, to elucidate the sequence of depositional and deformational events recognised within a Late Devensian, glaciolacustrine sequence overlain by diamicton, exposed in the Strathspey area of Scotland. Detailed microstructural analysis is used to provide information on the geometry, kinematics and relative age relationships between ductile (fold, shear) and brittle (fault) structures, and in the development of a polyphase deformation model for this sequence.

Micromorphological evidence for the variation in pore water pressure during deformation has also been recognised. The presence of deformation structures (particularly related to compaction) at relatively deep stratigraphical levels within the glaciolacustrine sequence also has have important implications for establishing the true thickness of the deforming layer during subglacial deformation.

## Quaternary geology of the Raitts Burn area, Speyside

The Speyside area of Scotland displays an exceptionally well-developed suite of landforms and deposits that formed during the later stages of the last major Late Devensian (Dimlington Stadial) ice-sheet glaciation of northern Britain. Most of the glacial sediments on the lower flanks of the Spey Valley are products of deposition by valley glaciers and their associated meltwaters (Young 1978). This 'valley glaciation' occurred after the initial deglaciation of the surrounding mountains, where erosional features (such as glacially polished rock surfaces, striae and roche moutonée) predominate. The landforms and sediments within the valley of Raitts Burn [NH 760 047 – NH 796 036] (Fig. 1), a tributary

*From*: MALTMAN, A. J., HUBBARD, B. & HAMBREY, M. J. (eds) *Deformation of Glacial Materials*. Geological Society, London, Special Publications, **176**, 279–292. 0305-8719/00/$15.00 © The Geological Society of London 2000.

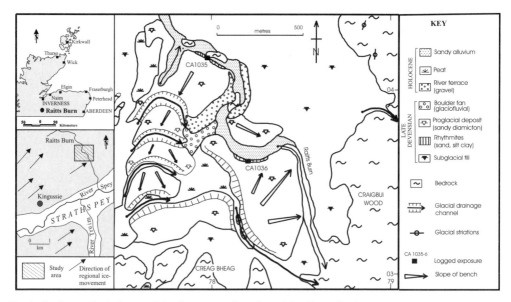

**Fig. 1.** Outline geological map of the Quaternary deposits and landforms in the Raitts Burn area, Strathspey, Scotland. Insets, location of the study area.

stream of the River Spey, were first described by Hinxman & Anderson (1915), in the Geological Survey Memoir account of the geology of mid-Strathspey and Strathdearn. The valley contains particularly good examples of glacigenic depositional landforms (including eskers and moraines) and sediments (matrix- and clast-supported diamictons, poorly sorted ice-proximal gravels and rhythmically laminated sands, silts and clays) all of which can be ascribed to the 'valley glaciation'.

Raitts Burn flows southeastwards across the northwestern flank of Strathspey, to join the valley of the River Spey *c.* 3 km downstream of Kingussie. It drains an ice-scoured hollow (*c.* 1.5 km$^2$ in diameter) in hard, resistant, mainly metasandstone bedrock of the Neoproterozoic Grampian Group. Regionally the bedrock shows evidence of glacial scouring, plucking and striation, indicating a general SW to NE direction of ice-movement (Fig. 1). This movement is reflected in the overall pattern of streamlining of rock exposures in the area (Young 1978; Auton 1998). However, on the interfluve north of Craigbui Wood (Fig. 1) glacial striae indicate NNE-directed ice movement. They may suggest divergent flow within the ice sheet, governed by perturbations of the bedrock, or a later advance of a glacier ice onto the interfluve from the Spey Valley.

Geological mapping (1:10 000 scale) of the Raitts Burn area, identified six gently sloping

benches surrounding a peat bog on the valley floor, and seventh bench on the northern side of the burn (Fig. 1). These features typically stand 10–15 m above the level of the floodplain of the stream. They resemble gently sloping glaciofluvial terraces, or deltas, but exposures in their frontal bluffs show that they are mainly composed of weakly stratified, sandy diamicton and are therefore interpreted as being principally of glacigenic rather than deltaic origin. The six western benches are dissected by arcuate glacial drainage channels that grade to successively lower elevations southwards. The morphology of the channels suggests that they formed at successive still-stand positions of the margin of the ice that laid down the sandy diamicton.

Lithological logging of sediments exposed in two cliff sections cut by the stream (CA1035 and CA1036, Fig. 2) has shown that the diamicton locally overlies a thin deposit of rhythmically interlaminated sand, silt and clay. These fine-grained rhythmites are interpreted as glaciolacustrine sediments which were deposited in an ice-dammed lake that occupied the topographic hollow now drained by the burn. The boggy, flat-lying peat covered ground, to the west of Raitts Burn, marks the former site of a subsequent lake that was impounded by ice at the mouth of the valley (to the SE) during the final stages of deglaciation. The glaciolacustrine sediments exposed in the cliffs, rest on a thin

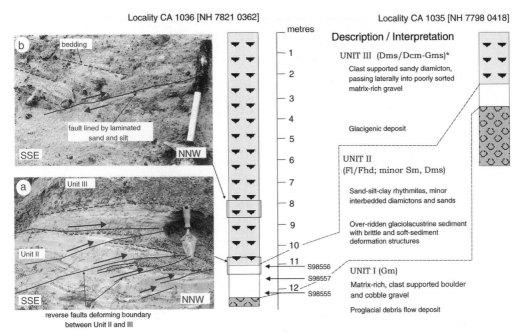

**Fig. 2.** Simplified lithological logs for the Quaternary deposits exposed at locations CA1035 and CA1306, Raitts Burn. (**a**) Small-scale reverse fault cutting silt-clay rhythmites, top of Unit II; trowel 26 cm long. (**b**) Sandy stratified diamicton, Unit III, hammer 40 cm long. * Lithofacies codes after Miall (1978) and Eyles *et al.* (1983).

unit of poorly sorted matrix-rich boulder, cobble and pebble gravel and contain a complex suite of sedimentary and tectonic structures. These structures were examined in detail from one exposure (CA1036, Fig. 1) using three orientated thin sections, with the aim of understanding the detailed sequence of depositional and deformational events that occurred during the last glaciation and deglaciation of the Raitts Burn basin.

## Stratigraphy and macroscopic deformation structures recorded in the logged sections

Lithological logs of the deposits exposed at localities CA1035 and CA1036 are summarized in Fig. 2. Three units are recognized at both sites. At the eastern end of exposure CA1035, poorly stratified, clast-supported, boulder and cobble gravel (2.8 m thick) at the base of the Quaternary sequence (Unit I) rests directly on reddened and decomposed metasandstone cut by granite sheets. The gravel comprises angular to subangular cobbles and boulders (>40 cm in diameter) within a hard, fine- to coarse-grained silty sand matrix. Cobbles and boulders of the

underlying metasandstone and granite predominate, but clasts of highly decomposed schistose semipelite are also present. This matrix-rich boulder and cobble gravel is interpreted as an ice-proximal debris-flow or glaciofluvial fan deposit. At CA1036, Unit I is a finer grained sandy gravel, containing rounded pebbles and cobbles of metasandstone. It is interpreted as being the more distal part of the deposit, which has undergone minor fluvial reworking. This suggests that the gravels of Unit I were laid down beyond an ice margin which retreated northwards (towards the head of the valley of Raitts Burn) during the initial deglaciation of the area.

The basal gravels in both exposures are overlain by a sequence of rhythmically laminated, graded, fine-grained sediments (Unit II). This unit also includes occasional thin interbeds of sand and sandy diamicton. At CA1035, Unit II comprises a 1 m thick interbedded sequence, of sands and upward-fining, sand–silt–clay rhythmites. Sedimentary structures in the rhythmites include climbing-ripple lamination and small-scale water-escape (flame) structures. Sand beds (0.3–0.4 m thick) occur near the top and bottom of the sequence. Some are cut by a number of steeply dipping, clay- and silt-lined fractures.

A number of more thinly laminated, graded beds, occurring in the middle of the unit, are boudinaged. Some of the thicker beds of sand have thin gravel lags, suggesting periods of more rapid water movement and sediment influx, perhaps associated with occasional drainage of the lake through or beneath its ice-dam. However, there is no evidence of desiccation or weathering indicating that subglacial drainage probably resulted in periodic lowering of lake level rather than complete emergence and drying out of the lake sediments.

At CA1036, the basal sandy gravel is overlain by thinly laminated rhythmites, which comprise a greater proportion of fine-grained sediment than at CA 1035. The intensity of faulting and folding visible in the exposures increases upwards through Unit II. In the basal 0.5 m the sand–silt–clay laminae (typically 1–2 mm thick) show little macroscopic evidence of deformation and are only locally displaced by small-scale, high-angle, NE–SW-trending normal faults. In the middle of the unit these laminated sediments are deformed by small-scale ENE–WSW-trending reverse faults (displacements of $c.$ 20 cm) and folds which yield an overall N/NNW-directed sense of movement. Individual graded silt-clay couplets are broken (boudinaged) into discrete 'rafts' within a matrix of homogenous fine-sand. This relationship is indicative of fluidisation and water escape; an interpretation supported by the presence of flame-structures within the middle of Unit II in exposure CA1035.

In the upper part of Unit II (CA1036), a thin (0.2 m) bed of diamicton is overlain by 0.2 m of finely laminated sand–silt–clay rhythmites with drop-stone pebbles. The upper 0.1 m of the laminated unit is deformed by small-scale, low-angle reverse faults (Fig. 2a) and gently inclined to recumbent, asymmetrical folds, which result in localized overturning of the lamination. The inclination of several of the fault planes, which cut through the fold hinges, is sub-parallel to the basal contact of the overlying sandy diamicton (Unit III). In section CA1035, this sandy diamicton passes laterally into clast-supported, poorly sorted, silty sandy gravel. At locality CA1036, the diamictic sediments (Fig. 2b) are 11 m thick. Small-scale, low-angle reverse faults, trending ENE–WSW, are developed within the lower half of the unit. These faults are lined with laminated, normally graded fine-grained sand and silt. The diamicton is stratified. It contains abundant sub-rounded pebbles of metasandstone and includes planar laminated interbeds of yellow-grey fine-grained sand and stringers of gravel. The deposit becomes more sandy and friable towards the top.

## Micromorphology of the deformed glaciolacustrine deposits

Orientated samples of the laminated, glaciolacustrine deposits (Unit II, Fig. 2) were collected for detailed examination of the range of microstructures developed and to assess changes in the style and intensity of deformation through this unit. Sample preparation (total time $c.$ 10 months) involved the replacement, in a vacuum, of the pore water by acetone. The latter was then replaced by a resin–acetone solution, progressively increasing the percentage of resin (up to 100%). The resin was then allowed to cure. Three large format ($75 \times 110$ mm) thin-sections were taken from the centre of the prepared samples. Each thin-section was cut orthogonal to the main deformation structures avoiding any artefacts associated with sample collection. The terminology used to describe the various planar fabrics developed within these sediments largely follows that proposed by van der Meer (1987, 1993; also see Menzies this volume). Successive generations of fabrics ($S_1, S_2 \ldots S_n$) and folds ($F_1, F_2 \ldots F_n$) are distinguished by the nomenclature normally used in structural geological studies ($S_1$ earliest fabric to $S_n$ latest). For a more detailed description see Phillips & Auton (1998).

### Sample S98555

This sample occurs near to the base of Unit II and is composed of finely laminated silt (estimated composition 60% silt, 40% clay) and clay ($\leq 5\%$ silt, $\geq 95\%$ clay) with a few thin (0.2 to 2.0 mm in thickness), normally graded (right-way-up), fine- to medium-grained sandy laminae (Figs 3 and 4a). The clay layers ($c.$ 50 laminae in 10 cm of thin section) possess sharp, planar tops. The sandy laminae (25% fine sand, 30% silt, 20% clay) represent periods of increased clastic input into a lake. The scattered, coarse sand to small pebble-sized clasts (Figs 3 and 4a), composed of metasandstone (psammite), biotite-granite and rhyolite rock fragments, are dropstones. Many are draped by the overlying sediment.

Two planar (plasmic) fabrics have been recognized within the clay-rich laminae: (1) a well developed bedding-parallel fabric ($S_1$); (2) a weaker, second ($S_2$) fabric developed at approximately 15° to bedding (in the plane of section). The bedding-parallel $S_1$ fabric is locally deformed/re-orientated adjacent to the dropstones, which are enclosed within a distinct 'pressure shadow'. Within the thicker clay laminae, $S_1$ is deformed by a set of lenticular to sigmoidal kink bands (dextral sense of off-set) which occur

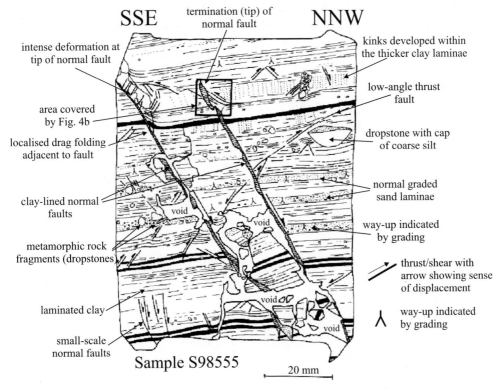

**SSE**  termination (tip) of normal fault  **NNW**

kinks developed within the thicker clay laminae

intense deformation at tip of normal fault

low-angle thrust fault

area covered by Fig. 4b

localised drag folding adjacent to fault

dropstone with cap of coarse silt

clay-lined normal faults

normal graded sand laminae

void

void

way-up indicated by grading

metamorphic rock fragments (dropstones)

thrust/shear with arrow showing sense of displacement

laminated clay

void

way-up indicated by grading

small-scale normal faults

**Sample S98555**

20 mm

**Fig. 3.** Thin section drawing of sample S98555 (see text for details). Also shown is the area of the thin section covered by Fig. 4b. The black filled layers are shaded to highlight the overall geometry of the microstructures developed within this sample.

at approximately 80° to 90° to bedding (Fig. 4b). Bedding and the kink bands (*sensu* Bordonau & van der Meer 1994) are deformed by a set of microfaults, most of which are compressional, and small-scale shears (Fig. 3) in which the shear plane is defined by a well-developed unistrial plasmic fabric (Fig. 4b). Immediately adjacent to these small-scale structures the lamination and $S_1$ fabric are folded/crenulated. In some cases, the tips of the small-scale shear planes appear to 'link' into bedding surfaces, suggesting that there has been some minor bedding-parallel movement.

The lamination and deformation structures described above are off-set by two, steeply inclined clay-lined normal faults. This faulting resulted in localized drag folding and brecciation of the sediments in the immediate footwall and hanging wall (Figs 3 and 4a). The amount of displacement on these structures varies progressively along the length of the fault, with maximum observed displacements of some 10 mm in their central parts, decreasing towards the fault-tip (Figs 3 and 4a). The clay-lining to the faults

possesses a well developed plasmic fabric. At least one further phase of movement occurred along these faults. It resulted in the brecciation of the clay-lining (Fig. 4b) and kinking/crenulation of its internal fabric. At the tip of one of the faults both the clay-lining and adjacent 'wall-rock' are highly brecciated. This breccia possess a fine clay matrix and expands away from the fault tip to form a 'v' shaped zone. The faults, as well as the earlier developed kink bands and shears, are locally sinistrally off-set along bedding (Figs 3, 4a and b), probably in response to later movement along bedding-parallel fractures.

*Sample S98557*

The second sample was taken from the middle of Unit II (Fig. 2). The lower part of the thin section is composed of laminated clay and silt (A on Fig. 5) which possesses a well-developed, bedding-parallel $S_1$ fabric. This fabric is deformed by variably developed, lenticular to sigmoidal, kink bands and small-scale shears

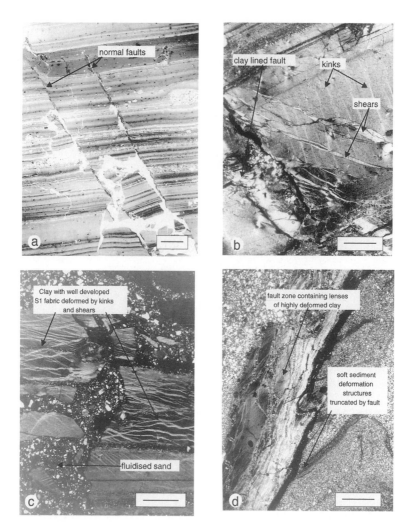

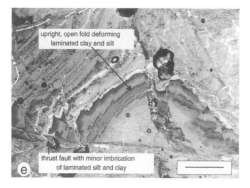

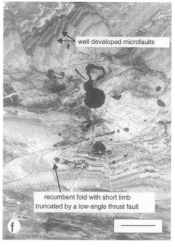

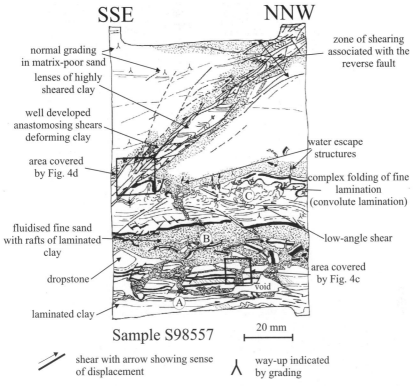

**Fig. 5.** Thin section drawing of sample S98557. Also shown is the area of the thin-section covered by Fig. 4c and d. A, laminated clay and silt; B, fluidized sand containing rafts of laminated clay and silt derived from A; C, laminated fine sand and silt deformed by soft-sediment deformation structures. The black (clay-silt) and stippled (sand) layers are coloured to highlight the overall geometry of the microstructures developed within this sample.

(Fig. 4c). The clay and silt are also deformed by a number of later fractures. These are filled by fine-grained, matrix-poor, homogeneous sand containing angular to weakly rounded fragments of clay, broken from the side-walls (Figs 4c and 5). This sand was derived from an overlying bed (10–15 mm thick). The lack of internal stratification within the sand bed (B on Fig. 5) and presence of elongate rafts of laminated silt and

clay indicate post-depositional fluidisation. The angular rafts of silt and clay preserve the $S_1$ fabric as well as the kinks and shears (Fig. 4c), indicating that both fluidization of the sand and the fracturing of the clay, post-dated $S_1$ and the development of the later ductile structures. The laminated silt and sand (C on Fig. 5), which overlies the fluidized sand layer, displays a well developed convolute lamination and is cut by a

**Fig. 4.** Photomicrographs. (**a**) Laminated clay and silt deformed by two clay lined extensional faults with minor drag folding and brecciation adjacent to these structures (sample S98555, plane polarized light, scale bar = 10 mm). (**b**) Thin clay lamina with well developed bedding-parallel ($S_1$) fabric. This fabric is deformed by later lenticular kink bands and shears. All of these microstructures are deformed by brittle deformation associated with the reactivation of the clay-lined extensional microfault (sample S98555, crossed polarised light, scale bar = 1 mm). (**c**) Fractured laminated clay with fractures filled by matrix-poor fine sand. Also note the presence of well developed kink bands and anastomosing shears within the clay which predate brittle deformation (sample S98557, crossed polarized light, scale bar = 1 mm). (**d**) Highly deformed clay within compressional microfault. Individual shear planes within the clay are defined by a well developed unistrial plasmic fabric (brighter birefringence). Also note the fault truncates an earlier fold developed during soft-sediment deformation (sample S98557, crossed polarized light, scale bar = 1 mm). (**e**) Upright, open fold developed in hanging wall of low-angle micro-thrust fault. Note that the lamination is deformed by a number of small-scale extensional and compressional microfaults (sample S98556, plane polarized light, scale bar = 2 mm). (**f**) Recumbent folds, thrusts and extensional microfaults developed within the laminated silt and clay at the top of the glaciolacustrine sequence (sample S98556, plane polarized light, scale bar = 2 mm).

number of sand-filled fluid escape conduits and microfaults (Fig. 5).

The upper part of the thin section comprises a medium-grained, matrix-poor sand which is deformed by a well developed complex reverse microfault zone (Fig. 5). This fault zone comprises an anastomosing network of discrete shear planes (showing a dextral sense of movement) that enclose broadly lenticular lenses of laminated clay and silt (Figs 4d and 5). Poorly defined shears also occur within the hanging wall of the fault (Fig. 5). The lenses are also deformed internally by anastomosing, fine-scale shears, which are defined by a very well developed unistrial plasmic fabric (Fig. 4d). In detail, the shears separate or enclose lenticular domains of relatively undeformed sediment, with the clay possessing a planar plasmic fabric. This fabric is comparable to $S_1$ within the laminated clay and silt in the footwall of the fault. The sedimentary lamination preserved within the clay-silt lenses was also crenulated during deformation associated with this reverse faulting. There is no evidence of injection of fluidized sand along the fault. It also clearly truncates the earlier developed convolute lamination (Figs 4d and 5), which shows that faulting post-dated soft-sediment deformation.

## Sample S98556

This sample represents the most intensely deformed part of Unit II (Fig. 2). In the thin section (Fig. 6a, b) inter-laminated clay and silt is overlain by coarse-grained, poorly sorted diamicton (Unit III). The most striking feature of this slide is the complexity of deformation recorded within the laminated silt and clay, when compared with the simple pattern preserved within the overlying diamicton (see below). A range of both ductile and brittle structures are developed in the silt and clay. These include: upright to recumbent microfolds, discrete shear planes, thrusts, normal and reverse microfaults, as well as a locally intense planar to crenulation-style fabric within the clay laminae (Figs 4e, f and 6). Dark-coloured laminae within the clay and silt provide useful marker horizons for unravelling the complex structure of this sample (see Fig. 6). Small-scale, normal grading within the silty layers indicates that, in general, the sediments are the right way up. However, localized overturning of the laminae occurs on the short, steep limbs of recumbent folds. The well developed, bedding-parallel $S_1$ fabric within the clay laminae is deformed by later kink bands, shears, microfaults and folds, which all relate to the main phase of deformation recorded in this sample.

Microfaults are common (Figs 4e, f and 6a, b) and are defined by very narrow shears that possess a variably developed plasmic fabric. Four principal types of microfaults are recognized: (1) Low-angle to bedding-parallel thrust faults (dextral sense of movement) resulting in localized, small-scale imbrication of the laminated silt and clay (Fig. 6a, b). Movement along these thrusts apparently locally reactivated and intensified $S_1$. Hence a composite $S_1/S_n$ fabric occurs within the adjacent clay laminae. (2) Moderate to shallowly dipping normal faults, which appear to link into the thrusts. (3) High-angle, locally closely spaced, extensional faults (most common). These divide the laminated clay and silt into a number of small, tilted fault blocks (Figs 4e, f and 6). These microfaults appear to terminate at the thrusts. (4) Moderate to steeply dipping reverse faults, which are locally associated with small-scale asymmetrical folds. These faults cut the steeply dipping short limbs of the folds. Many off-set the earlier developed thrusts and they are relatively common within the middle to upper part of the laminated silt and clay. The faults are particularly numerous close to the contact with the overlying diamicton.

Two styles of folds are recognized: (1) recumbent, close to tight and locally isoclinal structures; and (2) upright to gently inclined, open to moderately tight folds. Both sets of folds deform the $S_1$ fabric: The recumbent folds result in localized overturning of bedding within the laminated silt and clay (Figs 4f and 6a, b). The overturned short limbs are cut out by thrusts and their sense of asymmetry is consistent with the dextral sense of displacement along the thrusts. A second plasmic fabric occurs axial planar to the recumbent folds with small-scale shears developed in the cores of the folds. The open to moderately tight folds are deformed by closely spaced microfaults (types 3 and 4) which locally fan around the fold hinge (Figs 4e and 6a, b). These upright folds are also cut by the low-angle thrusts (Fig. 4e), indicating that they predate at least some of the displacement along the thrusts. The relative age relationships between the two styles folding is uncertain as no refolded fold relationships have been observed in thin section.

The diamicton at the top of the thin section (Fig. 6a) contains angular to subangular clasts ($\leq$7.0 mm in diameter) of metasandstone and granite supported within a finer grained sandy matrix. The boundary with the underlying laminated silt and clay is sharp, but irregular in form.

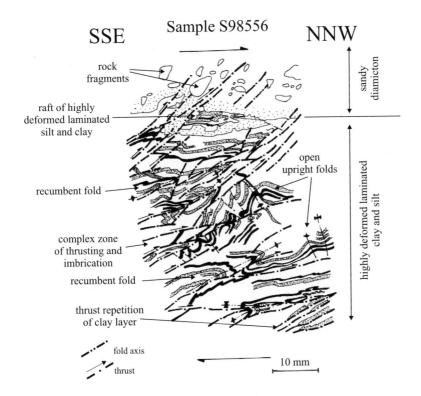

**(a)**

SSE — Sample S98556 — NNW

rock
fragments

raft of highly
deformed laminated
silt and clay

sandy
diamicton

recumbent fold

open
upright folds

highly deformed laminated
clay and silt

complex zone
of thrusting and
imbrication

recumbent fold

thrust repetition
of clay layer

fold axis

thrust

10 mm

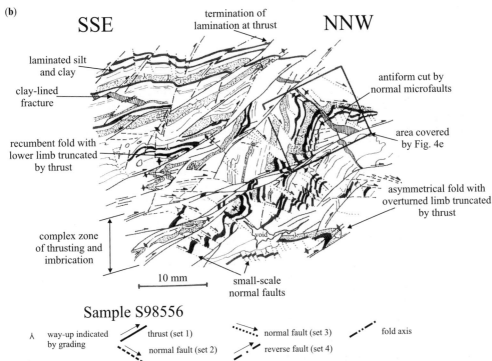

**(b)**

SSE — termination of
lamination at thrust — NNW

laminated silt
and clay

antiform cut by
normal microfaults

clay-lined
fracture

area covered
by Fig. 4e

recumbent fold with
lower limb truncated
by thrust

asymmetrical fold with
overturned limb truncated
by thrust

complex zone
of thrusting and
imbrication

void

10 mm

small-scale
normal faults

Sample S98556

way-up indicated
by grading

thrust (set 1)

normal fault (set 3)

fold axis

normal fault (set 2)

reverse fault (set 4)

**Fig. 6. (a)** Diagram summarizing the main microstructures developed within sample S98556. **(b)** Thin-section drawing showing, in detail, the range of microstructures developed within the laminated clay and silt (see text for details). Also shown is the area of the thin section covered by Fig. 4e. The black (clay) and stippled (silt) layers are shaded to highlight the overall geometry of the microstructures developed within this sample.

In places 'flame-like' apophyses of plastically deformed silt and clay extend into the diamicton. This boundary is off-set by at least two, moderate- to low-angle reverse faults (Fig. 6a). The faults can be traced into the diamicton where they are defined by a planar (omnisepic) fabric within its matrix. The sandy matrix of the diamicton appears to have been injected downward along these faults, detaching an elongate 'raft' of silt and clay near the base of the diamicton (Fig. 6a). Planar and lattice-like (lattisepic) fabrics are heterogeneously developed within the diamicton, reflecting the variation in clay content of its matrix. These fabrics locally appear to have nucleated upon the margins of the larger detrital grains (cf. cleavage development adjacent to porphyroblasts in metamorphic rocks, Vernon 1989; Passchier & Trouw 1996). Similar fabrics are also developed within wispy, highly deformed clay fragments within the diamicton. A concentric, circular (skelsepic) fabric is weakly developed within the matrix of the diamicton adjacent to some of the larger lithic clasts. Similar circular fabrics have been interpreted by van der Meer (1993) as having formed due to rotational deformation (simple shear).

## Interpretation of the microstructures and deformation history

### Origin of rhythmic lamination

The micromorphology of the fine-grained, rhythmically laminated sediments in Unit II, notably the systematic fluctuations in grain size and thickness indicate cyclic variations in sediment flux. Such fluctuations commonly result from daily, meteorological or annual variations in sediment discharge into lakes (Church & Gilbert 1975). Short-term (daily/meteorological) cycles commonly produce thin, macroscopically normally graded laminae with sharp basal contacts, forming gradational fining-upward units. In contrast, annual cycles tend to produce distinct sand-silt-clay rhythmites (varves) with macroscopically planar, sharp contacts between the coarse-, medium- and fine-grained components. These result from marked differences in sediment supply between summer and winter (De Geer 1912; Ashley 1975; Benn & Evans 1998). The macroscopically planar, sharp contacts between the sand, silt and clay laminae in sample S98555 suggests that these lacustrine sediments are true annual varves.

Analysis of the varves in thin section S98555 (which have only suffered minor post-deposi-

tional disturbance) indicates an approximate frequency of 50 varves within 10 cm of rhythmically laminated sediment. Using this frequency as representative of the sequence as a whole, then 1.2 m of varved sediment preserved in section CA1036 may equate to c. 600 years of glaciolacustrine deposition. This simplistic figure is probably an underestimate, however, as depositional hiatuses (caused by erosion during periods of more rapid water movement, associated partial drainage of the lake) are present in the sequence and its upper surface is truncated by the overlying diamicton.

### Deformation structures

Examination of the microstructures present in Unit II shows that the style of deformation changes and its intensity and complexity increase upwards through the glaciolacustrine sequence. The relationships between the microstructures described above indicate that these sediments have undergone polyphase deformation. The relative timing of the development of each type of microstructure is illustrated in Fig. 7. At all stratigraphic levels within the glaciolacustrine unit, the earliest recognized phase of deformation ('$D_1$') imposed a bedding-parallel $S_1$ fabric within the silt–clay laminae (Fig. 7). An absence of bedding-parallel or low-angle ductile shearing associated with $S_1$, suggests that this pervasive fabric developed in response to compaction/loading (pure shear) rather than simple shear. In the lower and middle parts of the unit, this early fabric was subsequently deformed by kink bands and ductile shears (assigned to '$D_2$', Fig. 7). These kinks and shears have not been recognized in the upper part of the unit, probably due to the intensity of subsequent deformation at this level.

The most prominent deformation structures toward the base of Unit II are the clay-lined normal microfaults. These post-date the development of the $S_1$ fabric, the '$D_2$' kink bands and shears (Fig. 4b) and are, therefore, '$D_3$' in age (Fig. 7). The presence of clay-linings show that faults were open due to essentially layer-parallel extension. This allowed water to flow along the fractures, their clay-linings to be deposited and also indicates that the sediments were not frozen during '$D_3$'. Brecciation of the clay-linings and the adjacent side-walls show that several phases of fault reactivation occurred during '$D_3$'. This may have been due to near vertical flow of overpressured subglacial groundwater through the sediments in a manner analogous to that described by Boulton & Caban (1995). The flow

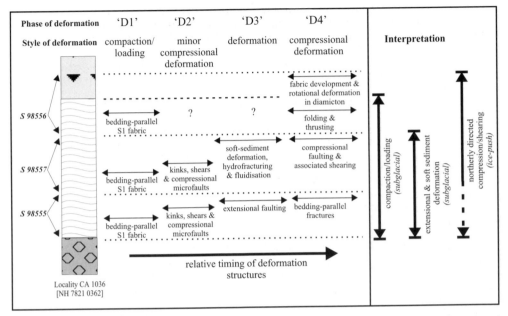

**Fig. 7.** Diagram summarizing relative age relationships (based upon microtextural evidence) and changes in style of deformation within the Raitts Burn glaciolacustrine sequence.

of overpressured water may have also resulted in the development of the 'v' shaped breccia zone directly above the termination of one of the normal faults (sample S98555). An increase in pore water pressure during fluid flow (syn-'D$_3$') would have exceeded the cohesive strength of the side-walls resulting in failure at the fault tip and brecciation. The morphology of this breccia zone is possibly comparable to the macroscopic 'funnel shaped' plumes described by Rijsdijk *et al.* (1999). Extensional movements on the faults were then followed by the development of bedding-parallel brittle fractures ('D$_4$', Fig. 7) which off-set all earlier structures.

The cohesive laminated clay and silt in the middle part of Unit II is fractured and the open fractures filled with fluidized matrix-poor sand. This phase of brittle deformation and associated fluidisation post-dates the development of S$_1$, as well as 'D$_2$' kink bands and shears (Fig. 4c). This is shown by the presence of clay–silt fragments, possessing all of the earlier 'D$_1$'–'D$_2$' microstructures, within the fluidised sand layer (B on Fig. 5). These fragments are derived from the underlying inter-laminated clay and silt (A on Fig. 5). In contrast the overlying, finely interbedded sands and silts (C on Fig. 5) were undergoing contemporaneous soft-sediment deformation, but only partial fluidisation (Fig. 7). Consequently, the change in style of deformation can be directly

related to variation in sediment composition and effective water pressure. Pore-water pressure increased until it exceeded the minimum cohesive stress of the sediment. This led to disaggregation and fluidisation of the matrix-poor sand and hydrofracturing (brittle deformation) of the underlying cohesive laminated clay and silt. In contrast, the overlying inter-laminated clayey silt and sand underwent plastic, soft-sediment deformation rather than disaggregation.

The microtextural evidence presented above suggests that hydrofracturing, fluidization and soft-sediment deformation within the middle of Unit II occurred at the same time as normal faulting within the lower parts of the unit ('D$_3$' on Fig. 7). Reverse faulting (designated as 'D$_4$') observed in the middle part of Unit II post-dates soft-sediment deformation and fluidisation (Fig. 7). The geometry of the reverse (compressional) fault and associated minor structures are consistent with an overall northerly directed sense of shear. Faulting and associated shearing of the matrix-poor sand and laminated clay indicates that there was an increase in the effective stress of the sediments after 'D$_3$'. This would have occurred due to a decrease in pore-water pressure in response to the coupled processes of hydrofracturing and fluidization.

As noted above, the most intense deformation occurs in the upper parts of Unit II. It produced upright to recumbent folds and associated thrusts ('D$_4$'). The sense of asymmetry of the folds and off-set on the thrusts are once again consistent with a northerly directed sense of shear. Continued movement along the thrusts led to the dissection of the upright folds and cut-out of the overturned limbs of recumbent structures. Ductile folding and thrusting during the initial stages of 'D$_4$', were followed by more brittle deformation. This formed the complex array of normal and reverse microfaults. The geometry of these microfaults is consistent with them having formed as Riedel shears (Passchier & Trouw 1996), with the bulk of displacement occurring along the approximately layer-parallel thrusts. These microfaults can be interpreted as accommodation structures resulting from continued shortening within the footwalls and hanging walls of the thrusts. The thrusts propagated upwards towards the boundary between the laminated clay and silt, and overlying diamicton.

The differences in style and apparent intensity of deformation recorded by the diamicton (Unit III) and underlying laminated silt and clay (Unit II) are interpreted as reflecting the gross lithological contrasts between the two units. The overall compressional regime and northerly directed sense of shear during folding and thrusting at the top of Unit II can be correlated with compressional faulting in the middle part of the unit.

## Deformation model for the Raitts Burn area

Analysis of the micromorphology of the glaciolacustrine sequence exposed in Raitts Burn has led to the development of a polyphase deformation history for these sediments.

'D$_1$' affected all levels within the glaciolacustrine sequence and is characterized by the development of a bedding-parallel S$_1$ fabric in response to compaction/loading.

'D$_2$', an apparently weak, ductile shearing event (compression) resulting in the development of kink bands and shears which deform S$_1$.

'D$_3$' was dominated by soft-sediment deformation, hydrofracturing and fluidization within the middle part of Unit II accompanied by extensional faulting in the lower part of the unit. Changes in the style of deformation are related to the variation in sediment composition and pore-water pressure. No evidence of 'D$_3$' soft-sediment deformation has been recognized in the upper part of the glaciolacustrine sequence.

'D$_4$' is characterized by locally tight folding, thrusting and brittle microfaulting in the upper part of the glaciolacustrine sequence caused by

northerly directed shear. In the middle part of the sequence this phase of deformation resulted in reverse faulting.

Elements of this 'D$_1$'–'D$_4$' deformation sequence can also be recognized within the macroscopic structures seen in outcrop. At both macro- and microscopic scales it is evident that: (1) the most intense deformation is partitioned into the upper most 30–40 cm of Unit II in the manner indicated by Boulton & Hindmarsh (1987); (2) that subglacial deformation is not homogeneous; (3) that it extends to depths in excess of 3 m below the presumed ice–sediment interface. Microstructural analysis also indicates that the initial stages of subglacial deformation involve compaction/loading (pure shear) and, therefore, can not be modelled by simple shear alone.

The deformation sequence outlined above can be directly applied to modelling the glacial and deglacial events within the Raitts Burn basin which followed initial regional ice-sheet decay. The changes in the style and intensity of deformation can be explained in terms of advance of ice across the lake basin (Auton 1998; Phillips & Auton 1998).

The simplest interpretation of the 'D$_1$' to 'D$_4$' structures is that they all developed during the same deformation event and that its intensity increased with time. Only the early weaker deformation structures ('D$_1$'–'D$_2$') are preserved in the lower parts of Unit II; late stronger deformation structures ('D$_3$'–'D$_4$') are preserved higher in the sequence. In the lower and middle parts of Unit II the deformation sequence is: 'D$_1$' compaction/loading (resulting in initial dewatering of the clays), minor 'D$_2$' shearing, followed by 'D$_3$' soft-sediment deformation (see Fig. 7). 'D$_1$' to 'D$_2$' can be attributed to initial glacial overriding of flat-lying lacustrine sediments and onset of diamicton deposition. The early stages of ice movement across the basin probably occurred on a water lubricated surface, which resulted in only a minor amount of shearing being translated into the underlying material. Further dewatering, extensional faulting and hydrofracturing occurred in the laminated clays and silts. This was accompanied by soft-sediment deformation and fluidisation of the interbedded sands as pore-water pressure increased during 'D$_3$'. Soft-sediment deformation of the silty sands and water flow along open normal faults further indicates that the glaciolacustrine sediments were not frozen during 'D$_1$' to 'D$_3$'.

The fall in pore-water pressure which followed 'D$_3$' hydrofracturing and sediment fluidization, increased the strength of the subglacial sediment. This enabled reverse faults to develop in the middle part of Unit II during 'D$_4$'. Continued northerly directed glacial overriding led to folding and thrusting in the upper part of the Unit II. The intensity of deformation in the upper part of Unit II suggests there was a reduction in lubrication within the ice-sediment boundary layer during 'D$_4$'. This allowed shearing to be translated into the underlying sediment. As 'D$_4$' continued, shearing also propagated upwards, leading to the deformation of the basal part of the diamicton, whilst the upper part of the unit was deposited. The deformation of the lower part of Unit III indicates that the pore-water pressure was relatively low during 'D$_4$' and that the unit was relatively coherent during its deposition.

All of the glacitectonic structures in Unit II and the lower parts of Unit III relate to the advance of ice northwards across the lake basin. The friable sandy diamictons in the upper parts of Unit III, are interpreted as ice-contact, subaerial debris-flows that have undergone a degree of winnowing and sorting by meltwater at an ice margin. They are thought to represent the glacigenic material, deposited at the front of the ice as it retreated southwestwards across the basin.

Morphologically the Raitts Burn glacigenic landforms do not readily fit into any of the categories of ice-marginal deposits described in the literature. Their gently sloping (c. 3–7°) planar upper surfaces are probably, in part, an artefact of erosion by meltwater at an ice front. However, the gross morphology of the landforms resembles that of composite ridges and thrust block moraines described by Evans (1991) and by Benn & Evans (1998). In particular, they display a rudimentary arcuate shape, with intervening depressions (glacial drainage channels) arranged in an 'concave up-glacier pattern'. The direction of slope of the upper surfaces of the features, together with the associated glacial drainage pattern, are also critical to understanding the pattern of retreat. The arcuate form of the western drainage channels (Fig. 1) and the southwestward slope of the western group of features, suggests southwesterly retreat of the western portion of the ice-front. The slopes of the eastern benches, which occur within an arc between NE and SE, indicate a pivoting and splitting apart of eastern portion of the ice front as it retreated.

The glacitectonic structures associated with advance of the ice across the basin are probably preserved because they are developed towards the base of a thick sedimentary sequence deposited within a topographic hollow. The small-scale structures described above are only

well developed within fine-grained deposits. No comparable structures, formed during ice-retreat, have been recognized in the upper parts of the sequence, in which sandy diamictons and gravels predominate. These coarser grained deposits are generally less well consolidated and poorly exposed. The model of ice-retreat in the basin is therefore mainly based on the morphology of the moraines and the associated pattern of meltwater drainage in the basin.

This model of glacial advance and retreat, established for the Raitts Burn basin, is probably also applicable to the nearby valleys of Glen Gynack and Allt na Baranachd, where similar glacigenic sediments are preserved. Comparable patterns of meltwater ponding, ice advance and retreat at other sites has the potential to elucidate the complex local variations in the regional pattern of deglaciation in the Central Highlands of Scotland towards the end of the Dimlington Stadial.

## Conclusions

Detailed examination of a range of micro-structures principally developed in glacio-lacustrine deposits exposed at Raitts Burn, Strathspey has led to recognition of polyphase deformation that was associated with glaciation and deglaciation of this small former ice-dammed lake basin. The relative ages of the tectonic microstructures have been established, providing important information concerning the glacitectonic evolution of the overridden lacustrine sequence. The recognition of deformation structures at a relatively deep stratigraphical level within the Raitts Burn glaciolacustrine sequence also has important implications for recognition of partitioning during subglacial deformation and for establishing the thickness of the deforming layer. Structures present in the lower parts of the sequence are associated with initial compaction of the lake sediments by wet-based ice. The intense ductile folding, shearing and associated brittle microfaulting at the top of the lacustrine unit records deformation occurring as a result of subsequent north-directed ice-push during deposition of the thick overlying sandy diamicton. All of the microstructures recorded relate to progressive deformation during the ice advance, whereas the pattern of ice retreat is evident only from the gross morphology of the glacigenic landforms.

The concept, applied here, of establishing the relative ages of micro- and/or macroscopic structures is of critical importance when attempting to elucidate the deformation history of all glacigenic sediments. This combination of field observations and micromorphological analysis is most usefully applied to interstratified sequences (e.g. tills and interbedded finer grained sediments), which are widespread throughout the Quaternary. In the finer-grained intercalations it is often possible to differentiate between little-deformed laminated units and their intensely deformed equivalents. This allows a more complete deformation sequence to be established than can be determined by concentrating on the structures developed in the tills alone. It is clear, therefore, that macro- and micromorphological examination of entire sequences is required to provide the most detailed interpretation of subglacial deformation.

This work forms part of the Monadhliath Regional Mapping programme. We would like to thank J. Merritt, J. Mendum and M. Smith for their comments and discussion of an earlier draft of the manuscript. J. Rose, J. van der Meer and A. Maltman are acknowledged for their constructive reviews of this paper, which is published with the permission of the Director of the British Geological Survey (NERC).

## References

ASHLEY, G. M. 1975. Rhythmic sedimentation in glacial Lake Hitchcock, Massachusetts-Connecticut. *In*: JOPLING, A. V. & McDONALD, B. C. (eds.) *Glaciofluvial an Glaciolacustrine Sedimentation*. SEPM, Special Publications, **23**, 304–320.

AUTON, C. A. 1998. *Aspects of the Quaternary Geology of 1:50 000 Sheet 74W (Tomatin)*. British Geological Survey, Technical Report, **WA/98/21**.

BENN, D. I. & EVANS, D. J. A. 1998. *Glaciers and Glaciation*. Arnold, London.

BORDONAU, J. & VAN DER MEER, J. J. M. 1994. An example of kinking microfabric in Upper Pleistocene deposits from Llavorsi (central southern Pyrenees, Spain). *Geologie en Mijnbouw*, **73**, 23–30.

BOULTON, G. S. & CABAN, P. 1995. Groundwater flow beneath ice sheets: Part II – its impact on glacier tectonic structures and moraine formation. *Quaternary Science Reviews*, **14**, 563–587.

—— & HINDMARSH, R. C. A. 1987. Sediment deformation beneath glaciers: rheology and geological consequences. *Journal of Geophysical Research*, **92**, 9059–9082.

CHURCH, M. & GILBERT, R. 1975. Proglacial fluvial and lacustrine sediments. *In*: JOPLING, A. V. & McDONALD, B. C. (eds) *Glaciofluvial and Glaciolacustrine Sedimentation*. SEPM, Special Publications, **23**, 22–100.

DE GEER, G. 1912. A geochronology of the last 12 000 years. *In*: *11th International Geological Congress, Stockholm, Compte Rendu*, **1**, 241–258.

EVANS, D. J. A. 1991. Canadian Landform Examples – 19: High Artic thrust block moraines. *The Canadian Geographer*, **35**, 93–97.

EYLES, N. E., EYLES, C. H. & MIALL, A. D. 1983. Lithofacies types and vertical profile models; an alternative approach to the description and environmental interpretation of glacial diamict and diamictite sequences. *Sedimentology*, **30**, 393–410.

HINXMAN, L. W. & ANDERSON, E. N. 1915. *The geology of mid-Strathspey and Strathdearn, explanation of Sheet 74*. Memoir of the Geological Survey, Scotland.

MENZIES, J. 2000. Brittle, ductile and polyphase deformation in glacial sediments using micromorphological analyses. *This volume*.

MIALL, A. D. 1978. Lithofacies types and vertical profile models in braided rivers: a summary. *In*: MIALL, A. D. (ed.) *Fluvial Sedimentology*. Canadian Society of Petroleum Geology Memoirs, **5**, 597–604.

PASSCHIER, C. W. & TROUW, R. A. J. 1996. *Microtectonics*. Springer.

PHILLIPS, E. R. & AUTON, C. A. 1998. *Micromorphology and deformation of a Quaternary glaciolacustrine deposit, Speyside, Scotland*. British Geological Survey Technical Report **WG/98/15**.

RIJSDIJK, K. F., OWEN, G., WARREN, W. P., McCARROLL, D. & VAN DER MEER, J. J. M. 1999. Clastic dykes in over-consolidated tills: evidence for subglacial hydrofracturing at Killiney Bay, eastern Ireland. *Sedimentary Geology*, **129**, 111–126.

SPRY, A. 1969. *Metamorphic textures*. Pergamon Press, Oxford.

VAN DER MEER, J. J. M. 1987. Micromorphology of glacial sediments as a tool in distinguishing genetic varieties of till. *In*: KUJANSUU, R. & SAARNISTO, M. (eds.) *INQUA Till Symposium, Finland 1985*. Geological Survey of Finland, Special Paper, **3**, 77–89.

—— 1993. Microscopic evidence of subglacial deformation. *Quaternary Science Reviews*, **12**, 553–587.

—— & LABAN, C. 1990. Micromorphology of some North Sea till samples, a pilot study. *Journal of Quaternary Science*, **2**, 95–101.

—— & WARREN, W. P. 1997. Sedimentology of Late Glacial Clays in lacustrine Basins, Central Ireland. *Quaternary Science Reviews*, **16**, 779–791.

——, RABASSA, J. O. & EVENSON, E. B. 1992. Micromorphological aspects of glaciolacustrine sediments in northern Patagonia, Argentina. *Journal of Quaternary Science*, **1**, 31–44.

——, VERBERS, A. L. L. M. & WARREN, W. P. 1994. The micromorphological character of the Ballycroneen Formation (Irish Sea Till): A first assessment. *In*: WARREN, W. P. & CROOT, D. (eds) *Formation and Deformation of Glacial Deposits*. Balkema, Rotterdam, 39–49.

VERNON, R. H. 1989. Porphyroblast-matrix microstructural relationships: recent approaches and problems. *In*: DALY *ET AL*. (eds) *Evolution of Metamorphic Belts*. Geological Society, London, Special Publications, **43**, 83–102.

YOUNG, J. A. T. 1978. The landforms of the Upper Strathspey. *Scottish Geographical Magazine*, **94**, 76–94.

# Large-scale glaciotectonic thrust structures in the eastern Danish North Sea

MADS HUUSE & HOLGER LYKKE-ANDERSEN

*Department of Earth Sciences, University of Aarhus, Finlandsgade 8, 8200 Århus N, Denmark (e-mail: mhuuse@geo.aau.dk)*

**Abstract:** The distribution of shallow, detached thrust fault complexes in the eastern Danish North Sea has been mapped based on 6400 km of high-resolution multichannel seismic data. Individual thrust segments are 200–1000 m long and 100–250 m thick with up to 200 m of horizontal displacement. Thrusting mainly affected upper Middle Miocene to lower Pleistocene strata and predated an extensive system of now buried subglacial valleys. The thrust structures are located along a 200 km long NW–SE trend. The general thrust direction is towards the SW, parallel to inferred directions of ice movement during the Elsterian and Saalian glaciations. This direction is, however, also parallel to the dip of the underlying detachment surface. The detachment surface corresponds to the mid-Miocene unconformity in the SE and the base of the Quaternary to the northwest. Detachment is generally located at depths of 130–270 m below sea level.

The structures are interpreted as glaciotectonic thrust structures of Elsterian and/or Saalian age. The driving mechanism is interpreted to be gravity spreading in front of glaciers advancing from the north, NE and east. Thrusting was facilitated by the presence of SW-dipping detachment surfaces.

Glaciotectonic deformation structures have been widely recognized in NW Europe and in the southern North Sea (Ussing 1907; Jessen 1931; Gry 1940; Figge 1983; Eissmann 1987; Aber *et al.* 1989; Ehlers 1990*a, b*; Brodzikowski 1995; Laban 1995; van der Wateren 1995). Apart from their spectacular influence on many formerly glaciated landscapes on the northern hemisphere, the main scientific interest in glaciotectonic structures is due to their usefulness for establishing directions and maximum extents of glacier advances during the Pleistocene (Gry 1940; Berthelsen 1978; Houmark-Nielsen 1987, 1988; Ehlers 1990*a, b*; van der Wateren 1995). Also of interest is their resemblance with thin-skinned deformation in foreland thrust and fold belts (e.g. Pedersen 1987; Aber *et al.* 1989).

The scale of deformation structures ranges from microscopic (Bruns 1989; van der Meer & Laban 1990) over m-scale (Berthelsen 1978; Houmark-Nielsen 1987; Aber *et al.* 1989) to several hundreds or even thousands of metres (Aber *et al.* 1989; Laban 1995; van der Wateren 1995; Petzold *et al.* 1998). The latter category is most suitable for high-resolution multichannel seismic investigations as outcrops are rarely of sufficient vertical and lateral extent to allow detailed mapping of more than parts of one such thrust sheet. A notable exception is that of the extensive browncoal pits in eastern Europe where large-scale glaciotectonic structures are revealed through progressive excavation of glacially deformed lignitic sands and clays (e.g. Eissmann 1987; Brodzikowski 1995; Eissmann *et al.* 1995; Hannemann 1995; Petzold *et al.* 1998).

Glaciotectonic structures have previously been reported from parts of the North Sea by Figge (1983), Cameron *et al.* (1989), Laban (1995), and Leth (1998). These studies employed single-channel, high-frequency and shallow-penetration seismic equipment, and generally provided little insight into the detailed architecture of the structures.

The objective of the present paper is to document the occurrence and architecture of a system of large-scale detached thrust structures in the eastern Danish North Sea. The study is based on high-resolution multichannel seismic data, which, to our knowledge, provide the first detailed images of large-scale glaciotectonic structures in the North Sea area. Moreover, a qualitative account of the origin of the structures is attempted.

## Geological setting

The eastern Danish North Sea is situated within the NW European lowland, which is characterized by lowland and shallow shelf areas (Fig. 1).

*From*: MALTMAN, A. J., HUBBARD, B. & HAMBREY, M. J. (eds) *Deformation of Glacial Materials.* Geological Society, London, Special Publications, **176**, 293–305. 0305-8719/00/$15.00 © The Geological Society of London 2000.

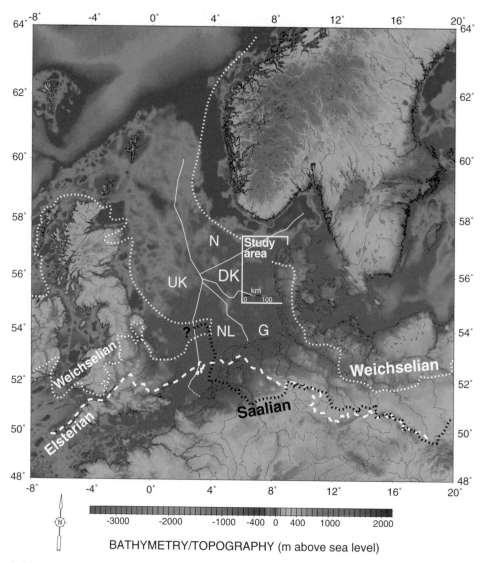

**Fig. 1.** The study area is located in the eastern Danish North Sea, which is part of the northwest European lowland. Topographic/bathymetric data have been extracted from the GTOPO30 database supplied with the GMT (3.0) package. Maximum extents of the Elsterian, Saalian and Weichselian ice sheets are indicated.

These areas are underlain by a thick Mesozoic and Cenozoic succession (see Ziegler 1990). Tectonic activity in the North Sea region generally ceased during the Cretaceous, leaving a passively subsiding basin to be filled by sediments supplied from the surrounding land areas during the Cenozoic (Nielsen *et al.* 1986; Ziegler 1990; Michelsen *et al.* 1998). During Oligocene and Miocene time the eastern North Sea was progressively filled by large deltas prograding from southern Norway (Oligocene),

Sweden (early Miocene) and finally from northern Germany (late Miocene). The progradational pattern was interrupted by a regional transgression in the middle Miocene, leading to widespread deposition of fine-grained clays in the North Sea, Denmark, and in NW Germany (Hodde and Gram Formations; Gramann & Kockel 1988; Koch 1989; Michelsen *et al.* 1998; Huuse 2000).

In Pliocene time, the eastern North Sea mainly constituted a bypass area for sediments filling-in the Central Trough further to the west (Gramann & Kockel 1988; Sørensen *et al.* 1997; Michelsen *et al.* 1998; Clausen *et al.* 1999; Huuse 2000). In the southwestern parts of the study area, a thin Pliocene succession is followed by an up to 300 m thick lower Pleistocene succession consisting of shallow to marginal marine gravel and sand, with subordinate amounts of silty clay (King 1973; DGU 1975; Toogood 1988; Huuse 2000).

During the Elsterian and Saalian glaciations the eastern North Sea was overridden by vast ice sheets that continued into northern Germany and the Netherlands, whereas the Weichselian ice sheets only reached the northern part of the study area (Fig. 1; Ehlers *et al.* 1984; Ehlers 1990*a, b*; Huuse & Lykke-Andersen 2000). Buried and overdeepened valleys analogous to the Mid-Late Pleistocene valleys known from other parts of the North Sea and onshore NW Europe have been observed in most of the Danish North Sea (Salomonsen 1993, 1995; Huuse & Lykke-Andersen 2000).

## Database

A total of 6400 km of high-resolution multi-channel seismic data was available for the study,

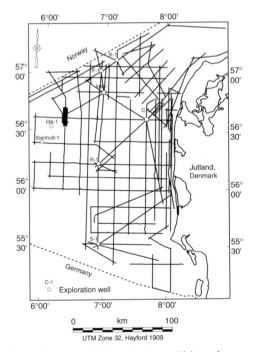

**Fig. 2.** Database. High-resolution multichannel seismic data (DA94, DA95, DA96, GR97, GR98) and available exploration wells with cuttings descriptions, biostratigraphy, and geophysical logs.

covering most of the eastern Danish North Sea (Fig. 2). Most of the high-resolution multi-channel seismic profiles (*c.* 5000 km) were acquired by the University of Aarhus (AU) in 1994–1996 (6 fold; DA94, DA95 and DA96), employing the Danish Research Vessel DANA. In 1997–98 a total of 1400 km were acquired by AU and the Geological Survey of Denmark and Greenland (24 fold; GR97 and GR98), as part of a five year mapping program, employing the naval Survey Vessel GRIBBEN.

Post-stack bandwidth of the migrated seismic data is 40–180 Hz (DA94) or 40–250 Hz (DA95, DA96, GR97, GR98). Penetration is of the order of 1–2 s two-way traveltime (TWT) corresponding to depths of 1–2 km. Vertical resolution of the data is of the order of 5 m and horizontal features wider than *c.* 20 m may be discerned in the upper 500 m subsurface.

Well logs (GR, sonic, velocity and resistivity) and unpublished stratigraphic reports, mainly based on cuttings analyses, were available from eight exploration wells in the study area (Fig. 2). Supplementary biostratigraphic analyses were carried out by Laursen (1995).

## Thrust structures

### Architecture of thrust fault complexes

In the western part of the study area, the thrust structures are detached at the WSW-dipping base Quaternary surface (Figs 3, 4), whereas further to the east, in the main part of the area, the detachment corresponds to the SW- to west-dipping mid-Miocene unconformity (Figs 5, 6).

The thrust structures in the NW are truncated at a high angle at the seabed (Fig. 3b), whereas the structures further to the east are fully preserved, except where truncated by Quaternary valleys (Fig. 5). As also indicated by boomer and multi-beam echo sounder data, the structures have no expression at the seabed (Fig. 3b), and may thus be considered inactive at present. The depth of detachment generally varies between 150–300 ms TWT (≈130–270 m) below sea level (Figs 4, 6), but slightly shallower and deeper levels have been observed in a few locations (Fig. 6). Thrusting appears to affect the entire sequence between the detachment surface and seabed, although the uppermost 50 ms (≈40 m) subbottom are often masked by seabed multiples, thus precluding confident interpretations of the shallowest parts (Fig. 5). Water depths are everywhere between 25 and 60 ms TWT (≈20–45 m), and the thickness of individual thrust sheets (depth to detachment minus water

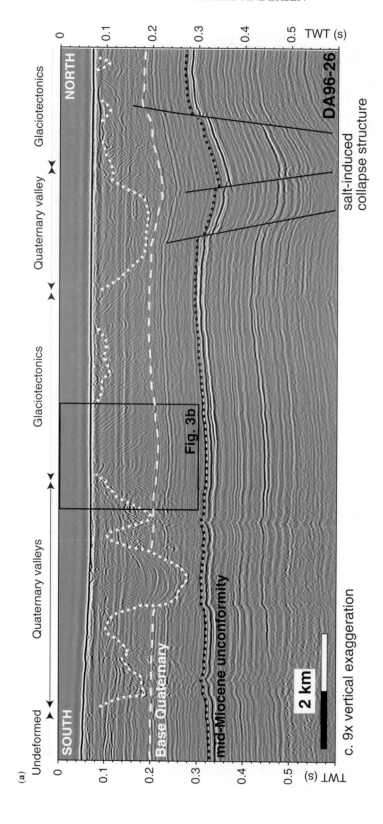

**(b)**

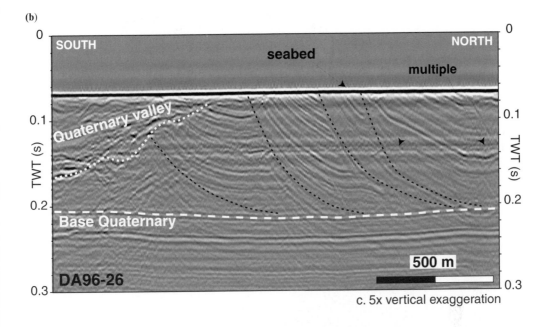

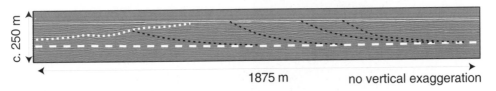

**Fig. 3.** (**a**) High-resolution multichannel seismic profile (DA96–26) showing a series of detached thrust structures intersected by deep Quaternary valleys. The detachment surface roughly coincides with the base Quaternary. The Quaternary succession is largely sandy (permeable), whereas the underlying Pliocene and Miocene succession consists mainly of silty clay (impermeable). (**b**) Zoom-in on a part of the thrusted succession. See Fig. 4 for location.

depth) is thus between 100 and 250 m. The length of individual thrust sheets, measured in the transport direction, may be fairly regular (Fig. 3a, b) or highly variable (Fig. 5). Generally, the length of individual thrust sheets is within 200–1000 m although slightly shorter and longer segments do occur. Typical angles of the thrust faults are between 10° and 25° where the faults have not been subsequently rotated due to continued deformation along more distal (younger) thrust faults (Figs 3b, 5). Generally, the thrusts steepen towards the surface before flattening at the tip of the thrust plane, where preserved (Fig. 5), thus producing a listric concave-upward morphology of the thrust plane followed by a slightly convex-upward curvature (see Figs 3b, 5).

Preliminary interpretations of the structures shown in Fig. 3 indicate that the two-dimensional (cross-sectional) geometry of the thrust planes may be approximated by the arc of a circle. Calculations based on this approximation yield an average horizontal displacement towards the SSW of up to 200 m per thrust. However, further studies are needed in order to test this approximation. Assuming a horizontal displacement of the order of 200 m per thrust fault and an average length of the thrust sheets of c. 500 m this yields a rough estimate of the amount of shortening accommodated by thrusting of c. 40%. For the example shown in Fig. 3a, this should leave a proximal depression due to removal of the thrusted sequence in the northern part of the profile, as in a hill and hole pair (see Aber *et al.* 1989). The lateral extent of such a depression should correspond to the amount of shortening, which may be roughly calculated as the amount of thrusts present, multiplied by the average horizontal displacement. Eight thrusts may be discerned in the central part of Fig. 3a, thus yielding an overall shortening of the order

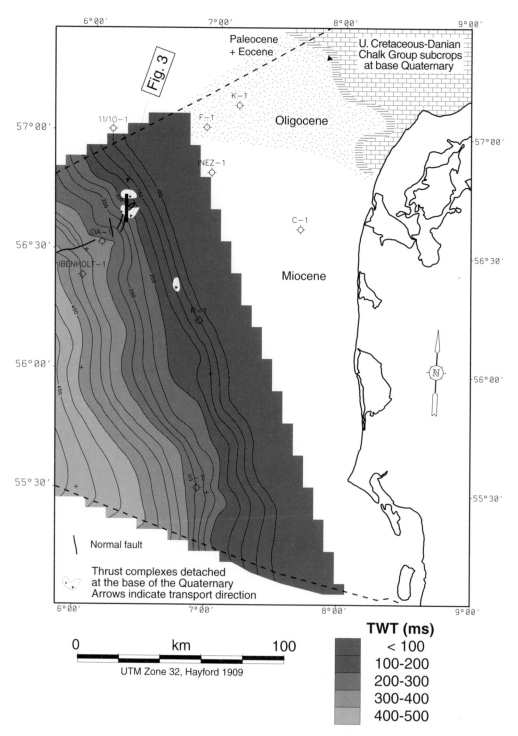

**Fig. 4.** Base Quaternary time-structure and subcrop map (Huuse & Lykke-Andersen 2000) with locations of thrust fault complexes detached at the base of the Quaternary.

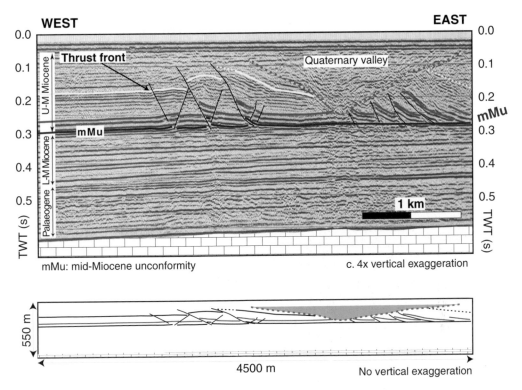

**Fig. 5.** High-resolution multichannel seismic profile (GR98-01) showing decreasing intensity of thrusting from east to west. The detachment surface is the mid-Miocene unconformity (mMu), which is characterized by fine-grained clays (Hodde and Gram Formations). Note that the upper 100 ms TWTT is masked by seabed multiples. See Fig. 6 for location.

of 1.5 km. This corresponds to the width of the Quaternary valley present above the salt-induced collapse structure to the north (Fig. 3a). However, the presence of deformed sediments north of the valley could indicate that the total amount of shortening was significantly greater than 1.5 km, and that the extensional part of the system could be located further to the north, beyond the extent of the seismic line. Unfortunately, the widespread occurrence of deep Quaternary valleys at the proximal parts of the deformed successions (see Fig. 7) generally precludes verification of the calculated amount of distal shortening by matching it with the amount of proximal extension.

The overall architecture of the thrusted succession (Figs 3, 5) shows a marked resemblance with the geometry of thrust structures associated with gravity spreading at the distal parts of thrust- and fold belts and at the toes of major deltaic wedges (see Boyer & Elliott 1982; Pedersen 1987; Aber 1988; Croot 1988). Based on similar considerations, gravity spreading has

previously been invoked in order to explain glaciotectonic thrust structures onshore Denmark (e.g. Pedersen 1987, 1996). The intensity of deformation associated with the structures described in this paper resembles that of large nappes rather than imbricate push moraines, which typically show much more closely spaced thrust planes and a generally shallower depth of detachment (see Klint & Pedersen 1995; van der Wateren 1995, figs 3.48, 3.49).

## Geographical extent of thrusting

Thrust structures detached at the base of the Quaternary are located in the NW of the study area, where the base Quaternary surface is situated at depths of 125–225 ms TWT (≈100–200 m) below sea level (Fig. 4). Further to the SE, thrust structures detach at (or near) the mid-Miocene unconformity at depths of 150–300 ms TWT (≈130–270 m) along a NW–SE trend (Fig. 6). Taken together, the thrust structures

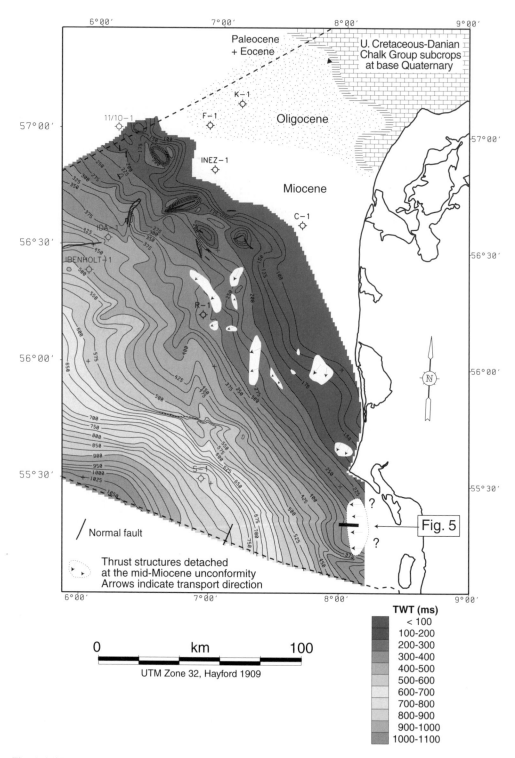

**Fig. 6.** Mid-Miocene unconformity time-structure and subcrop map (Huuse 1999). Locations of thrust fault complexes detached at the mid-Miocene unconformity (i.e. at the clays of the Hodde Formation) are indicated.

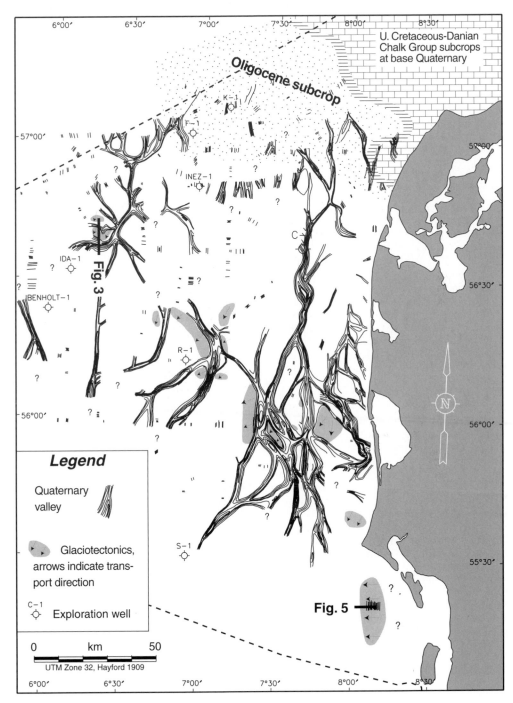

**Fig. 7.** Quaternary valleys (Huuse & Lykke-Andersen 2000, fig. 7) and distribution of glaciotectonic thrust fault complexes in the eastern Danish North Sea.

are distributed along a 200 km long NW–SE trend, from 56°40'N, 6°20'E to 55°10'N, 8°20'E (Fig. 7). However, important variations in the direction of thrusting occur between the south-easternmost part where thrusting is mainly towards the west, and the main part where thrusting is towards the SSW–SW (Fig. 7). The thrust direction of the structures detaching at the base of the Quaternary is somewhat oblique to the dip of the detachment (Fig. 4), whereas the thrust directions of the structures detaching at the mid-Miocene unconformity are virtually parallel to the dip of the detachment (Fig. 6).

## Relationship with subglacial valleys

The widespread occurrence of deep, buried Quaternary valleys in the Danish North Sea was demonstrated by Salomonsen (1993, 1995) and Huuse & Lykke-Andersen (2000). Salomon-sen (1993, 1995) attributed the valleys to fluvial incision during sea-level lowstand and subse-quent over-deepening by subglacial meltwater. However, regional considerations preclude an interpretation of the buried valleys as former river channels (Huuse & Lykke-Andersen 2000). Based on the morphology and distribution of the valleys, Huuse & Lykke-Andersen (2000) mainly attributed the valleys to subglacial erosion by meltwater.

The incision of the deep valleys clearly post-date thrusting (Figs 3a, 5). Thus, if the timing of valley incision was known it would place a con-straint on the timing of thrusting. However, the timing of valley incision is poorly constrained due to the scarcity of well control in the area (Fig. 2). The age of the valleys is therefore mainly constrained by comparisons with valleys onshore NW Europe, which are mainly Elsterian (e.g. Ehlers et al. 1984), and by the distribution of the valleys in relation to the maximum extents of the Pleistocene glaciations (Huuse & Lykke-Andersen 2000, fig. 4). Based on these con-siderations, Huuse & Lykke-Andersen (2000) suggested that most valleys were likely to be of Elsterian age, although a Saalian age is also possible. A Weichselian age can probably be excluded due to the limited extent of the Weich-selian ice sheets (Fig. 1).

## Substrate lithology

The depth and character of valley incision within the study area may have been influenced by variations in substrate porosity and permeability (Huuse & Lykke-Andersen 2000). The beha-viour of relatively unconsolidated sediments,

subject to rapid differential loading, is largely dependent upon grain size, porosity, permeabil-ity, shear strength, and pore pressure (e.g. Boulton & Hindmarsh 1987; Croot 1988; van der Wateren 1995). It is thus anticipated that failures will be preferentially located along bed boundaries where these factors change abruptly.

Offshore well control (DGU 1975; Toogood 1988), indicate that the interval affected by thrust-ing detached at base Quaternary (Figs 3, 4) is characterized by relatively permeable sand and gravel, whereas the underlying pre-Quaternary succession is composed mainly of less permeable silty clay. Onshore well and outcrop data (e.g. Rasmussen 1961; Michelsen et al. 1998) and offshore well data (S-1 well, DGU 1975) indi-cate that the structures detached at or near the mid-Miocene unconformity (Figs 5, 6) consist of upper Middle Miocene to Lower Pleistocene sands and silty clays. These structures detach into the fine-grained clays of the middle to upper Miocene Hodde and Gram Formations that overlie the pro-deltaic and deltaic Arnum, Ribe and Odderup Formations (see Rasmussen 1961). Amplitude anomalies related to gas accumula-tions below the mid-Miocene unconformity are widespread on seismic data from the eastern Danish North Sea, indicating that the Hodde Formation and probably the lower part of the Gram Clay constitute regional permeability barriers. Marked variations in pore-fluid pres-sures may thus be expected to occur across the mid-Miocene unconformity. Moreover, the shear strength of the Hodde and Gram clays may be further reduced in places by the presence of gas.

## Timing of thrusting

The location of the thrust structures beneath the floor of the North Sea precludes a direct dating of thrusting as only few exploration wells, yielding rather poor cuttings samples, have been drilled in the area (Fig. 2). The timing of thrust-ing is, however, constrained by the age of the youngest sediments affected by thrusting and by the age of unconformities truncating the struc-tures, e.g. subglacial valleys. It should be noted that, despite their morphological similarity, all thrust structures (or valleys) in the area are not necessarily coeval, and thrust complexes with different transport directions probably belong to different ice advances.

The youngest sediments affected by thrust-ing are of (undifferentiated) Pleistocene age (Figs 3, 4), indicating that thrusting occurred some time during the Pleistocene. Almost all of the areas affected by thrusting have been cut by subglacial valleys that post-date thrusting. These valleys are interpreted to be mainly of Elsterian and possibly Saalian age, thus constraining the youngest episode of thrusting to be of Elsterian or Saalian age.

## Mechanics of thrusting: discussion

The theoretical considerations regarding the mechanics of thrusting are still at an early stage and it is here only intended to give a qualitative account of how thrusting may have proceeded. As previously noted, the architecture of the detached thrust structures resembles structures formed by gravity spreading in thrust- and fold belts and at delta fronts. This resemblance, the close association of the structures with subglacial valleys of (inferred) Elsterian and possibly Saalian age, and the thrust-direction (sub-)parallel to the slope of the underlying detachment, point towards an interpretation of thrusting caused by gravity spreading in front of advancing glaciers.

The lithological variations of the substrate provide at least two discrete intervals suitable for detachment: the base of the Quaternary and the mid-Miocene unconformity. These intervals are characterized by abrupt variations in permeability and grain size. Note that both detachment surfaces slope towards the WSW, coinciding with directions of ice advance from the Scandinavian highland across Skagerrak and Jutland and from the Baltic. Advance of an ice sheet would subject the substrate to significant differential stresses, both vertically and horizontally at the ice front. These stresses would be analogous to stresses acting at the distal parts of thrust- and fold belts and at the toes of major deltas. In these settings gravity spreading may result in the development of thrust structures of geometries similar to those observed in this study (see Boyer & Elliott 1982; Pedersen 1987; Aber 1988). Gravity spreading in front of the advancing glaciers may have been facilitated by the presence of over-pressured meltwater at the base of the ice, causing build-up of over-pressure and decrease of shear strength focused at impermeable horizons. Increased pore pressure due to over-pressured meltwater could have been significant some distance into the foreland of the glacier, depending on the presence of permeability barriers such as clay beds or permafrost. In the case of sufficient differential stresses, over-pressured clay horizons could then develop into detachment surfaces such as observed at the base of the Quaternary and at the mid-Miocene unconformity. A detachment surface dipping away from the load imposed by the glacier would facilitate thrust faulting by letting the weight of the thrust sheet itself do a greater portion of the work than in the case of a flat-lying detachment.

For detached thrust structures (push moraines) observed elsewhere, it has been stated that 'the thickness of original sediments above the decollement can be a guide to the thickness of contemporary permafrost' (see Berthelsen 1978; Boulton *et al.* 1999). The presence of permafrost could explain the fact that stratification appear to be preserved within the thrust sheets (Figs 3b, 5). However, structures similar to those observed here have been observed in completely thawed environments (Pedersen 1987; Aber 1988; van der Wateren 1995). It thus appears that permafrost is not a requirement for producing the observed geometries in the presence of a sufficient differential load and a dipping detachment (see Croot 1988; van der Wateren 1995). On the other hand, permafrost would probably enhance the preservation potential of original stratification within the thrust sheets. Permafrost would probably be more important in the event that deformation was due to bulldozing by an advancing glacier, rather than gravity spreading.

## Conclusion

A large number of detached thrust fault complexes have been mapped in the eastern Danish North Sea. The individual thrust segments are 200–1000 m long and 100–250 m thick, and thrusting affect the entire sequence between seabed and the detachment surface, which is located at depths of 130–270 m below sea level. Preliminary studies of one thrust complex indicate that the amount of horizontal shortening may amount to 40%. Thrusting predated an extensive system of now buried subglacial valleys of Elsterian and possibly Saalian age.

The thrust complexes are generally located along a 200 km long NW–SE trend. The main thrust directions are towards the SW with minor components towards the south and west, and are thus parallel to inferred directions of ice movement during the Elsterian and Saalian glaciations. The thrust direction was also down-slope of the underlying detachment surface. The detachment surface corresponds to the mid-Miocene unconformity in the southeast and the base of the Quaternary towards the northwest, and the age of the affected sediments ranges from middle Miocene to Pleistocene.

The structures are interpreted as glaciotectonic thrust structures of Elsterian and possibly

Saalian age. Based on a comparison with other compressional settings, such as thrust and fold belts and toes of major deltas, the structures are interpreted as due to gravity spreading in front of advancing ice sheets. The driving mechanism is thought to be a combination of ice loading from the north, NE and east, and the presence of SW- to west-dipping detachment surfaces. Excess meltwater discharge during glacier advance may have facilitated thrusting by increasing the pore pressure and thus decreasing the shear strength of impermeable horizons. Permafrost may or may not have been present during thrusting; it is not thought to be a requirement for the formation of the structures described here or structures of similar scale observed elsewhere.

The financial support of the Danish Natural Science Research Council (grant nos. 9401161 and 9502760) is greatly appreciated. Thanks to S. A. S. Pedersen and G. Williams, who reviewed the manuscript and made many valuable comments and suggestions.

# References

ABER, J. S. 1988. Ice-shoved hills of Saskatchewan compared with Mississippi Delta mudlumps – Implications for glaciotectonic models. *In*: CROOT, D. G. (ed.) *Glaciotectonics: Forms and Processes*. Balkema, Rotterdam, 1–9.

——, CROOT, D. G. & FENTON, M. M. 1989. *Glaciotectonic landforms and structures*. Kluwer, Dordrecht.

BERTHELSEN, A. 1978. Methodology of kineto-stratigraphy as applied to glacial geology. *Bulletin of the Geological Society of Denmark*, **27**, 25–38.

BOULTON, G. S. & HINDMARSH, R. C. A. 1987. Sediment deformation beneath glaciers: Rheology and geological consequences. *Journal of Geophysical Research*, **92(B9)**, 9059–9082.

——, VAN DER MEER, J. J. M., BEETS, D. J., HART, J. K. & RUEGG, G. H. J. 1999. The sedimentary and structural evolution of a recent push moraine complex: Holmstrømbreen, Spitsbergen. *Quaternary Science Reviews*, **18**, 339–371.

BOYER, S. E. & ELLIOTT, D. 1982. Thrust systems. *American Association of Petroleum Geologists Bulletin*, **66**, 1196–1230.

BRODZIKOWSKI, K. 1995. Pre-Vistulian glaciotectonic features in southwestern Poland. *In*: EHLERS, J., KOZARSKI, S. & GIBBARD, P. (eds) *Glacial deposits in north-east Europe*. Balkema, Rotterdam, 339–359.

BRUNS, J. 1989. Stress indicators adjacent to buried valleys of Elsterian age in North Germany. *Journal of Quaternary Science*, **4**, 267–272.

CAMERON, T. D. J., SCHÜTTENHELM, R. T. E. & LABAN, C. 1989. Middle and Upper Pleistocene and Holocene stratigraphy in the southern North Sea between 52° and 54°N, 2° to 4°E. *In*: HENRIET, J. P. & DE MOOR, G. (eds) *The Quaternary*

*and Tertiary Geology of the Southern Bight, North Sea*. University of Ghent, 119–135.

CLAUSEN, O. R., GREGERSEN, U., MICHELSEN, O. & SØRENSEN, J. C. 1999. Factors controlling the Cenozoic sequence development in the eastern parts of the North Sea. *Journal of the Geological Society, London*, **156**, 809–816.

CROOT, D. G. 1988. Morphological, structural and mechanical analysis of neoglacial ice-pushed ridges in Iceland. *In*: CROOT, D. G. (ed.) *Glaciotectonics: Forms and Processes*. Balkema, Rotterdam, 33–47.

DGU 1975. *Dansk Nordsø S-1x, Lithologisk og biostratigrafisk rapport*. Geological Survey of Denmark.

EHLERS, J. 1990a. Reconstructing the dynamics of the north-west European Pleistocene ice sheets. *Quaternary Science Reviews*, **9**, 71–83.

——1990b. *Untersuchungen zur Morphodynamik der Vereisungen Norddeutschlands unter Berücksichtigung benachbarter Gebiete*. Bremer Beiträge zur Geographie und Raumplanung, **19**. Universität Bremen.

——, MEYER, K.-D. & STEPHAN, H.-J. 1984. Pre-Weichselian glaciations of north-west Europe. *Quaternary Science Reviews*, **3**, 1–40.

EISSMANN, L. 1987. Lagerungsstörungen im Lockergebirge. *Geophysik und Geologie, Geophysikalische Veröffentlichungen der Karl-Marx Universität, Leipzig*, **III(4)**, 7–77.

——, LITT, T. & WANSA, S. 1995. Elsterian and Saalian deposits in their type area in central Germany. *In*: EHLERS, J., KOZARSKI, S. & GIBBARD, P. (eds) *Glacial deposits in north-east Europe*. Balkema, Rotterdam, 439–464.

FIGGE, K. 1983. Morainic deposits in the German Bight area of the North Sea. *In*: EHLERS, J. (ed.) *Glacial deposits in north-west Europe*. Balkema, Rotterdam, 299–304.

GRAMANN, F. & KOCKEL, F. 1988. Palaeogeographical, lithological, palaeoecological and palaeoclimatic development of the Northwest European Tertiary Basin. *In*: VINKEN, R. (ed.) *The Northwest European Tertiary Basin*. Geologisches Jahrbuch, **A100**, 428–441.

GRY, H. 1940. De istektoniske forhold i molersområdet. *Meddelelser Dansk Geologisk Forening*, **9**, 586–627.

HANNEMANN, M. 1995. Über Intensität und Verbreitung glazigener Lagerungsstörungen im tiefen Quartär und im Tertiär Brandenburgs. *Brandenburgische Geowissenschaftliche Beiträge*, **2**, 51–59.

HOUMARK-NIELSEN, M. 1987. Pleistocene stratigraphy and glacial history of the central part of Denmark. *Bulletin of the Geological Society of Denmark*, **36**, 1–189.

——1988. Glaciotectonic unconformities in Pleistocene stratigraphy as evidence for the behaviour of former Scandinavian ice sheets. *In*: CROOT, D. G. (ed.) *Glaciotectonics: Forms and Processes*. Balkema, Rotterdam, 91–99.

HUUSE, M. 1999. *Cenozoic evolution of the eastern North Sea Basin – new evidence from high-resolution and conventional seismic data*. PhD thesis, University of Aarhus.

——2000. Late Cenozoic palaeogeography of the eastern North Sea Basin: climatic vs. tectonic forcing of deltaic progradation? *Bulletin of the Geological Society of Denmark*, **47**, in press.

—— & LYKKE-ANDERSEN, H. 2000. Overdeepened Quaternary valleys in the eastern Danish North Sea: morphology and origin. *Quaternary Science Reviews*, **19**, 1233–1253.

JESSEN, A. 1931. *Lønstrup Klint*. Danmarks Geologiske Undersøgelse, **II49**.

KING, C. 1973. *R-1x Stratigraphical/paleontological final report*. Paleoservices England. DGU Report File no. **3069**.

KLINT, K. E. S. & PEDERSEN, S. A. S. 1995. *The Hanklit Glaciotectoic Thrust Fault Complex, Mors, Denmark*. Geological Survey of Denmark, Series A, **35**.

KOCH, B. E. 1989. *Geology of the Søby-Fasterholt area (text)*. Geological Survey of Denmark, Series A, **22**.

LABAN, C. 1995. *The Pleistocene glaciations in the Dutch sector of the North Sea: A synthesis of sedimentary and seismic data*. PhD thesis, University of Amsterdam.

LAURSEN, G. V. 1995. Foraminiferal biostratigraphy of Cenozoic sections in five wells from the Danish area. *EFP-92 project: Basin development of the Tertiary of the Central Trough with emphasis on possible hydrocarbon reservoirs*. Report No. 20. University of Aarhus/Geological Survey of Denmark.

LETH, J. O. 1998. *Late Quaternary geology and Recent sedimentary processes of the Jutland Bank region, NE North Sea*. PhD thesis, University of Aarhus.

VAN DER MEER, J. J. M. & LABAN, C. 1990. Micromorphology of some North Sea till samples, a pilot study. *Journal of Quaternary Science*, **5**, 95–101.

MICHELSEN, O., THOMSEN, E., DANIELSEN, M., HEILMANN-CLAUSEN, C., JORDT, H. & LAURSEN, G. V. 1998. Cenozoic Sequence Stratigraphy in the Eastern North Sea. *In*: HARDENBOL, J., DE GRACIANSKY, P. C., JACQUIN, T. & VAIL, P. R. (eds) *Mesozoic and Cenozoic Sequence Stratigraphy of European Basins*. Society for Sedimentary Geology (SEPM), Special Publications, **60**, 91–118.

NIELSEN, O. B., SØRENSEN, S., THIEDE, J. & SKARBØ, O. 1986. Cenozoic Differential Subsidence of North Sea. *American Association of Petroleum Geologists Bulletin*, **70**, 276–298.

PEDERSEN, S. A. S. 1987. Comparative studies of gravity in tectonic Quaternary sediments and sedimentary rocks related to fold belts. *In*: JONES, M. E. & PRESTON, R. M. F. (eds) *Deformation of sediments and sedimentary rocks*. Geological Society, London, Special Publications, **29**, 165–180.

——1996. Progressive glaciotectonic deformation in Weichselian and Palaeogene deposits at Feggeklit, Mors, Denmark. *Bulletin of the Geological Society of Denmark*, **42**, 153–174.

PETZOLD, H., BAHRT, W., BAUER, M. & SEITZ, R. 1998. Seismic investigation of glacigenous macro structures at the border of a lignite deposit. *In*: *European Association of Geoscientists and Engineers 60th Conference and Technical Exhibition, Leipzig, Abstract Volume*, P130.

RASMUSSEN, L. B. 1961. *De miocæne Formationer i Danmark*. Geological Survey of Denmark, Series IV, **4(5)**.

SALOMONSEN, I. 1993. *Quaternary buried valley systems in the eastern North Sea*. PhD thesis, University of Copenhagen.

——1995. Origin of a deep buried valley system in Pleistocene deposits of the eastern central North Sea. *In*: MICHELSEN, O. (ed.) *Proceedings of the 2nd Symposium on: Marine Geology – Geology of the North Sea and Skagerrak*. Geological Survey of Denmark, Series C, **12**, 7–19.

SØRENSEN, J. C., GREGERSEN, U., BREINER, M. & MICHELSEN, O. 1997. High-frequency sequence stratigraphy of Upper Cenozoic deposits in the central and southeastern North Sea areas. *Marine and Petroleum Geology*, **14**, 99–123.

TOOGOOD, J. F. 1988. *Biostratigraphy of Danish North Sea well Phillips Norway 5605/20-1 (Ibenholt-1)*. Robertson Research Report No. **3755/Ia**.

USSING, N. V. 1907. Om floddale og randmoræner i Jylland. *Oversigt over Det Kongelige danske Videnskabernes Selskab Forhandlinger*, **4**, 161–213.

VAN DER WATEREN, F. M. 1995. *Structural Geology and Sedimentology of Push Moraines*. Mededelingen Rijks Geologische Dienst, **54**.

ZIEGLER, P. A. 1990. *Geological Atlas of Western and Central Europe*. Shell Internationale Petroleum Maatschappij BV, The Hague.

# An instability mechanism for drumlin formation

## A. C. FOWLER

*University of Oxford, Mathematical Institute, 24–29 St Giles', Oxford OX1 3LB, UK*
*(e-mail: fowler@maths.ox.ac.uk)*

**Abstract:** Drumlins are subglacial bedforms that are formed by the interaction of ice flow with an erodible basal topography. The mechanism of their formation bears resemblance to similar processes that cause the formation of dunes and anti-dunes in rivers, and sand dunes in deserts. In 1998 Hindmarsh showed that the interaction of a shearing ice flow with a deformable basal till layer could cause an instability which promotes the growth of basal topography, though he was unable to give analytic criteria for the instability. Here we analyse Hindmarsh's model, and by using certain approximations, we are able to give concise analytical parametric criteria for this instability. The resultant instability occurs if the basal shear stress is larger than a critical value which increases with increasing basal effective pressure, and which also depends on the basal till thickness. It is hypothesized that this instability is the basic mechanism involved in the formation of Rogen moraine and drumlins.

Theories for the origin of drumlins (Fig. 1) have been debated in the geological literature for over a hundred years. Early authors (Upham 1892; Tarr 1894; Russell 1895; Millis 1911) associated their formation with erosion or deposition of basal till. A summary of the early work is given by Gravenor (1953), together with an assessment of the relative merits of the erosional and depositional theories. An extensive survey of

the literature is that of Everett (1987). The books by Sugden & John (1976) and Embleton & King (1975) give useful discussions, both of the physical nature of drumlins, and of the various theories concerning their formation. As Sugden & John point out, the dynamical theories are conspicuously unsatisfactory.

More recently, the erosion/deposition type theories have been fused by Boulton (1987),

**Fig. 1.** Drumlins in the Ards Peninsula, County Down, Northern Ireland. Typical elevations are 30 metres and lengths several hundred metres. The photograph is taken from the top of one drumlin.

*From*: MALTMAN, A. J., HUBBARD, B. & HAMBREY, M. J. (eds) *Deformation of Glacial Materials*. Geological Society, London, Special Publications, **176**, 307–319. 0305-8719/00/$15.00 © The Geological Society of London 2000.

who suggests that the mechanism of drumlin formation is associated with the deformational behaviour of subglacial tills, an idea previously advanced by Smalley & Unwin (1968) and Menzies (1979), and evidentially supported by Boyce & Eyles (1991).

A radical alternative is the phenomenological flood theory of Shaw and co-workers (Shaw 1983, 1994; Shaw & Kvill 1984; Shaw et al. 1989), who suggest, on the basis of a perceived similarity with fluvially eroded bed forms, that drumlins are formed similarly, in massive sub-glacial floods. This idea has been supported by Sharpe (1987), for example. The plausibility of this theory relies primarily on the existence and nature of such large floods; and while it is undeniable that floods, such as jökulhlaups, do occur, their erosive effects remain hypothetical.

The present paper offers an orthogonal approach to the dynamics of drumlin formation. Although the opposing till deformation or flood-derived theories are closer to a quantitative theory than earlier more discursive ideas, they are still a long way short of a genuine dynamical description. Such a description was initiated by Hindmarsh (1996, 1998b); in particular, the latter paper offers an extended effort to bridge

the conceptual gap between the geomorphologist and the applied mathematician. While such bridge-building is essential, we do not seek to embroider the present content in this way. In the custom of the dynamicist, we strip the phenomenon to its dynamical essence, and will, for example, have nothing to say about drumlin stratigraphy.

## Stability of bedforms

To a dynamicist, the most obvious source of explanation for drumlins, such as those shown in Figs 1 and 2, is that they represent the evolved form of an instability. The image in Fig. 2 is suggestive of an evolved instability generated by local variations in topography. Drumlins (rock-cored or not) are formed of till (Fig. 3), and the basic dynamical explanation is that a uniform layer of subglacial till is unstable. Fluid mechanics abounds with such spatial instabilities, and the idea that the basic instability can lead, under different circumstances, to Rogen moraine, drumlins or megaflutes, carries little in the way of surprise. (Rogen moraine consists of semi-regular washboard-like ridges of till aligned

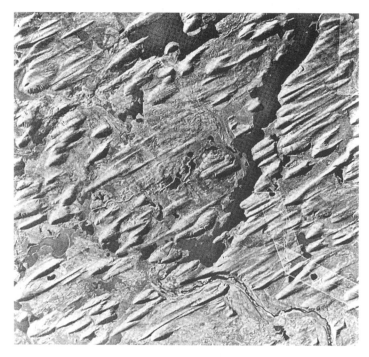

**Fig. 2.** A field of drumlins in Saskatchewan. Ice flow was from the bottom left (Boulton 1987, p. 66). This image is suggestive of a spatial instability, but one which is strongly influenced by inhomogeneity in the underlying substrate (whence the patchiness in the waveforms).

**Fig. 3.** Subglacial till, from a coastally exposed drumlin at Killough, County Down, Northern Ireland. Irregular rock fragments are dispersed in a matrix of finer material.

transverse to the ice flow.) If a uniform till layer is unstable, it is natural to expect a transition, as a critical parameter changes, from a stable state (no bedforms) to two-dimensional waves (Rogen moraine), which are themselves transversely unstable (drumlins), leading in extreme conditions to megaflutes. Indeed, such ideas have already been more or less expressed (Lundqvist 1989; Menzies & Shilts 1996). Such sequences of transitions are very familiar in convection (Busse 1978), shear flow (Maslowe 1985), and more germanely in fluvial erosion forms (Richards 1982), and the analogy between drumlins and fluvial dunes is of long standing (Kinahan & Close 1872, and particularly Davis 1884).

More controversial is the origin of the instability. The flood theory of Shaw assumes that bedforms (particularly drumlins) occur through subglacial floods. We do not examine this possibility further. The deformation theory assumes that the instability is analogous to fluvial bedforms (in the sense that dunes are caused by the interaction of an overlying shear flow with an erodible and mobile substrate), but

is due to the ice flow. Our purpose in this paper is to show that such an instability does in fact exist, and thereby elucidate what the primary controlling mechanics are. The model presented here is essentially due to Hindmarsh (1998a, b), who showed that the classical Nye–Kamb sliding theory allied with basal till deformation would produce instability in many cases. Hindmarsh used a viscous till rheology, but the resulting algebraic complication prevented him from deriving concise criteria for instability. Our purpose here is thus to derive analytic criteria for Hindmarsh's instability.

The starting point of the theory is the observation of Hindmarsh (1996, 1998b) that a layer of till of thickness $s$ depending, let us say, on one (flow-line) spatial direction $x$, satisfies an evolution equation of the form

$$s_t + q_x = 0 \qquad (1)$$

where subscripts denote partial derivatives, $t$ is time, and $q$ is the till flux. This equation neglects erosion or deposition rates, which would be represented by sink or source terms on the right hand side.

Evidently the till flux $q$ will depend on the till thickness $s$ and also the basal shear stress $\tau$. The dependence is mediated by the till rheology, and while there are different possibilities for this (Boulton & Hindmarsh 1987; Kamb 1991; Hooke et al. 1997), it seems clear that till flow decreases as the effective pressure $p_e$, defined as the difference between the confining pressure and the pore water pressure $p_w$, increases.

We will suppose for simplicity that the till layer rests on a flat substrate, e.g. bedrock. (More generally, bedrock irregularities will promote the instability we discuss, and Fig. 2 may be suggestive of this.) If the till layer is perturbed to have variable thickness, then as the ice flows over a protuberance (where $s$ is larger), the overburden normal stress will increase. This causes an increase in $p_e$ and thus a decrease in till flow. In turn, this can cause the till to stall and further enhance the growth of $s$. It is this feedback between growth of $s$ and increased normal stress in the ice which causes instability (Hindmarsh 1998a).

## Subglacial bedforms

We now consider two-dimensional flow of an ice sheet over a deformable substrate, as illustrated in Fig. 4. We take the vertical axis to be $z$, with $z = 0$ defining the ice–till interface in the uniform reference state, and we suppose that $z = -\bar{s}$ represents the (flat) base of the deforming till,

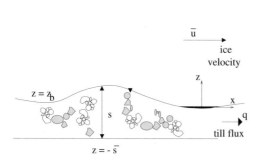

**Fig. 4.** Schematic geometry of the ice sheet and till flow.

whether this be bedrock or immobile till (Boulton 1987), and the ice surface is at $z = z_i$.

*Effective pressure*

Within the till, we let $p_s$ be the average pressure within the solid phase, and $p_w$ is the volume averaged pore water pressure. The total confining pressure within the till is therefore

$$P = (1 - n)p_s + np_w \qquad (2)$$

where $n$ is the porosity, and Terzaghi's effective pressure is then defined (see Bear & Bachmat 1990, pp. 153ff., for a lucid discussion) by

$$p_e = P - p_w = (1 - n)(p_s - p_w). \qquad (3)$$

Crucial to the determination of till velocity is the effective pressure at the ice–till interface, denoted by $N$, and defined as $-\sigma_{nn} - p_w$, where $-\sigma_{nn}$ is ice normal stress at the ice–till interface, and $p_w$ is pore water pressure.

Consider first a uniform shearing flow of ice of thickness $z_i$ over a layer of till of thickness $\bar{s}$ (we avoid the complication that in an ice sheet there is a small but crucially non-zero surface slope). In this reference state, the ice pressure is

$$p_i = p_a + \rho_i g(z_i - z) \qquad (4)$$

where $p_a$ is atmospheric pressure, $\rho_i$ is ice pressure, and $g$ is the acceleration due to gravity. If the pore pressure $p_w$ is equal to the pressure $p_c$ of a supposed channelized drainage system, then the interfacial effective pressure is given by

$$\bar{N} = p_a + \rho_i g z_i - p_c. \qquad (5)$$

Now consider the more general situation shown in Fig. 4 where the ice–till interface is at $z = z_b$. The ice flows over the irregular bed, and

this generates non-cryostatic pressure and normal stresses at the bed. If we define the reduced pressure $\Pi$ by

$$p_i = p_a + \rho_i g(z_i - z) + \Pi \qquad (6)$$

and let $\tau_{nn}$ be the deviatoric stress in the ice normal to the interface, then the effective pressure at the interface is given by

$$N = \bar{N} + \Delta\rho_{wi} g z_b + \Pi - \tau_{nn} \qquad (7)$$

where

$$\Delta\rho_{wi} = \rho_w - \rho_i \qquad (8)$$

$\rho_w$ is the density of water, and we suppose that the hydraulic potential within the till is constant and equal to the drainage channel pressure, i.e.

$$p_w + \rho_w g z = p_c. \qquad (9)$$

The drainage effective pressure $\bar{N}$ is assumed to be determined via a subglacial drainage theory such as that of Walder & Fowler (1994), and will be assumed to be constant in this study.

Below $z_b$, we suppose that the till and pore water pressures increase hydrostatically, and thus

$$p_e = N + (1 - n)\Delta\rho_{sw} g(z_b - z) \qquad (10)$$

where

$$\Delta\rho_{sw} = \rho_s - \rho_w \qquad (11)$$

and $\rho_s$ is the density of sediments.

*Till rheology*

To determine the till velocity, we must choose a rheology for till. The conceptual candidates are a viscous law (Boulton & Hindmarsh 1987), a plastic failure law (Kamb 1991; Iverson *et al.* 1998), or some more convoluted generalization (Clarke 1987). In addition, sliding probably occurs at ice–till and till–bed interfaces, the latter promoting erosion. Partly because our purpose is didactic, we select a viscous rheology. As pointed out by Boulton (1996), the plastic rheology can be accommodated within this framework by an appropriate parametric limit. This is discussed further in Appendix A.

We suppose that the shear rate of till is

$$\frac{\partial v}{\partial z} = A \exp(\alpha\tau/p_e) \qquad (12)$$

where $\tau$ is the shear stress, $v$ is the horizontal till velocity and we suppose that $\alpha$ is large. This is essentially Kamb's (1991) choice, and is promoted by Mitchell (1993, pp. 359ff.); but it can also be related in a practical sense to Boulton & Hindmarsh's (1987) one (see Appendix A).

We use (10) and denote $z_b - z = \zeta$, so that

$$-\frac{\partial v}{\partial \zeta} = A \exp\left[\frac{\alpha \tau}{N + \Delta\rho_{sw}g(1 - n)\zeta}\right]$$

$$\approx A^* \exp(-\zeta/\zeta^*) \qquad (13)$$

where

$$A^* = A \exp\left(\frac{\alpha \tau}{N}\right)$$

$$\zeta^* = \frac{N^2}{\alpha \tau \Delta\rho_{sw}g(1 - n)} \qquad (14)$$

using the assumption that $\zeta \ll \alpha\tau\zeta^*/N$ (mainly for algebraic convenience). Assume that at the till–bed interface, slip does not occur, that is, $v = 0$ at $\zeta = s$. If, in addition, slip does not occur at the ice–till interface, then $v = u$ at $\zeta = 0$, where $u$ is the basal ice velocity and

$$u - v = A^*\zeta^*[1 - e^{-\zeta/\zeta^*}] \qquad (15)$$

whence

$$u = A^*\zeta^*[1 - e^{-s/\zeta^*}] \qquad (16)$$

and thus the till flux $q = \int_0^s v\, d\zeta$ is given by

$$q = A^*\zeta^{*2}\left[1 - \left(1 + \frac{s}{\zeta^*}\right)e^{-s/\zeta^*}\right]. \qquad (17)$$

Figure 5 shows a sequence of graphs of $q$ as a function of $s$ with $\alpha = 10$, for various values

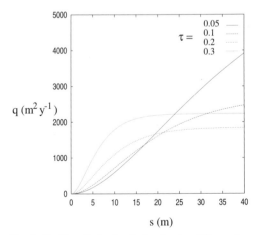

**Fig. 5.** Profiles of $q(s, \tau)$ calculated using (17); a value of $\alpha = 10$ is used, and units are $m^2\, a^{-1}$ for till flux, while $s$ is given in metres. Graphs are shown for $\tau = 0.05, 0.1, 0.2$ and $0.3$ bars. Other constants are taken as $\Delta\rho_{sw}g(1 - n) = 0.1\,bar\,m^{-1}$, $N = 1\,bar$, $A = 10\,a^{-1}$.

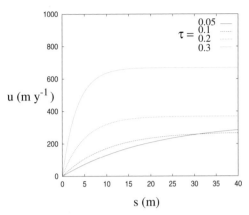

**Fig. 6.** Ice velocity as a function of till thickness, same parameters as in Fig. 5.

of $\tau$. It can be seen that $q$ is an increasing function of $s$, and also generally increases with $\tau$, except at low stresses and high till thicknesses. Figure 6 shows the basal ice velocity computed for the same parameter values. Note that $u$ also increases with both $\tau$ and $s$, except at low stresses and high till thicknesses.

Till conservation is expressed by the equation (1), where we ignore the small effects of basement erosion and depositional lodgement from the overlying ice.

### Basal sliding

The basic mechanism for the instability we shall discuss is the following. If a uniform layer of till is perturbed (for example by a bedrock knoll), then the flow of the ice over the bump causes an excess normal stress on it, and hence an increased effective pressure at the ice–till interface. Since the till flux (Fig. 5) is generally increasing with $\tau$ and thus decreasing with $N$, this causes a slowing down and thus accumulation of the till. In order to model this stability mechanism, we need to describe the deformational flow of the ice, essentially using the theory of basal sliding.

The classical theory of sliding was initiated by Weertman (1957), and developed principally by Nye (1969, 1970) and Kamb (1970). This is the form of the theory we use here (see also Hindmarsh 1998a). We ignore regelation and consider the slow flow of ice over the basal till layer; we take the ice viscosity $\mu$ to be constant. In the uniform (undisturbed) state, with the ice–till interface at $z = 0$, a steady shearing flow in the basal ice due to an imposed basal shear stress $\bar{\tau}$ is $(\bar{u} + \bar{\tau}z/\mu, 0)$, where the cartesian coordinates are $(x, z)$. In the theory presented here, we will suppose that the ice depth is very large, so that this velocity field is obtained far from the bed. The assumption of large depth is valid provided the wavelength of basal undulations is much less than the depth, and for drumlins this may not be highly accurate. We follow the assumption here for simplicity in exposition, and also because the instability we find does not rely on having a finite ice depth. In classical parlance, $\bar{\tau}$ is the basal shear stress and $\bar{u}$ is the effective sliding velocity. Sliding theory thus computes $\bar{\tau}$ as a function of $\bar{u}$ and other variables. (Notice that the uniform basic state is a shearing flow, which is not the case in classical sliding theory; Nye 1969.)

In terms of a stream function $\psi$ (so the ice velocity is $(\psi_z, -\psi_x)$), we have

$$\psi \sim \bar{u}z + \frac{\bar{\tau}}{2\mu}z^2 \quad \text{as} \quad z \to \infty. \tag{18}$$

We define the reduced pressure $\Pi$ via (6), so that

$$\Pi_x = \mu\nabla^2\psi_z$$
$$\Pi_z = -\mu\nabla^2\psi_x \tag{19}$$

with

$$\Pi \to 0 \quad \text{as} \quad z \to \infty. \tag{20}$$

At the ice–till interface $z = z_b$, we define the deviatoric shear stress to be $\tau$, the deviatoric normal stress is $\tau_{nn}$, and the tangential velocity (taken as continuous across the interface) is $u$, which we take to be given by (16). In terms of $\psi$, we have

$$-\tau_{nn} = \frac{2\mu}{1+s_x^2}[(1-s_x^2)\psi_{xz} + s_x(\psi_{zz} - \psi_{xx})]$$

$$\tau = \frac{\mu}{1+s_x^2}[(1-s_x^2)(\psi_{zz} - \psi_{xx}) - 4s_x\psi_{xz}] \tag{21}$$

Lastly, continuity of velocity at the interface implies

$$\psi_z = u$$
$$\psi_x + s_t + us_x = 0. \tag{22}$$

In order to see that the model is sensibly posed, we can argue as follows. $\psi$ satisfies

$$\nabla^4\psi = 0 \tag{23}$$

and requires two boundary conditions on each of $z = s$ and $z \to \infty$. As $z \to \infty$, (18) is equivalent to two conditions, and also serves to determine $\bar{\tau}$ in terms of $\bar{u}$. The equation (1 determines $s$ in principle, while (16) determines $u$. (22) then provide the two basal boundary conditions for $\psi$, while (21) serves to determine $\tau$ and $N$, which are involved in the determination of $q$ and $u$.

Of course, the boundary conditions do not actually uncouple in this way, but this way of thinking helps to show that the problem has the right number of boundary conditions.

## Linear stability analysis

The basic uniform state is that of a shear flow over a horizontal ice–till base at $z = 0$, and an underlying till layer of uniform thickness $\bar{s}$. We denote uniform state variables with an overbar, and consider perturbations to this state by writing

$$\psi = \bar{u}z + \frac{\bar{\tau}}{2\mu}z^2 + \Psi. \tag{24}$$

We linearize about the basic solution $\Psi = 0$, $s = \bar{s}$, $u = \bar{u}$, $\tau = \bar{\tau}$, on the basis that $s_x \ll 1$. We then have

$$\Pi_x = \mu\nabla^2\Psi_z$$
$$\Pi_z = -\mu\nabla^2\Psi_x \tag{25}$$

with

$$\Pi, \Psi \to 0 \quad \text{as} \quad z \to \infty \tag{26}$$

and

$$\Psi_z \approx u - \bar{u}$$
$$-\Psi_x \approx s_t + \bar{u}s_x$$
$$\tau - \bar{\tau} \approx \mu(\Psi_{zz} - \Psi_{xx})$$
$$N \approx \bar{N} + \Delta\rho_{wi}g(s - \bar{s})$$
$$\quad + \Pi + 2\mu\Psi_{xz} + 2\bar{\tau}s_x \tag{27}$$

on $z = 0$, approximately. For a mode of wavenumber $k$, $\Psi$ is given by

$$\Psi = (a + bz)e^{-kz}e^{ikx+\sigma t} \tag{28}$$

whence

$$\Pi = -2\mu ikbe^{-kz}e^{ikx+\sigma t} \tag{29}$$

and if

$$u = \bar{u} + \tilde{u}e^{ikx+\sigma t}$$
$$\tau = \bar{\tau} + \tilde{\tau}e^{ikx+\sigma t}$$
$$s = \bar{s} + \tilde{s}e^{ikx+\sigma t}$$
$$N = \bar{N} + \tilde{N}e^{ikx+\sigma t} \tag{30}$$

then the boundary conditions (27) give

$$b - ka = \tilde{u}$$
$$-ika = (\sigma + ik\bar{u})\tilde{s}$$
$$\tilde{\tau} = -2k\mu\tilde{u}$$
$$\tilde{N} = -2\mu ik^2 a + (\Delta\rho_{wi}g + 2ik\bar{\tau})\tilde{s}. \tag{31}$$

If we write the first term in $(31)_4$ in terms of $\tilde{s}$, then we can see that the relative sizes of the three terms in that equation are $2\mu k^2 u$, $\Delta\rho_{wi}g$ and $2k\tau$. If we use values $\mu \approx 6\,\mathrm{bar\,a}$, $u \approx 100\,\mathrm{m\,a^{-1}}$, $\Delta\rho_{wi} \approx 10^2\,\mathrm{kg\,m^{-3}}$, $g \approx 10\,\mathrm{m\,s^{-2}}$, $\tau \approx 0.2\,\mathrm{bar}$, and anticipating $k \approx 10^{-2}\,\mathrm{m^{-1}}$ (as a representative drumlin wave number), then the corresponding estimates in units of bar $\mathrm{m^{-1}}$ are 0.12, 0.01, and 0.004. On this basis we neglect the second and third terms in $(31)_4$. Strictly, this neglect is invalid when $k$ is very small, but this is of no significance, as we shall find that in that case the growth rate of unstable modes becomes very small. The basal ice velocity $u(s, \tau, N)$ is expanded as

$$\tilde{u} = u_s \tilde{s} + u_\tau \tilde{\tau} + u_N \tilde{N} \qquad (32)$$

where $u_s = \partial u/\partial s$, evaluated at $s = \bar{s}$, $\tau = \bar{\tau}$ and $N = \bar{N}$, and similarly if we define the perturbed till flux via $q = \bar{q} + \tilde{q}e^{ikx+\sigma t}$, then

$$\tilde{q} = q_s \tilde{s} + q_\tau \tilde{\tau} + q_N \tilde{N}. \qquad (33)$$

Finally, linearization of (1) gives the recipe for the eigenvalue $\sigma$ as

$$\sigma\tilde{s} + ik\tilde{q} = 0. \qquad (34)$$

The equations (31–34) provide seven relations for $a$, $b$, $\tilde{u}$, $\tilde{s}$, $\tilde{\tau}$, $\tilde{N}$, $\tilde{q}$ and $\sigma$. To simplify them, it is easiest to first use $(31)_2$ to write $\tilde{s}$ in terms of $a$, then combine (33) and (34) to eliminate $\tilde{q}$. (32) and the modified (33) then provide two homogeneous equations for $a$ and $\tilde{u}$, when $\tilde{s}$, $\tilde{\tau}$ and $\tilde{N}$ are written in terms of them. Cross multiplication then yields an expression for $\sigma$ in the form

$$\sigma = \rho - ikc \qquad (35)$$

where $\rho$ is the growth rate and $c$ is the wave speed, and these are given by

$$\rho = \frac{2\mu k^3 \Delta_1 \Delta_2}{(1 + 2\mu k u_\tau)^2 + 4\mu^2 k^4 \Delta_2^2} \qquad (36)$$

and

$$c = \bar{u} - \frac{\Delta_1(1 + 2\mu k u_\tau)}{(1 + 2\mu k u_\tau)^2 + 4\mu^2 k^4 \Delta_2^2} \qquad (37)$$

where

$$\Delta_1 = (1 + 2\mu k u_\tau)(\bar{u} - q_s) + 2\mu k q_\tau u_s$$
$$\Delta_2 = q_N + 2\mu k(u_\tau q_N - q_\tau u_N). \qquad (38)$$

Note that the ice normal stress effect on $N$ is manifested by the term $-2\mu ik^2 a$ in $(31)_4$, and the effect of ignoring this is equivalent to putting $u_N$ and $q_N$ to zero, in which case $\Delta_2 = 0$ and thus

the bed is neutrally stable. The ice normal stress effect is thus the mechanism for instability.

So far, no specific assumptions about the till flux and velocity have been used. Now we use the explicit expressions given by (16) and (17) in order to evaluate $\Delta_1$ and $\Delta_2$ explicitly; thus

$$u = A^*\zeta^*(1 - e^{-X})$$
$$q = A^*\zeta^{*2}[1 - (1 + X)e^{-X}] \qquad (39)$$

where

$$X = s/\zeta^*, \quad \zeta^* = \frac{N^2}{\alpha\tau\Delta\rho_{sw}g(1 - n)}$$

$$A^* = A\exp(\alpha\tau/N) \qquad (40)$$

Define

$$U(X) = 1 - e^{-X}, \quad W(X) = 1 - (1 + X)e^{-X}; \qquad (41)$$

after some algebra, which is outlined in Appendix B we find that the critical stability parameters $\Delta_1$ and $\Delta_2$ (defined in (38)) are given by

$$\Delta_1 = A^*\zeta^* W\left[1 + \frac{2\mu kA^*\zeta^*}{\tau}\left\{\frac{\alpha\tau}{N} - F(X)\right\}\right]$$

$$\Delta_2 = \frac{q}{N}\left[-\frac{\alpha\tau}{N} + 4Q(X) + \frac{2\alpha\mu kA^*\zeta^*}{N}J(X)\right] \qquad (42)$$

where (using $W' = U - W$ and $U' + U = 1$)

$$Q(X) = 1 - \frac{XW'}{2W}$$

$$J(X) = U\left[1 + X\left(\frac{U'}{U} - \frac{W'}{W}\right)\right]$$

$$F(X) = \frac{UW + 2WU' - XUU'}{W} \qquad (43)$$

and the variables are evaluated at the basic state. The functions $J$ and $Q$ increase montonically from 0 to 1, and are displayed in Fig. 7. The function $F(X)$ is well approximated by $1 - e^{-0.7X}$, and is shown in Fig. 8. Explicit definitions of the functions are given in Appendix B.

Clearly, instability occurs if $\Delta_1\Delta_2 > 0$. In practice, using typical values $\mu \approx 6\,\mathrm{bar\,y}$, $u \approx A^*\zeta^* \approx 100\,\mathrm{m\,a^{-1}}$, $\tau \approx 0.2\,\mathrm{bar}$, $N \approx 1\,\mathrm{bar}$, $\alpha \approx 10$, and anticipating drumlin wavenumbers $k \approx 0.01\,\mathrm{m^{-1}}$ (corresponding to wavelengths $2\pi/k \approx 600\,\mathrm{m}$), we find $2\mu ku/\tau \approx 60$, $2\alpha\mu ku/N \approx 30$, and these large values suggest that typically $\Delta_2 > 0$, so that the instability criterion is essentially that

$$\frac{\alpha\tau}{N} \gtrsim F(X). \qquad (44)$$

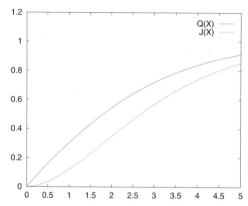

**Fig. 7.** The instability functions $Q(X)$ and $J(X)$ in the definition of $\Delta_2$.

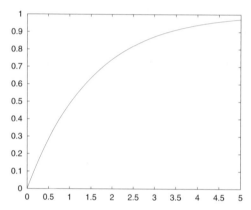

**Fig. 8.** The function $F(X)$ as used in (42). $F \to 1$ as $X \to \infty$.

We elaborate on this instability criterion in the following section.

## Discussion

We define the dimensionless wave number parameter $K$ by

$$K = \frac{2\mu\alpha Ak}{r} \qquad (45)$$

where

$$r = \Delta\rho_{sw}g(1-n) \approx 0.1 \,\text{bar}\,\text{m}^{-1} \qquad (46)$$

and the dimensionless stress $Y$ by

$$Y = \frac{\alpha\tau}{N}. \qquad (47)$$

Using the definitions of $A^*$ and $\zeta^*$ in (40), we have that the stability parameters $\Delta_1$ and $\Delta_2$ are given by

$$\Delta_1 = A^*\zeta^* W(X)\left\{1 + \frac{Ke^Y}{Y^2}[Y - F(X)]\right\}$$

$$\Delta_2 = \frac{q}{N}\left\{-Y + 4Q(X) + \frac{Ke^Y}{Y^2}J(X)\right\}. \qquad (48)$$

Using the values above (44), we have $K \approx 120$, and we will suppose that $K$ is large. The critical curves $\Delta_1 = 0$ and $\Delta_2 = 0$ are thus given by

$$(\Delta_1): \quad Y = F(X) + O\left(\frac{1}{K}\right)$$

$$(\Delta_2): \quad X = \sqrt{\frac{6}{K}}\,Ye^{-Y/2} + O\left(\frac{1}{K}\right) \qquad (49)$$

where for the latter we use the small $X$ approximations $Q \approx \frac{1}{3}X$, $J \approx \frac{1}{6}X^2$. The corresponding graphs are plotted in Fig. 9.

The region of instability lies between the two curves. A sufficient condition for instability is that

$$X = \frac{\alpha r\tau s}{N^2} \gtrsim \left(\frac{24}{K}\right)^{1/2} e^{-1} \approx \frac{1.8}{\sqrt{K}}$$

$$Y = \frac{\alpha\tau}{N} \gtrsim 1 \qquad (50)$$

as these values delineate the $X$ and $Y$ maxima of the two curves in Fig. 9. The practical implications are discussed in the conclusions.

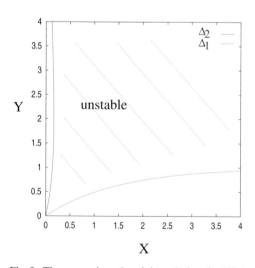

**Fig. 9.** The curves $\Delta_1 = 0$ and $\Delta_2 = 0$ given by (49) in terms of the parameters $X = \alpha r\tau s/N^2$ and $Y = \alpha\tau/N$, when $K = 120$. Instability occurs between the two curves.

## Growth rate

The growth rate $\rho$ is given by (36). In terms of $X$, $Y$, and $K$, we find

$$2\mu k u_\tau = \frac{Ke^Y}{Y^2}[UY - W]. \tag{51}$$

Using this together with (48) and assuming that $K$ is large, we derive the approximate expression for $\rho$,

$$\rho \approx \left(\frac{AN}{2\mu}\right)^{1/2}\left[\frac{D(k/k^*)^3}{B^2 + C^2(k/k^*)^4}\right] \tag{52}$$

where

$$k^* = \frac{r}{(2\mu AN)^{1/2}} \tag{53}$$

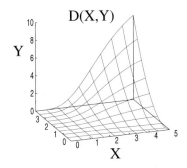

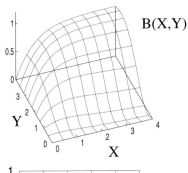

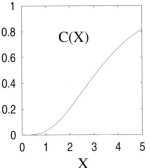

**Fig. 10.** The functions $D(X, Y)$, $B(X, Y)$ and $C(X)$ defined in (54).

and

$$D(X, Y) = YW^2(X)J(X)[Y - F(X)]$$
$$B(X, Y) = Y^2e^{-Y}[U(X)Y - W(X)]$$
$$C(X) = W(X)J(X). \tag{54}$$

These are $O(1)$ functions, and are shown in Fig. 10. For example, with values $\tau = 0.2$ bars, $s = 5$ m, $N = 1$ bar, $r = 0.1$ bar m$^{-1}$, $\alpha = 10$, we have $X = 1$, $Y = 2$, whence $D = 0.025$, $C = 0.032$, $B = 0.54$. We see that the growth rate tends to zero as $k \to 0$ and $k \to \infty$, and has a maximum at

$$k = k_{\max} = 3^{1/4}\sqrt{\frac{B}{C}}\,k^* \tag{55}$$

and this is $5.4k^*$ for $X = 1$, $Y = 2$. Clearly $k^{*-1}$ is a distinguishing length scale, as we would expect instabilities to occur at finite amplitude where the growth rate is largest. For values $\mu \approx 6$ bar a, $A \approx 10\,\text{a}^{-1}$, $N \approx 1$ bar, $r \approx 0.1$ bar m$^{-1}$, the length scale is 100 m and the preferred wavelength of instability would be $2\pi/k_{\max} \approx 116$ m. It is clear from the variation of $C$ in Fig. 10 that higher values of $X$ give higher wavelengths. From (52), the time scale for growth is $2\mu/AN$, and with the above values, this is a year. The explicit formula for the maximum growth rate is

$$\rho_{\max} = \left(\frac{AN}{2\mu}\right)^{1/2}\left(\frac{3^{3/4}D}{4B^{1/2}C^{3/2}}\right) \tag{56}$$

and this is plotted in Fig. 11. Note that the plot shows $\rho_{\max}$ for values $X > 0.5$. The approximation in (52) breaks down as $X \to 0$, and in fact

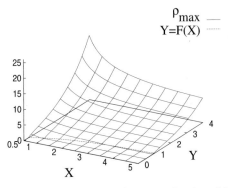

**Fig. 11.** The maximum growth rate as a function of $X$ and $Y$, given by the large $K$ approximation (56). $\rho_{\max}$ is put to zero where the system is stable, i.e. for $Y < F(X)$, and this curve is delineated on the $(X, Y)$ plane. The units of growth rate are $(AN/2\mu)^{1/2}$ and for representative values may be taken as a$^{-1}$. Note that $X$ lies in the range $0.5 < X < 5$.

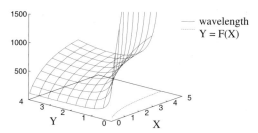

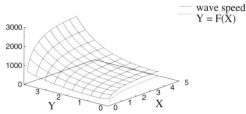

**Fig. 12.** The wavelength of the most rapidly growing mode. The units are $k*^{-1}/100 = (2\mu AN)^{1/2}/100r$, and for $k*^{-1} = 100\,\text{m}$, this means the units are m.

**Fig. 13.** The wave speed of the most rapidly growing mode. The units are $u*/100$, i.e. $(AN/r)/100$, and for $(AN/r) = 100\,\text{m a}^{-1}$, the units are $\text{m a}^{-1}$. The wave speed is less than the basal ice velocity, but the two are roughly comparable. The relatively high values are purely due to our choice of the rheological constant $A$. It should be noted that the $Y$ range in this figure and in Fig. 12 is $F(X) < Y < F(X) + 4$, and the surface extends beyond the line $Y = 4$.

the uniform state becomes stable again. The wavelength of the most rapidly growing mode is given by

$$L = \frac{2\pi}{3^{1/4}} \sqrt{\frac{C}{B}}\, k*^{-1} \tag{57}$$

and this is shown in Fig. 12 for $X > 0.5$ in the unstable region. We see that for our typical value $k*^{-1} = 100\,\text{m}$ the preferred wavelengths lie in the range 100–1000 m (the higher values near the stability threshold correspond to very slowly growing modes).

*Wave speed*

Using the same approximations as above, the wave speed of the perturbations is

$$c = u - u* \left\{ \frac{BD}{C[B^2 + C^2(k/k*)^4]} \right\} \tag{58}$$

where the velocity scale is

$$u* = \frac{AN}{r}. \tag{59}$$

With preceding values, we have $u* \approx 100\,\text{m a}^{-1}$, and the migration speed relative to the basal ice flow is of similar size to it. Since $Y > F$ for instability, and $W/U < F$ (in fact they are almost equal), then $B > 0$ when $\rho > 0$, and the wave speed is less than that of the ice.

In fact, we see from (39) that

$$u = u* \frac{e^Y}{Y} U \tag{60}$$

and therefore the wave speed at the wavenumber $k_{\max}$ of maximum growth in (55) is

$$c = u* V(X, Y) \tag{61}$$

where

$$V(X, Y) = \frac{e^Y U}{Y} - \frac{D}{4BC}. \tag{62}$$

Figure 13 plots the wave speed of the most rapidly growing mode as a function of $X$ and $Y$. These drumlins always move downstream, at a rate which is slightly less than the basal ice speed.

**Conclusions**

We have shown explicitly that Hindmarsh's (1998*a*) mechanism for the generation of bedform will realistically occur under ice sheets, given favourable conditions. Several preconditions must be met for the instability to occur. There must be a till layer; the basal ice must be at the melting point, and the resulting drainage pressure must be high, and presumably associated with a distributed drainage system. However, if it is too high, we might expect rapid sliding and therefore very low shear stresses, which mitigates against instability. Thus while the instability seems certainly to be accessible in general, conflicting external parameters may prevent it occurring; and indeed, drumlins do not occur everywhere.

Hindmarsh (1998*a*) found instability for many different sets of parameter values. Although our result here is based on a slightly different till rheology, the results should be comparable. The simplest presentation of the stability map in Fig. 9 is if we define

$$\xi = \frac{rs}{N} \tag{63}$$

so that $X = Y\xi$, and the instability region lies between the two curves $Y \approx F(Y\xi)$ and $\xi \approx (6/K)^{1/2} e^{-Y/2}$, as shown in Fig. 14. Drumlin forming instability is caused in this model by high basal shear stress, $\tau > N/\alpha$, and is likely to occur in practice if the till has a strong rate dependence on shear stress, and if, additionally, the effective pressure is low. This may be the case for a distributed drainage system as for example under ice stream B but for high values $N > 10$ bars appropriate to channelised drainage systems, instability is unlikely to occur (and indeed the till would be unlikely to deform).

We have found that the instability operates on the convective timescale, which is on the order of years; drumlins form rapidly. The instability has preferred wavelengths in the range 100–1000 m comparing favourably to observations, though this depends on our choice of values for $A$ and $N$. Like fluvial dunes, these drumlins move downstream, but unlike dunes, the wave speed is comparable to the ice speed.

An interesting feature of Fig. 14 is that it predicts instability independently of basal shear stress if the till thickness is between upper and lower limits of $1.5N/r$ and $0.22N/r$ (if $K = 120$). It is tempting to suppose that in this case drumlins might form from thickening till until they reach a sufficient thickness that they stabilize; however, examination of this possibility awaits the development of a non-linear theory (cf. Hindmarsh 1998b).

The theory advanced here does not preclude the possibility of other mechanisms, however those that have been promoted in the literature tend not to be built on sound mechanical principles. A recent example is the suggestion by Kleman & Hattestrand (1999) that Rogen moraine patterns occur when till accumulations at the

boundary between frozen and temperate basal ice are fractured. This notion sounds plausible even if hypothetical, but it is based on the underlying assumption that till accumulates at the frozen–temperate boundary because of the stress accumulation there as the ice switches from a no slip to a sliding basal boundary condition. This purely theoretical idea, based on an analysis of Hutter & Olunloyo (1982), is apparently not applicable to the frozen–temperate boundary when account is taken of realistic dependence of basal sliding on the temperature field (see Fowler & Larson 1980, p. 338).

In conclusion, drumlins most obviously represent the evolution of an instability, initially of transverse form (as in Rogen moraine), and this instability can be potentially explained via the interaction of ice flow and till flow. We expect these transverse waves to be susceptible to three-dimensional instabilities, in common with other similar wave transitions, for example fluvial dunes, and we have shown that Hindmarsh's (1998a) model can explain their initial evolution.

I thank W. Everett for correspondence and information concerning the early drumlin literature, and F. Ng for his ever-vigilant perusal of the manuscript. I am indebted to C. Schoof for spotting an error in an earlier version of the paper.

## Appendix A

### Till rheology

Till is a complex medium and, insofar as it is analogous to soil in its rheological behaviour, it can be expected that a detailed realistic description of it should include such phenomena as irreversible elasticity, secondary creep, failure, dilatation, effect of fabrics, and so on (Clarke 1987; Mitchell 1993). In selecting the simple viscous rheology in (12), we are partly motivated by the wish for simplicity, but also by the realistic hope that when considering long term strain in steady deformation, many of the above-mentioned phenomena may be practically irrelevant.

There has been much discussion in the literature (e.g. Kamb 1991; Hooke et al. 1997; Iverson et al. 1998) of whether an appropriate long-term rheology for till is a rigid–plastic one or a viscous one. Our point of view here is that the two choices can simply be viewed as representing different parametric limits, and are not conceptually distinct. Furthermore, we do not wish to be overly swayed by laboratory experiments where scaling up issues may be important.

The analogy between a power law rheology and the exponential rate law (12) can be drawn as follows. Define $\mu^*$ by

$$A = \varepsilon_{obs} \exp(-\alpha\mu^*) \qquad (A1)$$

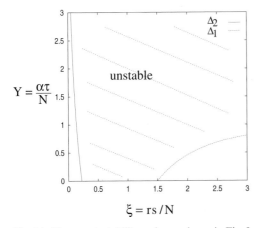

**Fig. 14.** The same instability region as shown in Fig. 9, but in terms of the primary instability parameters $\xi = rs/N$ and $Y = \alpha\tau/N$. See the comment on surface plotting in Fig. 12.

where $\varepsilon_{\text{obs}}$ is a typical observed ice shear strain rate, so that

$$\frac{\partial v}{\partial z} = \varepsilon_{\text{obs}} \exp\left[\frac{\alpha}{p_e}\{\tau - \mu^* p_e\}\right] \quad \text{(A2)}$$

and $\mu^* p_e$ is essentially a yield stress, with $\partial v/\partial z \ll U/d$ for $\tau < \mu^* p_e$. If we suppose $(\tau - \mu^* p_e)/\mu^* p_e$ is small (this is not a restriction if $\alpha$ is large, and if $\alpha$ is $O(1)$, then the exponential and power law rheologies are similar anyhow), (A2) is equivalent to

$$\frac{\partial v}{\partial z} \approx \varepsilon_{\text{obs}} \exp\left[\mu^* \alpha \ln\left(1 + \frac{\tau - \mu^* p_e}{\mu^* p_e}\right)\right]$$

$$= \varepsilon_{\text{obs}}\left(\frac{\tau}{\mu^* p_e}\right)^{\mu^* \alpha} \quad \text{(A3)}$$

which is one version of the Boulton–Hindmarsh rheology.

## Appendix B

In calculating the partial derivatives of $u$ and $q$, it is convenient to use the definition of the dimensionless stress, $Y = \alpha\tau/N$, given in (47). Then we have

$$A^* = Ae^Y, \quad \zeta^* = \frac{N}{rY} \quad \text{(B1)}$$

where $r$ is a constant given in (46), and the partial derivatives of $A^*$, $\zeta^*$ and $X$ are given by

$$\frac{A^*_\tau}{A^*} = \frac{Y_N}{\tau}, \quad \frac{\zeta^*_\tau}{\zeta^*} = -\frac{1}{\tau}, \quad \frac{A^*_N}{A^*} = -\frac{Y}{N}, \quad \frac{\zeta^*_N}{\zeta^*} = \frac{2}{N}$$

$$\frac{X_\tau}{X} = \frac{1}{\tau}, \quad \frac{X_N}{X} = -\frac{2}{N}, \quad \frac{X_s}{X} = \frac{1}{s}. \quad \text{(B2)}$$

Since

$$u = A^*\zeta^* U(X), \quad q = A^*\zeta^{*2} W(X) = A^*\zeta^* sW/X \quad \text{(B3)}$$

we find

$$\frac{u_\tau}{u} = \frac{1}{\tau}\left[Y - 1 + \frac{XU'}{U}\right]$$

$$\frac{q_\tau}{q} = \frac{1}{\tau}\left[Y - 2 + \frac{XW'}{W}\right]$$

$$\frac{u_N}{u} = \frac{1}{N}\left[-Y + 2 - 2\frac{XU'}{U}\right]$$

$$\frac{q_N}{q} = \frac{1}{N}\left[-Y + 4 - 2\frac{XW'}{W}\right]$$

$$\frac{u_s}{u} = \frac{XU'}{sU}, \quad \frac{q_s}{q} = \frac{XW'}{sW}. \quad \text{(B4)}$$

From these we find, using $U - W' = W$,

$$u - q_s = A^*\zeta^* W \quad \text{(B5)}$$

and the definitions of $\Delta_1$ and $\Delta_2$ in the text follow on substituting the expressions in (B4) into (38).

The instability functions defined in (43) are given explicitly by the following expressions, using the definitions of $U$ and $W$ in (41):

$$Q(X) = \frac{1 - (1 + X + \frac{1}{2}X^2)e^{-X}}{1 - (1 + X)e^{-X}}$$

$$J(X) = \frac{1 - (2 + X^2)e^{-X} + e^{-2X}}{1 - (1 + X)e^{-X}}$$

$$F(X) = \frac{1 - 2Xe^{-X} - e^{-2X}}{1 - (1 + X)e^{-X}}. \quad \text{(B6)}$$

## References

BEAR, J. & BACHMAT, Y. 1990. *Introduction to modeling of transport phenomena.* Kluwer, Dordrecht.

BOULTON, G. S. 1987. A theory of drumlin formation by subglacial sediment deformation. *In:* MENZIES, J. & ROSE, J. (eds) *Drumlin Symposium.* Balkema, Rotterdam, 25–80.

——1996. Theory of glacial erosion, transport and deposition as a consequence of subglacial sediment deformation. *Journal of Glaciology,* **42**, 43–62.

—— & HINDMARSH, R. C. A. 1987. Sediment deformation beneath glaciers: rheology and geological consequences. *Journal of Geophysical Research,* **92**, 9059–9082.

BOYCE, J. I. & EYLES, N. 1991. Drumlins carved by deforming till streams below the Laurentide ice sheet. *Geology,* **19**, 787–790.

BÜSSE, F. H. 1978. Nonlinear properties of convection. *Reports of Progress on Physics,* **41**, 1929–1967.

CLARKE, G. K. C. 1987. Subglacial till: a physical framework for its properties and processes. *Journal of Geophysical Research,* **92**, 9023–9036.

DAVIS, W. M. 1884. The distribution and origin of drumlins. *American Journal of Science,* **23**, 407–416.

EMBLETON, C. & KING, C. A. M. 1975. *Glacial geomorphology.* Edward Arnold, London.

EVERETT, W. 1987. *An analysis of the literature on drumlins and related streamlined forms.* MPhil thesis, University of London.

FOWLER, A. C. & LARSON, D. A. 1980. The uniqueness of steady state flows of glaciers and ice sheets. *Geophysical Journal of the Royal Astronomical Society,* **63**, 333–345.

GRAVENOR, C. P. 1953. The origin of drumlins. *American Journal of Science,* **251**, 674–681.

HINDMARSH, R. C. A. 1996. Sliding of till over bedrock: scratching, polishing, comminution and kinematic wave theory. *Annals of Glaciology,* **22**, 41–48.

——1998a. The stability of a viscous till sheet coupled with ice flow, considered at wavelengths less than the ice thickness. *Journal of Glaciology,* **44**, 285–292.

——1998b. Drumlinisation and drumlin-forming instabilities: viscous till mechanisms. *Journal of Glaciology,* **44**, 293–314.

HOOKE, R. LEB., HANSON, B., IVERSON, N. R., JANSSON & FISCHER, U. H. 1997. Rheology of till beneath Storglaciären, Sweden. *Journal of Glaciology*, **43**, 172–179.

HUTTER, K. & OLUNLOYO, V. O. S. 1980. On the distribution of stress and velocity in an ice strip, which is partly sliding over and partly adhering to its bed, by using a Newtonian viscous approximation. *Proceedings of the Royal Society of London*, **A373**, 385–403.

IVERSON, N. R., HOOYER T. S. & BAKER, R. W. 1998. Ring-shear studies of till deformation: Coulomb-plastic behaviour and distributed strain in glacier beds. *Journal of Glaciology*, **44**, 634–642.

KAMB, B. 1970. Sliding motion of glaciers: theory and observation. *Reviews of Geophysics and Space Physics*, **8**, 673–728.

——1991. Rheological nonlinearity and flow instability in the deforming bed mechanism of ice stream motion. *Journal of Geophysical Research*, **96**, 16 585–16 595.

KINAHAN, G. H. & CLOSE, M. H. 1872. *The general glaciation of Iar-Connaught and its neighbourhood, in the counties of Galway and Mayo*. Hodges, Foster and Co., Dublin.

KLEMAN, J. & HATTESTRAND, C. 1999. Frozen-bed Fennoscandian and Laurentide ice sheets during the Last Glacial Maximum. *Nature*, **402**, 63–66.

LUNDQVIST, J. 1989. Rogen (ribbed) moraine-identification and possible origin. *Sedimentary Geology*, **62**, 281–292.

MASLOWE, S. A. 1985. Shear flow instabilities and transition. *In*: SWINNEY, H. L. & GOLLUB, J. P. (eds) *Hydrodynamic instabilities and the transition to turbulence*, 2nd ed. Topics in Applied Physics, **45**, Springer-Verlag, Berlin, 181–228.

MENZIES, J. M. 1979. The mechanics of drumlin formation with particular reference to the change in pore-water content of the till. *Journal of Glaciology*, **22**, 373–384.

—— & SHILTS, W. W. 1996. Subglacial environments. *In*: MENZIES, J. (ed.) *Past glacial environments: sediments, forms and techniques*. Butterworth–Heinemann, Oxford, 15–136.

MILLIS, J. 1911. What caused the drumlins? *Science*, **34**, 60–62.

MITCHELL, J. K. 1993. *Fundamentals of soil behaviour*. 2nd ed. John Wiley, New York.

NYE, J. F. 1969. A calculation of the sliding of ice over a wavy surface using a Newtonian viscous approximation. *Proceedings of the Royal Society of London*, **A311**, 445–467.

——1970. Glacier sliding without cavitation in a linear viscous approximation. *Proceedings of the Royal Society of London*, **A315**, 381–403.

RICHARDS, K. 1982. *Rivers: form and process in alluvial channels*. Methuen, London.

RUSSELL, I. C. 1895. The influence of debris on the flow of glaciers. *Journal of Geology*. **3**, 823–832.

SHARPE, D. R. 1987. The stratified nature of drumlins from Victoria Island and Southern Ontario, Canada. *In*: MENZIES, J. & ROSE, J. (eds) *Drumlin Symposium*, Balkema, Rotterdam, 185–214.

SHAW, J. 1983. Drumlin formation related to inverted meltwater erosional marks. *Journal of Glaciology*, **29**, 461–479.

——1994. Hairpin erosional marks, horseshoe vortices and subglacial erosion. *Sedimentary Geology*, **91**, 269–283.

—— & KVILL, D. 1984. A glaciofluvial origin for drumlins of the Livingstone Lake area, Saskatchewan. *Canadian Journal of Earth Sciences*, **21**, 1442–1459.

——, —— & RAINS, B. 1989. Drumlins and catastrophic glacial floods. *Sedimentary Geology*, **62**, 177–202.

SMALLEY, I. J. & UNWIN, D. J. 1968. The formation and shape of drumlins and their distribution and orientation in drumlin fields. *Journal of Glaciology*, **7**, 377–390.

SUGDEN, D. A. & JOHN, B. S. 1976. *Glaciers and landscape*. Edward Arnold, London.

TARR, R. S. 1894. The origin of drumlins. *American Geologist*, **13**, 393–407.

UPHAM, W. 1892. Conditions of accumulation of drumlins. *American Geologist*, **10**, 339–362.

WALDER, J. S. & FOWLER, A. 1994. Channelised subglacial drainage over a deformable bed. *Journal of Glaciology*, **40**, 3–15.

WEERTMAN, J. R. 1957. On the sliding of glaciers. *Journal of Glaciology*, **3**, 33–38.

# Moraine-mound formation by englacial thrusting: the Younger Dryas moraines of Cwm Idwal, North Wales

DAVID J. GRAHAM[1] & NICHOLAS G. MIDGLEY[2]

[1] Centre for Glaciology, Institute of Geography and Earth Sciences, University of Wales, Aberystwyth, Ceredigion SY23 3DB, UK (e-mail: djg97@aber.ac.uk)

[2] School of Biological and Earth Sciences, Liverpool John Moores University, Byrom Street, Liverpool L3 3AF, UK

**Abstract:** The Younger Dryas (c. 11–10 ka BP) moraine-mound complex ('hummocky moraine') in the historically important site of Cwm Idwal, North Wales, has previously been interpreted using periglacial, subglacial, ice-marginal and englacial models. In this paper the morphology and sedimentology of these landforms is described and the competing hypotheses tested against this evidence. It is demonstrated that an englacial thrusting model, developed for polythermal glaciers in Svalbard, best fits the available evidence. Thrusting probably resulted from longitudinal compression against a reverse bedrock slope, although a frozen snout, downglacier of sliding ice, may also have been a trigger. It is suggested that the role of ice-deformation, especially thrusting, in landform development has been underestimated, and that the englacial thrusting model may find application in the interpretation of other sites in the palaeo-landform record.

Moraine-mound complexes with irregular topography have been described from a number of glacierized and glaciated regions (e.g. Eyles 1979; Sollid & Sørbel 1988; Benn 1990; Attig & Clayton 1993; Bennett & Boulton 1993; Johnson et al. 1995; Andersson 1998; Eyles et al. 1999). British moraine-mound complexes formed during the Younger Dryas Stadial (c. 11–10 ka BP) are commonly known as 'hummocky moraine' (e.g. Sissons 1974, 1979, 1980, 1983; Sissons & Sutherland 1976; Cornish 1981). These authors interpreted the moraines as evidence of widespread areal ice stagnation, brought about by rapid climatic amelioration at the end of the Younger Dryas. Subsequent work has challenged the general applicability of this model (e.g. Eyles 1983; Hodgson 1986; Bennett & Glasser 1991; Benn et al. 1992; Bennett & Boulton 1993; Hambrey et al. 1997; Bennett et al. 1998; Eyles et al. 1999) and it is now generally accepted that 'hummocky moraine' complexes are polygenetic (see reviews by Benn 1992; Bennett 1994). The genetic connotations that have become attached to the term 'hummocky moraine' mean that it is no longer useful for describing topographically complex moraines of undetermined origin (Benn 1992; Bennett 1994). The non-genetic term 'moraine-mound complex' (introduced by Bennett et al. 1996b) is, therefore, used here to describe these landforms.

During the last decade there has been a great deal of work on the genesis and significance of British Younger Dryas moraine-mound complexes (e.g. Bennett & Glasser 1991; Benn 1992; Bennett & Boulton 1993). Hambrey et al. (1997) proposed that an englacial thrusting model, originally developed for polythermal glaciers in Svalbard (e.g. Hambrey & Huddart 1995; Huddart & Hambrey 1996), may be applicable at some sites, particularly Coire a' Cheud-chnoic (Valley of a Hundred Hills, Torridon, Scotland) and Cwm Idwal (North Wales). Evidence in support of this model at Coire a' Cheud-chnoic was presented by Bennett et al. (1998). This current paper aims to demonstrate that englacial thrusting is the most likely origin of the Cwm Idwal moraine-mound complex, and addresses some of the wider implications of this conclusion.

## Cwm Idwal

Cwm Idwal (Fig. 1a) is one of Britain's classic glacial sites, figuring prominently in the development of Ice Age concepts in Britain. It was here that Darwin (1842) demonstrated glaciation in Wales for the first time. With an area of 1.37 km², Cwm Idwal is the largest cirque in the Glyderau range of North Wales (Addison 1988b). The cirque has an aspect of 038°

From: MALTMAN, A. J., HUBBARD, B. & HAMBREY, M. J. (eds) Deformation of Glacial Materials. Geological Society, London, Special Publications, **176**, 321–336. 0305-8719/00/$15.00 © The Geological Society of London 2000.

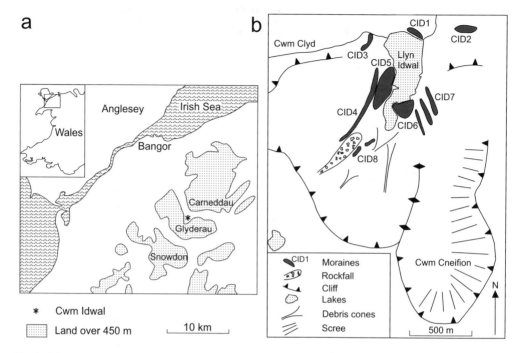

**Fig. 1.** (**a**) Location map for Cwm Idwal. (**b**) Generalized geomorphological map of the Cwm Idwal area drawn from an uncorrected aerial photograph, showing the location of the moraine groups (CID1 to 8) described in the text. Pits were located towards the northern end of CID4 and CID5.

(Addison 1988*b*), a floor at around 370 m OD and a backwall which rises 430 m almost vertically in places. Two smaller cirques drain into Cwm Idwal: Cwm Cneifion (with a floor height of around 640 m OD) from the SE; and Cwm Clyd (with a floor height of around 670 m OD) from the west. These cirques are carved out of strongly cleaved Ordovician Rhyolitic Tuffs and minor sandstones (BGS 1972).

The moraines of Cwm Idwal were first described by Darwin (1842) and subsequently by Jehu (1902), Seddon (1957), Escritt (1971), Unwin (1975), Addison (1988*a*, 1990) and Gray (1990). Small-scale maps of the moraines were presented by Unwin (1975) and Gray (1982), with detail being provided by Addison (1988*a*, 1990) and Gray (1990). Whilst recognizing many of the same general features, these maps differ in several aspects. Between them, Addison (1988*a*, 1990) and Gray (1990) identify eight moraine groups on the basis of morphology and spatial distribution (Fig. 1b) and the key features and previous interpretations of each are summarized in Table 1.

The moraines in the inner basin of Cwm Idwal (CID4, CID5, CID6, CID8, Fig. 1b and 2) have been assigned a Younger Dryas age on the basis

of pollen and lithostratigraphic evidence from within the moraine-mound complex (Godwin 1955; Seddon 1962; Tipping 1993). The age and extent of the moraine at the cirque lip (CID1) is unclear, with some authors arguing for a Late Devensian age on the basis of its subdued morphology (Unwin 1975; Addison 1988*a*; Gray 1990) and others a Younger Dryas age (Gray 1982). Despite numerous studies, the origin of the moraine-mound complex in the centre of Cwm Idwal (CID4, CID5, CID6) remains controversial and forms the subject of the remainder of this paper.

## Methods

A topographic survey of moraine groups CID4, CID5 and CID6 was undertaken using a Leica TC600 total station and the dataset used to construct a high-resolution digital elevation model. The model enabled three-dimensional visualization of the moraine-mound complex from any angle and the generation of cross- and long-profiles to aid landform interpretation. Data were collected by ground survey in preference to photogrammetry since significant

**Table 1.** *Outline of the key features and previous interpretations of the Cwm Idwal moraines*

| Moraine ID | Key features | Interpretation | Reference |
|---|---|---|---|
| CID1 | Dams Llyn Idwal. Extent contested (eastern part may be a thin veneer on bedrock). Subdued morphology. | End moraine | Seddon 1957 |
| CID2 | Encloses a small sedimentary basin outside Cwm Idwal. | | |
| CID3 | Arcuate ridge enclosing postglacial sediments. | End moraine (Cwm Clyd) | |
| CID4 | Long, discontinuous, sharp crested ridge (450 m). Height of west face disguised by postglacial infilling. | Pronival rampart<br>Lateral/end moraine<br>Lateral moraine<br>End moraine<br>Flute<br>'Bulldozed' end moraine (Cneifion Glacier)<br>Englacial thrust moraine | Escritt 1971; Unwin 1977<br>Darwin 1842<br>Jehu 1902<br>Seddon 1957<br>Gray 1982<br>Addison 1988a<br>Hambrey *et al.* 1997 |
| CID5 | Often imbricated, approximately linear ridges of varying lengths. Morphology confused by bedrock at surface in places. | Lateral/end moraine<br>Lateral moraine<br>End moraine<br>Flutes<br>'Bulldozed' end moraine (Cneifion Glacier)<br>Englacial thrust moraine | Darwin 1842<br>Jehu 1902<br>Seddon 1957<br>Gray 1982<br>Addison 1988a<br>Hambrey *et al.* 1997 |
| CID6 | Often imbricated, approximately linear ridges of varying lengths. Morphology confused by bedrock at surface in places. Generally smaller than CID5. | Lateral/end moraine<br>Terminal moraine<br>End moraine<br>'Bulldozed' end moraine (Cneifion Glacier)<br>Englacial thrust moraine | Darwin 1842<br>Jehu 1902<br>Seddon 1957<br>Addison 1988a<br>Hambrey *et al.* 1997 |
| CID7 | Indistinct ridges running up eastern flank of Cwm Idwal. | Lateral/end moraine<br>'Slope of drift' | Darwin 1842<br>Jehu 1902 |
| CID8 | Small linear ridge. | | |

**Fig. 2.** Photograph of the Cwm Idwal moraine-mound complex taken from the lip of Cwm Cneifion looking towards the NW. The lake (Llyn Idwal) is 50 m wide at the narrowest point.

errors are associated with photogrammetric elevation measurement on steep slopes, and there is a need to ensure optimum point-density to reduce processing time whilst maintaining model accuracy. Greatest priority was given to collecting point-data on the moraines, with points spaced so as to include all morphological characteristics that could be realistically represented using a 0.25 m contour interval. Areas surrounding and between the moraines had only sufficient data to generate a topographic context for the moraines in the digital elevation model. The resulting average point-spacing was 20 per 100 m$^2$ on the moraines and 5 per 100 m$^2$ in the surrounding area, with a total of 7800 points in the model.

A triangulated irregular network was generated from the dataset using the Delaunay (optimal) triangulation (McCullagh 1988), and the digital elevation model produced from this network using a projected distance-weighted average interpolator. By avoiding the need to interpolate onto a regular grid, the original data values in the model were preserved and the computer processing time minimised by reducing the number of calculations required in areas with a low density of data (McCullagh 1988).

A lack of natural sediment exposures of significant size, and restrictions on excavation, meant that investigations of the moraine-mound sedimentology were limited to four pits dug to a depth of between 1 m and 1.2 m. The pits were excavated approximately half-way down the eastern (rectilinear) faces of four adjacent mounds in moraine groups CID4 and CID5 (Fig. 1b). Clast samples were also collected for control from the outflow stream from Llyn Idwal, an alluvial fan to the south of Llyn Idwal and weathered scree on the western slope of the Cwm.

Sedimentary facies were described on the basis of laboratory analysis of grain-size and clast characteristics, and the field measurement of clast-fabric. Sediment texture was classified using the Moncrieff (1989) scheme for poorly sorted sediments as modified by Hambrey (1994). Matrix samples were collected from each facies below the soil horizon and the grain-size distribution assessed by a combination of dry-sieving and SediGraph analysis using a subsample of 70–100 g. Clast shape was analysed for samples

of 50–200 clasts of mixed lithology, following the method of Benn & Ballantyne (1994). This has been shown to discriminate well between glacigenic facies, even for samples of mixed lithology (Bennett et al. 1997). Surface features (striations and faceting) were recorded. Three-dimensional clast-fabric was measured based on samples of 50 prolate clasts (elongation ratio at least 3 : 2, Jones et al. 1999), selected from close to the bottom of each pit to minimize the effects of disturbance, and plotted on Schmidt lower-hemisphere equal-area stereographic projections.

## Morphology of the moraine-mound complex

The moraine-mounds east of Llyn Idwal (CID6, Figs 1b, 2, 3) have a maximum apparent relief of 12 m although peat deposits disguise their true size in places. They range in length from 8 to 50 m. They are generally discrete, straight crested ridges, with an orientation of between 013° and 037°E of N except in the constriction in Llyn Idwal where the morphology is controlled by the presence of glacially abraded bedrock at or just below the surface in many places. A few of the ridges have rectilinear faces on their eastern sides.

Moraine group CID5 (Figs 1b, 2 3) west of Llyn Idwal, consists of ridges that are generally larger than those east of the lake, reaching a maximum relief of 15 m. The ridges range in length from 25 to 80 m. Peat deposits disguise the size of the mounds in places. Most ridges are straight-crested, with an orientation ranging from 014° to 086°E of N. There is a continuum between whaleback and sharp-crested ridge forms. Most mounds have rectilinear eastern faces, with dips of between 22° and 36° (Fig. 4), whilst the western faces are generally less regular. Individual mounds appear to be stacked against one another in an imbricate fashion. Rafts of bedrock up to 6 m a-axis lie on the mounds in places, with their flat faces parallel to the rectilinear slopes. Exposed bedrock between some of the mounds has striations oriented along the axis of the cirque, roughly parallel to the ridge crests. At the western edge of the group, the moraines are stacked onto CID4 (Figs 1b, 2, 3), a ridge running for approximately 450 m along the west of Cwm Idwal. The

**Fig. 3.** Digital elevation model of the central part of Cwm Idwal showing the morphology of the moraine-mound complex. All heights are relative to an arbitrary datum. For location see Fig. 1b. (**a**) Model contoured with an interval of 1 m. Scale in metres. Arrow defines look direction for orthographic projection. Heavy lines define cross-sections. (**b**) Orthographic projection of part of the model. (**c**) Cross-sections through the digital elevation model showing the stacked nature of the moraine-mounds, prominent rectilinear slopes and long profiles.

a

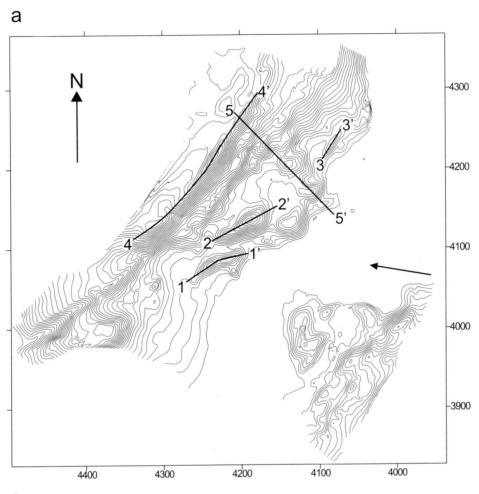

b

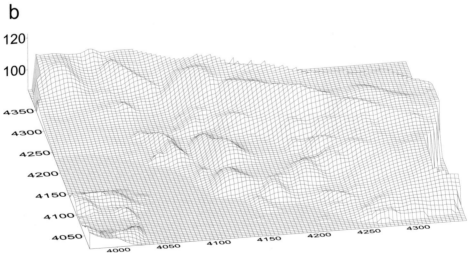

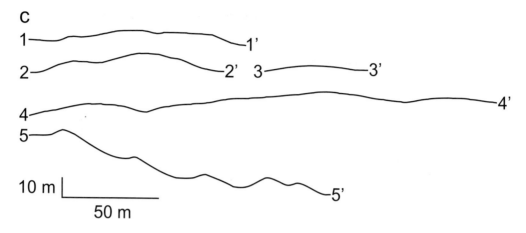

**Fig. 3.** (*Continued*).

ridge is breached by streams in a number of places and breaks into discrete subsidiary ridges at its southern end. At its northern end, it becomes more subdued until it can no longer be traced. The ridge crest is sharp and slightly curved, with a kink towards its southern end. The east face of the ridge has a rectilinear form and dips at between 26° and 40° (Fig. 4). The morphology of the west face is less regular than the east face. The volume of sediment in the mounds in CID4 and CID5 is uncertain, and may be less than the apparent volume, because the mounds are stacked against the bedrock slope of the cirque.

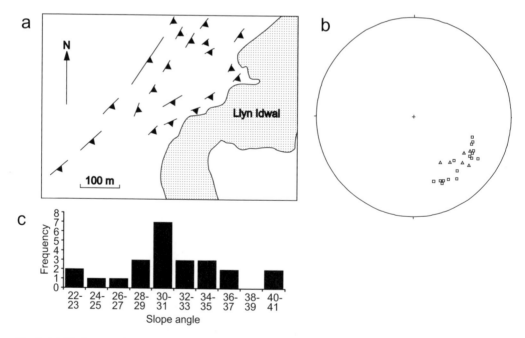

**Fig. 4.** (**a**) Glacio-tectonic map of the western part of the moraine-mound complex (moraine groups CID4 and 5). For location see Fig. 1b. The strike of ridge crests and rectilinear slopes are shown. (**b**) Schmidt lower hemisphere equal-area stereographic projection showing the dip and direction of dip of rectilinear slopes. Triangles refer to moraine group CID4, $n = 5$. Squares refer to moraine group CID5, $n = 16$. (**c**) Histogram showing the frequency of rectilinear slope angles in moraine groups CID4 and 5.

## Facies analysis

### Lithofacies descriptions

Two sedimentary facies were identified in the moraines: massive diamicton and massive gravel.

*Massive diamicton.* This facies occurred in all four of the excavations and texturally is clast-rich sandy diamicton. The proportion of gravel was estimated to be 30% in each excavation, with individual clasts up to 40 cm in diameter. The matrix is dominated by sand (73–80%), the remainder being mainly silt (Table 2). Samples of clasts from all pits showed peaks in the sub-angular category, the remainder being almost entirely within the angular and subrounded categories (Fig. 5a). The RA index (proportion of angular and very angular clasts) ranged from 11% to 38%. A co-variant plot of the RA and C40 indices (percentage of clasts with a $c/a$ axial ratio $\leq 0.4$) shows that the diamictons can be clearly differentiated from control samples of scree, alluvial fan and fluvial deposits collected from within Cwm Idwal (Fig. 5b). A small number of clasts in each sample were observed to be faceted (1–8%) and striated (1–3%). The clast-fabrics were weak (Fig. 6).

*Massive gravel.* This facies occurred only in pit 4 below the diamicton at a depth of 1 m. Texturally this facies is a sandy gravel, with the proportion of gravel estimated to be 60%. All the clasts were in the cobble- to granule-size classes, the largest being 70 mm in diameter, with the majority in the pebble and granule size classes. The matrix was dominated by sand (81%), the remainder being mainly silt (Table 2). Measurements of clast angularity peaked in the subangular category, the remainder being entirely within the angular and subrounded categories (Fig. 5a). The RA index was 25%. The gravel cannot be distinguished from the diamicton on

**Table 2.** *Grain-size data summary for the Cwm Idwal moraines*

| | Gravel (%) | Matrix (%) | | |
| --- | --- | --- | --- | --- |
| | | Sand | Silt | Clay |
| Diamicton (pit 1) | 30 | 77.7 | 16.2 | 6.1 |
| Diamicton (pit 2) | 30 | 79.8 | 13.2 | 6.9 |
| Diamicton (pit 3) | 30 | 74.3 | 14.9 | 10.8 |
| Diamicton (pit 4) | 30 | 72.6 | 15.5 | 11.9 |
| Gravel | 60 | 80.6 | 11.5 | 7.9 |

The proportion of gravel in the whole sediment was estimated to the nearest 10%. The proportions of sand, silt and clay are quoted for the matrix fraction.

a co-variant plot of the RA and C40 (Fig. 5b). Of the clasts observed, 24% were faceted and 9% striated.

### Lithofacies interpretation

Whilst some modification, especially from weathering processes, is likely to have occurred since deposition, the preservation of steep rectilinear slopes and striated clasts suggests that modification has not been significant. This is supported by evidence from a similar site in Scotland (Coire a' Cheud-chnoic, Hodgson 1982), where two-dimensional clast-fabric data from shallow excavations frequently showed preferred orientations along ridge crests, suggesting that down-slope movement during para- and post-glacial re-working was limited. Subsequent work at Coire a' Cheud-chnoic by Bennett *et al.* (1998) also utilized clast-fabric data from shallow exposures adjacent to footpaths. Nevertheless, care must be taken when interpreting sedimentological

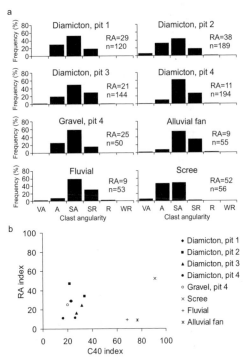

**Fig. 5.** Clast shape data from moraines and control samples. (**a**) Percentage frequency histograms of clast roundness. VA, very angular; A, angular; SA, subangular; SR, subrounded; R, rounded; WR, well rounded. (**b**) Co-variant plots of the RA index (percentage of angular and very angular clasts) and the C40 index (percentage of clasts with $c/a$ axial ratio $\leq 0.4$).

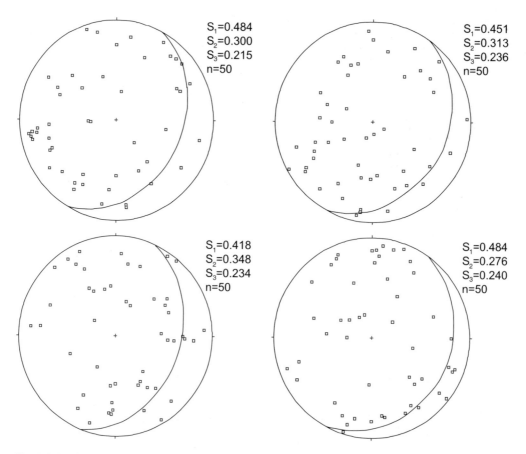

**Fig. 6.** Schmidt lower-hemisphere equal-area stereographic projections of clast fabric data, showing the angle and direction of dip of prolate clasts within the massive diamicton facies in four moraine mounds ($n = 50$). The curved 'line indicates the dip and orientation of the rectilinear face of each mound.

data from a small number of shallow excavations, particularly since the sample site may not give a representation of the sedimentary variation within the landform (Bennett 1995).

Both the gravel and diamicton facies are interpreted as basally derived glacial sediments. The presence of faceted and striated clasts in all the samples is strong evidence of traction at the glacier bed (Boulton 1978). The shape and angularity of the clasts contrasts sharply with control samples from fluvial, alluvial fan and scree deposits within Cwm Idwal (Fig. 5). The proportion of angular and very angular clasts is consistent with observations of basally transported material at contemporary glacier margins (Benn & Ballantyne 1993; Bennett *et al.* 1997; Hambrey *et al.* 1997) and other moraines in Britain interpreted as being formed largely of basally transported material (Benn 1989; Bennett *et al.* 1998). The grain-size distribution is

also consistent with a basal origin for the sediment (Hambrey 1994). Whether the sediments represent reworked material from an earlier glaciation is unclear.

## Discussion

### Previous interpretations of the moraine-mound complex

Four types of origin have been proposed for the Cwm Idwal moraine-mound complex and these are compared against the sedimentological and morphological evidence.

(1) *Formation of the long ridge (CID4) as a pronival (protalus) rampart against the western cliff of Cwm Idwal* (Escritt 1971; Unwin 1975, 1977). Although the traditionally accepted view

is that pronival ramparts contain few fines (e.g. Washburn 1979; Ballantyne & Kirkbride 1986) work on actively forming ramparts (Harris 1986; Ono & Watanabe 1986; Ballantyne 1987; Shakesby *et al.* 1995, 1999) indicates that fine material may be supplied by incremental snow-push of pre-existing sediment, avalanches and debris-flows. These processes may result in the sediment on the distal faces of ramparts being difficult to distinguish from glacigenic sediment on the basis of texture, especially within the matrix component (Harris 1986; Ballantyne 1987). Although Harris (1986) reported the proportion of gravel in a Norwegian rampart to be more than twice that in local till and reported in this study from the Cwm Idwal ridge, the scarcity of quantitative estimates of the relative proportions of gravel, sand and mud in active ramparts in the literature makes the use of texture to discriminate between moraines and ramparts problematic.

Recent work has stressed the importance of subnival processes in providing sediment to ramparts (Shakesby *et al.* 1995, 1999). Nevertheless, clast roundness has been shown to generally have a mode in the angular or very angular classes in both actively forming (Harris 1986; Ballantyne 1987; Shakesby *et al.* 1995, 1999) and relict (e.g. Unwin 1975; Ballantyne & Kirkbride 1986) features. This contrasts with the Cwm Idwal roundness data with a mode in the subangular class, an RA index of 29 and no clasts in the very angular class (Fig. 5). Postglacial weathering could account for some of this difference, but it seems unlikely that this could explain the whole difference given the clast roundnesses described for relic features of a similar age in Scotland (Ballantyne & Kirkbride 1986).

Shakesby *et al.* (1999) reported that clasts on the proximal faces of ramparts had strong preferred orientations oblique to ridge crests and dips parallel to the surface slope, whilst Harris (1986) found orientations perpendicular to ridge crests with dips downslope on both proximal and distal faces. The strong fabrics constrast with the very weak fabrics in Cwm Idwal, but the conflicting results of these studies and the possibility of postglacial modification in Cwm Idwal makes this a poor diagnostic tool. The morphology of pronival ramparts is also a poor diagnostic feature, their planform, slope-angles, crest-sharpness and size being very variable (Shakesby 1997). Ballantyne (1987) noted that the distal faces of Norwegian ramparts are approximately rectilinear, as in Cwm Idwal.

(2) *Formation as ice-marginal moraines* (Darwin 1842; Seddon 1957; Escritt 1971; Addison

1988*a*). The first possibility is that the moraine-mound complex represents lateral dump moraines associated with a glacier flowing along the axis of Cwm Idwal (Darwin 1842; Seddon 1957). Lateral moraine ridges are formed principally by the dumping of supraglacial material from a glacier surface, although they may also contain significant quantities of debris carried englacially and subglacially (Boulton & Eyles 1979; Small 1987). The RA index has been found to range from 64 in the upper part of a Norwegian glacier to 4 close to the terminus, with the mode changing from the angular to subangular class downglacier, reflecting an increase in the proportion of actively transported clasts downglacier (Benn & Ballantyne 1994). In Cwm Idwal, the RA index was found in the lower range described by Benn and Ballantyne (1994) for lateral moraines, with values between 11 and 38, and a mode in the subangular class.

The bedding in lateral moraines is often scree-like, dipping away from the glacier at 10°–40° (Small 1987). However, ice-proximal faces tend to be very steep (up to 70°, Small 1983) and prone to slumping and gullying when ice support is removed (Ballantyne & Benn 1994), so clast-fabrics taken on this face are poor diagnostic features for lateral moraines. The steep and unstable ice-proximal faces of lateral moraines contrast with the rectilinear, apparently unmodified, ice-proximal faces of the Cwm Idwal moraine-mounds, dipping at 22–40°.

Lateral moraines are generally continuous, although the crestline may be anastomosing. The longitudinal profile of the crest-line reflects the ice-surface slope of the glacier, and therefore may be expected to dip downglacier. In Cwm Idwal, only CID4 is continuous and its crestline, even allowing for postglacial breaching by streams, is undulating (Fig. 3c). The crestlines of other mounds, although oriented in approximately the same direction, cannot readily be joined to form ice-marginal positions and their longitudinal profiles are generally whalebacked.

An alternative model places a glacier in Cwm Cnefion, with the Cwm Idwal moraine-mound complex representing ice-marginal frontal moraines (Escritt 1971; Addison 1988*a*). The range of models for the formation of terrestrial ice-marginal moraines is large, and the sedimentology and morphology described from contemporary glaciers is diverse. In essence, however, transverse ice-marginal moraines on land may form by ice-marginal pushing of pre-existing sediment, dumping from the glacier, or ablation from the glacier surface. Complexity is increased by the operation of a variety of other processes such as basal lodgement, freezing-on of rafts of debris, mass movement and fluvial activity. Since frontal dump moraines are formed in the same way as lateral dump moraines, they are not discussed separately.

Ice-marginal moraines frequently form by the annual pushing of sediment at the glacier margin (Benn & Evans 1998). The sedimentology of such ridges is highly variable, depending on the sediments present in the glacier foreland, so sedimentary facies are poor diagnostic characteristics. These features rarely reach a height of more than 10 m at modern glaciers (Benn & Evans 1998) and are usually continuous (often anastomosing) ridges, broadly arcuate in planform, but reflecting the local ice-margin morphology (Matthews *et al.* 1979; Sharp 1984). In cross-section they generally have relatively gentle proximal slopes, resulting from glacier over-riding, and steep distal slopes at the angle of repose of the sediment (Sharp 1984; Boulton 1986). These characteristics contrast with the discrete mounds up to 15 m high, with steep rectilinear faces, in Cwm Idwal.

Ablation moraines result where the debris cover on a glacier surface is sufficient to retard ablation. The resulting landform is a chaotic association of mounds formed by repeated resedimentation as the ice-cores melt (Boulton 1972; Eyles 1979). The sediments in ablation moraines may be actively or passively transported. The sedimentology is frequently complex due to frequent reworking during deposition, and may contain interbedded debris flows and laminated lacustrine sediments (Benn & Evans 1998). This contrasts with the relatively simple sedimentology observed in Cwm Idwal. The gross morphology of ablation moraines is controlled by the debris distribution within and on the glacier, however the form of individual mounds is controlled by the history of mass movement processes as the ice-core melts. The resulting morphology contrasts with the rectilinear faces and imbricate nature of the mounds in Cwm Idwal.

(3) *Formation as subglacial bedforms* (Gray 1982). Gray (1982) called these flutes, but subglacial bedforms of this size are generally classified as drumlins or megaflutes (Rose 1987; Bennett 1995). Drumlins and megaflutes of Younger Dryas age have been reported from Scotland (Rose 1981; Hodgson 1986; Benn 1992; Bennett 1995), where they appear to be limited to valley systems and lowland areas and be absent in cirques. The textural and clast characteristics of the sediments from Cwm Idwal are consistent with descriptions of the sedimentology of bedforms of a similar size in the modern

and ancient landform record (Krüger & Thomsen 1984; Menzies & Maltman 1992; Nenonen 1994; Newman & Mickelson 1994). In contrast to the poorly developed clast fabric observed in Cwm Idwal, fabrics close to the surface in modern and ancient bedforms are expected to reflect ice-flow direction around the bedform (Andrews & King 1968; Krüger & Thomsen 1984). Although fabrics may be weakened by postglacial reworking, the presence of steep rectilinear faces in Cwm Idwal suggests this has not been significant and some indication of the original fabric may be expected to be preserved.

Subglacial bedforms of the scale observed in Cwm Idwal occur at contemporary glacier margins, but are generally smoothly streamlined and often taper down-glacier (e.g. Krüger & Thomsen 1984) and relic features in Scotland are generally low and broad (Bennett 1995). These descriptions contrast with the sharp ridge crests, commonly symmetrical long profiles and rectilinear faces of the Cwm Idwal mounds (Fig. 3).

(4) *Formation by the melt-out of sediment entrained in englacial thrusts in a glacier crossing Cwm Idwal from Cwm Cneifion* (Hambrey *et al.* 1997). This hypothesis was developed following the observation of similar landforms forming at the margins of Svalbard glaciers. The sediments in Svalbard thrust-moraines are very variable, being determined by the sediments overridden by the glacier, but include diamicton and gravel (e.g. Hambrey *et al.* 1997; Bennett *et al.* 1998). Clast-angularity within the moraines is also variable, but the diamictons and gravels are frequently subangular and subrounded, and striated and faceted clasts are common (Hambrey *et al.* 1997). Clast fabrics are highly variable and include those with no clear preferred orientation both within englacial thrusts (Hambrey *et al.* 1999) and the resulting moraine mounds (Bennett *et al.* 1999). The mounds, which are variable in length and size, are imbricately stacked with rectilinear slopes facing up-glacier approximately transverse to ice-flow and irregular down-glacier faces (e.g. Bennett *et al.* 1999). The sedimentary and morphological characteristics of Svalbard thrust-moraine complexes are similar to those observed in Cwm Idwal and a similar moraine-mound complex in Scotland (Bennett *et al.* 1998) (Table 3), although the variety of facies is apparently more limited.

## Depositional model for Cwm Idwal

Aspects of the sedimentology and morphology of the Cwm Idwal moraine-mound complex are

**Table 3.** *Comparison of moraine-mound complexes in Svalbard and Coire a' Cheud-chnoic (Valley of a Hundred Hills, Scotland) (Bennett* et al. *1998) with Cwm Idwal*

| Characteristic | Svalbard | Coire a' Cheud-chnoic | Cwm Idwal |
|---|---|---|---|
| *Morphology* | | | |
| Broad moraine-mound belt | ✔ | ✔ | ✔ |
| Rectilinear mound slopes | ✔ | ✔ | ✔ |
| Consistent dip and orientation | ✔ | ✔ | ✔ |
| Imbricate/stacked morphology | ✔ | ✔ | ✔ |
| Variable ridge/mound length | ✔ | ✔ | ✔ |
| *Sedimentology* | | | |
| One mound: one facies/facies association | ✔ | NE | NE |
| Facies variability | ✔ | ✔ | ✔ |
| Basal sediment | ✔ | ✔ | ✔ |
| Sharp tectonic facies contacts | ✔ | NE | NE |
| Fabrics: weak, girdle or random | ✔ | ✔ | ✔ |

NE, no evidence currently available.

consistent with all of the proposed origins. However, only the englacial thrusting hypothesis is consistent with *all* aspects of the sedimentology and morphology of the moraine-mound complex. As at Coire a' Cheud-chnoic, where a similar moraine-mound complex has been interpreted as a result of englacial thrusting (Bennett *et al.* 1998), the level of sedimentary exposure is insufficient to provide evidence of the extent of within- and between-mound facies variability, or the nature of the contacts between sediment facies. It is rarely possible to determine the genesis of features in the palaeo-landform record with certainty, and although it seems probable that the Cwm Idwal moraine-mound complex represents the landform record of englacial thrusting, without greater sedimentary exposure it will not be possible to substantiate this hypothesis.

The depositional model proposed for the Cwm Idwal moraine-mound complex places an ice accumulation area in Cwm Cneifion feeding a glacier that flowed obliquely across Cwm Idwal. Thrusts developed in the glacier, along which debris was entrained. As the glacier retreated during deglaciation, this debris melted out to form the moraine-mound complex (Fig. 7).

## Formation of thrust-moraine complexes

The formation and long-term preservation of a thrust-moraine complex requires that three conditions be met: thrusting must have occurred; significant quantities of sediment must have been incorporated into the thrusts; the sediment must have been distributed in the glacier and deposited in such a way that the landform it produces is preserved in the land-

form record. The implications of these conditions are examined below.

*Causes of thrusting.* Thrusting in glacier ice results from longitudinal flow compression that cannot be absorbed by ductile deformation. Flow compression may result from a cold-based ice margin preventing sliding, the passage of a surge front, compression against a reverse bedrock slope, a glacier confluence, or a combination of these factors.

Thrusting in the Cwm Cneifion Glacier was probably strongly influenced by the stacking of the moraines against the reverse bedrock slope on the western wall of Cwm Idwal. However, there is some evidence to suggest that the glacier may also have been polythermal. Striations on exposed bedrock surfaces on the west of Llyn Idwal are orientated along the axis of the valley, oblique to the direction of ice flow as required by the thrusting hypothesis. If the thrusting hypothesis is correct, the preservation of these striations implies that virtually no bedrock erosion took place during the Younger Dryas, a conclusion consistent with that of Sharp *et al.* (1989) for the nearby Cwm Llydaw on the Snowdon Massif. The glacier being cold-based close to the margin may most readily explain this. Thus thrusting was probably induced by a combination of flow compression resulting from the reverse bedrock slope, and the presence of cold-based ice at the glacier margin.

*Debris entrainment by thrusting.* The formation of thrust-moraines requires significant quantities of sediment to be entrained along englacial thrusts. Although the presence of basal debris in

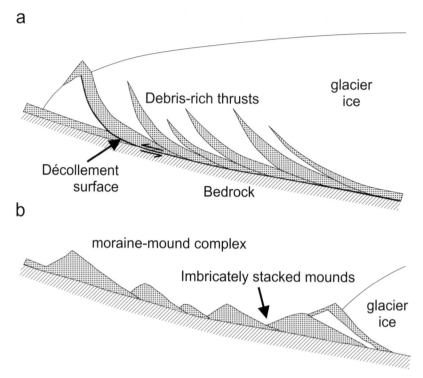

**Fig. 7.** Schematic model for the formation of the Cwm Idwal moraines. (**a**) Advancing glacier at the Younger Dryas limit. (**b**) Receding glacier with melt-out of thrust debris.

thrusts has been well documented (e.g. Sharp 1985, 1988; Knight 1997; Murray *et al.* 1997; Hambrey *et al.* 1999), the mechanism by which this debris is entrained remains controversial. Competing hypotheses stress the importance of debris-rich basal ice and direct incorporation of subglacial sediment.

Debris-rich basal ice-layers, up to several metres thick, may be formed at the glacier sole by a number of processes (Alley *et al.* 1997; Knight 1997). This debris-rich layer may be thickened by folding and, where ductile deformation cannot absorb the strain, may be elevated to higher levels within the glacier by thrusting (Boulton 1970).

If the décollement surface (sole thrust) is below the glacier/bed interface, slabs of basal sediment may be elevated by folding and thrusting (Clarke & Blake 1991; Hambrey 1994; Hambrey & Huddart 1995). The décollement may occur at the base of subglacial permafrost, where compression results in high porewater pressures (Hambrey & Huddart 1995; Boulton *et al.* 1999), the sediment–bedrock interface, or a less competent layer in subglacial sediments. In polythermal glaciers, sediment incorporation may be most effec-

tive at the boundary between warm and cold ice as the compressive regime favours thrust initiation, and sediment may be frozen onto the glacier sole (e.g. Sollid & Sørbel 1988). There is some evidence that rafts of sediment, containing primary depositional structures and not affected by permafrost, may be incorporated into thrusts, some without significant deformation occurring (Bennett *et al.* 1999). It is currently unclear, however, how widespread this phenomenon is.

The modes of debris entrainment in Cwm Idwal are uncertain, but a number of observations may be made. Given the limited distance available for a debris-rich basal ice-layer to form, it is unlikely that sufficient debris could have been entrained in the basal ice-layer to explain the volume of sediment in each mound. The presence of bedrock exposures between moraine mounds suggests that the sediment–bedrock interface acted as a décollement surface, enabling subglacial debris to be elevated into the glacier. Whether the subglacial sediment was frozen during entrainment is uncertain, but the preservation of oblique striations inferred to date from prior to the Younger Dryas suggests that the glacier was cold-based at the margin and, therefore, may have been underlain by sediment affected by permafrost.

*Debris deposition.* The preservation of a distinctive thrust-moraine morphology in the landform record requires limited syn- and post-depositional mass movement. Ablation moraine will be formed where considerable amounts of ice are incorporated into the moraine-mound complex during deposition. As the ice-cores melt, resedimentation is likely to result in the destruction of the tectonic structure (Bennett *et al.* 1996b). The tectonic structure is most likely to be preserved where the volume of sediment in thrusts is large compared with the spacing of the thrusts. Steeply dipping thrusts also favour preservation as it is the base of the thrust that forms the mound and in steep thrusts more material is left in the base of the mound and less ice is draped with sediment, promoting resedimentation (although above 45° slumping is likely, Bennett *et al.* 1999).

## Wider significance

The recognition of thrust-moraine complexes in the palaeo-landform record is of importance as it has potential implications for the reconstruction of glacier dynamics and palaeo-climate. (Hambrey *et al.* 1997; Bennett *et al.* 1998). In the case of Cwm Idwal, it appears that the palaeo-climatic implications are limited, as the presence of a reverse bedrock slope is likely to have been the dominant control on thrusting. The preservation of a tectonic structure does, however, demonstrate that glacier retreat was dynamic and not dominated by *in situ* stagnation and associated resedimentation. Because they form englacially, thrust-moraine complexes do not provide direct evidence of the position of the ice-margin. Where an ice-marginal position has been inferred from the distribution of moraine-mound complexes, as has often been the case in the reconstruction of British Younger Dryas glaciers (e.g. Sissons 1980; Gray 1982), the maximum size of the glacier may be underestimated. This illustrates the importance of careful interpretation of the geomorphological and sedimentological evidence before undertaking glacier reconstructions.

Although the deformation structures present in glacier ice are relatively well understood, the role of glacier ice-deformation in landform development has, until recently, been little studied. Recent work on contemporary glaciers in Svalbard (e.g. Bennett *et al.* 1996a, b, 1998, 1999; Hambrey *et al.* 1997, 1999; Glasser *et al.* 1998) has begun to correct this deficiency, but work on

evidence of ice-deformation in the palaeo-landform record has hardly begun. In the British Younger Dryas landform record, evidence of ice-deformation is limited to a moraine-mound complex interpreted as thrust-moraines in Torridon, Scotland (Bennett *et al.* 1998), and some sedimentological evidence of debris elevation in the morainic sediments on the Isle of Skye (Benn 1990). Although care must be taken to avoid forcing landforms to fit prevailing paradigms (Sugden 1980), it seems likely that if two englacial thrust-moraine complexes are present in the British Younger Dryas landform record, others remain to be discovered. Indeed, although thrust-moraine complexes have thus far only been recognised in association with cirque and valley glaciers, there is no reason to believe they may not also have been formed close to the margins of the Pleistocene ice-sheets where hummocky topography is common (e.g. Sollid & Sørbel 1988; Johnson *et al.* 1995; Andersson 1998; Eyles *et al.* 1999). If they do exist, their recognition may be problematic due to degradation of the tectonic structure. It is also likely that landforms resulting from other forms of ice-deformation may await discovery in the palaeo-landform record.

## Conclusions

(1) The only hypothesis that is able to satisfactorily explain all aspects of the morphology and sedimentology of the Cwm Idwal moraine-mound complex is the deposition of material entrained by englacial thrusting.

(2) The initiation of thrusting in Cwm Idwal probably resulted from compression against a reverse bedrock slope, although polythermal ice may also have played a role.

(3) It is tentatively suggested that debris entrainment was by the incorporation of sub-glacial sediment directly into thrusts, with the décollement at the sediment–bedrock interface. It is unclear whether the sediment was permafrost-affected during entrainment.

(4) Relict englacial thrust-moraine complexes have now been described from Cwm Idwal and Coire a' Cheud-chnoic (Bennett *et al.* 1998), suggesting that the role of ice-deformation, particularly thrusting, in landform genesis has been underestimated.

(5) The Cwm Idwal and Coire a' Cheud-chnoic thrust-moraine complexes are both of Younger Dryas age. Although it seems likely that similar features of greater antiquity exist in the landform record, degradation may make their recognition problematic.

We thank M. J. Hambrey, N. F. Glasser and M. R. Bennett for their valuable advice and reviewing this paper. Thanks also to H. R. Roberts and the Countryside Council for Wales for permission to undertake work in Cwm Idwal and P. L. Noble for fieldwork assistance. Digital elevation modelling was undertaken using an evaluation copy of Panacea32 donated by M. McCullagh. D.J.G. thanks the University of Wales, Aberystwyth, the Bill Bishop Memorial Fund and the Quaternary Research Association for contributions to the costs of fieldwork and field equipment. N.G.M. thanks Liverpool John Moores University for their contribution to the cost of fieldwork.

# References

ADDISON, K. 1988a. *The ice age in Cwm Idwal*. K. & M. K. Addison, Broseley, Shropshire.

——1988b. *The ice age in Y Glyderau and Nant Ffrancon*. K. & M. K. Addison, Broseley, Shropshire.

——1990. Cwm Idwal. *In*: ADDISON, K., EDJE, M. J. & WATKINS, R. (eds) *North Wales Field Guide*. Quaternary Research Association, Cambridge, 93–94.

ALLEY, R. B., CUFFEY, K. M., EVENSON, E. B., STRASSER, J. C., LAWSON, D. E. & LARSON, G. J. 1997. How glaciers entrain and transport basal sediment: physical constraints. *Quaternary Science Reviews*, **16**, 1017–1038.

ANDERSSON, G. 1998. Genesis of hummocky moraine in the Bolen area, southwestern Sweden. *Boreas*, **27**, 55–67.

ANDREWS, J. T. & KING, C. A. M. 1968. Comparative till fabrics and till fabric variability in a till sheet and a drumlin: a small scale study. *Proceedings of the Yorkshire Geological Society*, **36**, 435–461.

ATTIG, J. W. & CLAYTON, L. 1993. Stratigraphy and origin of an area of hummocky glacial topography, northern Wisconsin. *Quaternary International*, **18**, 61–67.

BALLANTYNE, C. K. 1987. Some observations on the morphology and sedimentology of two active protalus ramparts, Lyngen, northern Norway. *Arctic and Alpine Research*, **19**, 167–174.

—— & BENN, D. I. 1994. Paraglacial slope adjustment and resedimentation following recent glacier retreat, Fåbergstølsden, Norway. *Arctic and Alpine Research*, **26**, 255–269.

—— & KIRKBRIDE, M. P. 1986. The characteristics and significance of some Late-glacial protalus ramparts in upland Britain. *Earth Surface Processes and Landforms*, **11**, 659–671.

BENN, D. I. 1989. Debris transport by Loch Lomond Readvance glaciers in Northern Scotland: basin form and the within-valley asymmetry of lateral moraines. *Journal of Quaternary Science*, **4**, 243–254.

——1990. *Scottish Lateglacial Moraines: Debris Supply, Genesis and Significance*. PhD thesis, University of St Andrews.

——1992. The genesis and significance of 'hummocky moraine': evidence from the Isle of Skye, Scotland. *Quaternary Science Reviews*, **11**, 781–799.

—— & BALLANTYNE, C. K. 1993. The description and representation of particle shape. *Earth Surface Processes and Landforms*, **18**, 665–672.

—— & ——1994. Reconstructing the transport history of glaciogenic sediments – a new approach based on the co-variance of clast form indices. *Sedimentary Geology*, **91**, 215–227.

—— & EVANS, D. J. A. 1998. *Glaciers and glaciation*. Arnold, London.

——, LOWE, J. J. & WALKER, M. J. C. 1992. Glacier response to climatic change during the Loch Lomond Stadial and early Flandrian: geomorphological and palynological evidence from the Isle of Skye, Scotland. *Journal of Quaternary Science*, **7**, 125–144.

BENNETT, M. R. 1994. Morphological evidence as a guide to deglaciation following the Loch Lomond Readvance: a review of research approaches and models. *Scottish Geographical Magazine*, **110**, 24–32.

——1995. The morphology of glacially fluted terrain: examples from the Northwest Highlands of Scotland. *Proceedings of the Geologists' Association*, **106**, 27–38.

—— & BOULTON, G. S. 1993. A reinterpretation of Scottish 'hummocky moraine' and its significance for the deglaciation of the Scottish Highlands during the Younger Dryas or Loch Lomond Stadial. *Geological Magazine*, **130**, 301–318.

—— & GLASSER, N. F. 1991. The glacial landforms of Glen Geusachan, Cairngorms: a reinterpretation. *Scottish Geographical Magazine*, **107**, 116–123.

——, HAMBREY, M. J. & HUDDART, D. 1997. Modification of clast shape in High-Arctic environments. *Journal of Sedimentary Research*, **67**, 550–559.

——, ——, —— & GHIENNE, J. F. 1996a. The formation of a geometrical ridge network by the surge-type glacier Kongsvegen, Svalbard. *Journal of Quaternary Science*, **11**, 437–449.

——, ——, —— & GLASSER, N. F. 1998. Glacial thrusting and moraine-mound formation in Svalbard and Britain: the example of Coire a' Cheundchnoic (Valley of a Hundred Hills), Torridon, Scotland. *In*: OWEN, L. A. (ed.) *Mountain Glaciation*. Quaternary Proceedings, **6**. Wiley & Sons, Chichester, 17–34.

——, ——, ——, —— & CRAWFORD, K. 1999. The landform and sediment assemblage produced by a tidewater glacier surge in Kongsfjorden, Svalbard. *Quaternary Science Reviews*, **18**, 1213–1246.

——, HUDDART, D., HAMBREY, M. J. & GHIENNE, J. F. 1996b. Moraine development at the High-Arctic glacier Pedersenbreen, Svalbard. *Geografiska Annaler*, **78A**, 209–222.

BGS 1972. *Geological Special Sheet (Central Snowdonia) – Solid*. British Geological Survey.

BOULTON, G. S. 1970. On the origin and transport of englacial debris in Svalbard glaciers. *Journal of Glaciology*, **9**, 213–229.

——1972. Modern Arctic glaciers as depositional models for former ice sheets. *Journal of the Geological Society, London*, **128**, 361–393.

——1978. Boulder shapes and grain-size distributions of debris as indicators of transport paths through a glacier and till genesis. *Sedimentology*, **25**, 773–799.

——1986. Push-moraines and glacier-contact fans in marine and terrestrial environments. *Sedimentology*, **33**, 677–698.

—— & EYLES, N. 1979. Sedimentation by valley glaciers: a model and genetic classification. *In*: SCHLÜCHTER, C. (ed.) *Moraines and varves*. Balkema, Rotterdam, 11–23.

——, VAN DER MEER, J. J. M., BEETS, D. J., HART, J. K. & RUEGG, G. H. J. 1999. The sedimentary and structural evolution of a recent push moraine complex: Holmstrømbreen, Spitsbergen. *Quaternary Science Reviews*, **15**, 961–987.

CLARKE, G. K. C. & BLAKE, E. W. 1991. Geometric and thermal evolution of a surge-type glacier in its quiescent state: Trapridge Glacier, Yukon Territory, Canada 1969–89. *Journal of Glaciology*, **37**, 158–169.

CORNISH, R. 1981. Glaciers of the Loch Lomond Stadial in the western Southern Uplands of Scotland. *Proceedings of the Geologists' Association*, **92**, 105–114.

DARWIN, C. 1842. Notes on the effects produced by the ancient glaciers of Caernarvonshire, and on the boulders transported by floating ice. *Philosophical Magazine*, **3**, 180–188.

ESCRITT, E. A. 1971. Plumbing the depths of Idwal's moraines. *Geographical Magazine*, **44**, 52–55.

EYLES, N. 1979. Facies of supraglacial sedimentation on Icelandic and Alpine temperate glaciers. *Canadian Journal of Earth Sciences*, **16**, 1341–1361.

——1983. Modern Icelandic glaciers as depositional models for 'hummocky moraine' in the Scottish Highlands. *In*: EVERSEN, E. B., SCHLÜCHTER, C. & RABSON, J. (eds) *Tills and related deposits*. Balkema, Rotterdam, 47–60.

——, BOYCE, J. I. & BERENDREGT, R. W. 1999. Hummocky moraine: sedimentary record of stagnant Laurentide Ice Sheet lobes resting on soft beds. *Sedimentary Geology*, **123**, 163–174.

GLASSER, N. F., HAMBREY, M. J., CRAWFORD, K., BENNETT, M. R. & HUDDART, D. 1998. The structural glaciology of Kongsvegen, Svalbard, and its role in landform genesis. *Journal of Glaciology*, **44**, 136–148.

GODWIN, H. 1955. Vegetational plant history at Cwm Idwal: a Welsh plant refuge. *Svensk Botanisk Tidskrift*, **49**, 35–43.

GRAY, J. M. 1982. The last glaciers (Loch Lomond Advance) in Snowdonia, North Wales. *Geological Journal*, **17**, 111–133.

——1990. The Idwal moraines. *In*: ADDISON, K., EDJE, M. J. & WATKINS, R. (eds) *North Wales Field Guide*. Quaternary Research Association, London, 94–95.

HAMBREY, M. J. 1994. *Glacial Environments*. UCL Press, London.

—— & HUDDART, D. 1995. Englacial and proglacial glaciotectonic processes at the snout of a thermally complex glacier in Svalbard. *Journal of Quaternary Science*, **10**, 313–326.

——, BENNETT, M. R., DOWDESWELL, J. A., GLASSER, N. F. & HUDDART, D. 1999. Debris entrainment and transfer in polythermal valley glaciers. *Journal of Glaciology*, **45**, 69–86.

——, ——, GLASSER, N. F., HUDDART, D. & CRAWFORD, K. 1999. Facies and landforms associated with ice deformation in a tidewater glacier, Svalbard. *Glacial Geology and Geomorphology (http://ggg.qub.ac.uk/ggg)*.

——, HUDDART, D., BENNETT, M. R. & GLASSER, N. F. 1997. Genesis of 'hummocky moraines' by thrusting in glacier ice: evidence from Svalbard and Britain. *Journal of the Geological Society, London*, **154**, 623–632.

HARRIS, C. 1986. Some observations concerning the morphology and sedimentology of a protalus rampart, Okstindan, Norway. *Earth Surface Processes and Landforms*, **11**, 673–676.

HODGSON, D. M. 1982. *Hummocky and fluted moraines in part of North-West Scotland*. PhD thesis, University of Edinburgh.

——1986. A study of fluted moraines in the Torridon area, north-west Scotland. *Journal of Quaternary Science*, **1**, 109–118.

HUDDART, D. & HAMBREY, M. J. 1996. Sedimentary and tectonic development of a high-arctic thrust-moraine complex: Comfortlessbreen, Svalbard. *Boreas*, **25**, 227–243.

JEHU, T. J. 1902. A bathymetrical and geological study of the lakes of Snowdonia and eastern Caernarvonshire. *Transactions of the Royal Society of Edinburgh*, **40**, 419–467.

JOHNSON, M. D., MICKELSEN, D. M., CLAYTON, L. & ATTIG, J. W. 1995. Composition and genesis of glacial hummocks, western Wisconsin, USA. *Boreas*, **24**, 97–116.

JONES, A. P., TUCKER, M. E. & HART, J. 1999. Guidelines and recommendations. *In*: *The Description and Analysis of Quaternary Stratigraphic Field Sections*. Quaternary Research Association, London, Technical Guides, **7**, 27–62.

KNIGHT, P. G. 1997. The basal ice layer of glaciers and ice sheets. *Quaternary Science Reviews*, **16**, 975–993.

KRÜGER, J. & THOMSEN, H. H. 1984. Morphology, stratigraphy, and genesis of small drumlins in front of the glacier Mýrasjökull, south Iceland. *Journal of Glaciology*, **30**, 94–105.

MATTHEWS, J. A., CORNISH, R. & SHAKESBY, R. A. 1979. 'Saw-tooth' moraines in front of Bødalsbreen, southern Norway. *Journal of Glaciology*, **22**, 535–546.

MCCULLAGH, M. 1988. Terrain and surface modelling systems: theory and practice. *Photogrammetric Record*, **12**, 747–779.

MENZIES, J. & MALTMAN, A. J. 1992. Microstructures in diamictons – evidence of subglacial bed conditions. *Geomorphology*, **6**, 27–40.

MONCRIEFF, A. C. M. 1989. Classification of poorly sorted sedimentary rocks. *Sedimentary Geology*, **65**, 191–194.

MURRAY, T., GOOCH, D. L. & STUART, G. W. 1997. Structures within the surge front at Bakaninbreen,

Svalbard, using ground-penetrating radar. *Annals of Glaciology*, **24**, 122–129.

NENONEN, J. 1994. The Kaituri drumlin and drumlin stratigraphy in the Kangasniemi area, Finland. *Sedimentary Geology*, **91**, 365–372.

NEWMAN, W. A. & MICKELSON, D. M. 1994. Genesis of the Boston Harbor drumlins, Massachusetts. *Sedimentary Geology*, **91**, 333–343.

ONO, Y. & WATANABE, T. 1986. A protalus rampart related to alpine debris flows in the Kuranosuke Cirque, northern Japanese Alps. *Geografiska Annaler*, **68A**, 213–223.

ROSE, J. 1981. Field guide to the Quaternary geology of the south eastern part of the Loch Lomond basin. *Proceedings of the Geological Society of Glasgow*, **123**, 3–19.

——1987. Drumlins as part of a glacier bedform continuum. *In*: MENZIES, J. & ROSE, J. (eds) *Drumlin Symposium*. Balkema, Rotterdam, 103–116.

SEDDON, B. 1957. Late Glacial cwm glaciers in Wales. *Journal of Glaciology*, **3**, 94–99.

——1962. Late-glacial deposits of Llyn Dwythwch and Nant Ffrancon, Caernarvonshire. *Philosophical Transactions of the Royal Society*, **B244**, 459–481.

SHAKESBY, R. A. 1997. Pronival (protalus) ramparts: a review of forms, processes, diagnostic criteria and palaeoenvironmental implications. *Progress in Physical Geography*, **21**, 394–418.

——, MATTHEWS, J. A. & McCARROLL, D. 1995. Pronival ('protalus') ramparts in the Romsdalsalpane, southern Norway: forms, terms, subnival processes, and alternative mechanisms of formation. *Arctic and Alpine Research*, **27**, 271–282.

——, ——, McEWAN, L. J. & BERRISFORD, M. S. 1999. Snow-push processes in pronival (protalus) rampart formation: geomorphological evidence from Smørbotn, Romsdalsalpane, southern Norway. *Geografiska Annaler*, **81A**, 31–45.

SHARP, M. J. 1984. Annual moraine ridges at Skalafellsjökull, south-east Iceland. *Journal of Glaciology*, **30**, 82–93.

——1985. Sedimentation and stratigraphy at Eyjabakkajökull – an Icelandic surging glacier. *Quaternary Research*, **24**, 268–284.

——1988. Surging glaciers: geomorphic effects. *Progress in Physical Geography*, **12**, 533–559.

——, DOWDESWELL, J. A. & GEMMELL, J. C. 1989. Reconstructing past glacier dynamics and erosion from glacial geomorphic evidence: Snowdon, North Wales. *Journal of Quaternary Science*, **4**, 115–130.

SISSONS, J. B. 1974. A Late-glacial ice cap in the central Grampians, Scotland. *Transactions of the Institute of British Geographers*, **62**, 95–114.

——1979. The Loch Lomond Stadial in the British Isles. *Nature*, **280**, 199–203.

——1980. The Loch Lomond Advance in the Lake District, northern England. *Transactions of the Royal Society of Edinburgh, Earth Sciences*, **71**, 12–27.

——1983. Quaternary. *In*: CRAIG, G. Y. (ed.) *Geology of Scotland*. Scottish Academic Press, Edinburgh, 399–424.

—— & SUTHERLAND, D. G. 1976. Climatic inferences from former glaciers in the south-east Grampian Highlands, Scotland. *Journal of Glaciology*, **17**, 325–346.

SMALL, R. J. 1983. Lateral moraine of Glacier de Tsidjiore Nouve: form, development, and implications. *Journal of Glaciology*, **29**, 250–259.

——1987. Moraine sediment budgets. *In*: GURNELL, A. M. & CLARK, M. J. (eds) *Glacio-fluvial Sediment Transfer*. Wiley, Chichester, 165–197.

SOLLID, J. L. & SØRBEL, L. 1988. Influence of temperature conditions in formation of end moraines in Fennoscandia and Svalbard. *Boreas*, **17**, 553–558.

SUGDEN, D. E. 1980. The Loch Lomond Advance in the Cairngorms (a reply to J. B. Sissons). *Scottish Geographical Magazine*, **96**, 18–19.

TIPPING, R. 1993. A detailed early postglacial (Flandrian) pollen diagram from Cwm Idwal, North Wales. *New Phytologist*, **125**, 175–191.

UNWIN, D. J. 1975. The nature and origin of the corrie moraines of Snowdonia. *Cambria*, **2**, 20–33.

——1977. Cwm Idwal. *In*: BOWEN, D. Q. (ed.) *Wales and the Cheshire-Shropshire Lowland, International Union for Quaternary Research, X Congress, Guidebook for excursions A8 and C8*. GeoAbstracts, Norwich, 48–51.

WASHBURN, A. L. 1979. *Geocryology: A Survey of Periglacial Processes and Environments*. Edward Arnold, London.

# Index

Page numbers in italic, e.g. *221*, signify references to figures. Page numbers in bold, e.g. **60**, denote references to tables.